Space Technology Library

Volume 38

The *Space Technology Library* is a series of high-level research books treating a variety of important issues related to space missions. A wide range of space-related topics is covered starting from mission analysis and design, through a description of spacecraft structure to spacecraft attitude determination and control. A number of excellent volumes in the *Space Technology Library* were provided through the US Air Force Academy's Space Technology Series. The quality of the book series is guaranteed through the efforts of its managing editor and well-respected editorial board. Books in the *Space Technology Library* are sponsored by ESA, NASA and the United States Department of Defense.

More information about this series at http://www.springer.com/series/6575

Deng-Yun Yu · Ze-Zhou Sun · He Zhang

Technology of Lunar Soft Lander

国防工業出版社
National Defense Industry Press

Springer

Deng-Yun Yu
China Aerospace Science and Technology Corporation
Beijing, China

Ze-Zhou Sun
Beijing Institute of Spacecraft System Engineering
Beijing, China

He Zhang
Beijing Institute of Spacecraft System Engineering
Beijing, China

Space Technology Library
ISBN 978-981-15-6582-3 ISBN 978-981-15-6580-9 (eBook)
https://doi.org/10.1007/978-981-15-6580-9

Jointly published with National Defense Industry Press
The print edition is not for sale in China (Mainland). Customers from China (Mainland) please order the print book from: National Defense Industry Press.

This Springer imprint is published by the registered company Springer Nature Singapore Pte Ltd.
The registered company address is: 152 Beach Road, #21-01/04 Gateway East, Singapore 189721, Singapore

Foreword By Fanpei Lei

It is an exciting moment that the circumlunar return and reentry spacecraft, Chang'E-5T1, was launched and return safely just then. Now the book "*Technology of Lunar Soft Lander*" edited by YU Dengun et al. will be published. Congratulation!

The target of China's Lunar Exploration Program (CLEP) was Orbiting, Landing, and Return by three steps. With mission success of Chang'E-1, Chang'E-2, and Chang'E-3, the scheduled target of Orbiting and Landing had been implemented. The book "*Technology of Lunar Soft Lander*" was edited on the basis of summary of Chang'E-3 lunar lander development by Chinese aerospace scientists and engineers. It is not only the summary of engineering practice experience, but also important result of CLEP development, as well as cultural and spiritual wealth of Chinese aerospace undertaking.

Chairman Mao Zedong once said that human beings should constantly summarize experience, invent, discover, and advance. Exploring mystery of the universe and utilizing lunar resource will be long-term enterprise of human being. The publication of this book will have two effects. One is to summarize past technical practice systematically, refine experience or theory, so as to guide future development of lunar lander and even provide useful experience or theory for future exploration of the Moon and other celestial bodies. The other one is to advocate that Chinese aerospace scientists and engineers shall summarize experience in time and make it a fashion and habit for scientists and engineers to write what they have done.

Now Chinese aerospace enterprise is prompting to build a space power. There shall be not only excellent and reliable spacecraft, but also strong support of technology, science, and culture. It is undoubted that the publication of aerospace technology is an important part of our aerospace enterprise. The aerospace enterprise shall be inherited by generations and the aerospace technical books shall also be written by generations of scientists and engineers. We shall take the publication of "*Technology of Lunar Soft Lander*" as an opportunity to prompt the series publication of CLEP which covers all technology fields of CLEP development. I hope that the publication of this book will lead publication of related technology books of all technology fields involved in CLEP development.

In the end, I'd like to sincerely thank all personnel who participated edit and publication of this book, as well as all leading comrades who supported the approval of the book publication. Wish the publication of "*Technology of Lunar Soft Lander*" a great success!

November 2014

Fanpei Lei
President of Chinese Aerospace Science
and Technology Corporation
Beijing, China

Foreword By Peijian Ye

The Moon is the only natural satellite of the Earth. It is not only the start point of deep space exploration, but also the focus of all space countries. Along with development of China's Deep Space Exploration Program, the Chang'E-1 orbiting the Moon was implemented in 2007 and lunar soft-landing of Chang'E-3 was implemented in 2013. There will be lunar sample and return in the future.

After landed on surface of celestial bodies, exploration in short distance will help to learn the celestial body more clearly and deeply. At the same time, it is also possible to observe the solar system and other celestial bodies with unique location and environment on the body. There are lunar exploration, observation of the Earth and space in mission objectives of lunar soft-landing exploration.

It is not easy for a lunar lander to be launched on Earth and landed on lunar surface. The lunar lander shall be landed stably on lunar surface after a long journey of 389 thousand kilometers. It shall adapt to different lunar terrain and soft lunar soil, as well as environment on lunar surface such as lunar dust, temperature alternation in lunar day and night. A series of new technologies shall be broken through for the lunar lander. The implementation of lunar soft-landing is an important mark of aerospace technology level.

The book "*Technology of Lunar Soft Lander*" is the first book for systematic and deep introduction of lunar lander technology in China. The editors of this book are young and middle-aged scientists and engineers who I am familiar with and thinking highly. Some of them participated in the development of Chang'E-1 with me in the beginning of twenty-first century, and then participated in the development of Chang'E-2 and Chang'E-3. They are important ones working on front line of CLEP, who have solid theoretical foundation and rich engineering experience. The design methodology, typical technology, and verification methodology based on engineering practice of Chang'E-3 mission are systematically described in this book including environment, system design, subsystem design, AIT, and ground test.

The book "*Technology of Lunar Soft Lander*" can provide technology reference for aerospace scientists and engineers in lunar and deep space exploration, as well as related knowledge for researchers in colleges and institutes. The publication will help to summarize engineering experience in Chinese lunar exploration and provide strong support for sustainable development of deep space exploration.

November 2014

Peijian Ye
Academician of the Chinese Academy of Sciences
Chief Scientist of Chang'E-3 lunar probe
Beijing, China

Preface I

Soft-landing on celestial body is an important approach to explore the mystery of the universe by human beings. Compared with Earth artificial satellite and circum-lunar lunar spacecraft, there were a great many new technologies in system design, Guidance, Navigation, and Control (GNC), propulsion, landing cushion, and thermal design of lunar lander. There were also a lot of ground verification and validation tests to be conducted.

On the basis of mission success of Chinese Chang'E-3 lunar lander, the design and development technology of lunar lander are summarized in the book. There are 17 chapters. In chapters 1–3, there is system introduction for history of lunar exploration, mission and environment of lunar lander, system design of lunar lander. In chapters 4–11, there is design of important subsystems including structure and mechanism, thermal control, power supply, GNC, propulsion, On-Board Data Handling (OBDH), Tracking, Telemetry and Command (TT&C) and telecommunication and landing gear. In chapters 12–15, there is technology of Integration and Assembly, test, verification and validation, including Integration and Assembly, metrology, electrical test, and validation. In chapter 16, the instrumentation technology of lunar lander is introduced. In chapter 17, there is future prospect of lunar lander technology.

The book is edited by YU Dengyun, LIN Yiming, LIU Guoqing, SUN Zezhou, and ZHANG He. There are also experts involved in the compilation of this book including WU Xueying, WANG Chuang, DENG Xiangjin, CHAI Hongyou, LI Xinli, XU Yanling, LIU Zijun, ZHANG Ming, LEI Yingjun, CHENG Ming, JIN Guangming, CAO Wei, ZHAO Lei, HUANG Xiaofeng, LIU Shi, MAN Jianfeng, YANG Jianzhong, HE Xiaoyu, QUAN Shuang, WANG Jinrong, ZOU Xin, FAN Shichao, YI Wangmin, FENG Wei, YANG Zaihua, SHI Lixia, DU Chen, and GAO Qinghua. The final editor and author of the book is YU Dengyun and ZHANG He.

It costed two years to finish the book. It has been strongly supported by LEI Fanpei (President of Chinese Aerospace Science and Technology Corporation (CASC)), WU Yansheng and YUAN Jie (Vice President of CASC), YANG Baohua (Vice President of China Academy of Space Technology), YE Peijian (Academician of the Chinese Academy of Sciences). The manuscript was reviewed by experts including LI Chunlai, PENG Chengrong, FENG Jisheng, ZHAO Jiancheng, HUA Chengsheng, FAN Hanlin, HAN Guojing, HAN Hongyin, ZHANG Honghua, WANG Xianwen,

YAN Xinrong, ZHAO Heping, ZHANG Qingjun, BAO Xiaoping, RAO Wei, and CAI Zhenbo, and valuable opinions and suggestions are put forward. Experts helped in publication of the book including FAN Hanlin, LIU Xueguang, TAN Ying, GUO Fan, HUANG Jun, and ZHANG Chuanjun. Many sincere thanks to them!

Due to the limitation of the author's level, we sincerely ask the readers to criticize and correct the inevitable shortcomings of the book.

Beijing, China
November 2015

Deng-Yun Yu
Ze-Zhou Sun
He Zhang

Preface II

On December, 14, 2013, the Chinese Chang'E-3 lunar lander was landed on lunar surface successfully. Since the last lunar soft-landing in 1976, there was no probe made by human beings to be landed on the Moon. The success of Chang'E-3 lunar lander made China the 3rd country who had capability to implement lunar soft-landing. As major participants of the Chang'E-3 project, we began to write the book *"Technology for Lunar Soft Lander"* in Chinese in 2014, which could help to summarize the technology and experience of the project development. In 2016, the book was published by China National Defense Industry Press.

In 2017, the famous international publisher Springer paid attention to the book. They suggested to publish the English version. Then we began to translate the book into English. Doctor PENG Jing of China Academy of Space Technology proofread the book. And then, Prof. WU Bo of the Hong Kong Polytechnic University and Prof. HUANG Hai of Beijing University of Aeronautics and Astronautics reviewed the English version, especially for correctness and accuracy of the specific terms. According to their suggestions, we revised the book. The editors of the Springer made a lot of revise for the English version at the same time. We sincerely appreciate their work and devotion to the publication of the English version of the book.

The publication of *"Technology for Lunar Soft Lander"* in English will help more scientists and engineers in other countries to know lunar exploration technology in China, and provide useful reference for future lunar exploration.

Due to the limitation of the author's level, we sincerely ask the readers to criticize and correct the inevitable shortcomings of the book.

Beijing, China
July 2020

Deng-Yun Yu
Ze-Zhou Sun
He Zhang

Contents

Acronyms

A/D	Analog/Digital
AAAP	Automatic Aircraft Fuselage Assembly Procedure
ABBET	Broad-Based Environment for Test
AGV	Automated Guided Vehicle
AOS	Advanced Orbiting System
API	Application Programming Interface
APS	Active Pixel Sensor
ASIC	Application Specific Integrated Circuit
ASOS	Application Specific Operating Systems
ATML	Automatic Test Markup Language
ATS	Automatic Test System
BPSK	Binary Phase Shift Keying
CAD/CAM	Computer Aided Design/Computer Aided Manufacture
CAN	Controller Area Network
CCSDS	Consultative Committee for Space Data Systems
CCU	Central Control Unit
CLEP	China's Lunar Exploration Program
CMOS	Complementary Metal Oxide Semiconductor
COB	Chip On Board
CPU	Central Processing Unit
CTU	Central Terminal Unit
DB	Data Bus
DC	Direct Current
DDR	Digital Data Recorder
DFA	Design For Assembly
DIU	Data Interface Unit
DMU	Data Multiplexing Unit
DoD	Department of Defense
DOF	Degree Of Freedom
DPU	Digital Processing Unit
DSMC	Direct Simulation Monte Carlo
DSP	Digital Signal Processor

EDAC	Error Detection And Correction
EGSE	Electrical Ground Support Equipment
EM	Electrical Model
EMC	Electro Magnetic Compatibility
EOL	End Of Life
EPC	Electronic Power Controller
ESA	European Space Agency
ESTEC	European Space Research and Technology Center
EUV	Extreme Ultra-Violet
FM	Flight Model
FPGA	Field Programmable Gate Array
FSK	Frequency Shift Keying
GCR	Galactic Cosmic Ray
GEO	Geosynchronous Earth Orbit
GNC	Guidance, Navigation and Control
GSFC	Goddard Space Flight Center
HC	Hydrocarbons
I/F	Current to Frequency
I/O	Input/Output
IEEE	Institute of Electrical and Electronics Engineers
IMU	Inertial Measurement Unit
IMUE	Inertial Measurement Unit Electronics
ISRO	India Space Research Organization
IVI	Interchangeable Virtual Instrument
JAXA	Japan Aerospace eXploration Agency
LDPC	Low Density Parity Check
LEO	Low Earth Orbit
LET	Linear Energy Transfer
LHP	Loop Heat Pipe
LLO	Lunar Low Orbit
LLRF	Lunar Landing Research Facility
LLRV	Lunar Landing Research Vehicle
LOI	Lunar Orbit Insertion
LOX	Liquid Oxygen
LTO	Lunar Transfer Orbit
LTU	Local Terminal Unit
LVDS	Low Voltage Differential Signaling
LXI	LAN eXtensions for Instrumentation
MEO	Medium Earth Orbit
MGSE	Mechanical Ground Support Equipment
MMH	Mono Methyl Hydrazine
MON	Mixed Oxides of Nitrogen
MRVS	Microwave Range and Velocity Sensor
NASA	National Aeronautics and Space Administration
OBDH	On-Board Data Handling

OCOE	Overall CheckOut Equipment
OSR	Optical Sun Reflector
PCM	Pulse Code Modulation
PID	Proportional Integral Derivative
PRM	Pseudo Rate Modulator
PROM	Programmable Read-Only Memory
PSK	Phase Shift Keying
PUS	Packet Utilization Standard
PWM	Pulse Width Modulation
PXI	PCI eXtensions for Instrumentation
R&D	Research and Development
RAM	Random Access Memory
RCS	Reaction Control System
RF	Radio Frequency
RFSA	Russian Federation Space Administration
RHU	Radioisotope Heat Unit
RTG	Radioisotope Thermoelectric Generator
RTOS	Real Time multitask Operation System
RTU	Remote Terminal Units
RVS	Range and Velocity Sensor
SADA	Solar Array Drive Assembly
SAVOIR	Space Avionics Open Interface Architecture
SBI	Same Beam Interferometry
SCDE	Surface Charge and Discharge Effects
SCOE	Special CheckOut Equipment
SCR	Solar Cosmic Ray
SDB	Serial Data Bus
SEE	Single Event Effects
SEFI	Single Event Functional Interrupt
SEI	Solid Electrolyte Interphase
SEL	Single Event Latch-Up
SESL	Space Environment Simulation Laboratory
SEU	Single Event Upset
SMU	System Management Unit
SOC	System On Chip
SOIS	Spacecraft On-board Interface Service
SRAM	Static Random Accessed Memory
SSPA	Solid State Power Amplifier
TCU	Telecommand Unit
TID	Total Ionizing Dose
TPS	Test Program Set
TT&C	Telemetry Tracking and Command
TWTA	Traveling Wave Tube Amplifier
UDMH	Unsymmetrical Dimethyl Hydrazine
UHF	Ultra-High Frequency

USB	Unified S Band
USMN	Unified Space Metrology Network
VCHP	Variable Conductance Heat Pipe
VLBI	Very Long Baseline Interferometry
ΔDOR	Delta-Differential One-way Ranging

Chapter 1
Introduction

1.1 Significance of and General Approaches to Lunar Exploration

1.1.1 Significance of Lunar Exploration

The Moon is the only natural satellite of the Earth. It has attracted endless attention of human being. The Moon was the preferred target for deep space exploration since the space era. Since 1950s, The United States and the Soviet Union built complete aerospace industry system in early lunar exploration competition. A rapid growth of science and technology was prompted as well. At the same time, extended application of the lunar exploration technology in military and civil area cultivated a great many high-tech industrial enterprise groups and produced remarkable social and economic benefits.

For fifty years, China's space activities extended from Earth application satellites to manned space flight and deep space exploration. It is important for China to explore the Moon, and master technologies of lunar exploration, and research on lunar science and application and utilize lunar resources. China would protect its interest and devote to mankind sustainable development with lunar exploration. The significance of China lunar exploration includes as follows [1–3].

1. To Improve the Comprehensive National Strength and Enhance the Cohesion of the Nation

Lunar exploration is a major symbol of comprehensive national strength as well as science and technology of a country. Since the first China's man-made satellite was launched on 24 April 1970, aerospace technologies have made remarkable achievements in the launch vehicle, application satellites and manned spaceship in China. It will fill the gap in deep space exploration and lay a solid technology basis for future planetary exploration to implement China's Lunar Exploration Program (CLEP). It has great significance to improve the comprehensive national strength and enhance the cohesion of China.

D.-Y. Yu et al., *Technology of Lunar Soft Lander*, Space Technology Library 38,
https://doi.org/10.1007/978-981-15-6580-9_1

2. To Protect China Interests in Outer Space Activities and Future Lunar Resources Utilization

It was the major concern for the states engaged in lunar exploration to guarantee rights and obligations by the Agreement Governing the Activities of States on the Moon and Other Celestial Bodies, or well known as the Moon Treaty. It was the objective fact that only the states capable of lunar exploration could benefit from the Moon. China was a member of United Nations Committee on the Peaceful Uses of Outer Space. It is practical for China to be capable of guarantee rights and obligations by the Moon Treaty and sharing Moon benefits and protecting China's interests after implementation of lunar exploration program.

3. To Prompt Development of Advanced Technology in China

Many advanced technologies are integrated in space activities such as deep space exploration. Implementation of lunar exploration will prompt development of aerospace technologies and other advanced technologies, including launch vehicle, deep space telecommunication, new power supply, photoelectricity, robotics, artificial intelligence, remote science, new material and etc. The application of such advanced technologies in national economy and defense will improve China's science and technology as well.

4. To Prompt Innovation and Progress of Fundamental Science

The lunar surface absorbs the direct radiation from the Sun without interference such as absorption/scattering/refraction of sunlight and electromagnetic wave by the atmosphere and magnetic field like the Earth. It is ideal place to study astrophysics, space physics and material science on the Moon. It will profoundly prompt innovation and progress of fundamental science such as space astronomy, space physics, Earth observation and space material, which will prompt progress of other fundamental science at the same time.

5. To Utilize Lunar Resources and Prompt Sustainable Social Development of Human Being

There are special resources of minerals and energy on the Moon that the Earth doesn't have. Utilization of lunar resources will have profound effect on sustainable social development of human being. There is plentiful high-quality ilmenite in lunar mare basalts, which will be raw material for production of metal, oxygen and water. Other minerals resources like kalium, phosphorus, iron, uranium, thorium and other rare elements on Earth will be important supplement of Earth resources. China will devote to sustainable social development of human being by implementation of lunar exploration program.

6. To Prompt International Cooperation in Space Domain

Future space program is getting more complex and enormous investment is needed. Compared with Earth application satellites and manned space flight, lunar exploration is more open to international communication and cooperation. Attempts at international cooperation in lunar exploration will help China's space science, aerospace technology and space application to be more open, which will prompt international cooperation among China and other international parties in space domain.

1.1.2 General Approaches to Lunar Exploration

There are various approaches to explore the Moon along with development of space technology and growing requirements of exploration. To date, there had been various approaches to explore the Moon including flyby, impact, orbiting, soft-landing, sample and return, manned lunar mission. Generally we called spacecraft to the Moon 'Probe' compared to the man-made Earth satellite.

1. Flyby

Flyby the Moon means that the probe passes through the Moon at a distance from thousands to ten thousand kilometers. While flying by the Moon, the probe took photo or explored the environment. Usually the operation time was short and the accuracy was low for a lunar flyby mission. At the beginning of lunar exploration era 1960s, flyby was used because of the limit of space technology. For example, the very first photo of lunar far side was taken by the Soviet Union Luna 3 probe when flying by the far side of the Moon.

2. Impact

Impact is also called hard landing, which means that the probe impacts lunar surface at high speed without deceleration. Sometimes there might be a special impact probe onboard the lunar spacecraft. In the early stage, the probe impacted lunar surface in order to acquire high resolution image and environment data when approaching, as well as observe the matter due to the impact on the Earth or by the orbiter around the Moon. Recently, some special impact probes were developed to generate the moonquake in order to explore the internal structure of the Moon by detecting the propagation of the moonquake wave. Some impact probes might even penetrate lunar surface to study lunar interior geology, mechanics and thermal properties. It is hard to penetrate into the Moon interior so that no such mission has been accomplished up to date.

3. Orbiting

When the probe approaches the Moon, the velocity relative to the Moon is decelerated by propulsion system to make it captured by the Moon gravity, and then the probe will orbit around the Moon in circle or ellipse. Such probe is also called an orbiter. An orbiter can cover large area of lunar surface during long life cycle around the Moon, so remote sensing in visible light and multiple spectrum and exploration of space environment near the Moon can be implemented. Global survey of the Moon can be accomplished by the lunar orbiter, which makes it a common approach for lunar reconnaissance survey. Now the Soviet Union, the United States of America and China have successfully implemented lunar orbiting exploration.

4. Soft-Landing

When the probe approaches or circles the Moon, it descends by its own propulsion and lands on lunar surface at predefined attitude and low speed to support instruments operation on lunar surface. After landing, instruments onboard the lander can make different types of exploration including taking photos of lunar surface, survey of physical and chemical characteristics of lunar regolith, exploration of internal structure, radiation and magnetic field of the Moon, observing the space and the Earth by taking advantage of special condition of the Moon, and deploying the lunar rover to explore different locations on the Moon. It is more extensive and accurate in exploration of the Moon with comparison to orbiting the Moon. It is more difficult to implement soft-landing on the Moon. Only the Soviet Union, the United States and China succeeded in lunar soft-landing to date.

5. Sampling and Return

When the probe lands on the Moon, it takes lunar regolith or rock sample from surface or subsurface, and brings the sample back to the Earth after lift-off on lunar surface. It is important because the return sample can be analyzed by precise instruments in ground laboratory besides in situ exploration after soft-landing. The Soviet Union was the only country successfully implementing robotic lunar sample and return up to date. The total mass of lunar sample returned by three successful the Soviet Union missions was about 300 grams.

6. Manned Lunar Mission

A manned Moon mission is a mission that sends astronauts to the Moon orbit or surface. When astronauts arrive at the Moon orbit or surface, they can perform different kinds of exploration and research with tools and instrumentation. There are some scientific research and experiments that cannot be done by robotic missions. Usually a manned Moon mission can bring more lunar sample back to the Earth, as well as gain experience for future manned lunar base construction and manned deep space exploration beyond the Moon. In 1970s, the Apollo Program had sent 12 astronauts to lunar surface and brought back about lunar rock and regolith 380 kg.

The approach selection of lunar exploration is up to requirements of scientific exploration and feasibility of technology. The contents of this book focus on the lunar lander technology.

1.2 Status of Lunar Lander Technology Development Outside China

Only the Soviet Union and the United States succeeded in lunar soft-landing in 1960s and 1970s to date. Their success prompted development of lunar lander technology. There were 44 lunar soft-landing missions implemented by the Soviet Union and the United States before 2014, 18 of them were successful, where 7 robotic lunar landing missions by the Soviet Union and 11 by United State including 6 manned lunar soft-landing missions.

1.2.1 Development of Lunar Lander Technology in the Soviet Union [1, 14]

The Luna Program was the first Soviet Union lunar exploration program and 24 successful missions were launched since 1959–1976. There were different types of lunar exploration including flyby, orbiting, soft-landing, roving, sample and return in the Luna Program. Great achievement of the Luna Program made a lot of records in lunar exploration history of human being.

The Luna 1 was the first man made spacecraft flying by the Moon. The Luna 2 was the first spacecraft impact the Moon. The Luna 3 was the first spacecraft that took photos of far side of the Moon over 70% area. The Luna 9 was the first spacecraft that softly landed on a celestial body surface. The Luna 16 was the first spacecraft that brought lunar sample back to the Earth robotically. The Luna 17 was the first spacecraft that deployed a lunar rover on lunar surface.

There were 7 missions that successfully softly landed on lunar surface. The Luna 9 and the Luna 13 were landed by airbag. The Luna 17 and the Luna 21 were landed by legs and rovers were deployed after landing. The rovers of the Luna 17 and Luna 21 roved around the lander and took photos. The Luna 16, Luna 20 and Luna 24 brought lunar sample back to the Earth. Brief information of the Luna probes is listed in Table 1.1.

1. The Luna 9 Lander

The Luna 9 was the first man made spacecraft landing on lunar surface successfully. It was launched by Molniya rocket on 31 Jan. 1966. It landed on the 64.37°W, 7.13°N near Oceanus Procellarum of the Moon on 2 Feb. 1966. The mission goals of Luna 9 were including:

(1) To take photo near landing site.
(2) To explore the bearing capacity of lunar surface.
(3) To detect the properties of illumination, thermal and radiation on lunar surface.

The launch mass of Luna 9 was about 1500 kg. The probe was made up of landing module, instrument module and propulsion subsystem. The launch configuration is shown in Fig. 1.1 and major components are shown in Fig. 1.2.

Table 1.1 Brief information of the Luna probes

Name	Launch time	Landing Time	Brief information of the probes
Luna 1	1959-01-02	–	The first spacecraft approached the Moon, flew by the Moon and entered heliocentric orbit
Luna 2	1959-09-12	–	The first spacecraft impacted on lunar surface
Luna 3	1959-10-04	–	The first spacecraft that took photos of far side of the Moon over 70% area
Luna 4	1963-04-02	–	Designed to land on lunar surface, but entered geocentric orbit because of trajectory error
Luna 5	1965-05-09	–	Designed to land on lunar surface, but crashed on lunar surface
Luna 6	1965-06-08	–	Designed to land on lunar surface, but entered heliocentric orbit because of trajectory error
Luna 7	1965-10-04	–	Designed to land on lunar surface, but crashed on Oceanus Procellarum of the Moon
Luna 8	1965-12-03	–	Designed to land on lunar surface, but crashed on Oceanus Procellarum of the Moon
Luna 9	1966-01-31	1966-02-03	The first spacecraft that softly landed on lunar surface
Luna 10	1966-03-31	–	The first lunar orbiter
Luna 11	1966-08-24	–	Lunar orbiter
Luna 12	1966-10-22	–	Lunar orbiter that took photo of the Moon equator
Luna 13	1966-12-21	1966-12-24	Landed on the moon and analyzed lunar regolith
Luna 14	1968-04-07	–	Lunar orbiter that explored the Moon gravity
Luna 15	1969-07-13	–	Designed to take sample and return, but crashed on lunar surface
Luna 16	1970-09-12	1970-09-20	The first spacecraft that successfully took lunar sample and return to the Earth
Luna 17	1970-11-10	1970-11-17	Landed on lunar surface and deployed a rover Lunkhod 1
Luna 18	1971-09-02	–	Lunar orbiter that orbited the Moon 54 circles before losing contact with ground station
Luna 19	1971-09-28	–	Lunar orbiter whose performance was better than Luna 10
Luna 20	1972-02-14	1972-02-21	Successful robotic lunar sample and return mission
Luna 21	1973-01-08	1973-01-15	Landed on lunar surface and deployed a rover Lunkhod 2
Luna 22	1974-05-29	–	Lunar orbiter that took photos and explore lunar magnetic field
Luna 23	1974-11-28	–	Designed to took lunar subsurface sample, but failed when landing

(continued)

Table 1.1 (continued)

Name	Launch time	Landing Time	Brief information of the probes
Luna 24	1976-08-09	1976-08-18	Successful robotic lunar sample and return mission

Fig. 1.1 Luna 9 in launch vehicle

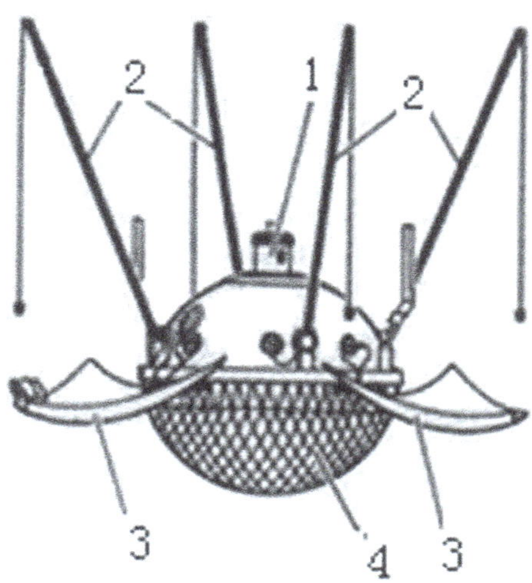

Fig. 1.2 Major components of Luna 9. **1** TV camera system; **2** Boom Antennae; **3** Petal antennae; **4** Bottom shell with thermal insulation layer

The landing module of Luna 9 looked like a ball that the diameter was 58 cm and the mass was 99 kg. The upper semisphere was TV camera and the lower semisphere was made up of battery, thermal controller and telecommunication subsystem. The lander was a sealed capsule that electronic devices were installed and the temperature was between 19 to 30 °C inside. There were four petal antennae on the top of the lander. The diameter was 160 cm and the height was 112 cm after petal antennae deployed.

After landing, the 4 petal antennae on the top of the lander were deployed and signal was transmitted to the Earth. The 4 petal antennae could help landing stability and serve as telecommunication subsystem reflector with four 75 cm boom antennae. Then the TV camera system made up of fixed and rotated lens took black-and-white photo of lunar surface near the lander. The first black-and-white photo was divided into seven parts and transmitted back to the Earth in 8 hours.

The Luna 9 stopped work because of exhaustion of the battery on 7 Feb. 1966. After the data send back by Luna 9 was analyzed, the Soviet Union scientists concluded that lunar surface was solid so that human being could land on lunar surface.

2. The Luna 16 Lander

After development of the Luna 9 and the Luna 13, new model of lunar lander was developed by the Soviet Union and was used in the Luna 16, Luna 17, Luna 20, Luna 21 and Luna 24 missions.

1) Specification of the lander

The launch mass of the lander was about 5800 kg, while the dry mass of the lander was about 1100 kg with the mass of the lunar rover or ascent vehicle was about 800 kg. It was a middle-class lander.

2) The landing process

The powered descent and landing of the Luna 16 was performed after lunar orbit insertion. The lander was inserted into a 110 km circular orbit around the Moon and then the perilune altitude was lowered to 18 km. The initial descent was performed at the perilune. When the altitude of the lander was about 2 km, the precise descent phase was initiated. The altitude to lunar surface was measured by the Doppler speedometer and radar altimeter. The attitude was controlled and the main engine was throttleable. When the altitude to lunar surface was 20 m, the main engine was shut off and small thrusters were powered on until the altitude to lunar surface was 3.5 m and the vertical velocity was less than 2.4 m/s. Then the small thrusters were power off and the lander freely fell to lunar surface. The landing process is shown in Fig. 1.3.

3) Basic configuration of the lander

The design principle of the lander was integrated structure with propulsion, i.e. the fuel tank served as bearing structure. There were four cylinder-and-spherical-dome fuel tanks for midcourse trajectory correction and lunar orbit insertion, while four spherical tanks for landing. The electronic devices were installed on the lateral wall

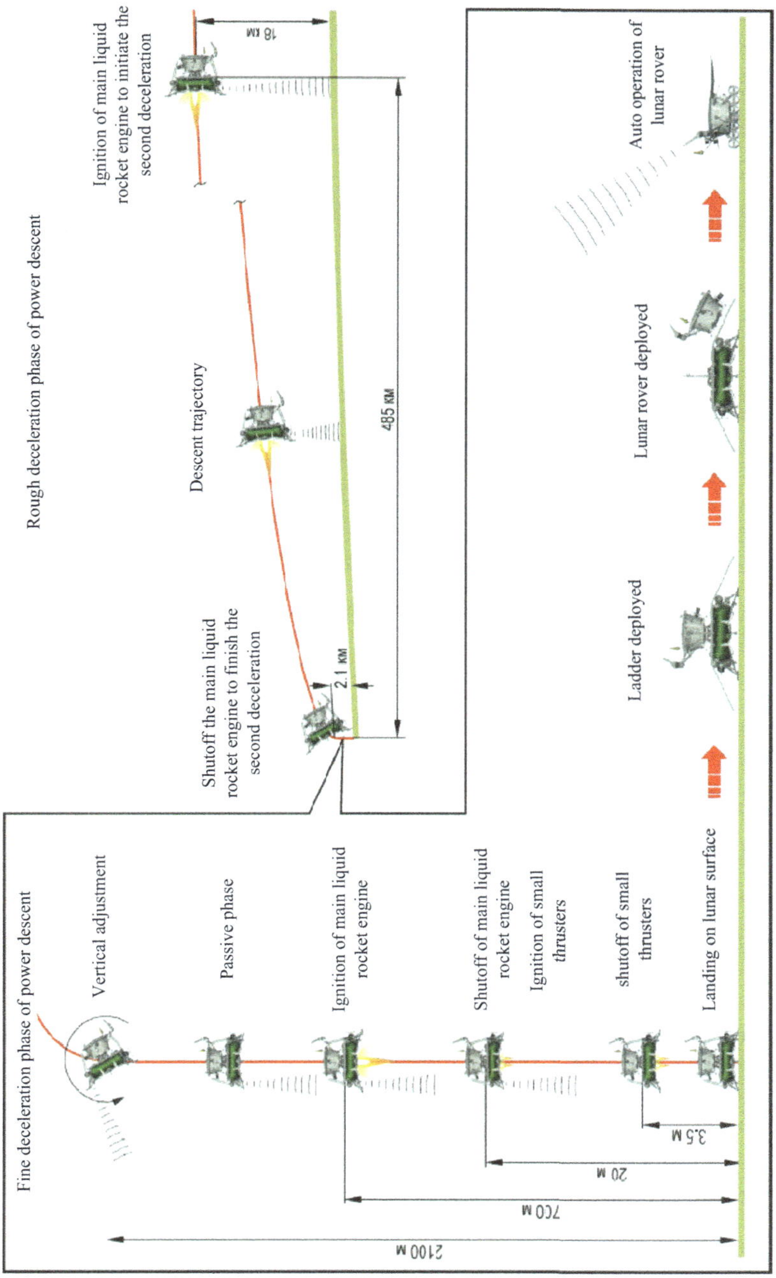

Fig. 1.3 Landing process of Luna 16

Fig. 1.4 Basic configuration of Luna 16 lander

of the tanks. There were four legs around the main structure for landing cushion. On the top of the lander, there was a platform for a rover or an ascent vehicle. A ramp for the lunar rover deployment was installed on the lateral. The basic configuration of the lander is shown in Fig. 1.4. The configuration of the lander was symmetric and balanced for dynamic environment, as well as the center of gravity was lower, which could benefit safe landing and deployment of the lunar rover.

4) Technology of navigation, control and propulsion for soft-landing

To meet with requirements of soft-landing mission, different types of sensors were used for navigation of landing control onboard the lander including star tracker, gyroscope, program controller, radio altimeter, Doppler speedometer and short range altimeter. But there was no sensor for terrain identification, so that the lander could not avoid obstacles at landing. The landing process was 'blind'.

There were two propulsion subsystems including throttleable main engine and small thruster by liquid bio-propellant for midcourse trajectory correction, lunar orbiter insertion, powered descent and attitude control. The throttleable main engine could produce 7500 N–19290 N of thrust with pump fuel feed system. There were one 2000 N thruster and several 200 N thrusters for attitude control.

5) Landing cushion

To meet the requirements of large landing mass, the landing cushion mechanics was powerful. It could work well under the condition that the vertical velocity was less than 5 m/s, the horizontal velocity was less than 1 m/s and the slope of lunar surface was less than 15 degree.

1.2.2 Development of Lunar Lander Technology in the United States [1, 6–13]

The lunar exploration program of Unite States was initiated in 1958. The ultimate goal of Unite States lunar exploration program was manned lunar landing. A series of robotic lunar missions were implemented in early stages including impact, orbiting, soft-landing and etc. Technologies for manned lunar mission were demonstrated and environment properties of the Moon were acquired by those missions.

The first generation of lunar probes of the United States was the Pioneer Program including 5 probes. The probes were designed to enter the lunar orbit, but all failed because of different reasons. After that, the second generation—the Ranger Program succeeded in lunar orbiting exploration. Nine Ranger probes were launched from Aug. 1961 to Mar. 1965: The Ranger 1 and the Ranger 2 were failed; The Ranger 3, Ranger 4 and Ranger 5 sent back images of lunar surface; The camera onboard the Range 6 failed; The Ranger 7 was the first probe that sent back photo close to lunar surface; The Ranger 8 and Ranger 9 sent back high resolution images of lunar surface before they impacted the Moon.

The third generation—the Surveyor Program probes were launched since 1965. Surveyor probes succeeded in lunar soft-landing.

1. Brief Information of Surveyor Landers

The Surveyor landers could land on lunar surface and perform scientific exploration and send back data to the Earth. Seven Surveyor landers were launched by the United States, where five of them succeeded and two failed. The first four of five successful probes landed on different locations near lunar equator, and the last one landed on the high land of south hemisphere. Four probes survived the coldest lunar night and the lifetime on lunar surface was longer than one month. All lifetime on lunar surface of the five probes was over 17 months and over 87,000 images were sent back. Six different types of chemical analysis were conducted to surface and subsurface sample. The composition of lunar regolith was tested by digging or other means. The mechanics, thermal and electric properties of lunar regolith were measured. The exploration results profoundly expanded human knowledge of the Moon. The mission goals of Luna 9 were including:

(1) Developing and validating the technology for landing softly on the Moon.
(2) Providing fundamental data for support of the Apollo Program.
(3) Adding to the scientific knowledge of the Moon by lunar exploration.

Brief information of the Surveyor landers is listed in Table 1.2.

2. Brief Introduction of the Lander Technology of Surveyor Landers

The configuration of Surveyor landers is show in Fig. 1.5. The payloads of each probe were different because of different goals of each mission.

1) Specification of the lander

The launch mass of Surveyor lander was about 1000 kg, while the dry mass was about 320 kg with 40 kg payload. There was no rover onboard the lander. It was a small class lander.

Table 1.2 Brief information of the Surveyor landers

Name	Launch time	Landing Time	Brief information of the probes
Surveyor 1	1966-05-30	1966-06-02	The first successful soft-landing on the Moon by the United States, over 11,000 images of lunar surface were sent back Proved that strength of lunar surface could support the lander Demonstrated the capability of soft-landing and telecommunication and acquired engineering data of the lander and data of Moon surface environment
Surveyor 2	1966-09-20	–	One thruster failed during midcourse trajectory correction to make the lander rotate, failed
Surveyor 3	1967-04-17	1967-04-20	Successfully landed on predefined Apollo landing site to the east of Surveyor 1 landing site and took images Acquired parameters of bearing capacity, radio reflection and thermal properties of lunar surface Dug the surface by sampler to observe influence of the lunar regolith properties to digging
Surveyor 4	1967-07-14	–	Lost contact during powered descent and landing, failed
Surveyor 5	1967-09-08	1967-09-11	Landed on Mare Tranquillitatis. Took images and conducted first chemical analysis of lunar regolith composition
Surveyor 6	1967-11-07	1967-11-10	Successfully landed and took images. Acquired parameters of dynamics, thermal and radio reflection properties of lunar surface Jumped once after landing to detect the mechanics properties of the lunar regolith
Surveyor 7	1968-01-07	1968-01-10	Successfully landed. Conducted detection of lunar regolith composition with alpha scattering device. Measured distance between the Earth and the Moon by laser ranging technology

Fig. 1.5 Configuration of Surveyor landers

2) The landing process

The Surveyor landers were launched directly into trans-lunar orbit and then landed on lunar surface without lunar orbit insertion. The trajectory of trans-lunar orbit intersected lunar surface. One or two midcourse trajectory correction was performed during trans-lunar flight. When the probe approached the Moon, powered descent and soft-landing was performed. The flight profile of the Surveyor landers was shown in Fig. 1.6.

It was 66 hours after launch for the Surveyor lander to perform soft-landing when the distance to the Moon was about 96 km. The first phase was deceleration and the second phase was powered descent. When the distance to the Moon was about 4.2 m, the engine was powered off and the final landing velocity was less than 6 m/s.

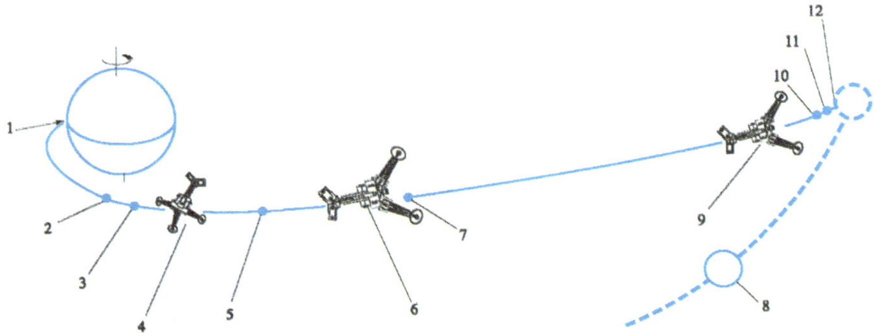

Fig. 1.6 Flight profile of the Surveyor landers. **1** Lift-off; **2~3** Launching; **3** Orbit insertion and separation; **4** Initial capture of TT&C; **5** Sun capture; **6** Sun oriented; **7** Midcourse correction; **8** Moon position when launched; **9** Ready for landing; **10** Initiation of main deceleration; **11** Descent with throttling engine; **12** Landing on lunar surface

It was short time for this type of flight process. But the precision of landing was low. There were few missions like the Surveyor in subsequent missions.

3) The basic configuration of the Surveyor lander

The configuration of the Surveyor lander was like a tripod. The main structure was truss type. The main engine was attached to the lower corners of the structure. Three landing legs were installed on three corners. There were deployable solar panel and telecommunication antenna. Instruments were installed on the truss. The truss type was light weight and easy for instruments installed so that it was suitable for small class lander.

4) Technology of navigation, control and propulsion for soft-landing

There were sun sensors, star sensors, Inertial Measurement Unit and etc. onboard the Surveyor lander in order to measure three axis attitude and acceleration. The radio altimeter and the Doppler speedometer were used to measure the distance and three-axis velocity to the Moon during final landing after the main engine was powered on. Similar to the Luna lander, there was no sensor for terrain identification onboard the Surveyor lander because of the limit of technology at that time. The landing process was also 'blind'. So the landing site for the Surveyor lander were all chose to be flat terrain. Propulsion and attitude control were provided by cold-gas (nitrogen) attitude control jets during cruise phases, three throttleable liquid bipropellant engines during powered phases including the landing, and the solid-propellant retrorocket engine during terminal descent. Each throttleable engine could produce 133 N to 463 N of thrust. The solid-propellant retrorocket engine could produce average 40000 N of thrust and was ignited for 40.5 second burn.

5) Landing cushion

The three landing legs of the Surveyor lander could absorb impact energy of final landing by crushable honeycomb aluminum inside. It could adapt to lunar surface condition that the slope was less than 15 degree, the height of hump was less than 10 cm and the coefficient of friction was more than 0.1.

6) Telecommunication

Telecommunications were achieved via S band low gain antenna and a large planar array high gain antenna to send back images and data back to the Earth.

3. Major Scientific Exploration Activities and Results of The Surveyor Lander

A great many scientific exploration was implemented by the Surveyor Program and solid basis was built up for subsequent lunar scientific research and engineering implementation including:

(1) A lot of high-resolution images (1 mm resolution). Tens of thousands of images were taken by the Surveyor landers which provided natural information of lunar terrain near the landers including numbers, size and distribution of rock and pits. Those photos also supported research of lunar regolith mechanics, magnetic properties and composition.
(2) Measurement of lunar surface mechanics properties. The lunar surface mechanics properties were studied through retrieval of telemetry data and analysis of images of landing process. There were strain gauges mounted on the lander.
(3) Research on lunar regolith mechanics properties. Samplers of the Surveyor 3 and Surveyor 7 were used to execute bearing capacity, impact and scoop experiments with monitoring of TV cameras to study the lunar regolith mechanics properties.
(4) Analysis of lunar surface temperature and thermal properties. Thermal sensors on outer surface and solar panel of the Surveyor landers were used to measure local thermal radiation environment.
(5) Analysis of magnetic properties of lunar surface. Small magnet attached to the bottom of the Surveyor 5 and Surveyor 7 as well as sampler of the Surveyor 7 was included to detect the magnetic field. The images of magnet covered with lunar dust showed that there were some magnetic particles in lunar regolith.
(6) Analysis of composition of lunar regolith. The data collected by alpha scatters on the Survey 5 and Surveyor 7 were used to determine the composition of lunar regolith for chemical analysis.

The Surveyor Program successfully implemented predefined scientific exploration and validated the technology for soft-landing on the Moon to support the subsequent Apollo Program.

1.3 China's Lunar Exploration Program

1.3.1 General Plan of CLEP

Since 1990s, scientists and aerospace experts began to study the scientific goals and key technologies of China's lunar exploration. At the beginning of 2004, the first phase of CLEP was approved by the Chinese government, i.e. the project of orbiting the Moon. The second and third phase of CLEP was listed in the "National Guideline on Medium- and Long-Term Program for Science and Technology Development (from 2006 to 2020)" as one the 16 National Science and Technology Major Programs. Proposal and preparation for the implementation of the second and third phase of CLEP was initiated then.

There were three steps in CLEP including orbiting the Moon, landing on the Moon, robotic sample and return from the Moon. The name of CLEP was named after a fairy tale goddess of Chinese conventional culture [2, 3, 15].

The first step was orbiting the Moon from 2004 to 2007. The mission goals were: developing and launch the first Chinese lunar probe to break through key technologies for lunar orbiting exploration; exploring the lunar terrain, element abundance and material distribution, lunar regolith, and space environment between the Earth and the Moon globally; establishing primary engineering system of CLEP.

The second step was soft-landing and roving on the Moon from 2008 to about 2013. The mission goals were: breaking through key technologies for lunar soft-landing, roving on lunar surface, deep space telecommunication and launching; developing lunar lander and rover for in situ exploration on lunar surface including research on lunar terrain of landing site, geological structure and material composition, and astronomy observation based on the Moon.

The third step was robotic sample and return from the Moon before 2020. The mission goals were: breaking through key technologies for lunar sampler such as driller and surface sampler, robotic operation on the Moon; studying the lunar sample on ground laboratory after recovery of return capsule on the Earth based on in situ analysis of the lunar sample in order to deepen the knowledge of the origin and revolution of the Earth-Moon system.

Till 2014, four missions of CLEP were implemented successfully including the Chang'E-1, the Chang'E-2, the Chang'E-3 and the Chang'E-5T, i.e. the circumlunar return and reentry spacecraft.

1.3.2 Brief Information on Chang'E Probes

1. Chang'E-1 Probe [4]

The Chang'E-1 probe was Chinese first lunar orbiter. The scientific objectives of the Chang'E-1 were: acquiring stereo 3-D images of lunar surface; analyzing the element abundance and material distribution; detecting the thickness of lunar regolith; exploring space environment between the Earth and the Moon. The Chang'E-1 was launched at 18:05 24, Oct. 2007 (Beijing Time) in Xichang satellite launch center. It was inserted into 200 km Lunar Low Orbit (LLO) and became a manmade lunar satellite at 7, Nov. 2007. Over 1.3 TBytes scientific data was transferred back including the Moon global images, some element abundance on lunar surface and etc. during the Chang'E-1 lifetime around the Moon. Scientific and engineering objectives were all achieved successfully. 494 days after launched, the Chang'E-1 impacted on the Mare Fecunditatis after 37 minutes deceleration from 15:36 1, Mar. 2009 by command.

2. Chang'E-2 Probe

The Chang'E-2 probe was originally a backup of the Chang'E-1. After success of the Chang'E-1, some new requirements of 2nd step of CLEP were integrated in the Chang'E-2 probe. The mission goals of the Chang'E-2 were acquiring high resolution images of predefined landing site—Sinus Iridum for the Chang'E-3 soft-landing and continuing scientific lunar exploration. The Chang'E-2 was launched at 19:00 1st, Oct. 2010 in Xichang satellite launch center. It orbited the Moon at 100 km circular orbit and 15 km × 100 km orbit to conduct lunar scientific exploration and high-resolution imaging. It escaped from Lunar Low orbit at 16:50 9, Jun. 2011 and entered a trajectory to the Sun-Earth Lagrange point 2 1.5 million kilometers away from the Earth. The Chang'E-2 probe then became the first Chinese manmade satellite of the Sun that escaped from the LLO. It flew by the Toutatis asteroid and acquired high resolution images of the Toutatis asteroid on Dec. 2012. The distance of the Chang'E-2 to the Earth was more 90 million kilometers and became the Chinese furthest spacecraft till now.

3. Chang'E-3 Probe

The Chang'E-3 probe was made up of a lander and a rover named Yutu or Jade Rabbit. The mission goals of the Chang'E-3 were first soft-landing, roving on lunar surface and conducting in situ exploration including observing the Moon, the Earth and space on the Moon. The Chang'E- 3 was launched at 01:30 2nd, Dec. 2013 in Xichang satellite launch center. The lunar orbit insertion was performed by the Chang'E-3 probe at 17:53:06, Dec. 2013 to enter into 100 km circular LLO. The lunar lander was landed successfully on the Moon at 21:11:14, Dec. 2013 and pretty images of landing process taken by the descent camera was transferred back. The Yutu rover was deployed at 04:35 15 Dec. 2013. After that, the lander and the lunar rover conducted scientific and engineering exploration respectively.

The Chang'E-3 was the Chinese first soft lander on celestial body. It broke through a lot of key technologies and served as a connecting link in the CLEP.

1.4 Technical Challenges to Development of Lunar Lander [5, 14, 16]

There are 6 major technical challenges from design to validation for the lunar lander.

The uncertainty of lunar terrain environment. The terrain of lunar surface is complex including mare and upland. The mare is relatively smooth terrain covering about 17 percentage of lunar surface, which is similar to the basin on the Earth. The upland is the area higher than the mare covering about 83 percentage of lunar surface, which is relatively rugged and undulatory. There are rock and craters in different size and shape. Although there is some macro statistics information about the distribution of craters, rock and slopes on lunar surface, it is still uncertain for a particular landing site. Those environmental factors must be considered carefully when a lunar lander is designed and simulation method for ground experiment and test is chosen, which are also important for a successful soft-landing. If the terrain of landing site is extremely rugged, the lunar lander might fail to have a stable land on lunar surface.

Power descent without assistance of atmosphere. There is no atmosphere on lunar surface. So powered descent of the lunar lander must be done by its own propulsion system without assistance of air brake. The velocity of the lunar lander relative the Moon must be decelerated from 1.7 km/s to several meters per second in order to have a soft-landing. It was challenging for the propulsion system to provide enough Delta-V and throttleable thrust at the same time. Because the Delta-V was big, a great many propellant must be fueled in propulsion system. Sometimes the wet mass (mass of the propellant) was bigger than 2/3 of the total launch mass of the lunar lander. Thus a series of technical difficulties were induced in system design and structure design of the lunar lander.

Autonomous navigation and control of landing process. The period of powered descent was about 10 minutes, which was relatively short. The dynamics of landing process varied fast so it was hard for ground operators to control the lander during powered descent. It must be done autonomously in close loop control. To complete autonomous control tasks including deceleration, descent along with attitude maneuver, hovering and obstacles avoidance, slow descent, autonomous navigation and control must be achieved by speedometer, ranging and terrain identification sensors onboard the lander. It was a big challenge because there was no such technology in the past spacecraft including accurate terrain identification, control of velocity and height, control of throttleable thrust, attitude control and obstacles avoidance maneuvers.

Cushion of landing impact. When the lander touched lunar surface, the velocity relative to the Moon was not zero. So the landing cushion system should absorb the impact energy of the lander to avoid possible damage to devices onboard the lander. At the same time, requirement of landing stability and envelop constraints of launch vehicle fairing must be satisfied when the landing cushion system was designed.

Thermal control on lunar surface. The temperature difference between lunar day and night is extreme because the rotation of the Moon is slow and there is no atmosphere. The highest temperature in lunar day is 120 °C, while the coldest temperature in lunar night might be −180 °C. The temperature difference of 300 °C induced big challenge in thermal control of the lunar lander on lunar surface. New type of thermal technologies and equipment must be used to control internal temperature of the lander, while strong adaptability to extreme temperature difference must be satisfied for the devices direct exposure to the Moon environment.

Ground simulation and test for the lunar environment. The environment of the Moon is totally different from the Earth and Earth orbit such as low gravity, extreme temperature, terrain, lunar regolith, lighting condition, lunar dust and radiation. To make sure the sufficiency and validity of ground test in lunar lander development, special facilities must be built up to simulate lunar environment and new theories and methods should be developed for ground experiment and test.

1.5 Summary

Lunar exploration of the lunar lander was 'immersive' compared to remote sensing. More precise and profound research fruits might be achieved by lunar lander exploration. It was not only an important approach of lunar scientific research and utilization of lunar resource by human being, but also crucial key technology for lunar base construction and manned Moon missions. It was very important to master technology for lunar soft-landing and capability to arrive at surface of celestial bodies in the Moon exploration and deep space exploration.

CLEP had attracted worldwide attention. According to the strategic plan of CLEP, the the Chang'E-3 probe was successfully landed on the Moon in Dec. 2013 and conducted long-term scientific exploration on lunar surface, which made China the 3rd country that was capable to soft-landing on the Moon in the world.

The significance of lunar exploration is introduced in this chapter. A brief review of successful lunar soft-landing of the Soviet Union and the United States in history is given. At the end, general information about CLEP and brief information of the Chang'E probes is described and major technical challenges of lunar lander is analyzed. It can provide primary background information for the readers to read subsequent chapters.

References

1. Chu G, Zhang H (2007) Technology of lunar probes[M]. China Science and Technology Press, Beijing
2. Ouyang Z (2004) Scientific objectives of Chinese Lunar Exploration Project and development strategy[J]. Advance in Earth Sciences 19(3):351–355

3. Ouyang Z (2005) Review of lunar exploration and introduction of Chinese Lunar Exploration Project[J]. Chinese Journal of Nature 27(4):187–190
4. Ye P, Sun Z, Rao W (2007) Research and development of Chang'E-1[J]. Spacecraft Engineering 16(6):9–15
5. Han H, Wang Y, Jiang Y et al (2012) Characteristics and application of deep-space explorer landing impact attenuation system[J]. Spacecraft Engineering 21(6):7–24
6. Jet Propulsion Laboratory (1966) Surveyor 1 mission report: Part I. Mission description and performance[R]. California Institute of Technology
7. Jet Propulsion Laboratory (1967) Surveyor 2 mission report[R]. California Institute of Technology
8. Jet Propulsion Laboratory (1967) Surveyor 3 mission report: Part I. Mission description and performance[R]. California Institute of Technology
9. Jet Propulsion Laboratory (1968) Surveyor 4 mission report[R]. California Institute of Technology
10. Jet Propulsion Laboratory (1968) Surveyor 5 mission report: Part I. Mission description and performance[R]. California Institute of Technology
11. Jet Propulsion Laboratory (1968) Surveyor 6 mission report: Part I. Mission description and performance[R]. California Institute of Technology
12. Jet Propulsion Laboratory (1969) Surveyor 7 mission report: Part I. Mission description and performance[R]. California Institute of Technology
13. Surveyor Program (1969) Lunar and planetary program division office of space science and applications. Surveyor program results[R]. NASA, USA
14. Wang C, Deng Z, Gao H et al (2006) Development status of lunar landers[J]. Missiles and Space Vehicles 284(4):31–36
15. Zheng W, Xu Z, Zhong M et al (2012) Progress in international lunar exploration programs[J]. Progress in Geophysics 27(6):2296–2307
16. Wu W, Yu D (2014) Key technologies in the Chang'E-3 soft-landing project[J]. Journal of Deep Space Exploration 1(2):105–109

Chapter 2
Environment Analysis of Lunar Soft Landing Exploration

2.1 Introduction

The lunar environment considered in the lunar soft-landing mission is generally the natural environment, of which the lunar probes are exposed to on lunar orbit or on lunar surface, such as the lunar radiation environment, the thermal environment on lunar surface, the topography of lunar surface, the lunar soil, the lunar dust, the gravitational field of the Moon, the nearly vacuum environment of the Moon and the variation of lighting condition on lunar surface [1]. These features have various effects on the normal function and safety of the lunar lander, and are also the important bases and constraints in the design of the lunar lander. In this chapter, engineering design parameters are extracted from the related references inside and outside of China and the data obtained in the Chang'E missions of China, which can be used as reference in the design of future lunar lander.

This chapter focuses on the analysis of the lunar environment itself, while its detailed effect on the lunar lander will be discussed in later chapters. In addition, some special designs of the lunar lander may introduce new and unusual environment effects. For example, the lunar lander is equipped with Radioisotope Heat Unit (RHU) to deal with thermal control in lunar night of 14 Earth days, while the RHU produces nuclear radiation environment. Another example is the lunar dust environment and the thermal environment that are caused by the interaction between the plume of the rocket engine and lunar surface in the landing process [2–5]. Such secondary environment is called induced environment. The effects of lunar environment on the lunar lander, as well as the analysis of induced environment and its effects on the lunar lander will be discussed in Chapter 3.

D.-Y. Yu et al., *Technology of Lunar Soft Lander*, Space Technology Library 38,
https://doi.org/10.1007/978-981-15-6580-9_2

2.2 Lunar Radiation Environment

On lunar orbit or on lunar surface, the major sources of charged particles radiation include the cosmic ray, the solar energetic particles and the plasma environment on lunar orbit.

2.2.1 *Galactic Cosmic Rays*

The Galactic Cosmic Rays (GCRs) come from the interstellar space of the Milky Way galaxy beyond the solar system. It is primarily composed of lower flux and extremely high-energy protons and heavy ions. The energy of the GCRs particles is generally ranging from 10^2 to 10^{12} MeV, while the energy of the majority of particles is between 10^3 MeV and 10^7 MeV. The free space flux of the particles is only 0.2~0.4 $(cm^2 \cdot sr \cdot s)^{-1}$. GCRs cover almost all the elements in periodic table of elements, while the energy spectrum of the particles is different from each other. Among them, 84.3% are protons, 14.4% are α particles and the rest 1.3% are the other heavy nuclei.

Due to the shielding effect of the interplanetary magnetic field on the GCRs, the energy spectrum of the GCRs that reach the Earth space has a close relation with intensity of interplanetary magnetic field. Since the solar activity affects on the interplanetary magnetic field, it also affects the GCRs. Generally, there is negative correlation between the intensity of the GCRs and the solar activity. The intensity of cosmic ray reaches its minimum in a solar maximum year and reaches its maximum in a solar minimum year.

2.2.2 *Solar Energetic Particles*

The solar particle events (also called the solar proton events since the most of particles are proton) happens during the solar storm. It is a paroxysmal turbulent of radiation environment and its intensity is related to the intensity and the position of solar burst activity. Large number of high-energy particles (primarily protons) are ejected in the solar burst activities (such as solar flare and coronal mass ejection). Due to the effects of various acceleration mechanisms, the ions and electrons are accelerated to a very high energy level from the lower layer of atmosphere of the Sun to the broad interplanetary space. They travel to the Earth along the solenoid magnetic field of the Sun under the effects of propagation and diffusion, which are eventually detected by the spacecraft in the space and the particle monitor on the ground.

The solar energetic particles are composed of all the elements that exist in nature. The basic properties of solar particle events are listed in Table 2.1. It is shown that most of the solar high-energy particles are protons, and the rest are α particles and a very small amount of heavy nuclei particles in Table 2.1. Although the proportion

Table 2.1 Basic characteristics of solar particle events

Proportion of particles	Energy	Integral flux of particles higher than 10 MeV/cm^{-2}	Peak of differential flux of particles higher than 10 MeV/(cm^{-2}·s-1)
96.4% of proton, 3.5% of α particle and about 0.1% of other heavy nucleus particle	As high as GeV	As much as 10^{10}	As much as 10^5

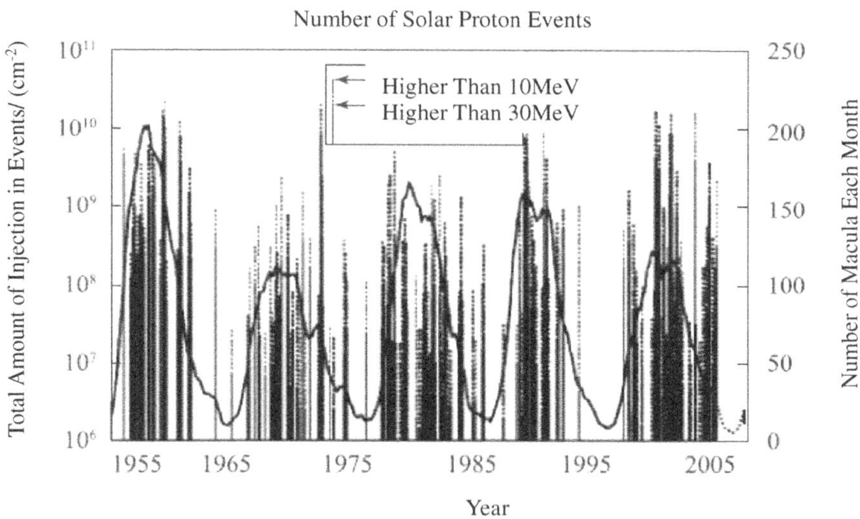

Fig. 2.1 Relationship between the integral flux of proton and the average value macula over months in the solar proton events during the No. 19~23 solar activity week

of heavy nuclei particles is low, its effect shall not be ignored. In fact, just like the proton and α particle, the heavy nuclei particles are an important source of single event effect on spacecraft.

The observation records to date show that the solar particle event can be classified as impulsive event and progressive event. The impulsive event is generally caused by solar flare eruption and is rich in electrons of energy ranging from 1 keV to 100 keV. The ratio of ^{3}He to ^{4}He is relatively high. But the duration of the event is short (usually several hours) and the integral flux of high-energy flux is generally low. The progressive solar particle event is primarily caused by coronal mass ejection and its primary component is proton. Such events usually last longer (several days) and the integral flux of high-energy particles may exceed 10^9 cm^{-2}. Therefore, the more intense solar particle event is generally the progressive event.

The solar particle event has a strong correlation with the intensity of solar activity. Figure 2.1 shows the relationship between the integral flux of proton and the average value macula over months in the solar proton event during the No. 19~23 solar activity week. It is shown that the larger solar proton events (with higher proton flux) often occur at the time of more intense solar activity (with more macula). Significant correlation exists between the intensity/numbers of solar proton event and the solar activity period. Around the peak year of solar activity, the number of solar proton events may be as many as 10 times, while in the valley year of solar activity the number of solar proton events may reduce to three or four times, or even less [6].

2.2.3 *Plasma Environment in Lunar Orbit*

While the Moon is orbiting the Earth in different phases, it is either inside or outside the magnetic tail of Earth. The relationship between the orbit of the Moon and the magnetic tail of the Earth is shown in Fig. 2.2. When the Moon is outside the magnetic tail, it is exposed to the solar wind. The energy and component of particles are very different under the plasma environment of the solar wind and the magnetic tail.

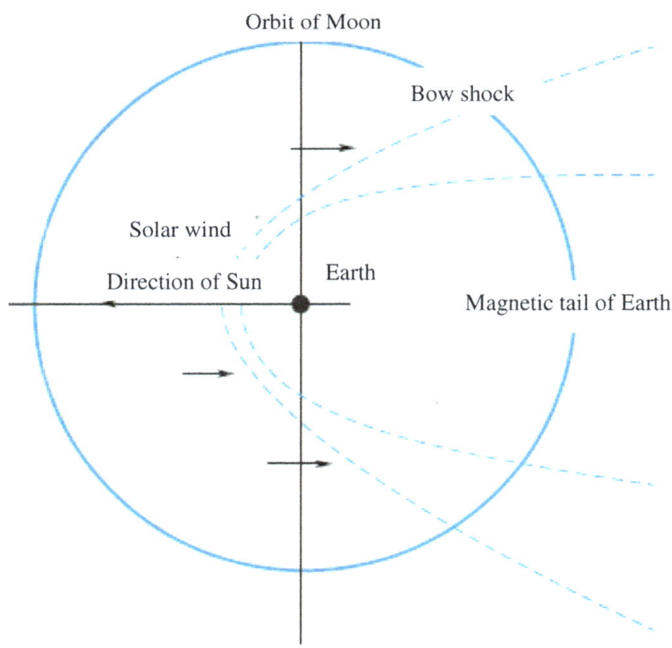

Fig. 2.2 Relationship between the orbit of the Moon and the magnetic tail of the Earth

Table 2.2 Basic parameters of the solar wind

Basic parameter	Value	Basic parameter	Value
Average energy/(keV/u)	1	Average velocity/(km/s)	450
Density of particle/cm^{-3}	1~20	Energy of electron	Several electron volts
Density of electron/cm^{-3}	7.1	Energy of proton/keV	0.21~3.3
Density of proton/cm^{-3}	6.6	Magnetic field/nT	7

1. Solar Wind

The solar wind is rare plasma originated in high temperature corona. When the temperature (kinetic energy) of high temperature corona is over the control of solar gravity, the solar wind emits in all direction. On average, the solar wind particles are composed of about 95% of protons, 4% of α particles and about 1% of heavy ions. Among the heavy ions, the abundance of carbon, nitrogen, oxygen, neon, magnesium, silicon and ferrum is relatively high. The basic parameters of solar wind are shown in Table 2.2.

Since the energy of proton in the solar wind is only a few thousand electron volts, and its penetrating range inside the material of spacecraft is only about 0.1 μm, it does not have any effect on spacecraft beside that the very thin optical coating suffers a certain amount of damage caused by total ionizing dose [7].

In the space plasma environment, electrons and ions have similar density and energy (in electroneutral state). However, the mass of ion is 1000 times of the mass of electron, therefore its speed is far lower than that of electron. The electronic current is in leading role and the negative current is larger than positive current. In other words, the charging effect in spacecraft is mainly triggered by electron. It is shown that the energy of electrons in solar wind is as low as a few electron volts in the Table 2.2. Its influence on the charging effect in spacecraft can be ignored.

2. Plasma in Magnetic Tail

The Moon passes the remote magnetic tail region of the Earth magnetic field while orbiting the Earth. In each revolution period, the Moon stays in the magnetic tail of Earth for about 4 days (before and after the full moon). While passing the magnetic tail, the Moon encounters two plasma regions that have very different properties. The two regions are plasma sheet and plasma mantle (which is also called tail lobe). The density of plasma in plasma sheet is very low and is usually lower than 0.1 cm^{-3}. The energy of plasma is lower than several electron volts in general. Therefore, its effect on spacecraft can be ignored. Plasma mantle distributes over the magnetic equator of Earth with a thickness of about 4Re (Re is radius of the Earth). It is primarily composed of thermal plasma with thousand electron volts of energy.

2.3 Lunar Thermal Environment

The natural thermal environments that the lunar lander encounters in lunar orbit or on lunar surface are the solar radiation, the lunar reflection and the lunar radiation [8].

2.3.1 *Solar Radiation*

Solar radiation is the primary heat source of the lander in operation. It is mainly comprised of the visible light, the infrared light and the X ray of the electromagnetic radiation of the Sun. The relative distance and angle between the Sun and the Earth vary according to different position of Earth on the ecliptic plane. Therefore, the solar radiation energy that reaches Earth varies with the distance from Earth to the Sun when Earth is at different position on the ecliptic plane.

The direct solar radiation is generally expressed in solar constant S. The solar constant is a flux density in the unit of W/m^2 measuring mean solar electromagnetic radiation (solar irradiance) per unit area. It is measured on a surface perpendicular to the rays, one astronomical unit (average distance from the Sun to Earth is 1AU, which is about 1.5×10^8 km) from the Sun.

For the lander on lunar surface, since the Moon may be at the closest or furthest point of the Earth-Sun line, the distance between the Sun and the Moon is ranging between $1\text{AU} - 3.8 \times 10^5$ km and $1\text{AU} + 3.8 \times 10^5$ km. Therefore, the average distance between the Moon and the Sun is approximately 1AU in analyzing the solar radiation energy. Meanwhile, since the lunar atmosphere is extremely thin, the attenuation of solar radiation energy by lunar atmosphere can be ignored, similar as Earth spacecraft beyond Earth atmosphere.

Therefore, the value of solar radiation energy for the lunar lander is the same as it for the Earth spacecraft that are beyond Earth atmosphere. The solar radiation energy at 1AU is denoted by 1S and the value is 1367 W/m^2. S varies in different seasons. It reaches its maximum of 1414 W/m^2 in winter solstice (perihelion) and reaches its minimum of 1322 W/m^2 in summer solstice (aphelion).

2.3.2 *Lunar Albedo*

For the solar radiation that reaches the Moon, part of it is absorbed and the other part is reflected. The reflected part is called lunar albedo. The albedo of solar radiation from the lunar surface is another important thermal environment for the lunar lander in lunar orbit or on lunar surface. The albedo of Moon is closely related to the thermal property (absorption and reflection) of lunar substance (such as lunar soil and lunar rock). The authoritative lunar albedo is documented in the design criterion SP-8023

Table 2.3 Nominal albedo of lunar surface

Regions on the Moon	Nominal Albedo		
	Min	Max	Average
Dark region	0.07	0.12	0.095
Upland region	0.108	0.24	0.150

"Lunar Surface Models, NASA Space Vehicle Design Criteria (Environment)" of NASA. Based on the data from this criterion, a certain margin in mission design is required to offset the uncertainty and limitation. The nominal albedo of lunar surface given by the design criterion SP-8023 of NASA is shown in Table 2.3.

2.3.3 Lunar Radiation

The infrared radiation of lunar surface is one of the heat sources for the lunar lander. Lunar radiation data is also given in the design criterion SP-8023 of NASA: In a unit area with average albedo of 0.110, the thermal radiation from lunar surface is about 310 W/m^2.

Based on the "Explorer" A-21A model, the infrared emissivity of Moon $\varepsilon = 0.875$ and the sunlight absorptivity $= 0.93$. These data can be applied in design of the lunar lander.

2.3.4 Earth Albedo

For lunar landers having orbiting Earth phases, heat flux from Earth albedo should be considered in thermal analysis. The heat flux density of Earth albedo varies with the surface morphology of Earth, the season, the geographical latitude and the orbit inclination. However, the flight altitude and speed of the lunar lander are high, and the energy of Earth albedo is relatively low compared to the energy of direct radiation from Sun. Therefore, an average albedo of 0.35 is usually used in the thermal analysis.

2.3.5 Earth Radiation

For the lunar lander passing the Earth, heat flux from Earth radiation should be considered in thermal analysis. The heat flux density of Earth radiation varies with the surface morphology of Earth, the season, the climate and the geographical latitude. In thermal analysis, an average value is usually used. The average infrared heat flux density of Earth is about 220 W/m^2.

2.4 Lunar Landform and Topography

2.4.1 Lunar Landform

The lunar surface can be generally classified in two kinds of landforms, that is, the lunar mare and the upland. The lunar mare is the broad plain on lunar surface, which occupies about 17% of the surface area of the Moon. Most of the lunar mares are in circular shape and surrounded by ridges (the slender and stretched mountains). It is similar to the basin on Earth. Upland is the region that is 2~3 km higher in altitude than lunar mare. It occupies about 83% of the surface area of the Moon [9]. The distribution of lunar mare and upland on the near side and far side of the Moon is shown in Figs. 2.3 and 2.4.

Fig. 2.3 The distribution of lunar mare and upland on the near side of the Moon (https://cseligman.com/text/moons/moonnear.htm)

Fig. 2.4 The distribution of lunar mare and upland on the far side of the Moon (https://cseligman.com/text/moons/moonfar.htm)

2.4.2 Lunar Topography

Lunar mare and upland are covered with rocks and craters in different size and shape. Craters are the ring pits that can be found everywhere on the Moon, including the ring mountain, the radiating ridge and the rise feature. Figure 2.5 is the first picture taken by the Chang'E-3 lunar lander after landing. Large amounts of rocks and craters can be found in the picture. The slope, the rocks and the craters of the landing site may have impact on the soft-landing and operation on lunar surface.

1. Distribution of Slope on Lunar Surface

The changes of average slope of lunar surface according to the variation of baseline for different topography is shown in Fig. 2.6. There are two major characteristics of the distribution of slope: the first is that the standard deviation of the average distribution

Fig. 2.5 The first picture taken by the ChangE'-3 lander after landing on the Moon

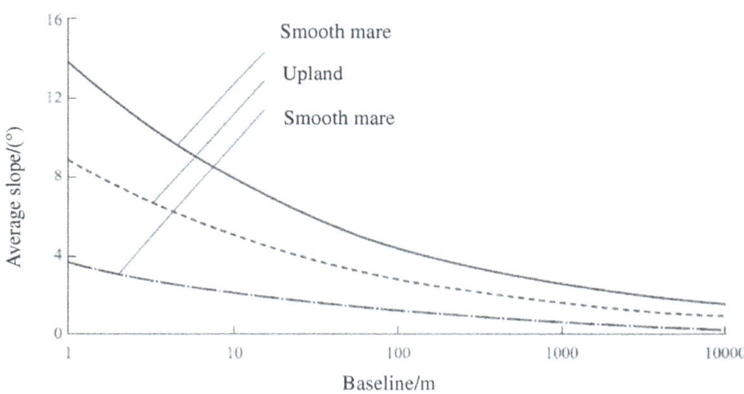

Fig. 2.6 The changes of average slope of lunar surface according to the variation of baseline for different topography

of slope is generally 30% of average slope; the second is that the maximum slope is usually no more than 4 times of the average slope.

The cumulative distribution of slope of the smooth mare and the rough mare is shown in Fig. 2.7. With the increase of slope, the variation of cumulative percentage of area becomes slower, which means that the larger the slope, the smaller the percentage. Besides, as the baseline increases, the maximum of slope becomes

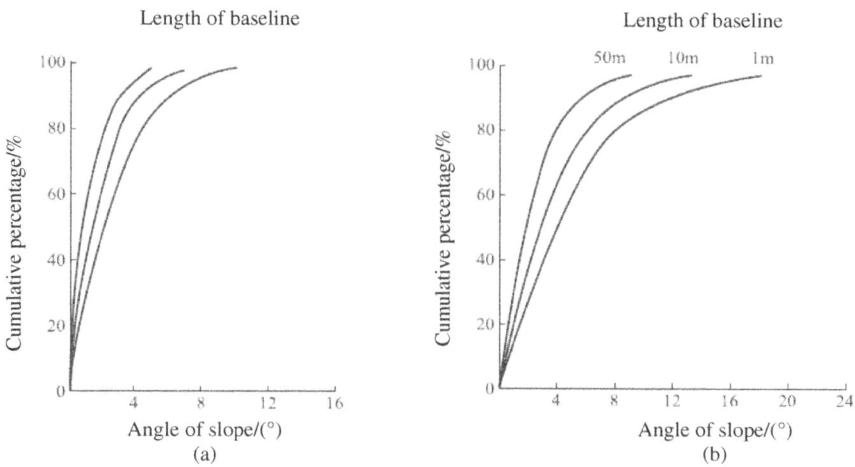

Fig. 2.7 The cumulative distribution of the slope of the smooth mare and rough mare. **a** Smooth mare; **b** Rough mare

smaller (in Fig. 2.7, the cumulative area below the maximum slope is about 95% of the total area). Take the smooth mare as an example, the maximum slope is about 10° when the baseline is 1 m, and the maximum slope is about 5° when the baseline is 50 m.

Table 2.4 shows the cumulative percentage of the area that is occupied by the slope over a certain angle, with the slope ranges from 0 to 15° and the baseline length is 10 m and 25 m.

It shows that the longer the baseline, the smaller the percentage of area that is occupied by the slope over a certain angle by analysis of the data in Table 2.4. For the smooth mare, the percentage of area that is occupied by the slope greater than 7° is 0 when the baseline length is 10 m.

2. Distribution of Rocks on Lunar Surface

The cumulative distribution of rocks in different topography is shown in Fig. 2.8.

By applying fitting method to Fig. 2.8, the cumulative number of rocks with diameter greater than D in a unit area (1 m^2) in smooth mare, rough mare and upland can be estimated. The functions are shown below.

In the smooth mare, the cumulative number of rocks $N_{b,s}$ (m^{-2}) larger than a certain diameter in a unit area (1 m^2) can be expressed as

$$N_{b,s}(D) = 10^{-3.3955} \times D^{-2.589} \tag{2.1}$$

where D is the diameter of rock.

In the rough mare, the cumulative number of rocks $N_{b,r}$(m^{-2}) larger than a certain diameter in a unit area (1 m^2) can be expressed as

Table 2.4 The cumulative percentage of the area that is occupied by the slope over a certain angle

Slope/(°)	Smooth mare		Rough mare	
	10 m baseline (%)	25 m baseline (%)	10 m baseline (%)	25 m baseline (%)
1	62.8	53.4	82.8	79.2
2	42.8	31.1	63.3	54.4
3	19.4	13.5	48.8	39.0
4	14.4	8.5	37.2	27.0
5	6.1	2.3	27.4	19.1
6	3.3	1.2	20.9	13.5
7	0	0	15.3	9.0
8	0	0	11.2	6.4
9	0	0	7.4	4.3
10	0	0	5.1	2.7
11	0	0	3.3	1.4
12	0	0	2.3	0.8
13	0	0	1.4	0.5
14	0	0	0.4	0.1
15	0	0	0	0

$$N_{b,r}(D) = \begin{cases} 10^{-3.3955} \times D^{-2.589} & (D < 0.003\,\text{m}) \\ 10^{-1.4845} \times D^{-1.8315} & (0.003\,\text{m} \le D \le 0.245\,\text{m}) \\ 10^{-2.233} \times D^{-3.057} & (D > 0.245\,\text{m}) \end{cases} \tag{2.2}$$

where D is the diameter of rock.

In the upland region, the cumulative number of rocks $N_{\mathrm{b,T}}(\mathrm{m}^{-2})$ larger than a certain diameter in a unit area (1 m^2) can be expressed as

$$N_{b,t}(D) = \begin{cases} 10^{-3.3955} \times D^{-2.589} & (D < 0.01\,\text{m}) \\ 10^{-2.5239} \times D^{-2.1532} & (0.01\,\text{m} \le D \le 0.763\,\text{m}) \\ 10^{-2.625} \times D^{-3.014} & (D > 0.763\,\text{m}) \end{cases} \tag{2.3}$$

where D is the diameter of rock.

3. The distribution of craters on lunar surface

The cumulative distribution of craters on lunar surface is shown in Fig. 2.9. Number of craters is nearly inversely proportional to its diameter. Craters with diameter greater than 1 km cover 7~10% of lunar surface area. Generally, the depth of crater does not exceed 25% of its diameter, and the height of its edge is less than 6% of its diameter. Therefore, the round crater is normally gentle.

By applying fitting method to Fig. 2.9, the cumulative number of craters with diameter greater than D in a unit area (1 m^2) in smooth and rough lunar mare can be calculated. The functions are shown below.

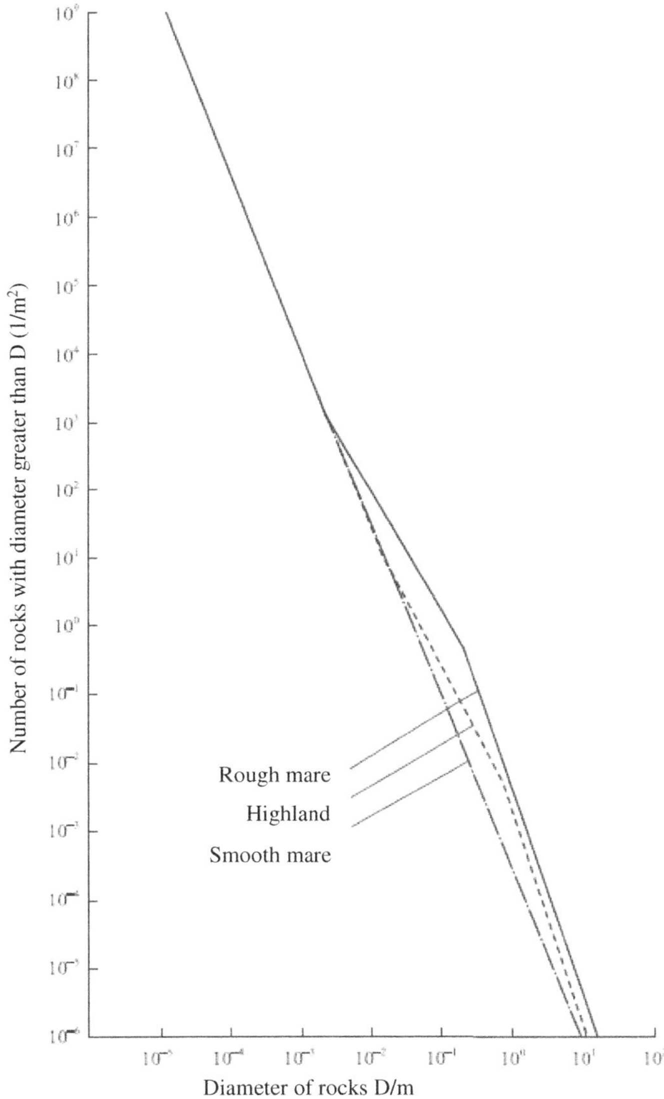

Fig. 2.8 The cumulative distribution of rocks on lunar surface

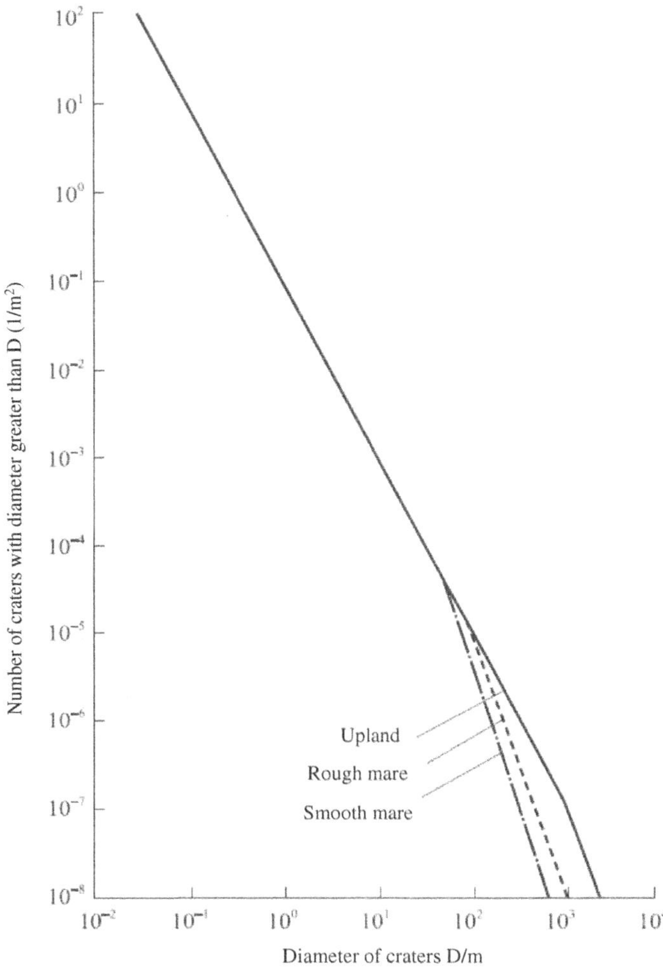

Fig. 2.9 The cumulative distribution of craters on lunar surface

In the smooth mare, the cumulative number of craters $N_{c,s}(m^{-2})$ larger than a certain diameter in a unit area (1 m^2) can be expressed as

$$N_{c,s}(D) = \begin{cases} 10^{-1} \times D^{-2} & (D < 40\,\text{m}) \\ 10^{0.602} \times D^{-3} & (40\,\text{m} \leq D \leq 100\,\text{m}) \\ 10^{-2.038} \times D^{-1.68} & (100\,\text{m} < D \leq 200\,\text{m}) \\ 10 \times D^{-3} & (D > 200\,\text{m}) \end{cases} \tag{2.4}$$

where D is the diameter of crater.

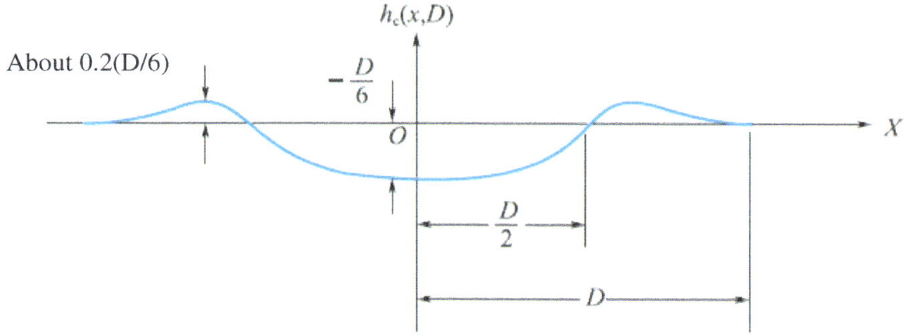

Fig. 2.10 The profile of typical crater

In the rough mare, the cumulative number of craters $N_{c,r}(m^{-2})$ larger than a certain diameter in a unit area (1 m^2) can be expressed as

$$N_{c,r}(D) = \begin{cases} 10^{-1} \times D^{-2} & (D \leq 100\,\mathrm{m}) \\ 10 \times D^{-3} & (D > 100\,\mathrm{m}) \end{cases} \tag{2.5}$$

where D is the diameter of crater.

The depth of crater is basically proportional to its diameter. The ratio is about 1:6. The model of crater is:

$$h_c = \frac{D^2}{12\Pi x} \sin \frac{2\Pi x}{D} \tag{2.6}$$

where h_c is the height of crater and D is the diameter of crater.

The profile of typical crater is shown in Fig. 2.10.

The depth to diameter ratio and the height of crater lip to diameter ratio of the crater vary with the different age (fresh, young, mature and old) of crater. The older is the crater, the smaller is the ratio, and shallower is the crater. Therefore, the variation of shape according to different age of crater should be taken into consideration in constructing the model of crater. The ideal shapes of craters in various ages are shown in Table 2.5.

According to the age of crater stated above, the ratio between the number of craters with diameter over D and the diameter of crater in different ages is shown in Fig. 2.11.

Table 2.5 The ideal shape of crater in various ages

Age of crater	Typical profile	Depth to diameter ratio	Height of crater lip to diameter ratio
Fresh	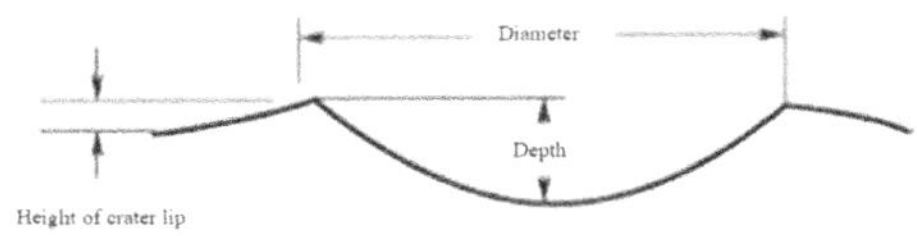	0.23–0.25	0.022–0.06
Young		0.17–0.19	0.016–0.045
Mature		0.11–0.13	0.008–0.03
Old		-	-

2.5 Lunar Soil and Dust

Lunar soil is consisted of rock fragment, powder, breccia and impacting melting glass. It is loose mixture that covers the lunar surface. Lunar soil is loose, non-consolidation and fine-particle. It is generally ecru or dark gray. The lunar soil is mainly formed by the crush of rocks due to thermal expansions and contractions caused by the combined action of the impact of meteorite and micrometeorite, the cosmic rays, the continuous bombardment of solar wind and the large diurnal temperature difference.

Lunar dust is the fine particle (diameter less than 1 mm) among the lunar soil, which is pretty loose and will be raised under certain impact of force. The basic physicochemical properties of lunar dust are the same with those of lunar soil. The two kinds of activation mechanism of lunar dust are natural mechanism and artificial mechanism. The natural mechanism includes the two-stage eruption and micro dust electrostatic floating caused by the collision of meteor and micrometeoroid. There are three kinds of artificial mechanism and are ranked by the degree of influence from the smallest to the largest as: the walk of astronaut, the rotation of the wheels of rover and the landing and life-off of spacecraft. Up to present, the lunar dust activated by the landing and life-off of spacecraft is the majority [10].

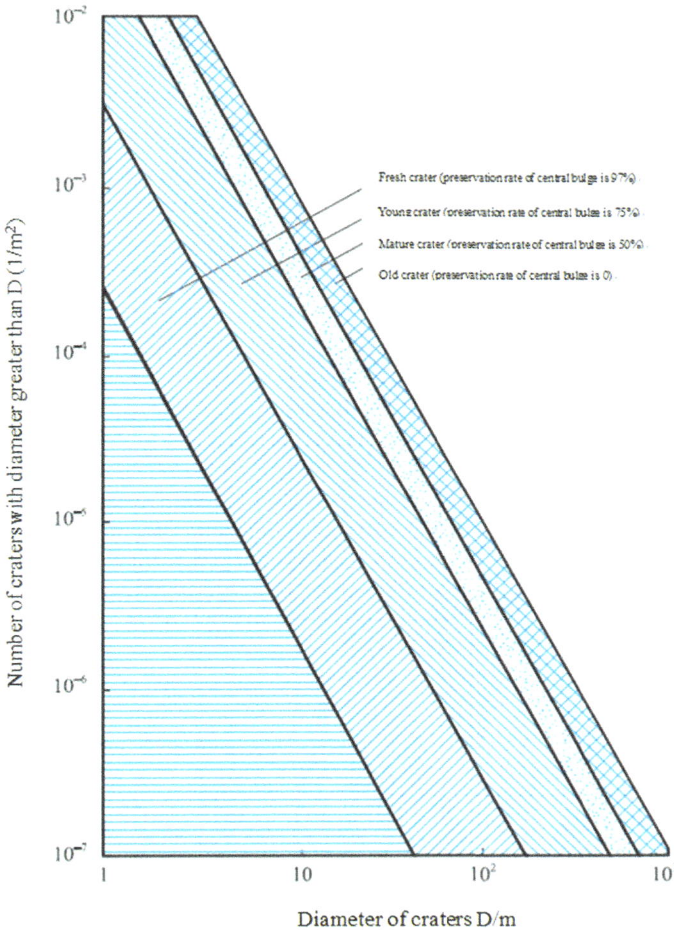

Fig. 2.11 The ratio between the number of craters with diameter over *D* and the diameter of crater

2.5.1 Physical Properties

1. Composition of Particles

Generally lunar soil is hardly sorted and the distribution range of its granularity is very wide. The diameter of most of the particles is smaller than 1 mm. The diameter of the majority of particles is 30~1000 μm. The medium of diameter is 40~130 μm and the average is 70 μm.

The fine particle of lunar soil with diameter smaller than 20 μm takes 10~20% of the lunar soil. These particles may easily float and attach to the equipment. The shape of the particle varies extremely. Shapes ranging from spherical to extreme angular shape show up, while the rod-like, sub angular and angular shape appears more frequently.

2. Density

Density of the unspoiled lunar soil is the mass of lunar soil in a unit volume. The analysis of existing lunar soil sample shows that the relation between the density ρ (g/cm^3) and depth z (cm) can be expressed as follows equation:

$$\rho = 1.92 \times \frac{z + 12.2}{z + 18} \tag{2.7}$$

The analysis result of lunar soil density is shown in Fig. 2.12. In the depth 0~40 cm, the range of the density of lunar soil is 1.3~1.7 g/cm^3. The best estimation of the average density of lunar soil in different depth of crater is listed in Table 2.6.

3. Porosity

Degree of porosity includes void ratio and porosity. The void ratio of lunar soil means the ratio of the void volume and the particle volume and is denoted as e. The natural void ratio of lunar soil may be used to evaluate the degree of compaction of lunar soil. When $e < 0.6$, it is dense and low compressibility lunar soil. When $e > 1.0$, it

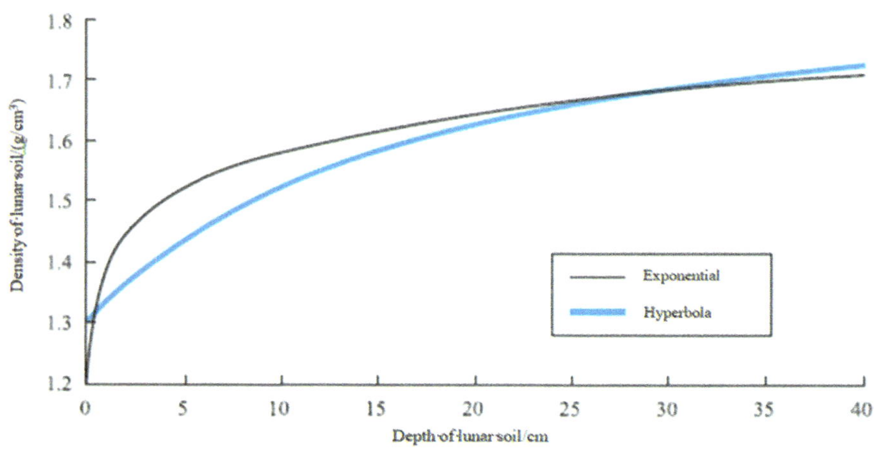

Fig. 2.12 The analysis result of lunar soil density

Table 2.6 The best estimation of the average density of lunar soil in different depth of crater

Depth/cm	Average density/(g/cm^3)
0~15	1.45~1.55
0~30	1.53~1.63
0~60	1.61~1.71
30~60	1.69~1.79

Table 2.7 The best estimation of average void ratio and porosity in different depth of lunar soil

Depth/cm	Average void ratio	Average porosity (%)
0~15	1.00~1.14	50~54
0~30	0.89~1.03	47~51
0~60	0.80~0.94	44~48
30~60	0.71~0.85	42~46

Table 2.8 The recommended value of cohesion and internal friction angle of lunar soil in different depth

Depth/cm	Cohesion/kPa		Internal friction angle/(°)	
	Variance	Average	Variance	Average
0~15	0.44~0.62	0.53	41~43	42
0~30	0.74~1.1	0.92	44~47	45.5
0~60	1.3~1.9	1.6	48~51	49.5
30~60	2.4~3.8	3.1	52~55	53.5

is loose and high compressibility lunar soil. Porosity is the ratio of the void volume and the total volume and is denoted as A. The best estimation of average void ratio and porosity in different depth of lunar soil is shown in Table 2.7.

2.5.2 *Mechanical Properties*

1. Shear Resistance

Position Alternation of particles will happen while lunar soil is under external force, which causes relative motion between two parts of lunar soil. The capability of lunar soil particle in resisting this kind of slide motion is called shear resistance of lunar soil. It is determined by cohesion c and internal friction angle φ. The shear resistance of lunar soil is expressed as

$$\tau = c + \sigma \tan \varphi \tag{2.8}$$

where σ is vertical stress.

The recommended value of cohesion and internal friction angle of lunar soil in different depth is shown in Table 2.8.

2. Bearing Capacity

Bearing capacity is the capacity of carrying load of lunar soil. The relationship between the static bearing capacity and the depth of lunar soil in different regions is shown in Fig. 2.13.

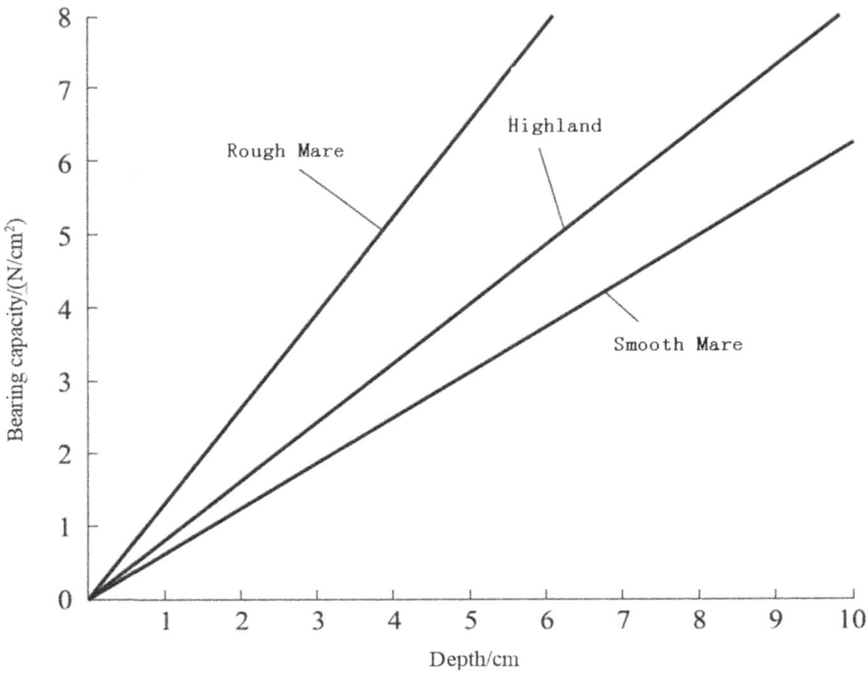

Fig. 2.13 The relationship between the static bearing capacity and the depth of lunar soil in different regions

The comparison between the ultimate bearing capacity of lunar soil and the vertical stress design of the various probes (such as the foot pad of "Apollo" lunar lander, the foot pad of "Surveyor" lunar lander and the wheel of "Apollo" lunar rover) and the one-foot-stand astronauts is shown in Fig. 2.14. As shown in the figure, the vertical stress value applied in engineering design is much smaller than the ultimate bearing capacity on the same contact width. The margin is enough.

2.5.3 *Electromagnetic Properties*

Conductivity and dielectric property are the important parameters that characterize the scattering and transmission of electromagnetic energy inside the lunar soil and rock layer. It is the response characteristics of lunar substance to electric current.

1. Conductivity of Lunar Substance

Lunar substance is mainly composed of silicate and is typically low-conductivity material. Due to loss of water, the conductivity of lunar substance with frequency lower than 1 Hz is equal to DC conductivity. Its value is extremely low and primarily up to temperature.

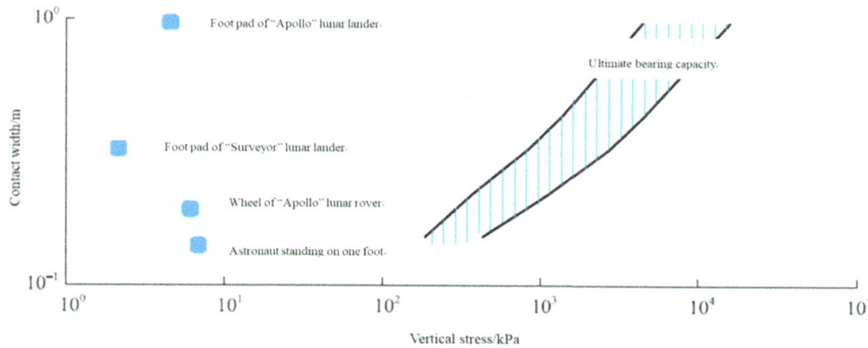

Fig. 2.14 The relationship between the ultimate bearing capacity of lunar soil and the ground contact pressure

The variation of the conductivity of lunar soil with respect to temperature shows amorphous characteristics, which means the lunar soil on lunar surface suffers intense radiation damage. For example, the variation of DC conductivity σ with respect to temperature of the "Apollo-15" lunar soil (15301, 38) satisfies the exponential relationship shown below:

$$\sigma = 6 \times 10^{-18} \mathrm{e}^{0.0237\mathrm{T}} \tag{2.9}$$

where T is temperature (K).

The characteristic of low frequency conductivity of lunar rock is similar to that of anhydrous silicate minerals on Earth. The variation of DC conductivity σ with respect to temperature of the "Apollo-16" lunar soil (65015, 6) is similar to that of lunar soil:

$$\sigma = 3 \times 10^{-4} \mathrm{e}^{0.0230\mathrm{T}} \tag{2.10}$$

However, the variation of dc conductivity with respect to temperature of the other lunar rock sample satisfies the expression below:

$$\sigma = \sigma_0 e^{E_0/kT} + \sigma_1 e^{E_1/kT} \tag{2.11}$$

where k is the Boltzmann constant, E_0 and E_1 is the activation energy, as shown in Table 2.9.

The conductivity of lunar substance is not only up to temperature, but is also affected by solar radiation. The solar radiation may greatly change the conductivity of lunar substance, and therefore causes great charge movement nearby the terminator. Measurement shows that the dc conductivity of lunar substance varies from 10^{-14}S/m of lunar soil to 10^{-9}S/m of lunar rock at night; while exposed to sunlight, the lunar

Table 2.9 Conductivity of lunar rock sample

Lunar rock	σ_0/(S/m)	E_0/eV	σ_1/(S/m)	E_1/eV
10048	3.5	0.896	2.66×10^{-3}	0.559
12002, 85	13	1.09	1.80×10^{-4}	0.48
15058	134	1.347	2.78×10^{-3}	1.593
15418	137	1.509	9.84×10^{-2}	0.971
15555	36.8	1.040	1.27×10^{-2}	0.604
68415	1.27×108	2.640	0	0
68815	14.2	1.366	0	0

soil conductivity of infrared region increases 10 times and that of ultraviolet region increases 10^6 times when comparing to the conductivity at night.

2. Complex Dielectric Constant of Lunar Substance

Lunar substance is typical silicate medium. The two complex numbers that scale the characteristics of medium are complex dielectric constant ε^* and complex permeability μ^*:

$$\varepsilon^* = \varepsilon_0\varepsilon_r = \varepsilon_0(\varepsilon' - \mathrm{j}\varepsilon'') \tag{2.12}$$

$$\mu^*\mu = \mu_0\mu_r = \mu_0(\mu' - \mathrm{j}\mu'') \tag{2.13}$$

where: ε_0 is the vacuum medium constant and μ_0 is the vacuum magnetic permeability.

δ_ε and δ_μ are defined as the electric loss angle and magnetic loss angle of the medium material, respectively:

$$\tan\delta_\varepsilon = \frac{\varepsilon''}{\varepsilon'} \tag{2.14}$$

$$\tan\delta_\mu = \frac{\mu''}{\mu'} \tag{2.15}$$

where ε' is the real part of the complex dielectric constant and is often called as dielectric constant or capacity. It is the major macroscopic physical quantity that comprehensively expresses the polarization of medium and the capability of substance to store magnetic energy. ε'' is the imaginary part of the complex dielectric constant and is related to conductivity. It is expressed as

$$\varepsilon'' = \frac{\sigma}{\omega\varepsilon_0} \tag{2.16}$$

where σ is the conductivity and ω is the angular frequency of electromagnetic wave.

Table 2.10 Comparison of dielectric properties between lunar sample and basalt on Earth

Sample	ρ/(g/cm^3)	ε'	tanδ	low temperature conductivity/(S/m)
lunar rock 10020	3.18	10~15	0.09~0.2	10^{-9}~10^{-11}
lunar rock 10057	2.88	9~13	0.09~0.2	10^{-9}~10^{-11}
lunar rock 10046	2.21	6~9	0.05~0.09	10^{-10}~10^{-11}
lunar rock 12002	3.30	8~10	0.02~0.09	10^{-10}~10^{-11}
lunar rock 12022	3.32	7~14	0.002~0.2	10^{-9}~10^{-11}
basalt on Earth (Hawaii Lake)	2.68	7	0.03	10^{-10}~10^{-12}
basalt on Earth (Nedick Cape)	2.60	6	0.05	10^{-10}~10^{-12}
semiconductor medium LMT-J	3.01	6	0.002	10^{-10}~10^{-12}

Complex dielectric constant is the physical quantity that measures the capability of substance to keep the distance between electric charges (polarization of charges). The dielectric constant ε' of almost every nonmetallic mineral is 4~13, while the dielectric constant of metallic mineral is 17~74 or even higher. The dielectric loss of metallic mineral is much higher than that of nonmetallic mineral. Therefore, for the absolute anhydrous lunar rock, the composition, structure and conformation are the major factors that determine the resistivity and complex dielectric constant; for the loose lunar soil, the influence factors of complex dielectric constant include the test frequency, the density of sample, the test temperature and the chemical composition.

The high frequency dielectric constant of lunar igneous rock is 7~14 under room temperature, while dielectric constant of the same type of dry rocks on Earth is normally lower than lunar rock (as shown in Table 2.10). The reason is that the lunar igneous rock is rich in high conductance material (such as ilmenite) and its low temperature conductivity is generally greater than the igneous rock on Earth.

2.6 Other Lunar Environments

2.6.1 Lunar Gravity

The average gravitational acceleration on lunar surface is 1.62 m/s^2 and is 1/6 of the gravitational acceleration of Earth, which is 9.8 m/s^2. When more detailed data of lunar gravitational field is required in the orbit design of lunar probe, lunar gravitational field model should be applied. The lunar gravitational field model refers to the set of spherical harmonic function coefficient. The lunar gravitational potential expressed in the Moon Fixed Coordinate in the series form of spherical harmonic coefficient is as below

$$V = \frac{GM}{r}\left\{1 + \sum_{n=1}^{\infty}\sum_{m=0}^{n}\left(\frac{R}{r}\right)^{n}\overline{P}_{nm}(\sin\varphi)\left[\overline{C}_{nm}\cos(m\lambda) + \overline{S}_{nm}\sin(m\lambda)\right]\right\} \tag{2.17}$$

where $\bar{C}_{nm}$ and $\overline{S}_{nm}$ are the normalized potential coefficient; $\overline{P}_{nm}$ is the standardized Legendre adjoint function; $R = 1738$ km is the average radius of lunar equator; $M = 7.353 \times 10^{22}$ kg is the mass of Moon; r, λ and φ are the Moon-center distance, the longitude and the latitude of space flow point, respectively.

The fairly precise lunar gravitational field model is the Konopliv method. This method analyzes all the measurement data obtained by lunar probes and achieves the LP165P model. Its effective order is 110 for the near side of the moon and 60 for the far side of the moon.

2.6.2 Near Vacuum Environment of the Moon

The atmospheric pressure at lunar surface is 10^{-10}~10^{-6}Pa. The molecular number density and the altitude range of air at lunar surface are shown in Table 2.11. The lunar atmosphere is mostly consisted of noble gas, H_2, CO_2, CH_4 and NH_3, while it does not contain the atomic oxygen, which has great influence on the spacecraft on the low-Earth orbit [11].

In order to understand the lunar atmosphere for aerospace engineering, the status of the atmosphere of Earth on the near-Earth orbit with an altitude range of 100~2500 km is analyzed. The result is shown in Table 2.12.

It is shown in Table 2.12 that the atmospheric pressure (vacuum degree) on lunar surface is similar to that of near-Earth orbit with an altitude range of 500~2500 km. The altitude of the Earth orbit spacecraft is generally higher than 500 km. Therefore, the design of the lunar lander in the material outgas, the evaporation, sublimation and decomposition of material, the thermal design, the vacuum cold welding and dry

Table 2.11 The molecular number density and the altitude range of air on lunar surface

Gas	Molecular number density/cm^{-3}		Altitude range/km	
	Daytime	Night	Daytime	Night
20Ne	4×10^3~10^4	10^5	100	25
He	8×10^2~4.7×10^3	4~7×10^4	511	128
H2	2.5~9.9×10^3	10+~1.5×10^5	1022	256
Ar	2×10^3	<10^2	55	–
CH4	1.2×10^3	–	–	–
CO2	10^3	–	–	–
NH3	4×10^2	–	–	–

Table 2.12 The result of analysis of the atmosphere of Earth on the near-Earth orbit with an altitude range of 100~2500 km

Height/km	Density/(g/cm^3)	Pressure/Pa	Molecular number density/cm^{-3}					
			O_2	N_2	O	Ar	He	H
100	5.99×10^{-10}	3.47×10^{-2}	2.25×10^{12}	9.72×10^{12}	7.15×10^{11}	1.16×10^{11}	1.61×10^{8}	0
200	3.52×10^{-13}	1.23×10^{-4}	4.18×10^{8}	4.27×10^{9}	4.95×10^{9}	0	7.68×10^{6}	0
300	2.91×10^{-14}	1.36×10^{-5}	1.11×10^{7}	1.76×10^{8}	7.65×10^{8}	0	4.63×10^{6}	0
500	8.55×10^{-16}	4.82×10^{-7}	1.80×10^{4}	6.34×10^{5}	3.05×10^{7}	0	2.05×10^{6}	1.95×10^{4}
800	1.41×10^{-17}	1.53×10^{-8}	2.54×10^{0}	2.70×10^{2}	3.62×10^{5}	0	6.77×10^{5}	1.47×10^{4}
1000	2.89×10^{-18}	5.45×10^{-9}	1.03×10^{-2}	2.17×10^{3}	2.31×10^{4}	0	3.40×10^{5}	1.24×10^{4}
1500	4.86×10^{-19}	1.15×10^{-9}	3.72×10^{-8}	3.73×10^{-5}	4.38×10^{1}	0	7.09×10^{4}	8.34×10^{5}
2500	4.21×10^{-20}	1.39×10^{-10}	3.36×10^{-17}	4.51×10^{-13}	1.32×10^{-3}	0	5.24×10^{3}	4.33×10^{3}

friction and the vacuum discharge of electronic devices is similar to the design of Earth spacecraft.

It is shown that the order of magnitude of the density of lunar atmosphere is equivalent to that of Earth atmosphere with an altitude over 1500 km by comparison of Tables 2.11 and 2.12. So, the atmospheric drag of the Moon is considered as the same as that of the Earth with an altitude over 1500 km in the design of circumlunar lunar orbit.

2.6.3 *Illumination Environment on the Lunar Surface*

Similar to Earth, the Sun rises from east and sets in west while viewing on lunar surface due to rotation around its own axis. The rotation period of Moon is 1 month, and one lunar day is approximately 28 Earth days. This period is the lunar cycle. It is almost identical to the lunation in the traditional Chinese calendar.

Except the lunar night as long as 14 Earth days, short-time invisibility of Sun may happen on the Moon due to the block of the Earth. It is the solar eclipse of the Moon. The solar eclipse of the Moon is the lunar eclipse in the view of Earth. Since the ecliptic (orbit of Earth) intersects the Moon's path (orbit of Moon) with an incline angle of about 5°, the Moon moves into the Earth shadow when the Sun approaches closely to the two intersection points of these two orbital planes and the connection line between Earth and Moon points to the Sun at the same time. Then the lunar eclipse of the Earth, which is also the solar eclipse of the Moon, will happen. Earth shadow is sorted as umbra and penumbra. The penumbra is beside umbra, as shown in Fig. 2.15. The lunar eclipse may only occur on the 15th or 16th day of each month of the traditional Chinese calendar.

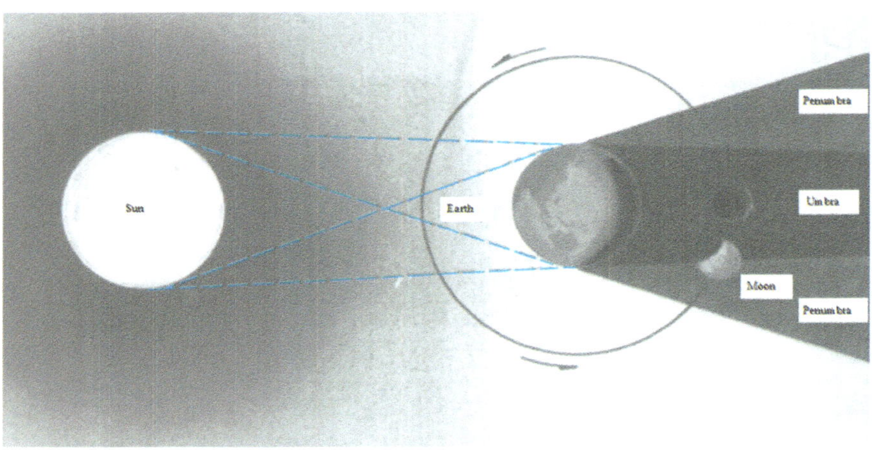

Fig. 2.15 Demonstration of lunar eclipse

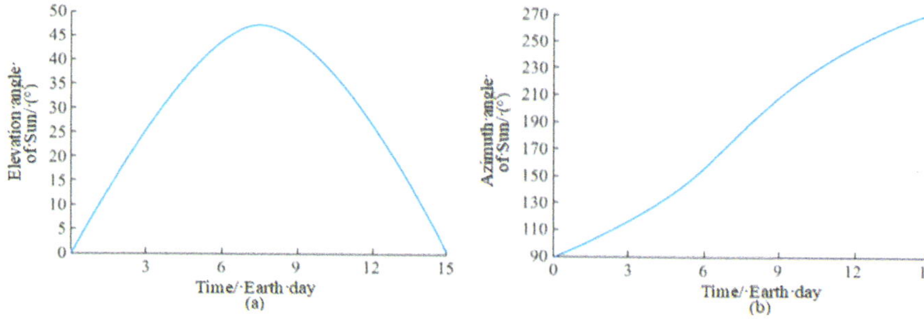

Fig. 2.16 The variation of elevation angle and azimuth angle of Sun in one lunar daytime

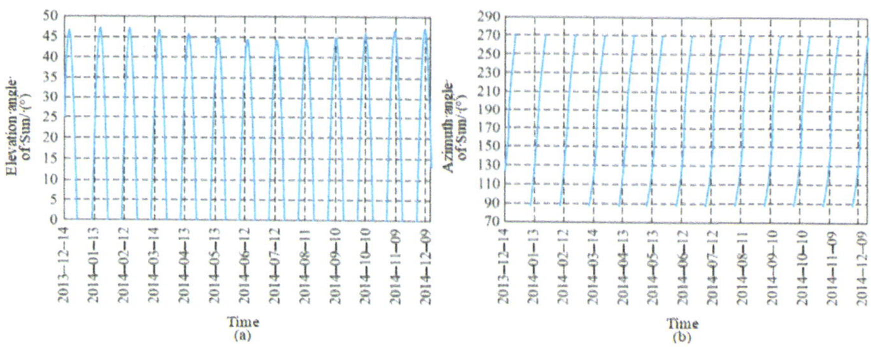

Fig. 2.17 The variation of elevation angle and azimuth angle of Sun in one Earth year

Similar to Earth, the elevation angle and the azimuth angle of Sun vary with the different longitude and latitude of locations on the lunar surface. And the rotational period of the Moon is about 28 times of that of the Earth. Therefore, the rate of variation of solar angle on lunar surface is about 1/28 of that of the Earth. Figure 2.16 shows the variation of elevation angle and azimuth angle of Sun in one lunar daytime regarding the local horizontal plane of the landing site of the Chang'E-3 probe. Figure 2.17 is the variation of elevation angle and azimuth angle of Sun in one Earth year (December of 2013–December of 2014) at local level.

2.7 Summary

The lunar environment faced by lunar soft-landing mission on lunar orbit or on lunar surface is introduced in this chapter. Lunar natural environment including the radiation environment, the thermal environment on lunar surface, the landform and topography of lunar surface, the lunar soil and dust, the lunar gravitational field, the near vacuum environment of the Moon and the illumination law on lunar surface are presented.

Lunar radiation environment mainly includes the solar cosmic ray, the galactic cosmic ray and the solar wind. Since the distance from Earth to Moon is relatively small comparing to the diurnal distance, the lunar radiation environment faced by the lunar lander is similar to the radiation environment of spacecraft on Earth orbit.

The thermal environment of Moon is an important factor in the design of the thermal control system of the lunar lander. It includes solar radiation, reflection of sunlight by lunar surface, lunar radiation and terrestrial albedo. The value of solar radiation, which is the majority of external heat flow, is equivalent to that for normal spacecraft on Earth orbit. Lunar radiation is the secondary external heat flow, while the external heat flow caused by terrestrial albedo usually can be ignored.

The landform and topography on lunar surface are the environment that requires serious consideration in the lunar soft-landing mission. The lunar surface is generally classified in two kinds of topography, i.e., the lunar mare and the highland. The lunar mare and the highland are both covered by rocks and craters in different sizes and shapes. The slopes, rocks and craters at landing site are important factors for safe landing. Distribution of slope, rock and crater on the different topography of lunar surface is presented.

Lunar soil is consisted of rock fragment, powder, breccia and impacting melting glass. It is loose mixture that covers the lunar surface. Lunar dust is the small size particle among the lunar soil, which is pretty loose and will be raised under certain impact of force. The basic physicochemical properties of lunar dust are same with those of lunar soil. This chapter discusses the physical property, the mechanical property and the electromagnetic property of lunar soil. These properties are the important factors that affect the landing stability, the cushion capability, the navigation sensor and the telecommunication on lunar surface of the lunar lander.

The last section of this chapter discusses the gravitational field, the near vacuum space and the illumination of the Moon. The design of the lunar lander has to take the 1/6 g gravity into consideration. The precise model of lunar gravitational field is an important basis in the design of the lunar orbit and the guidance rule in powered descent process. The vacuum degree on lunar surface is similar to that of near-Earth orbit with an altitude range of 500~2500 km. The design of the lunar lander in the material outgas, the evaporation, sublimation and decomposition of material, the thermal design, the vacuum cold welding and dry friction and the vacuum discharge of electronic devices is similar to the design of Earth spacecraft. One lunar day is approximately 28 Earth days. Except the lunar night as long as 14 Earth days, short-time invisibility of Sun may happen on the Moon due to the block of the Earth. Similar to Earth, the elevation angle and the azimuth angle of Sun vary with the different longitude and latitude of locations on the lunar surface. Besides, the rotational period of the Moon is about 28 times of that of the Earth. Therefore, the rate of variation of solar angle on lunar surface is about 1/28 of that of the Earth.

This chapter only analyzes the characteristics of the lunar natural environment. The analysis of induced environment is not included and the effects of the induced environment on the lunar lander are not discussed too.

References

1. Chu G, Zhang H (2007) Technology of lunar probes. China Science and Technology Press, Beijing
2. Ouyang Z (2005) Introduction to lunar science. China Astronautic Publishing House, Beijing
3. Guibai CHU, Shijun MA (2002) Introduction to astronautic technology. China Astronautic Publishing House, Beijing
4. Weichi TAN, Jingang HU (2009) Spacecraft system engineering. China Science and Technology Press, Beijing
5. Chengrong PENG (2011) Spacecraft system design. China Science and Technology Press, Beijing
6. Gross FC, Park JJ et al (1972) Analysis of Surveyor III television cable after residence on the Moon. NASA, USA
7. Marvin DC, Corney DJ (1991) Solar proton events of 1989: effects on spacecraft solar arrays. J Spacecraft Rockets 28(6):11–12
8. NASA (1969) Lunar surface models. NASA, USA
9. Murcray, DG (1969) Lunar surface studies. AD0685729
10. Gromov V (1998) Physical and mechanical properties of lunar and planetary soils. Earth Moon Planet 80:51–72
11. Lucas WC, Vaughan OH (1966) The lunar atmosphere. Marshall Space Flight Center

Chapter 3
System Design of Lunar Lander

3.1 Introduction

The system design of lunar lander usually included system-level design and analysis for the lander during the life cycle of the lander development. The objective of the system design was to provide an optimized spacecraft system solution meeting with specific requirements by integration of advanced technologies. The system engineering methodology was the key to achieve optimization in system-level. Major tasks of system design include mission analysis, system design, system simulation and analysis, definition of system and subsystem specification and interface, design and implementation of system integration, test and validation for the lander.

To achieve lunar soft landing, the function of lunar lander should include:

(1) To perform Lunar Transfer Orbit (LTO) and lunar orbiting after separation from the launch vehicles;
(2) To perform powered descent and soft landing;
(3) To be landed safely at predefined area on lunar surface with proper velocity, attitude and accuracy, and providing a stable platform after landing;
(4) To perform in situ exploration after landing and send scientific data back to Earth, i.e. to support scientific payloads working on lunar surface with suitable condition such as installation, view, orientation, thermal condition and power supply.

Detail information of system design for lunar lander is given in this chapter including principles of system design, mission analysis, trajectory design, system specification, spacecraft configuration design, telecommand and telemetry, powered descent and landing, landing stability analysis and simulation, and etc.

D.-Y. Yu et al., *Technology of Lunar Soft Lander*, Space Technology Library 38,
https://doi.org/10.1007/978-981-15-6580-9_3

3.2 Principles of System Design

The majors involved in system design of lunar lander included theory and methodology of system engineering, aerospace dynamics, engineering cybernetics, propulsion, heat transfer theory, structure mechanics, computer science, mechanical engineering, electronic engineering, microwave telecommunication, environment engineering and etc. The basic principles should be followed to achieve a good lunar lander as followings [1].

(1) Satisfying requirements of mission goals. According to the requirements of mission goals, mission analysis should be conducted at first to determine a set of quantitative functional specification by user's requirements. Those specifications could be transferred to detail specification of the design.
(2) Implementing system integrity. Lunar lander was a complex system which was a functional unit made up of several components that interacted with and rely on each other. Function and performance of each component should be studied. Furthermore, the system function and properties of lunar lander as an integrated system should be studied under interaction among all components in system design. The system integrity should be considered when the subsystems were designed. If there was incompatibility and contradiction among subsystems,
(3) Regarding system hierarchy. There should be a hierarchy system of all elements in a system from simplicity to complexity and from low to high. The engineering system of lunar lander included system, subsystem, equipment, assembly, component, parts and etc. in hierarchy. For the system design of a lunar lander, requirements and validation should be analyzed from top to bottom. Then it was validated to satisfy the requirements from bottom to top. Finally the system was integrated.
(4) Pursuing optimization in system level. The optimization of system design of a lunar lander should be done to determine the lander type, system specifications and some partial specifications by digital models and comprehensive evaluation with computer. It should be considered including orbital environment, performance, schedule, cost, risk, benefit and constraints such as launch vehicle launch site, ground telecommunication station and ground application in the system optimization. The optimization criteria might be the performance, or lightest weight or lowest costs. The system design aimed at the optimization of system performance, not optimization of some parts, assembly or subsystems.
(5) Adapting to specific environment. The specific environment for lunar lander was different from satellites on Earth orbit and lunar orbiters including engine plume, vibration, lunar dust during landing and lunar terrain, lunar regolith, lunar low gravity, extreme high and low temperature. Those environmental effects should be considered for solution during system design of lunar lander.
(6) Assuring accordance of engineering system. The engineering system for the lunar soft landing mission included lunar lander, launch vehicle, launch site, ground telecommunication station system and ground application system. The accordance of lunar lander interface with other engineering system should be ensured to achieve the successful landing.

(7) Assuring reliability and safety. The reliability was one of the most important key performance indicators of lunar lander as well as safety in system design. The development and launch of lunar lander was a high investment and high risk engineering project. In space history, the failure probability of lunar lander as well as other deep space probes was higher than Earth satellites because of the specific environment, uniqueness and irreversibility of critical control point. So it should be paid more attention to design of reliability and safety.

To achieve the system optimization, the fundamental principles for conventional spacecraft should be followed and specific requirements of lunar lander should be carefully considered as well.

3.3 Mission Analysis

Mission analysis was the primary task of the system design of spacecraft. The input of the mission analysis was mission goals and user's requirements. The output of the mission analysis was a series of specifications of function and performance in qualitative or quantitative form.

3.3.1 Analysis of Mission Characteristics

There are several differences from lunar soft landing to conventional Earth satellites as followings [2, 7]:

(1) Powered descent and deceleration. There is barely no atmosphere near lunar surface, i.e. the pressure is 10^{-14} atm (1 atm = 1.013×10^5 Pa). So the lunar lander could not be decelerated by airbraking. The velocity of lunar lander relative to the Moon should be decelerated to several meters per second, which meant that near 1.7 km/s Delta V be provided by propulsion system of lunar lander. At the same time, the attitude of lunar lander should be controlled precisely during powered descent to make sure that the vertical and horizontal velocity of lunar lander relative to the Moon meet the requirements of soft landing, and lunar lander was landed safely at the predefined location. The thrust of the engine should be throttlable as well. Those factors should be considered when the propulsion system was designed.

(2) Autonomous navigation and control during powered descent and landing. Because the time was short and the Delta V was big, guidance, navigation and control could not be realized by ground control for the powered descent and landing. The sensors such as speedometer, ranging and terrain identification of the guidance, navigation and control subsystem could help to realize autonomous navigation and control for powered descent and landing of lunar lander. The capability of the guidance, navigation and control subsystem to

identify lunar terrain and avoid obstacles could help to make sure lunar lander not to turn over and sink after landing because of the complexity of lunar terrain.

(3) Cushion of landing impact. There would be large impact load when lunar lander finally touches lunar surface. The landing cushion subsystem should be designed to absorb the impact load to make sure lunar lander not to turn over and sink at a predefined range of attitude and provide solid support for operation on lunar surface after landing.

(4) Thermal control and power supply for survival on lunar surface. The illumination condition and external heat flux changed greatly from lunar daytime to lunar night on lunar surface. It was a challenge to scatter heat under high temperature during lunar day. It was also a challenge to provide enough power supply and control the temperature without sun light during lunar night lasting for 14 days.

(5) Ground simulation test for lunar environment. It was different in gravity, temperature, illumination, radiation condition on lunar surface from Earth orbit. So the theory and methodology for ground test should be studied, as well as ground facilities and simulation equipments should be constructed to make sure the sufficiency of ground test and validation.

3.3.2 Analysis of Landing Site

The landing site of lunar lander should be determined at first for the system design. The difference between the landing sites had essential effects on the system design of trajectory, telecommunication and configuration for lunar lander.

The basic principles for landing site selection were both engineering implementability and satisfying requirements of scientific exploration on the basis of safe landing.

1. Engineering requirements

The implementability of engineering of landing site selection included lunar terrain, access of trajectory, telecommunication condition of powered descent and operation on lunar surface, sun light condition, thermal environment and etc. Effects on the system design of lunar lander included by those factors as following:

(1) Lunar terrain. There are almost craters and rock everywhere on lunar surface. It is relative smooth on the mare, while it is rugged on the highland. The lunar terrain properties including slopes, pits and shape of rock were important to safe landing and stable attitude after landing of lunar lander.

(2) Access of trajectory. The latitude of landing site should be less than the orbit inclination of lunar lander before landing. Theoretically if the orbit was polar orbit, lunar lander could be landed at any place on the Moon. But the repeat period and telecommunication condition for specific landing site should be considered at the same time. So the access of trajectory should be studied.

(3) Telecommunication condition. Because the rotation period was almost same as the period of revolution of the Moon around the Earth, the near side of the Moon is always in front of the Earth while the far side of the Moon could not be seen on the Earth. If the landing site was on the near side of the Moon, the telecommunication condition to the Earth was good. All lunar landers were landed on the near side of the Moon up to now (before 2017).
The telecommunication condition during powered descent should be considered besides the condition on lunar surface. During powered descent, the height and velocity would be change greatly, and there would be a lot of operation of sensors and thrusters of lunar lander. The risk of powered descent was the highest one of the flight profile of lunar lander. Just before powered descent process of lunar lander was initiated, a lot of command and data should be sent to lunar lander by ground control, which meant that the ground station should be able to access lunar lander at that time. For the safety of powered descent, it would be better if the ground station could access lunar lander during powered descent in case of possible malfunction for the ground operators to try possible operation to save lunar lander.

(4) Sun light condition. There would be adequate sun light during lunar daytime on lunar surface, so the solar cells were able to provide energy source for operation. The inclination of the Moon equator to the ecliptic is only about 1.5°, so the sun light condition is the best on the Moon equator. The sun elevation is getting smaller when the latitude is getting higher on lunar surface. For the landing site selection, the variation pattern of the sun elevation and azimuth should be studied for the design of the size and the orientation of the solar panel. Generally the landing time should be on the lunar day to get enough energy supply for the landing site. At the same time, the effects of the sun elevation on the optical sensors should be considered, because it was hard for the optical sensor to obtain good images when the sun elevation was too small or too big.

(5) Thermal environment. There is nearly no atmosphere around the Moon. So the temperature on lunar surface is up to the solar radiation. When it is lunar daytime, the difference of the temperature on the equator and polar region of the Moon is relative big. The highest temperature on higher latitude is low and it is higher when the latitude is getting to zero. When it is lunar night, the difference of the temperature on the equator and polar region of the Moon is relative small. The heat flux on different location of the Moon should be considered when the thermal control of lunar lander was designed.
The effects of all above constraints on the system design of lunar lander should be analyzed and weighed. The optimum selection of the landing site should be determined after comprehensive evaluation.

2. Scientific requirements

The principles for landing site selection could be summarized as three aspects for scientific exploration.

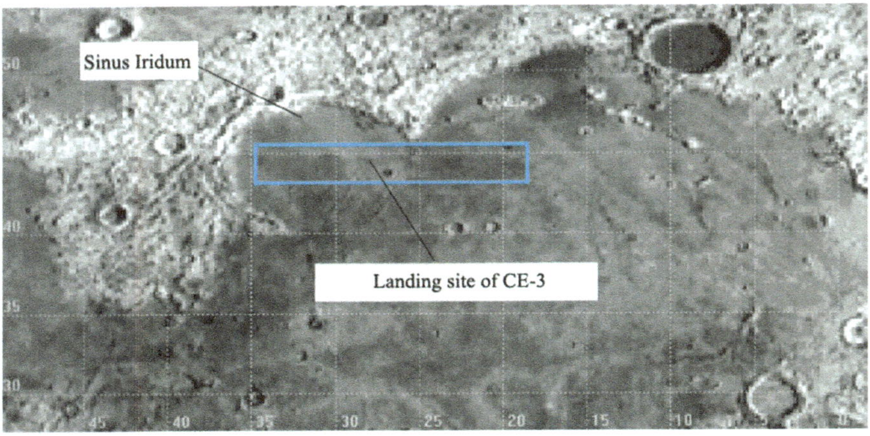

Fig. 3.1 Landing site of CE-3 and the Sinus Iridum on the Moon

(1) The geology appearance should be diverse. It could satisfy scientific requirements for lunar geology research in regions such as the border of the mare and the highland, the mountain and typical craters.
(2) The lunar regolith should be mature and the mineral resource should be abundant to satisfy the requirements for lunar resource utilization and research.
(3) The geology and location of landing site should be different from previous missions which may provide unique and additional information for lunar science.

After synthesis of scientific requirements and engineering implementability, several candidates of landing sites for China's first lunar landing mission was studied and the Sinus Iridum on the south-west of near side of the Moon was finally selected. The longitude was 18.2°W to 34.6°W and the latitude was 44.1° ± 1.5°N. The location of landing site is shown in Fig. 3.1.

3.3.3 Analysis of Flight Profile

The mission profile of lunar lander could be divided into four parts: flight process, landing process, operation on lunar surface and survival on the lunar night.

1. Requirements analysis of flight process.

The design of the flight process should aim at satisfying requirements of soft landing at the first of all. So the analysis of the flight process should be from the end to the start point.

(1) Flight process before landing. There were two approaches to achieve soft landing: one was direct landing and another was orbiting the Moon before landing. For the direct landing, the trajectory of lunar transfer orbit (LTO) was designed to intersect with lunar surface directly. When the probes approached the Moon, powered descent was performed at the end of approach and the probe was landed on the Moon at relative low speed, such as Soviet Luna 9 and Luna 11, the U.S. Surveyor landers. For the orbiting before landing, the trajectory of LTO would not intersect with lunar surface. When the probe approached the Moon, orbit insertion was performed for the probe to enter into lunar orbit. The altitude would be adjusted at appropriate time and then the powered descent was performed to land on the Moon, such as Soviet Luna 16, Luna 17, Luna 21 and Luna 24, the U.S. Apollo program.

(2) Design of lunar orbit. There were constraints including latitude of landing site, Delta V of lunar orbit insertion (LOI), safety of LOI and long term safety of lunar orbit in lunar orbit design. The altitude of perilune and semi-major axis should be determined as well as inclination in lunar orbit design.
The capability of navigation and control, propellant consumption, thrust for LOI, and safety of trajectory maneuver should be considered when the perilune altitude before powered descent was determined. For example, if the altitude of perilune was higher, the time for powered descent would be long so that the propellant consumption and error of landing would increase. If the altitude of perilune was lower, the accuracy of trajectory maneuver should be higher and the risk of impacting lunar surface would increase. The inclination of the lunar orbiter should be greater than the latitude of the landing site. Generally the lander could be landed at anywhere on the Moon if the lunar orbit was polar orbit.

(3) Selection of launch trajectory type. There were two types of launch trajectory for lunar landing mission including phasing orbit and direct LTO. It would take full advantage of the launch vehicle capability and decrease the requirement of propellant consumption for lunar lander to be launched directly into LTO, which was also common in deep space exploration.

2. Requirements of powered descent.

The powered descent process of soft landing was the key to the safe landing of lunar lander. There were a series of new requirements different from conventional spacecraft.

(1) Requirements for navigation and control. Guidance, navigation and control of lunar lander during powered descent could not be provided by ground control because of short time, great variation of velocity, unpredictable altitude and velocity relative to lunar surface. Autonomous guidance, navigation and control could be achieved by the speedometer and ranging sensors of GNC subsystem and cooperation among GNC, propulsion and other subsystems. There might be slopes, rock and pits in lunar terrain which could not be predicted, so lunar lander should be capable of autonomous identifying lunar terrain and avoiding obstacles to ensure not to turn over after soft landing.

(2) Requirements for propulsion. The propulsion subsystem of lunar lander should provide near 1.7 km/s Delta-V during soft landing process. The thrust weight ratio should be higher to lower the gravity loss. The thrust of main engine should be throttlable at a specific range to ensure the stability of powered descent because the mass of lunar lander was changing during powered descent. If the thrust weight ratio was higher, the time of powered descent would be shorter and the gravity loss was smaller. But the overload would be higher and the dynamics would be worse for the powered descent. Thus the total impulse of the propulsion subsystem, thrust and throttlable range of the main engine should be synthetically traded off.

(3) Requirements for landing cushion. The landing cushion absorbing impact load was necessary although the relative velocity was small when lunar lander touches lunar surface finally. The landing cushion subsystem would avoid damage to instruments onboard lunar lander due to the final impact and ensure that lunar lander would not turn over and sink. It could also provide solid support for subsequent operation of lunar lander on lunar surface.

3. Selection of landing cushion.

There were two types of landing cushion for lunar soft landing: airbag and landing gear. For the airbag, gas in the airbag was used to absorb landing impact. For the landing gear, different types of absorbing energy mechanisms inside the legs were used to absorb landing impact.

(1) Airbag. There might be one single airbag or multiple airbags complex for the airbag. Lunar lander was packed inside the airbag. The airbags were used for Luna 9 and Luna 13. The advantage of airbag was strong cushion capability and not sensitive to lunar terrain. The disadvantage of airbag was that airbag could only be used in small scale lander. The bouncing time after landing was long and the landing site could not be controlled precisely for the airbag. The airbag was usually used in Mars landing. The airbag used in the U.S. Mars lander 'Spirit' is shown in Fig. 3.2.

Fig. 3.2 Airbag for the U.S. Mars lander 'Spirit'

Fig. 3.3 Landing gear for landing cushion

(2) Landing gear. The landing gear for landing cushion (shown in Fig. 3.3) was widely used in lunar, Mars and asteroid exploration. The attitude after landing was stable and would not bounce after landing by design for the landing gear.

In order to ensure the landing stability, the distance between landing legs should be proportional to the height of mass center of lunar lander. If the size of the landing gear was beyond the envelope of rocket fairing, the landing gear should be designed to be unfolded before launch and deployed after launch.

4. Analysis of operation on lunar surface.

After lunar lander was landed on lunar surface safely, the rover would be deployed. All payloads onboard lunar lander would implement exploration on lunar surface and scientific data would be transferred back to Earth.

(1) Requirement for energy. The solar angle varies periodically on lunar day. So the solar array should be able to be pivoted in at least one direction to be pointed toward the Sun to satisfy the requirement for power source on lunar surface.
(2) Requirement for data transmission. Lunar lander should be able to transmit data to the Earth at higher rate. Because the orientation to the Earth on lunar surface was shifting, the directional antenna onboard should be pivoted in two directions.
(3) Requirement for rover deployment. A lunar rover onboard lunar lander could rove on lunar surface for exploration. One of the important tasks of lunar lander was the deployment of the lunar rover. There was special mechanism for connection, locking/unlocking and deployment to implement separation of the lunar rover and move the lunar rover to lunar surface. Ground telecommunication support was necessary to the deployment process of the lunar rover under constraints of temperature, power consumption and etc.
(4) Requirement for thermal control. The thermal environment was changing by the solar elevation. There are both solar radiation heat flux and infrared heat flux on lunar surface. Thermal control of lunar lander should be designed to block out the external heat flux and dissipate heat generated by instruments operation in lunar lander. Because the infrared heat reflection of lunar surface

was strong, the heat radiating surface should be orientated toward zenith. The effects of the attitude on the heat radiation efficiency should be considered.

(5) Requirement for exploration on lunar surface. For the requirements of payloads onboard lunar lander, lunar lander should satisfy requirements such as installation room, orientation, power supply, data transmission, temperature and etc.

5. Analysis of survival on lunar night.

The rotation period of the Moon is about 27.3 days. The lunar daytime is about 14 Earth days while the lunar night is also 14 Earth days. There is no sun light and the extreme temperature may be −180 °C on lunar night.

There were big differences for lunar lander on lunar night from spacecraft in shadow of the Earth or the Moon, which are listed as following.

(1) Long duration. The lunar night is about 14 Earth days without sun light, which is much longer than the time of shadow on the Earth or lunar orbit.
(2) Lack of space infrared radiation heat flux. For the spacecraft on Low Earth Orbit, there is specific infrared radiation heat flux from the Earth in the shadow. But such heat flux is little for the Moon because the lowest temperature of lunar surface may be −180 °C on lunar night.
(3) Profound effects of lunar surface. The lunar terrain might be regarded as infinite for lunar lander. There was not only heat radiation exchange but also irregular conduction of heat because lunar lander touched lunar surface. Because the temperature on lunar surface drops rapidly on lunar night, the temperature of lunar lander was influenced profoundly.

Because of the special properties of surface environment on lunar night, the power supply was the key of survival on lunar night for lunar lander. If energy storage device such as battery was used to provide power and heat, the capacity should be big enough. If lunar lander was designed to hibernate on lunar night, the requirement for power supply could be reduced greatly because only storage temperature should be maintained. But the autonomous wake up of lunar lander should be solved at the same time. When the power supply of lunar lander was off, the criterion for waken up may be illuminance of sun light and temperature, which means lunar lander was awakened up autonomously when the illuminance of sun light and temperature on lunar surface satisfied the requirements for operation after the dawn.

The instruments in lunar lander might be damaged or failure if there was no thermal control on lunar night. So there must some kind of power source to ensure the safety of lunar lander on lunar night that was adapted to lunar environment and able to work without sun light and special thermal control. The radioisotope heat source was one of nuclear energy, which was longevity, adaptable to environment, reliable, no external energy required and capable of providing heat and power source simultaneously. So Radioisotope Thermoelectric Generator (RTG) and heat unit could provide lunar lander with heat and power energy under extreme environment to support survival on lunar night.

3.3.4 Analysis of Environment Effects

It was necessary for the system design of lunar lander to analyze lunar terrain, regolith on landing site.

1. Analysis of landing site effects.
1) Analysis of elevation and its effect along powered descent path.

During the powered descent, the height relative to lunar surface should be measured by sensors onboard lunar lander. So the elevation along the powered descent path would greatly affect the safety of powered descent, control scheme and propellant consumption. The typical elevation diagram along powered descent path of the CE-3 lunar lander is shown in Fig. 3.4 by image data obtained by the CE-1 lunar orbiter, where the straight line in the middle of the figure is an example for a specific launch window. The pattern of elevation variation is shown in Fig. 3.5, where the reference plain is the surface of sphere of 1737400 m radius.

The elevation along powered descent path might be summarized as four categories: monotone flight path without craters/mountains, flight path over crafters, flight path over mountains, and flight path over rugged terrain like craters/mountains. When the circumlunar orbit before powered descent was designed, the effects of elevation variation along powered descent path should be considered when the altitude of the start point of powered descent was determined. And the capability of navigation and control should be able to adapt to the elevation variation along powered descent path.

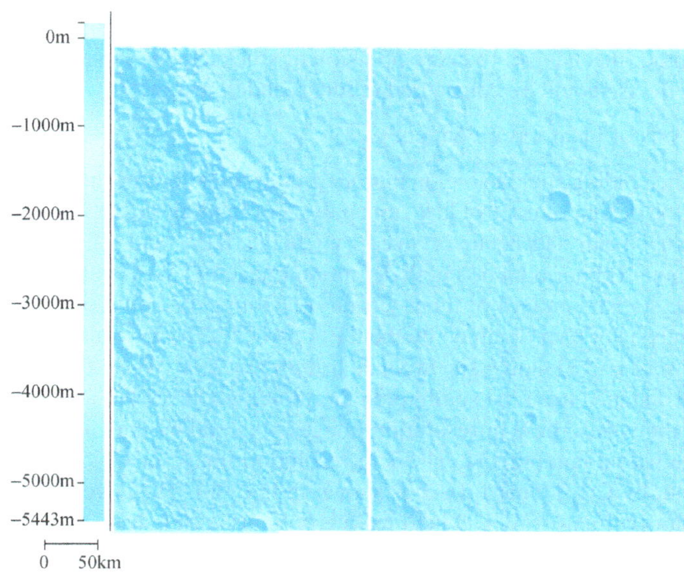

Fig. 3.4 Flight path diagram of typical powered descent process

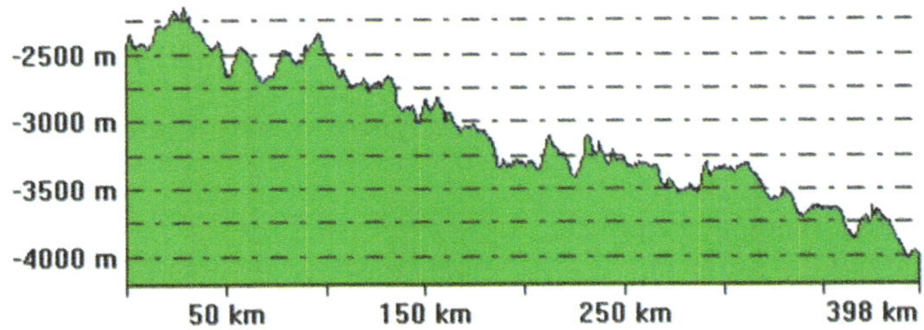

Fig. 3.5 Pattern of elevation variation along typical powered descent path

2) Analysis of lunar regolith/dust effects [3, 11]

The basic properties of lunar regolith and dust are analyzed in Sect. 2.5. The effects on design of lunar lander are analyzed in this section.

(1) Analysis of lunar regolith effects. The major effects of lunar regolith on lunar lander are landing impact and stability. The mechanical properties of lunar regolith varied in a specific range for the whole Moon surface. The worst case should be simulated and validated for landing impact and stability in ground test.
(2) Analysis of lunar dust effects. The lunar dust might cause large error or failure of sensors such as speedometer, ranging and optical sensors. The lunar dust might also pollute sun sensor, optical lens, solar array, thermal paints and etc. to degrade the efficiency of light transmission, power supply and thermal performance. The fine lunar dust particles might erode unsealed bearing, gear and other mechanism to cause abrasion or stuck. So the lunar dust properties and effects should be considered in lunar lander design.
(3) Analysis of lunar terrain effects.

For a specific landing site, the regional properties of lunar terrain should be analyzed to support the design of control law and analysis of landing safety.

The lunar terrain had been surveyed by China's CE-1 and CE-2 lunar orbiter. The CE-2 lunar orbiter obtained high definition image of CE-3 landing site in Sinus Iridum. The statistic pattern of lunar terrain in landing site could be figured out by detailed analysis of data obtained by the CE-1 and CE-2 lunar orbiter.

The range of CE-3 landing site in Sinus Iridum was 18.2°W–34.6°W and 44.1° ± 1.5°N. The distance from the south to the north was about 91 km and 356 km from the east to the west. The 7 m resolution images obtained by the CE-2 lunar orbiter covered the whole landing site. The 7 m resolution image of CE-3 landing site in Sinus Iridum by CE-2 lunar orbiter is shown in Fig. 3.6. The coverage diagram of 1.5 m resolution image of CE-3 landing site in Sinus Iridum by CE-2 lunar orbiter is shown in Figure 3.7.

The statistic pattern may be obtained for the slopes, size and contribution of rock and pits by analysis of high definition images of lunar surface, which could support design of landing cushion mechanism and determination of capability of lunar terrain identification and obstacles avoidance.

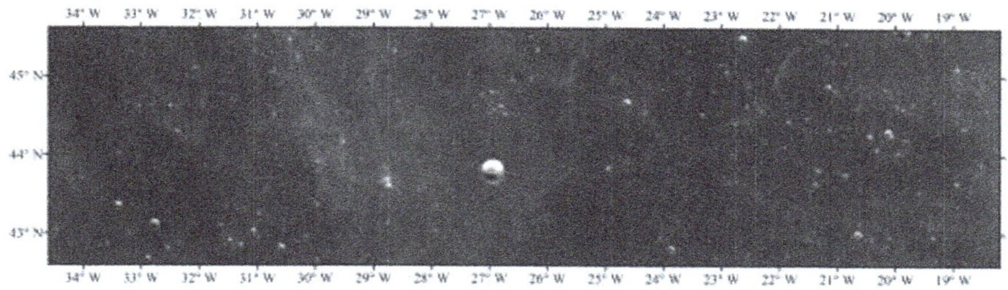

Fig. 3.6 7 m resolution image of CE-3 landing site in Sinus Iridum

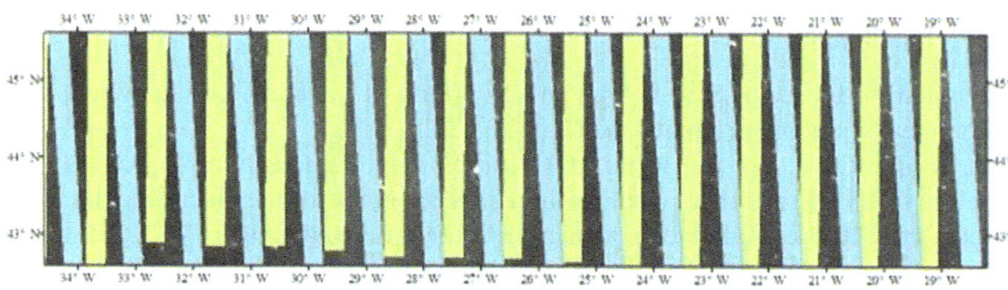

Fig. 3.7 Coverage diagram of 1.5 m resolution image of CE-3 landing site in Sinus Iridum (*Notes* Where the darker stripe for data of first access, the brighter stripe of data of second access)

2. Analysis of space environment

The space environment of charged particles should be considered including electron and proton captured by the Earth radiation belt, Galactic Cosmic Ray(GCR), Solar Cosmic Ray (SCR, only when there is solar flare), and solar wind [6].

1) Total Ionizing Dose(TID) effect

The radiation source of space charged particles that cause TID effect is electron and proton captured by the Earth radiation belt, and solar flare proton. So the TID effect encountered by lunar lander is within 5 h in the Earth's magnetosphere during LTO flight. Then in the life time on lunar surface, it would increase TID effect of solar flare proton for lunar lander if there is solar flare. There would not be more TID effect on lunar lander if there is no solar flare during life time on lunar surface. If lunar lander is launched directly into LTO, the TID effect caused by particles of the Earth radiation belt is far below the level of the CE-1 lunar orbiter during phasing orbit and LTO. For the CE-3 lunar lander, the requirement for TID effect in one year on lunar surface was far below the level of GEO spacecraft.

2) Radiation damage to solar cell

The radiation damage to the solar cell is caused by particles captured by the Earth radiation belt and solar flare proton. So it may occur during the LTO and thereafter.

Table 3.1 Equivalent 1 MeV electron damage dose for the single junction gallium arsenide solar cell

The cover		Lunar lander		
Density/(g/cm^2)	Depth/mm	Open circuit voltage/V	Max. power consumption/W	Short circuit current/A
0.0308	0.12	5.14×10^{13}	3.68×10^{13}	1.69×10^{13}

For the worse case, there might be a super solar flare like the event in Oct. 1989 in the life time of lunar lander. The equivalent 1 MeV electron damage dose for the single junction gallium arsenide solar cell is listed in Table 3.1.

3) Single Event Effects (SEE)

The radiation source of SEE is high energy proton of the Earth radiation belt, high energy heavy ion and proton of the GCR and SCR.

It is up to not only the value of Linear Energy Transfer (LET), but also the capability of SEE tolerance, whether there is SEE or not for the electronic parts and components. There might be frequent SEE if the threshold value of LET is low and event cross area is big which means very sensitive to SEE, even there is no solar flare. Therefore, methods including control of electronic parts and components, anti-SEE design in system level and etc. must be adopted to avoid influence of SEE.

4) Surface Charge and Discharge Effects (SCDE)

The surface charge and discharge on lunar lander might be caused by hot ion in the Earth's magnetosphere and solar wind. The surface charge and discharge effect caused by hot ion is only encountered when lunar lander is through hot ion region in the Earth's magnetosphere (20,000–40,000 km) during LTO. It is obvious when there is sub storm of the Earth magnetic storm. So the surface charge potential of lunar lander should be analyzed, while the electrical resistivity of surface material and ground connection should be controlled strictly to lower the surface electric potential of lunar lander in order to rule out the effect of surface charge and discharge effect.

The solar wind particles encountered may cause surface charge effect when the electronic devices of lunar lander work on lunar surface. Because the surface charge and discharge effect in life time of lunar lander is weak, the surface electric potential could be lowered by strict control of the electrical resistivity of surface material and ground connection distance to rule out the effect of surface charge and discharge effect.

5. Internal Charging Effects

The internal charging effect is due to strong geomagnetic activity. There would be charging within the spacecraft dielectric materials when the spacecraft is radiated continuously by long term high energy and high-flux electrons in Earth's magnetosphere. Because it is only 5 h for lunar lander to get through the Earth's magnetopause (about 65,000 km height), there is hardly no strong geomagnetic activity encountered by lunar lander. After then, lunar lander would no encounter the environment of high energy, high-flux and long term electrons that may induce the Internal Charging Effects during LTO, circumlunar orbit, powered descent and operation on lunar surface. So it is unnecessary to consider the Internal Charging Effects in the life time of lunar lander.

The summary of the space radiation environment of charged particles and their effects on lunar lander are listed in Table 3.2.

3. Analysis of the environment effects induced by soft landing.
1) Analysis of plume effects

Generally there would be at least one orbit control engine and several thrusters for attitude control in order to satisfy the requirements of orbit and attitude control in lunar lander. The plume of engine and thrusters in vacuum when they are ignited would cause force, thermal and pollution effects on surrounded equipments. Especially, when lunar lander was approaching lunar surface, there might be worse effects because the plume of orbit control engine was reflected by lunar surface. So it's important to analyze and validate effects of engine plume.

Table 3.2 Summary of the space radiation environment of charged particles and their effects on lunar lander

Flight phase	Earth's radiation belt	GCR	SCR	Solar wind	Radiation effects
LTO	Yes in Earth's magnetopause No beyond Earth's magnetopause	Yes	Yes when there is solar flare, otherwise no	Yes	TID: yes for material in Earth's magnetosphere, no for electronic parts and components SEE: yes SCDE: yes, but no influence on lunar lander Damage to solar cell: ignored
Circumlunar and powered descent	No				TID: yes when there is solar flare, otherwise no Damage to solar cell: yes when there is solar flare, otherwise no SEE: yes SCDE: weak

For example, there was a 7500 N throttlable engine for orbit control of the CE-3 lunar lander. The distribution of density, static pressure, dynamic pressure, stagnation pressure, static temperature, dynamic temperature, stagnation temperature, axial velocity, radial velocity and stream line of the flow field inside and outside the nozzle was analyzed for 7500 N engine plume. The plume angle was set 45° for the 7500 N engine. The relative position of the plume angle to all navigation sensors is shown in Figs. 3.8, 3.9, and 3.10. The effects of plume field should be considered when the navigation sensors are designed.

During powered descent, the throttlable engine was ignited continuously. The engine plume interacts with lunar surface directly. The thermal status of lunar lander was influenced by reflux of engine plume on lunar surface. On the other hand, lunar dust activated by engine plume may affect on function of lunar lander. The flow field of throttlable engine plume at 1500 N was analyzed by axial symmetric Direct Simulation Monte Carlo (DSMC) method. The distribution of stream line is shown in Figs. 3.11 and 3.12 when the distance of the bottom of engine nozzle to lunar surface was 3 m and 5 m.

It is shown in the analysis results that the remote field of engine plume spread freely. It is also shown that along with gas free spread, the density, pressure and temperature decreased while the Mach number increased. Where the distance to engine nozzle was 25 m, the density was about 2×10^{-7} kg/m^3, the pressure was

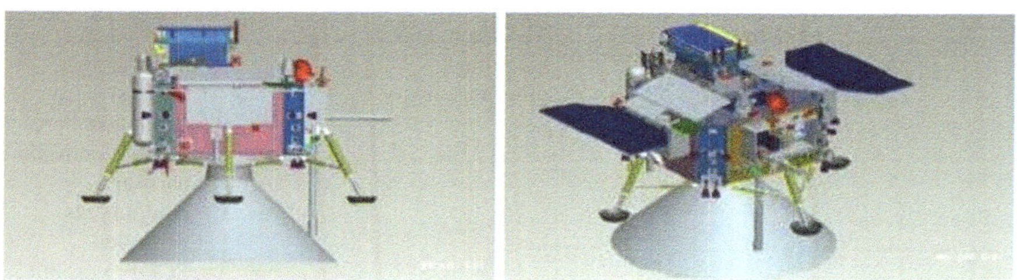

Fig. 3.8 Relative position of 7500 N engine plume angle to laser ranging sensor

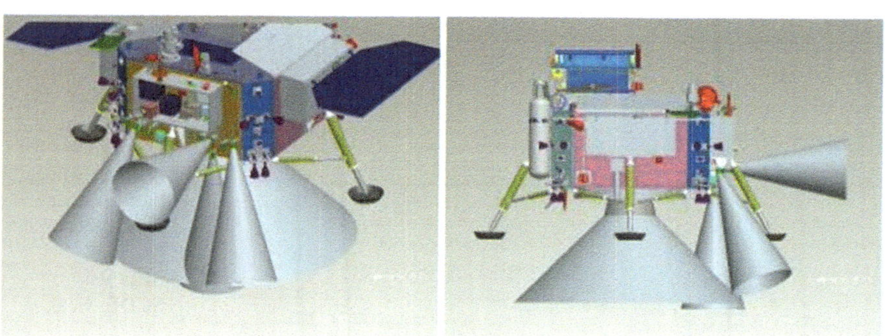

Fig. 3.9 Relative position of 7500 N engine plume angle to microwave ranging sensors and speedometers

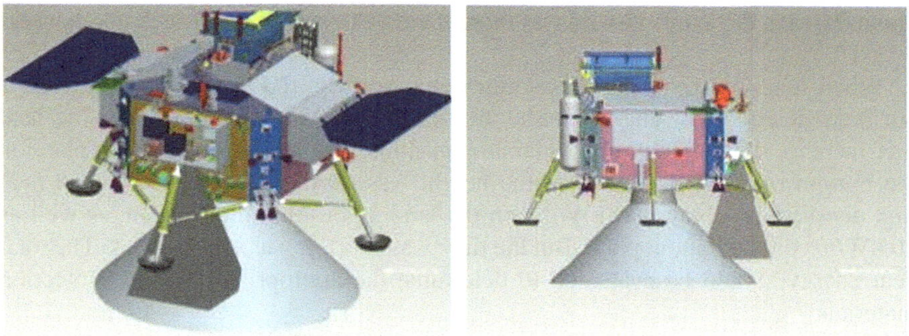

Fig. 3.10 Relative position of 7500 N engine plume angle to laser 3D imaging sensor

Fig. 3.11 Stream line of continuous flow of throttlable engine plum at 1500 N when the distance to lunar surface was 5 m

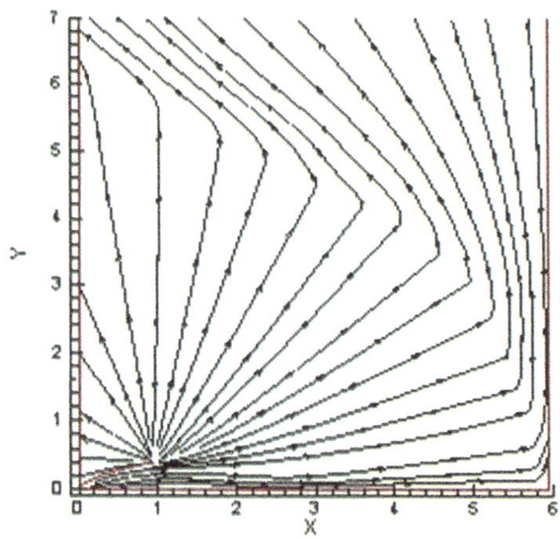

Fig. 3.12 Stream line of continuous flow of throttlable engine plum at 1500 N when the distance to lunar surface was 3 m

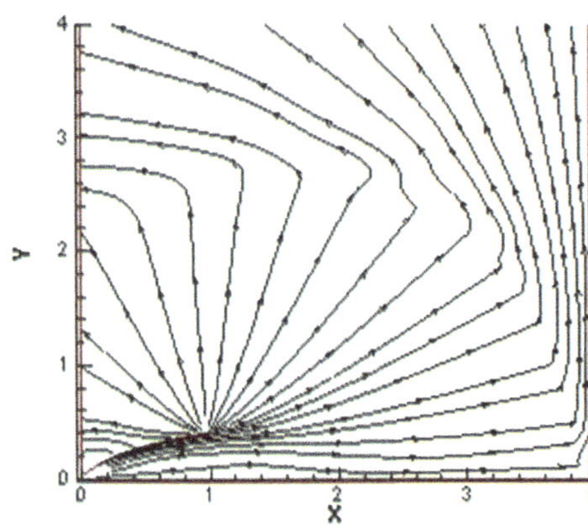

about 0.01 Pa, the axial velocity was more than 3000 m/s, and the Mach number was over 15.

When lunar lander was approaching lunar surface, the most important effects are thermal effects of engine plume and its reflux. Analysis shows that the heat flux increase dramatically on the landing cushion mechanism and bottom plate on the bottom of lunar lander when the height was decreasing. The maximum heat flux density was about 400 kW/m^2 on the landing cushion mechanism, as well as 80 kW/m^2 on the bottom plate. But the time was short. It was only 1 to 2 s. The total heat energy should be evaluated to determine the appropriate thermal protection materials.

2) Analysis of lunar dust effects

(1) Lunar dust activated by engine plume. The physical simulation model of engine plume and lunar dust on smooth lunar surface was established by actual thrust output during soft landing, variation of velocity and attitude for lunar lander. The interaction between engine plume and lunar dust during soft landing was digitally simulated for two phases coupled with gas phase (engine jet stream) and solid phase (lunar dust particles). The example of analysis results is given as following for the CE-3 lunar lander.

① When the height of lunar lander to lunar surface was higher than 20 m, the effects of engine plume on lunar surface was small and no lunar dust would be activated.

② When the height of lunar lander to lunar surface was lower than 20 m, the effects of engine plume on lunar dust increase along with decreasing of the height, as following.

a. Along with decreasing of the height of lunar lander, the pattern of stream line of the engine plume on lunar surface would change from spreading along lunar surface to upturning. The upturning angle was bigger when the height was lower.

b. Along with variation of the height of lunar lander and the stream line of engine jet stream, lunar dust particles carried by jet stream would cause growth of particle density on landing site. But the height of lunar dust particles was obviously lower than the jet stream gas. Even at the 1 m height to lunar surface, there was no continuous big lunar dust particle on the surface of lunar lander.

c. If the engine was always ignited during soft landing, there would be hardly any lunar dust particle under lunar lander and on all surface of lunar lander because of the filling effect of jet stream gas.

d. If the engine was powered off at 3 m height and lunar lander freely falls on lunar surface, small lunar dust particles under lunar lander would move upward along the inertial movement in V shape while the upward angle was about 20°. There would be small lunar dust particles stream on the bottom surface and no small particles reflux on top surface of lunar lander when it was free falling. The effect

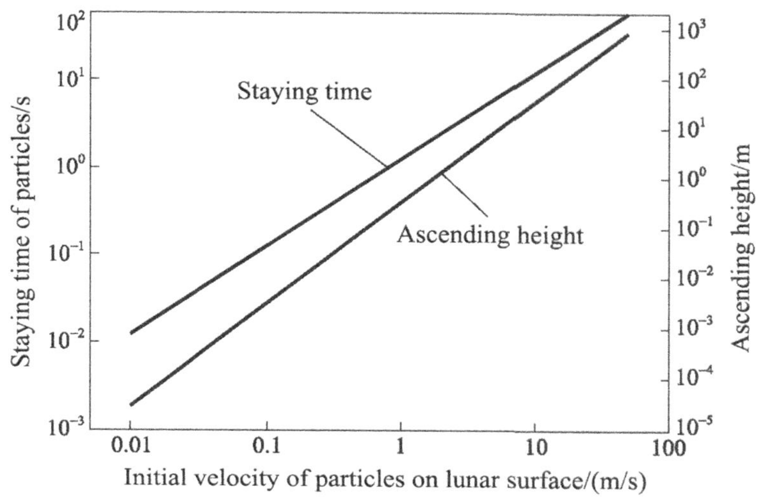

Fig. 3.13 Staying time and ascending height in vacuum for lunar dust particles at different initial velocity

on lunar lander was not obvious because the time was short and the concentration of small particles was lower.

(2) Lunar dust activated by landing impact. The activated lunar dust by touching lunar surface of landing cushion mechanism could be analyzed by digital simulation and test to obtain its space distribution and deposition. Fox example, analysis result of staying time and ascending height for lunar dust particles at different initial velocity is shown in Fig. 3.13 for the CE-3 lunar lander.

The analysis results implied that the effects of lunar dust activated by engine plume and landing impact on lunar lander were not obvious. Appropriate measures could be adopted to mitigate its effects as much as possible.

3) Analysis of vibration dynamic environment of engine thrust for orbit control

There was usually a main engine with large thrust for orbit control and soft landing. When the main engine was ignited, there would be high frequency vibration of thrust because of the inherent combustion instability in the combustion chamber. When the vibration force acts on lunar lander, there might be abnormal output of high accuracy sensors such as accelerometer and gyroscope which are sensitive to middle and high frequency vibrations. Therefore the vibration dynamic environment should be studied to evaluate its effects on sensors. The 7500 N throttlable engine was the largest engine on China's spacecraft. The vibration dynamic environment of 7500 N throttlable engine was measured and the result of typical working condition is shown in Fig. 3.14.

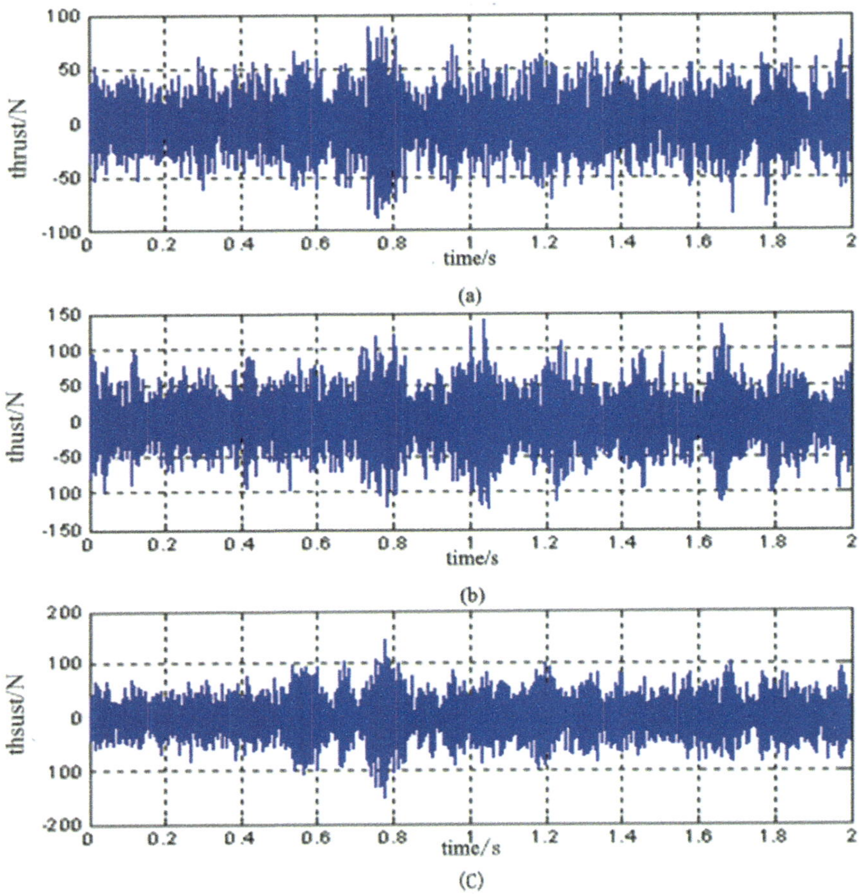

Fig. 3.14 Thrust vibration of typical working condition. **a** Diagram of thrust vibration at equivalent load point 1. **b** Diagram of thrust vibration at equivalent load point 2. **c** Diagram of thrust vibration at equivalent load point 3

The response of acceleration and angular displacement to middle and high frequency vibration caused by thrust vibration was analyzed by Finite Element Analysis software to evaluate effects on sensitive equipments on lunar lander.

4) Analysis of radiation effects of Radioisotope Thermal Unit (RTU)

The RTU was adaptable to different environment and reliable. It was not sensitive to space radiation damage. It could survive under different worse environment such as impact, acceleration, vibration, space vacuum and extreme temperature in space application. But the radioactivity of raw material of RTG would have effects on surrounding environment. Now the ^{238}Pu was the major raw material of RTU in space application, which could generate heat by its own decay. There were several spacecraft of the U.S. and Russia with power supply by RHU. Those spacecraft has flown for 20 years and might continue to fly another 20 years.

The half life period of ^{238}Pu is 87.74 years. When ^{238}Pu is decaying, it will give off alpha particles and gamma ray in 99.89 keV and 43.491 keV, whose intensity is lower 2–3 order of magnitude than alpha particles. The major form of radioisotope fuel is ^{238}Pu and PuO_2 in application, whose specific power is about 0.55 W/g and density is 5.09 g/cm^3.

The core of RHU was usually $^{238}PuO_2$ in ceramics container. The advantages of ceramics structure are high temperature resistant, high hardness, dust free and low leach rate of radioisotope element. Therefore all space RTG was in such form. The outer wall of RTG was made up of multiple materials and multilayered structure to safeguard, seal and avoid leakage of radioactive materials. There were four layers from inside to outer including: seal layer, high-strength protection layer, heat insulation layer and carbon fiber braided outer layer. The structure of RTG source was compact with higher safety and reliability.

The radiation dose around RHU of the CE-3 lunar lander was analyzed by point-source method. The distribution of radiation dose on different height at cylindrical surface 4 m distance to the radioisotope core is shown in Fig. 3.15. The analysis result showed that the effects of RHU on electronic devices could be ignored.

5) Analysis of dynamic environment for landing impact

There was additional dynamic environment such as landing impact besides dynamic environment of launch phase for lunar lander which was different from conventional

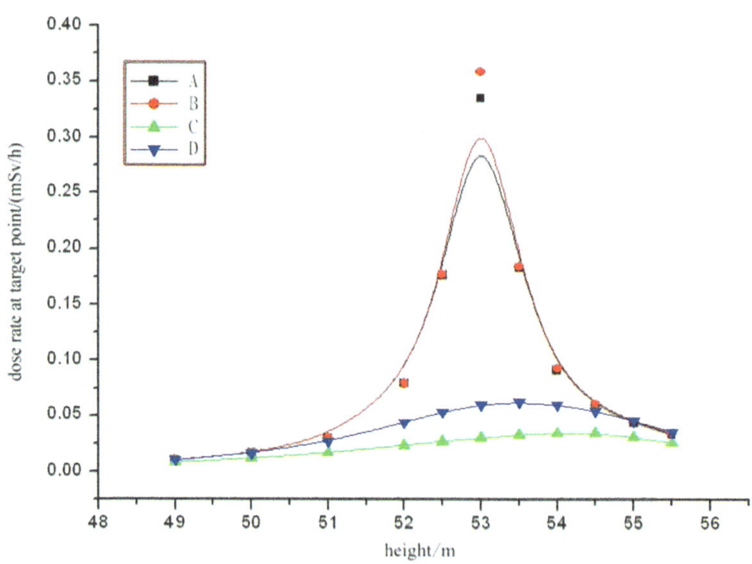

Fig. 3.15 Distribution of radiation dose at cylindrical surface 4 m distance (*Note* A, B, C, D for different angular position on same height)

spacecraft. The effects of landing impact on equipments of lunar lander should be analyzed to ensure the safety and reliability of equipments after landing. The load of landing impact was transferred to structure by landing cushion mechanism. Most of energy was absorbed by landing cushion mechanism while the residual energy would be transferred to the main structure and equipments of lunar lander. The pattern of force on the connection between landing cushion and main structure is shown in Figs. 3.16, 3.17, and 3.18 for a specific condition by analysis and test.

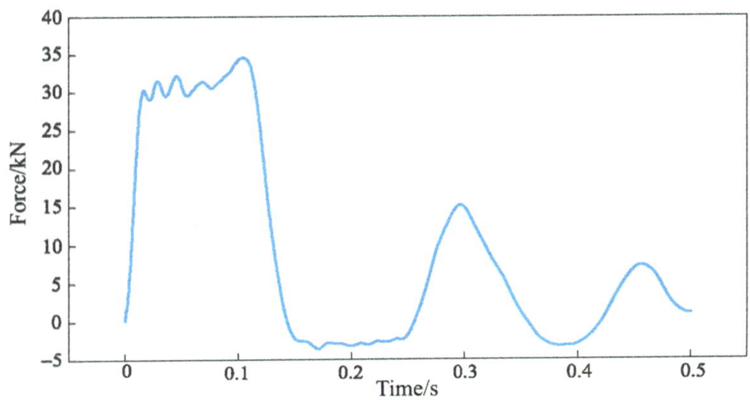

Fig. 3.16 The *X* axis force to time curve on the connection between the main pillars of landing legs with main structure when four landing legs touch the lunar surface simultaneously

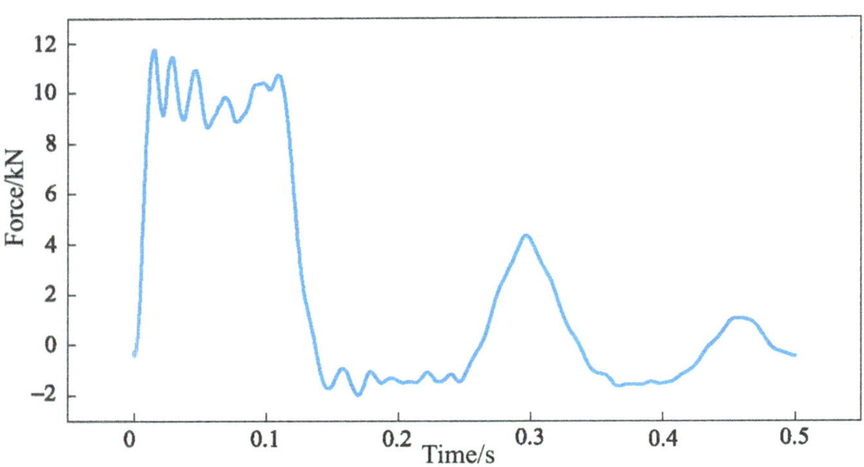

Fig. 3.17 The *Y* axis force to time curve on the connection between the main pillars of landing legs with main structure when four landing legs touch the lunar surface simultaneously

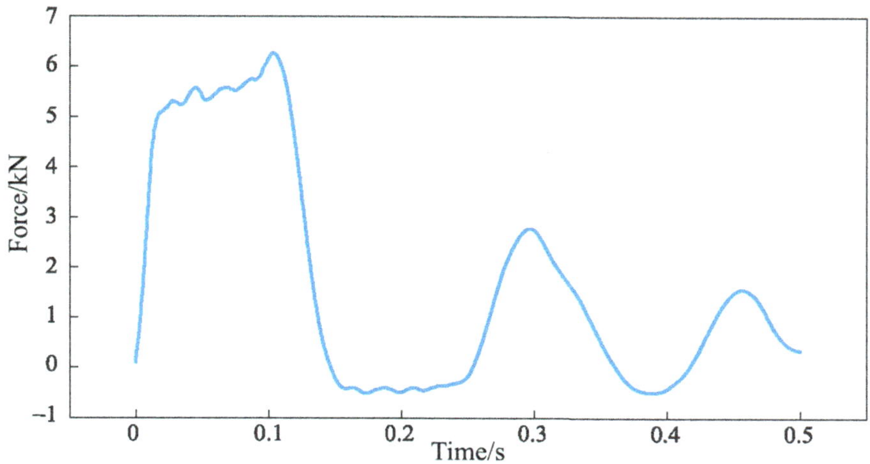

Fig. 3.18 The *Z* axis force to time curve on the connection between the main pillars of landing legs with main structure when four landing legs touch the lunar surface simultaneously

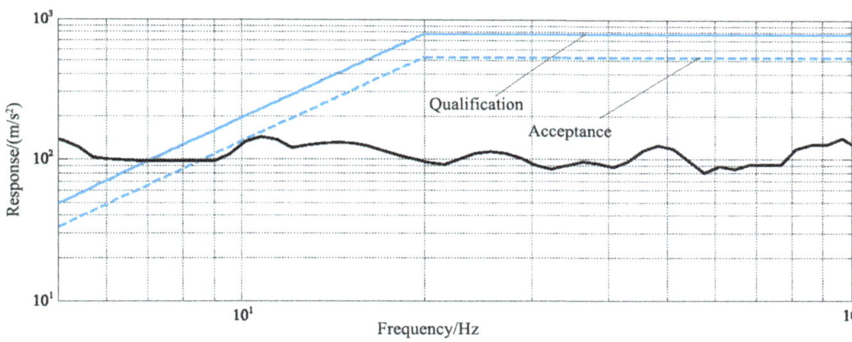

Fig. 3.19 Impact response spectrum of typical location on lunar lander (in comparison with sinusoidal vibration condition)

The dynamic response of crucial locations on lunar lander was analyzed by the model of the lander and input of the force on the connection of main structure. The impact response spectrum on typical location was obtained by impact spectrum transform shown in Fig. 3.19. The result showed that the spectrum energy of landing impact load was in uniform distribution at 5–100 Hz and the acceleration was small. For the frequency higher than 10 Hz, the spectrum of landing impact was not all in the envelope of launch phase environment. Because the frequency below 10 Hz was relative lower, the load could be regarded as static load. The landing impact environment could be treated as equivalent acceleration energy to validate.

3.3.5 *Analysis of Payload Configuration*

Lunar soft landing exploration extended depth and breadth of scientific research. Scientific exploration on lunar surface might be summarized as three categories.

(1) To the Moon: survey of lunar terrain and geological structure, research on chemical composition and material on lunar surface, subsurface structure of shallow lunar internal structure.
(2) To the Earth: observation of the Earth's ionosphere on near side of the Moon that always faces the Earth for research on variation mode of space weather of the Earth and the Sun.
(3) To space: lunar base observatory as a dream of astronomers which is ideal for astronomy because there is no atmosphere and induced twinkling and flickering of stars as well as slow rotation rate of the Moon relative to the Earth.

During powered descent of lunar lander, descent camera and other scientific instruments could obtain images of large area of lunar surface when approaching the Moon in high definition.

3.4 Trajectory Design

The aim of trajectory design for lunar lander was: to satisfy requirements of launch and telecommunication constraints, the start point of powered descent satisfying requirement of soft landing, the end of powered descent at predefined landing site, illumination condition after landing satisfying requirement of operation on lunar surface, and all crucial flight events in access time of ground station inside China. The basic idea of mission trajectory design for lunar lander is shown in Fig. 3.20.

For example, the basic idea of trajectory design for the CE-3 lunar lander was: analyzing illumination and telecommunication condition at first to determine the

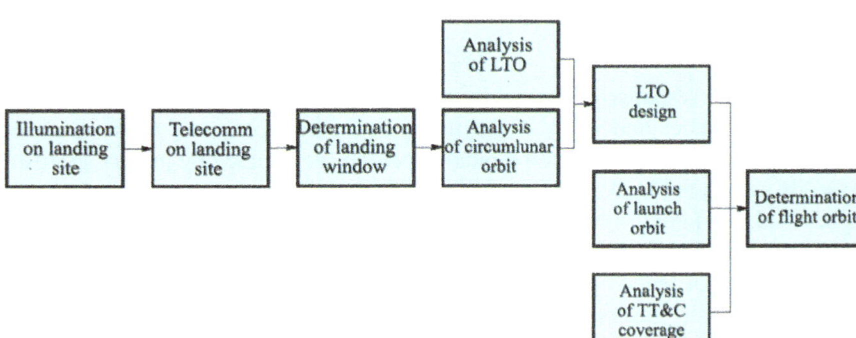

Fig. 3.20 Basic idea of mission trajectory design for lunar lander

window of landing; defining the initial condition for the powered descent to get parameters of LLO then; analyzing characteristics of LTO to determine the range of key trajectory parameters; determining the launch window under constraints of launch vehicle and ground telecommunication finally.

3.4.1 Overview of Flight Profile

The flight profile of the CE-3 lunar lander is shown in Fig. 3.21 [4].

(1) Launch phase: from ignition of launch vehicle to separation of lunar lander with launch vehicle. Lunar lander was launched directly into the high elliptical LTO with 200 km perigee and 380,000 km apogee.
(2) LTO: from separation of lunar lander with launch vehicle to LOI.
(3) Circumlunar orbit: from LOI to the start point of powered descent.
(4) Powered descent: from start point of powered descent to the final soft landing on lunar surface.

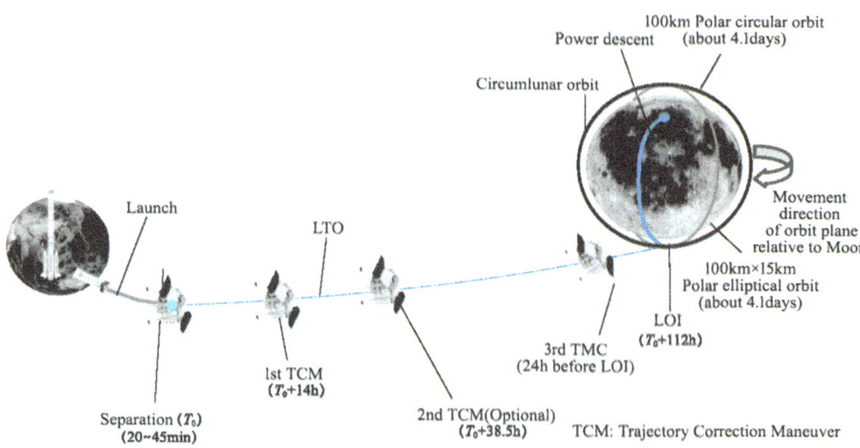

Fig. 3.21 Flight profile of the CE-3 lunar lander

3.4.2 Constraints of Trajectory Design

Major constraints of trajectory design for lunar lander by mission requirements and engineering system constraints were including [5]:

(1) Requirement for launch window: there should be more and wider launch windows for implementation feasibility.
(2) Launch site: there were launch site location, launch azimuth and etc. in constraints of launch trajectory design.
(3) Launch vehicle: there were constraints like launch mass, accuracy of launch trajectory, limitation of launch trajectory and etc.
(4) Landing site: range of longitude and latitude of landing site on lunar surface.
(5) Ground station: location and distribution of the ground stations.
(6) Illumination condition: illumination condition of all procedures in mission profile.
(7) Accuracy of orbit control: the error of Delta V control for orbit maneuvers.
(8) Target orbit before powered descent: altitude, inclination, argument of perilune and etc.
(9) Requirements of telecommunication access: requirements for telecommunication access time especially for key events.
(10) Accuracy of orbit determination and prediction: Accuracy of measurement and prediction for the trajectory position and velocity of all procedures in mission profile.

All those above factors should be considered to define optimum solution when the trajectory is designed.

3.4.3 Determination of Landing Window

The determination of landing window was up to requirements of solar elevation and telecommunication condition when lunar lander was landed on lunar surface. For example, it ws analyzed for the Sinus Iridum (31.5°W, 44.1°N) in lunar day in Aug. 2013 for the CE-3 lunar lander. The variation of solar elevation is shown in Fig. 3.22 in one lunar daytime (14 Earth days) assuming that the local plane was horizontal.

The solar elevation and telecommunication access for Chinese domestic ground station is shown in Fig. 3.23. The landing window selection should satisfying requirements for both illumination and telecommunication access by analysis of illumination and telecommunication condition.

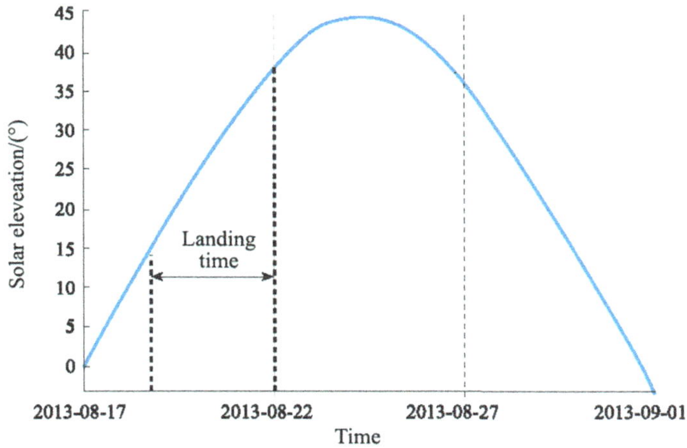

Fig. 3.22 Variation diagram of solar elevation in a lunar day in the Sinus Iridum

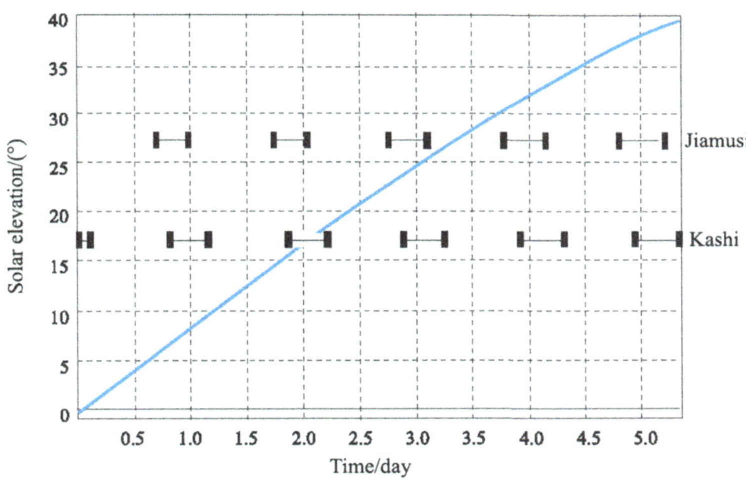

Fig. 3.23 Solar elevation and telecommunication access for Chinese domestic ground station from 19 Aug. 2013 to 22 Aug. 2013 in the Sinus Iridum

3.4.4 Analysis of Circumlunar Orbit

Major concerns of determination of circumlunar orbit before powered descent were including:

(1) Inclination of circumlunar orbit. It was better for the powered descent in orbit plane of circumlunar orbit. If it was not, a massive amount of propellant would be requirement to implement orbit maneuver outside the orbit plane. Therefore, the landing site should be on track of subsatellite point of circumlunar orbit before powered descent. It would be helpful to increase landing opportunities for the polar orbit with 90° inclination to cover global lunar surface.
(2) Altitude of perilune. The altitude of perilune of circumlunar orbit was defined as 100 km after synthesis of circumlunar orbit safety and Delta V requirement for LOI.
(3) Altitude of start point of powered descent. It should be considered for orbital safety and adaptability to lunar terrain undulatory along powered descent path when the altitude of start point of powered descent was determined. The altitude of the highest mountain is 10 km. So the range of altitude of start point of powered descent was usually set to 12–18 km after trade-off.
(4) Time on circumlunar orbit. The time of circumlunar orbit was up to the relative position of landing site to the orbit plane of circumlunar orbit after LOI. The orbit shadow and power balance of lunar lander should be considered when the time on circumlunar orbit was determined.
(5) Illumination condition of circumlunar orbit. The illumination condition on lunar surface had direct relation to the β angle of circumlunar orbit (the angle of orbit plane to solar vector) before powered descent for lunar lander. If the solar elevation on landing site was 15° when lunar lander was landed, the β angle of circumlunar orbit was 68.9°. If the solar elevation on landing site was 40° when lunar lander was landed, the β angle of circumlunar orbit was 26.5°.

3.4.5 Determination of LTO

The aim of LTO determination is to decrease Delta V requirement of LOI for lunar lander. The LTO with 110–140 h transfer time is optimum in energy for the lunar polar orbit after analysis. At the same time, the constraints should be satisfied such as argument of perigee for launch trajectory, illumination condition of LTO, telecommunication access of LOI, launch window requirements, and etc. The major parameters of LTO for the CE-3 lunar lander are defined after analysis as following.

(1) Altitude of perigee for LTO: 200 km.
(2) Altitude of perilune: 100 km.
(3) Inclination of LTO: 28.5°.
(4) Argument of perigee for LTO: 143.579°–225.530°
(5) Flight time of LTO: about 112 h or 114 h, 136 h or 138 h.

3.4.6 Analysis of Launch Window

When the launch window is determined for lunar lander, there are factors to be considered including illumination and telecommunication condition on landing site, time on circumlunar orbit, flight time of LTO and constraints of launch trajectory.

When the solar elevation is greater than 10°, usually the optical imaging sensor would obtain fine pictures. So the solar elevation on landing site was designed to be greater than 15° when the CE-3 lunar lander was landed. The telecommunication access of powered descent process should be available for Chinese domestic ground station to monitor powered descent.

The telecommunication access with satisfying requirements of illumination condition was analyzed between 2012 and 2013 for the CE-3 lunar lander. There were 20 telecommunication accesses that satisfied requirements of both illumination and telecommunication condition in Jul. 2012 to Dec. 2013, while the duration of each access was about 3–4 days. The total time of LTO and circumlunar orbit was about 12–15 days. Then the launch time was determined. The argument of perigee of the launch trajectory was determined by the launch time. The final launch window was determined by constraints of argument of perigee of the launch trajectory.

On above analysis, the launch windows satisfying all requirements and constraints of the CE-3 lunar lander in 2013 were in May, June, July, August and December.

3.5 System Components and Specifications

3.5.1 System Components

The components of lunar lander included subsystems such as structure and mechanism, landing cushion, GNC, propulsion, thermal control, Telemetry, Tracking and Command (TT&C), directional antenna, power supply, general circuits, On Board Data Handling (OBDH), payloads, engineering parameters measure and etc. The basic function of subsystems of the CE-3 lunar lander is listed as following.

(1) Structure and mechanism subsystem was made up of structure and mechanisms. There were main bearing structure and auxiliary structure in structure, whose function was: providing installation surface and room for instruments of the lunar probe; satisfying requirements of installing accuracy for instruments; assuring the integrity of the structure during all phases including AIT, launch, soft landing and operation on lunar surface; enduring load condition during assembling, parking, suspending, transporting, integration, test, launch, flight, landing and operation on lunar surface. There were mechanisms such as solar panel pivoting mechanism, lunar rover connection, release and deployment mechanism, camera pointing mechanism, cover mechanisms

for small modules, whose function were implementation of deployment and pivoting [8].

(2) Landing cushion subsystem was a special subsystem in lunar lander. There are four landing legs while each leg is made up of main pillar, auxiliary pillar, foot pad, holding and release mechanism. The function of landing cushion subsystem was: absorbing impact energy of soft landing, adapting different lunar terrain and maintaining stable attitude after landing, providing solid support for long-term operation on lunar surface.

(3) GNC subsystem was abbreviation of Guidance, Navigation and Control subsystem, which was made up of star trackers, sun sensors, Inertial Measurement Units (IMU), Central Control Unit (CCU), laser ranging sensor, microwave ranging sensors and speedometer, laser three-dimensional imaging sensors and etc. The function of GNC subsystem was: attitude measurement and maintenance, orbit control, providing information of ranging and speedometer relative to the Moon to satisfy requirements of powered descent, autonomous navigation and control of powered descent.

(4) The bi-propellant propulsion subsystem was made up of control unit, fuel tank, gas tank, pipe and valves, thrust for attitude control and one 7500 N throttlable engine. Its function was to provide force required for attitude control and powered descent of lunar lander.

(5) Thermal control subsystem was made up of innovative variable conductance heat pipe, two phase heat loop, RHUs besides conventional heat pipe, thermal paints and radiators. Its function was to ensure thermal management and temperature of all instruments in whole life cycle of lunar lander.

(6) TT&C subsystem was made up of S/X/UHF band antenna, transponder, solid amplifier, unified frequency source, data transmission modulator. Its function was to provide radio channel between lunar lander and the ground station, cooperate with the ground station to implement TT&C and data transmission of lunar lander, provide radio channel between lunar lander and the lunar rover to receive telemetry and scientific data from the lunar rover and send the data to OBDH for processing. There are S band and X band telemetry and command link with X band for data transmission in order to satisfy the requirement of the ground station. There was UHF telecommunication link between lunar lander and the lunar rover. There was also mid-gain data transmission channel for near real time data transmission during powered descent.

(7) OBDH subsystem was an integrated synthetic electronic subsystem. It could provide comprehensive service and management for lunar lander platform and payloads by a centralized computer system that could process measure and control tasks for lunar lander. It could implement orbital operation management and synthetic information processing by a kernel data system for lunar lander. It could manage and control every task with high efficiency and reliability, monitor health status and coordinate different tasks for lunar lander. It could manage and process the data of payloads. It could process and share information of lunar lander centrally. It was made up of

System Management Unit (SMU) and Data Interface Unit (DIU) whose integrated function was to provide management of onboard uplink and downlink data, thermal control, mechanism control, power supply distribution and autonomous routine management.

(8) Power supply subsystem was made up of solar panels, battery and power supply controller. Its function was to be responsible for first-hand power supply acquirement, modulation and supply.

(9) General circuit subsystem was made up of cable, connectors and separation switchers. Its function was to transmit electricity in lunar lander.

(10) Payloads subsystem was made up of four scientific instruments including descent camera, lunar terrain camera, lunar optical telescope and ultra-violet camera to obtain scientific data.

(11) Engineering parameters measure subsystem was made up of monitor camera, lunar dust detector for accumulating engineering data, monitoring implementation of different tasks and acquiring further data of lunar environment.

3.5.2 Major System Specifications

The major system specifications of the CE-3 lunar lander are listed in Table 3.3.

3.5.3 Propellant Budget

For lunar lander, the propellant for the powered descent was the emphasis in propellant budget besides propellant for orbit control, attitude control, bias of mixture ratio, expulsion efficiency of propellant tank, propellant remaining in pipe and etc. as for conventional spacecraft.

For the CE-3 lunar lander, the amount of propellant consumption varied in ± 30 kg because of different lunar terrain along powered descent flight path in different launch windows and error of measurement and control. So requirements of different launch windows should be considered for propellant budget.

The design of propellant budget margin should be traded off for lunar lander. For conventional spacecraft in orbit, the margin of propellant was relative large to maintain strong capability for orbit control and attitude correction so that all conventional or unexpected requirements could be satisfied. For the lunar lander, it was helpful to keep large propellant margin before powered descent in order to improve safety of soft landing in case of possible failure during powered descent. But the mass of propellant required to make the propellant margin be landed on lunar surface was two times of the mass of propellant margin. After soft landing, the mass of propellant margin became useless 'dumb mass'. The remaining of propellant should be depleted to avoid propellant leakage under extreme high and low temperature on

Table 3.3 Major system specifications of the CE-3 lunar lander

Item		Major specifications
System specifications	Mass	Launch mass less than 3780 kg Dry mass 1200 kg
	Dimension	Diameter 3650 mm, height 3600 mm
	Operation time on lunar surface	Longer than 12 months
	Reliability (End of Life)	$\geq$0.65
Structure and mechanisms	Configuration	Truss with panel, landing legs for cushion
Landing cushion	Condition for not to turn over	Slope less than 8°, height of hump or depth of pit less than 20 cm, vertical velocity less than 3.8 m/s, horizontal velocity less than 1 m/s, attitude within ±4°
	Angle between axis X of lunar lander with opposition of local Moon gravity after stable landing	$\leq$14°
GNC	Requirements for landing site selection	Height of hump or depth of pit: < 20 cm Slope: < 8°
	Requirements for landing velocity	Vertical velocity less than 3.8 m/s Horizontal velocity less than 1 m/s
	Control error of three axis angular rate	1°/s
	Control error of three axis attitude	1°
Propulsion	Propellant	Bi-propellant: dinitrogen tetroxide (MON-1), monomethylhydrazine (MMH)
	Engine for orbit control	Throttlable engine: 7500 N, 1500 N–5000 N Nominal specific: 305 s
	Thrusters for attitude control	16×150 N and 12×10 N Specific of 150 N thruster: $\geq$290 s Specific of 10 N thruster: $\geq$275 s
Thermal control	Operation temperature	−20–55 °C
	Storage temperature	−50–70 °C
	Output heat power of RHU	$\geq$120 W
TT&C	Frequency	S/X/UHF
	Data rate of tele-command	1000 b/s

(continued)

Table 3.3 (continued)

Item		Major specifications
	Data rate of telemetry	128 b/s, 1024 b/s (without channel coding) 256 b/s, 2048 b/s (with channel coding)
	Data rate of mid-gain antenna	500 kb/s (with channel coding)
	Data transmission rate	2.5 Mb/s (with channel coding)

lunar surface at last. Thus design of propellant budget margin should be traded off in a reasonable scope.

3.6 System Configuration Design

3.6.1 Design Principles

To implement soft landing, there was usually landing cushion mechanism which was folded before launch and unfolded after injection of launch trajectory. The mass of propellant might be 2/3 of the launch mass to meet the requirements of LOI and powered descent. There should be sensors for speedometer, ranging, terrain identification during powered descent to be installed oriented to lunar surface. The position of the lunar rover on lunar lander should be convenient for separation and deployment.

According to above analysis, the design of lunar lander configuration should follow principles as following:

(1) Compatible with mechanic interface between lunar lander and the launch vehicle.
(2) Lower vertical height of mass center to ensure landing stability.
(3) Assurance of safety and reliability of the lunar rover deployment.
(4) Satisfying requirements of installation accuracy, orientation, field of view, power supply and thermal control.
(5) Protection against lunar dust activated by landing impact.
(6) Satisfying requirements of accessibility for assembling, lifting, turning over, shipping, fueling, docking and test.

The basic idea for design of the CE-3 lunar lander configuration is listed as following:

(1) Cross strut structure with lateral plates. The main structure of CE-3 lunar lander was cross strut plates with the interface ring. There were four instrument modules on +Y, −Y, +Z, and −Z plates. The load could be transferred to the interface ring by cross strut. The dynamics of lunar lander was good because the path for force transfer was short and the bearing is reasonable. At the same

time, as the lateral surface of cross strut plate could bear load, the requirement of force transfer of landing cushion was satisfied.

(2) Centrosymmetric and paralleled installation of propellant tanks. Four quadrants provided by cross strut central module were used for installation of four propellant tanks in centrosymmetry and parallel. The bottom flanges of propellant tanks were connected with interface ring directly and the top flanges of propellant tanks were connected with cross strut plates by rods. Such design would provide good vibration environment for propellant tanks because the vibration load of the launch vehicle was transferred to propellant tanks by interface ring directly. The height of mass center of central module is lower as well as the height of mass center of the whole lunar lander. The mass center of lunar lander would keep in line with the vertical axis when the mass of propellant was changing.

(3) Connection between landing cushion mechanism with cross strut plates. When landing cushion mechanism touched lunar surface, the impact load upon landing cushion mechanism would act on lunar lander. Therefore reliability and strength of the connection between landing cushion mechanism and lunar lander should be ensured. The force transfer could be ensured when the four landing legs were connected to cross strut plates by triangle plates in centrosymmetry. The large span of landing cushion mechanism after deployed was helpful to increase landing stability.

(4) Configuration of equipments in different modules. There were two types of instruments aboard lunar lander: one need survive through lunar night and another one need not. Instruments of GNC and propulsion subsystems would not work on lunar surface after safe landing, therefore it was unnecessary to provide appropriate temperature. On the other hand, equipments of OBDH, TT&C, power supply and payload subsystem would work for a long term on lunar surface, therefore appropriate temperature should be maintained in lunar daytime and night. Different modules for different equipments were designed according to above characteristics then. There were two modules: +Y and −Y modules for equipments which should survive through lunar night, where RHU provided heat in lunar night. There were −Z and central modules for equipments which should not survive through lunar night. Such design could effectively save heat requirements in lunar night, which was good for design of thermal control.

(5) Solar panels on top and radiators in different modules. The solar panels were installed on top of lunar lander for repeated folding and unfolding as well as hold-down before soft landing. The −Z module radiator was on the lateral in which instruments should not survive through lunar night. The radiators of +Y and −Y modules were on the top regarding the characteristic of lunar external heat flux. The radiators were designed to tilt to adapt to tilt of lunar lander after landing.

(6) Lunar rover on top and deployed in Z direction. The lunar rover was installed on the top of lunar lander to provide good accessibility. The deployment mechanism of lunar rover was installed in +Z to avoid interference with other equipments during deployment.

3.6.2 System Configuration

1. Configuration design

The configuration of Chang'E-3 lunar lander is a box type as shown in Fig. 3.24.

2. Configuration in launch fairing

When the Chang'E-3 lunar lander was in launch vehicle fairing, the landing cushion mechanism and solar panels were folded as shown in Fig. 3.25.

3. Flight configuration

The two solar panels in lateral were unfolded and landing cushion mechanism was deployed, while the Chang'E-3 lunar lander was oriented toward Sun as shown in Fig. 3.26.

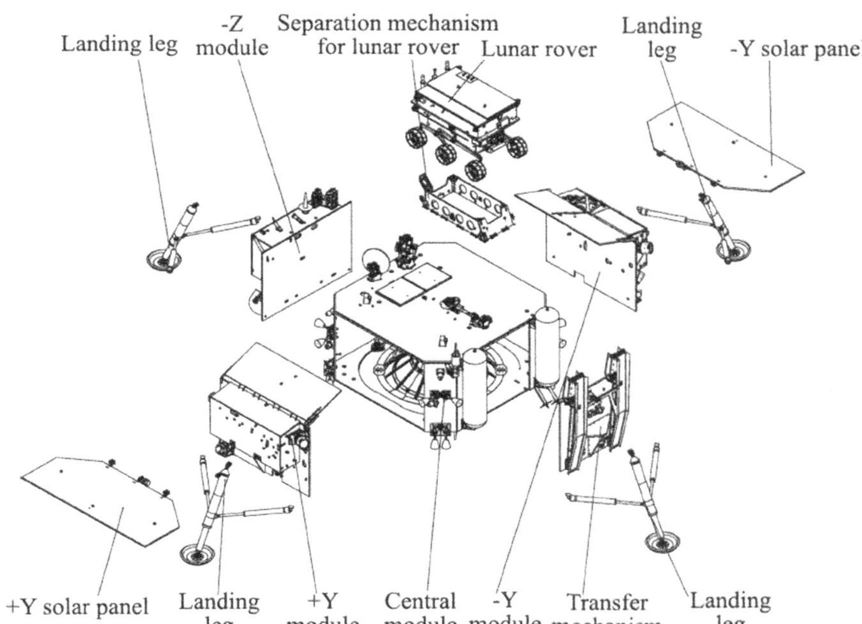

Fig. 3.24 Exploded view of Chang'E-3 lunar lander

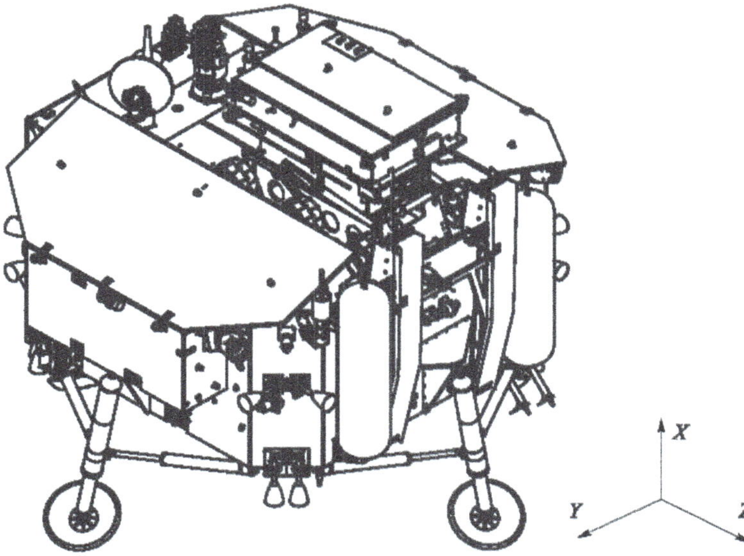

Fig. 3.25 Configuration of Chang'E-3 lunar lander in launch vehicle fairing

Fig. 3.26 Flight configuration of Chang'E-3 lunar lander

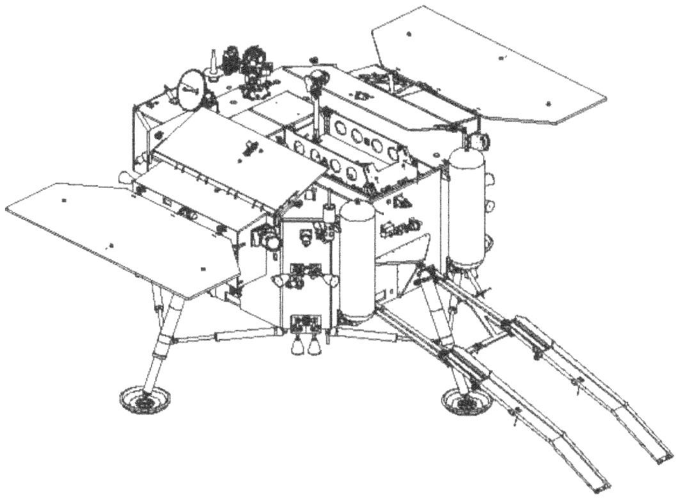

Fig. 3.27 Configuration of Chang'E-3 lunar lander on lunar surface operation

4. Configuration on lunar surface operation

After soft landing, the lunar rover was deployed through hold-down and release mechanism. The lunar rover moved to lunar surface by transfer mechanism. Then the +Y and −Y solar panels were unfolded and tilt of solar panels could be adjusted in order to obtain optimum illumination condition to get enough power supply. During operation on lunar surface, the pointing mechanism for the camera could pitch and yaw to take photos of lunar surface and the lunar rover. The lunar optical telescope began to work and the −Y module cover could be folded or unfolded to eliminate interference of parasitic light to the telescope. The directional antenna was oriented to Earth to download scientific data. The configuration of Chang'E-3 lunar lander on lunar surface operation is shown in Fig. 3.27.

5. Configuration in hibernation on lunar night

When the solar elevation decreased gradually and the requirement of power supply could not be satisfied, Chang'E-3 lunar lander was commanded to hibernate through lunar night. The tilt of +Y solar panel was set to satisfy the requirement of solar elevation and power supply when Chang'E-3 lunar lander was awakened. The −Y solar panel was folded to cover the radiator in order to maintain the temperature in −Y module.

3.7 Planning of TT&C Link

The TT&C of lunar lander with ground station should be implemented during flight. The data downlink and receiving from the lunar rover should be implemented during lunar surface operation [9].

The TT&C link of Chang'E-3 lunar lander with ground station and the lunar rover is shown in Fig. 3.28.

The major frequency was X-band for TT&C and data downloading of Chang'E-3 lunar lander. There were two phases for TT&C from separation with launch vehicle to approaching perilune. When the distance to Earth was within 2×10^4 km, the S-band TT&C link was setup. When Chang'E-3 lunar lander could be accessed by deep space ground station during LTO, the deep space ground station could support X-band TT&C and the X-band TT&C link is setup.

When Chang'E-3 lunar lander approached perilune, the X-band TT&C was maintained. During LOI and circumlunar orbit, the X-band TT&C was used while Chang'E-3 lunar lander could be accessed by deep space ground station during LOI. During powered descent, the TT&C link between Chang'E-3 lunar lander with deep space ground station was setup and data was downloaded by mid-gain antenna.

After soft landing, the X-band data downlink was setup when the directional antenna was operated normally. The downlink was switched to directional link. When Chang'E-3 lunar lander was hibernated in lunar night or awakened in lunar daytime, the downlink was switched to omnidirectional link.

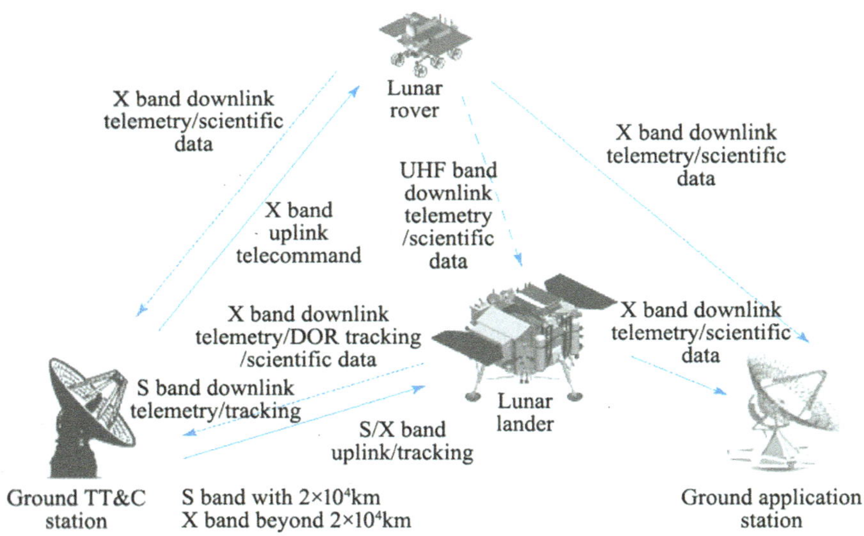

Fig. 3.28 TT&C link of Chang'E-3 lunar lander with ground station and the lunar rover

The telecommunication channel between Chang'E-3 lunar lander and the lunar rover could be setup when required. Chang'E-3 lunar lander could receive and transmit telemetry and scientific data of the lunar rover.

3.8 Design of Powered Descent and Soft Landing

The process of powered descent and soft landing was unique for the lunar lander. The major tasks of this process could be summarized as following.

(1) To decelerate the velocity relative to the Moon to a certain range by lunar lander's propulsion system.
(2) To obtain height and velocity relative to lunar surface by onboard speedometer and ranging sensors and IMU, and formulate control scheme autonomously by onboard computer.
(3) To identify lunar terrain of landing site by onboard optical and laser sensors, seeking safe landing site autonomously and formulate control scheme of obstacle avoidance.
(4) To shut off main engine autonomously at certain height before lunar lander touches lunar surface and ensure the falling velocity in a certain range.
(5) To absorb landing impact load by landing cushion mechanism, ensure stable attitude after landing and support subsequent scientific exploration.

Different control schemes should be formulated according to different GNC control scheme, configuration of navigation sensors and propulsion system.

3.8.1 Initial Status and Setup of Powered Descent

Generally, the start point of powered descent was design to be perilune of circumlunar orbit. The altitude of perilune for Chang'E-3 lunar lander was designed to be 15 km. The range of perilune altitude was determined by orbit control error, circumlunar orbit perturbation and orbit determination error. The GNC subsystem should ensure safe landing control within such range of initial status.

The setup before powered descent included folding of onboard mechanisms and setup of initial parameters and sensor status.

The solar panel should be folded before powered descent to avoid damage to solar panel by landing impact and power supply was provided by battery during powered descent.

The initial orbit parameters, orbital calibration results of sensors, compensation and correction parameters and etc. were uplinked by ground station. There was also uplink of delayed commands including power on of sensors, switch of telecommunication channels and power on of descent camera in initial status setup.

3.8.2 *Process of Powered Descent and Soft Landing*

There were seven phases in powered descent of Chang'E-3 lunar lander including preparation of ignition, main deceleration, rapid attitude adjustment, approaching, hovering, obstacle avoiding and slow descent.

(1) Preparation of ignition. The major tasks of preparation of ignition were to correct the time and orbit of ignition, calculate the attitude for ignition and adjust attitude as calculated before ignition. The attitude was determined by gyroscopes together with star sensors and controlled with 10 N thrusters during preparation of ignition.

(2) Main deceleration. The altitude relative to the Moon was from 15 km to 3 km during main deceleration. The major tasks were to decelerate velocity to the Moon from about 1.7 km/s to a predefined value and lower the altitude to about 3 km. The trajectory was a natural gentle descent trajectory caused by deceleration with orbit control engine according to the requirements of main deceleration. The drop of altitude was about 12 km and the range was about 430 km.
The initial pitch angle relative to lunar surface was 90° at the start point of main deceleration. The final pitch angle relative to lunar surface was about 60° at the end of main deceleration.

(3) Rapid attitude adjustment. The altitude relative to the Moon was from 3 km to 2.7 km during rapid attitude adjustment. The major tasks were to connect main deceleration with approaching, shift attitude to entry of approach rapidly and lower the thrust of main engine at the same time. The trajectory was a natural gentle descent trajectory caused by deceleration with orbit control engine according to the requirements of rapid attitude adjustment. The drop of altitude was about 300 m and the range was about 1 km.
The initial pitch angle relative to lunar surface was 60° at the start point of rapid attitude adjustment. The final pitch angle relative to lunar surface was about 9° at the end of rapid attitude adjustment.

(4) Approaching. The altitude relative to the Moon was from 2.7 km to 100 m during approaching. The major task was to avoid obstacle roughly. The trajectory of approaching target landing site was designed to satisfy the requirements of attitude and descent trajectory in order to align field of view of optical sensor with landing site according to requirement of rough obstacle avoiding. Lunar lander was approaching landing site gradually at 45° pitch angle in a nearly straight line. The large obstacles such as rock or pits whose diameter was greater than 1 m were identified by optical sensors to determine safe landing site and avoid obstacles. At last lunar lander approached the position above the landing site at 100 m height and near 0 relative velocity. The range of approaching along the flight path was about 3 km.
The initial pitch angle relative to lunar surface was 9° at the start point of approaching. The process of rough obstacle avoiding was implemented by adjusting attitude angle of pitch and yaw of lunar lander. When the necessary

acceleration in appropriate direction was generated, the pitch angle relative to lunar surface was adjusted to 9° and the yaw angle was adjusted reversely till the height and velocity of lunar lander satisfied requirements of hovering.

(5) Hovering. The height relative to lunar surface was about 100 m. The major task was to finely identify obstacles in landing site. The gravity of lunar lander was compensated by thrust of throttlable engine and lunar lander was kept hovering in order to observe landing site by 3-D imaging sensor to select safe landing site.
The roll, pitch and yaw angle relative to lunar surface was adjusted to about 0° during hovering.

(6) Obstacle avoiding. The height relative to lunar surface was from 100 m to 30 m. The major task was to finely avoid obstacles and descent. According to relative position of safe landing site given during hovering, lunar lander was descended to the position above the safe landing site, while the descent velocity relative to lunar surface was at predefined value and horizontal velocity is about zero. The trajectory during obstacle avoiding was an oblique line toward landing site.
The roll, pitch and yaw angle relative to lunar surface was adjusted to about 0° during obstacle avoiding.

(7) Slow descent. The height relative to lunar surface was from about 30 m to the height where the gamma sensor for engine shut off was triggered. The major tasks were to ensure steady and slow descent of lunar lander and the control accuracy of velocity and attitude satisfy requirements of soft landing. If the gamma sensor for engine shut off failed, the lunar surface sensor might provide backup signal for engine shut off.
The roll, pitch and yaw angle relative to lunar surface was adjusted to about 0° during slow descent.

3.8.3 Landing Cushion

When lunar lander was landed on lunar surface, the footpad of landing legs touched lunar surface at first. The ball joint could rotate freely to adapt lunar terrain. Then the footpad pushed the inner tube of main pillar to slide relative to the outer tube. The relative movement of inner tube to outer tube would compress aluminum honeycomb to absorb impact energy for cushion. When the horizontal load exceeded the limit, the two auxiliary pillars could provide cushion for both tension and compression. The slender rod could implement tension attenuation. The aluminum honeycomb could implement compression attenuation.

There were lunar surface sensors on footpad of landing legs. When the footpad touched lunar surface, the signal of touching lunar surface was transferred to GNC subsystem, which served as a backup method to shut off main engine in case of failure of the gamma sensor for engine shut off.

When the footpad touches lunar surface, the +X axis of lunar lander was nearly vertical to lunar surface while the +Z axis was oriented to the local north. The telemetry of lunar surface sensor was transmitted back to ground station as an indicator of soft landing.

3.8.4 Downloading of Descent Images

Lunar lander should be able to download images taken by descent camera during powered descent so images of lunar surface could be obtained for ground operation as soon as possible. The images of descent camera were used only for display, which would not be used for powered descent control. After safe landing, all images taken by descent camera could be downloaded through directional telecommunication channel.

3.9 Analysis and Simulation of Landing Stability

For lunar lander, the indicator of safe landing was that lunar lander was landed stably on lunar surface without turn over or sink, the stable attitude was maintained and there was no damage to onboard equipments caused by landing impact. Therefore analysis of landing stability should be performed for detail design of lunar lander, initial condition of soft landing and design of landing cushion mechanism with mechanical property and terrain of lunar surface. The sensitivity analysis of all factors should be performed to find the worst case in order to cover all possible condition in flight.

The influential factors for landing stability of lunar lander included features of lunar lander, initial condition of soft landing and lunar environment. The features of lunar lander include property of mass and moment of inertia, landing cushion features of mechanism and geometry parameters. The initial condition of soft landing included vertical velocity, horizontal velocity and attitude. The lunar environment includes mechanic property, slope, distribution of impact pits and rocks of lunar surface. In order to obtain probability of stable landing, those influential factors should be analyzed for landing stability by Monte Carlo simulation of different combinations. Thus the probability statistical model of stable landing should be established.

The impact response during soft landing, attenuation stroke and attitude after landing could be obtained to evaluate landing stability by establishing dynamic model of lunar lander, generating random lunar terrain and analyzing landing dynamics process under different combinations of initial condition [10].

In order to obtain quantitative indicator of landing stability, the criterion of stability could be given based on method of minimum stable distance. At first, concepts of turn over line and safe line are given as shown in Figs. 3.29 and 3.30. The turn over line was the closed curve that the line between centers of adjacent footpad is projected on horizontal plane as black frame shown in Fig. 3.30. The safe line was the closed

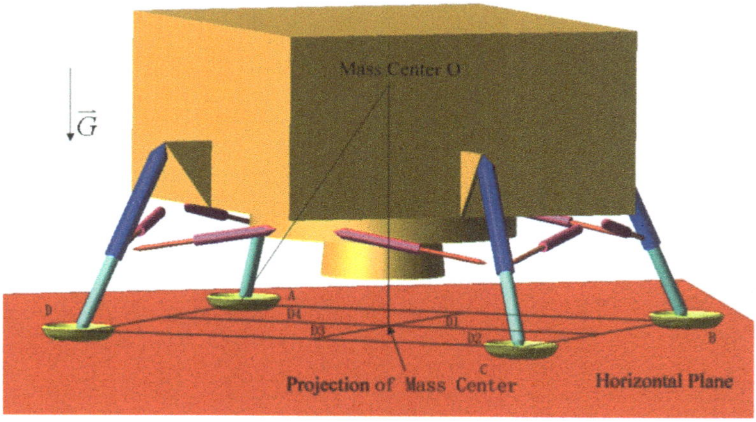

Fig. 3.29 Diagram of geometry criterion of landing stability

Fig. 3.30 Diagram of turn over line and safe line for lunar lander

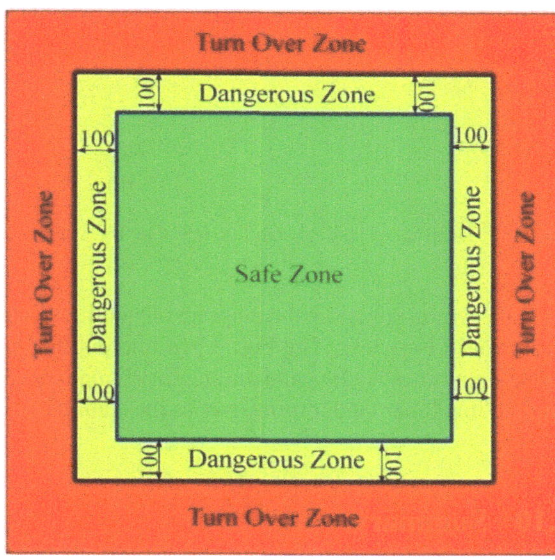

curve that the turn over line was translated *d* distance to inner side as red region shown in Fig. 3.30. The turn over line was determined by structure of lunar lander. When the dimension of lunar lander after deployment of landing cushion mechanism was determined, the turn over region was determined as well shown in Fig. 3.29. The translation distance *d* of the safe line was determined by the requirements of stability design of lunar lander. When the *d* is bigger, the safe region was smaller and safety factor of stability design was bigger.

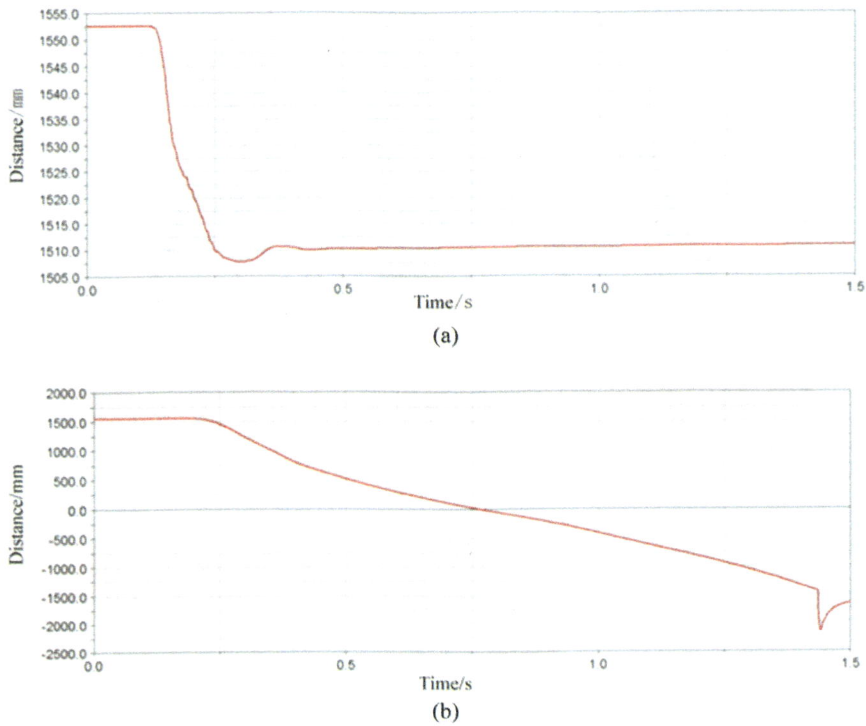

Fig. 3.31 Diagram of landing stability, **a** Stable. **b** Unstable

For a certain situation, the diagram of stability boundary was obtained by analysis as shown in Fig. 3.31. The Fig. 3.31a implied that lunar lander was maintained in stable status after landing and the stability boundary was not changed. The Fig. 3.31b implied that lunar lander turns over as the stability boundary was changed to minus.

3.10 Summary

System design of lunar lander is given in this chapter including mission analysis, environment effects analysis, trajectory design, definition of system and subsystem specification, design of system configuration, TT&C, powered descent and landing process, landing stability analysis. Special description of plume effects, landing impact, design of powered descent and landing process for lunar lander is given too.

System design was important to a system-level optimized solution. So methods such as comparison of different proposals and simulation analysis were used to determine the optimized design from pre-phase A to phase D.

Not all elements of system design of lunar lander are listed in this chapter. Special attention should be paid in system design of lunar lander to improvement of adaptability to complex terrain, accurate evaluation of plume and lunar dust effects and protection, miniaturized and integrated design, improvement of reliability and safety, with respect to changes of mission requirements.

References

1. Chu G, Zhang H (2007) Technology of lunar probes[M]. China Science and Technology Press, Beijing
2. Peng Chengrong (2011) Spacecraft system design[M]. China Science and Technology Press, Beijing
3. Ouyang Z (2005) Introduction to lunar science[M]. China Astronautic Publishing House, Beijing
4. Wu Weiren (2007) To the moon[M]. China Astronautic Publishing House, Beijing
5. Xi Xiaoning, Zeng Guoqiang, Ren Xuan et al (2001) Orbit design of lunar probe[M]. National Defense Industry Press, Beijing
6. Chu Guibai, Ma Shijun (2002) Introduction to astronautic technology[M]. China Astronautic Publishing House, Beijing
7. Tan Weichi, Hu Jingang (2009) Spacecraft systems engineering[M]. China Science and Technology Press, Beijing
8. Yuan Jiajun, Yu Dengyun, Chen Liemin et al (2004) Design and analysis of spacecraft structure[M]. China Astronautic Publishing House, Beijing
9. Zhou Xianwei (2009) Deep space communication[M]. National Defense Industry Press, Beijing
10. NASA (1969) Lunar surface models[R]. NASA, USA
11. Gromov V (1998) Physical and mechanical properties of lunar and planetary soils[J]. Earth Moon Planet 80:51–72

Chapter 4
Structure and Mechanism Technology of Lunar Lander

4.1 Introduction

The lunar lander structure is a general term for parts and components that support and transmit loads to provide support, stiffness, and dimensional stability for the lunar lander and its subsystem equipment. The lunar lander mechanism is the mechanical component that executes motion of the spacecraft accessories, such as solar panel mechanism, connection/unlocking and deployment mechanism for the lunar rover, and pointing mechanism for directional antenna.

The main functions of lunar lander structure are as follows:

(1) To provide a mechanical connection and separation interface for lunar lander with the launch vehicle.
(2) To provide interface for parking, lifting, packaging, and transportation of lunar lander.
(3) To provide fixed interface for installation of instruments on lunar lander with definite rigidity.
(4) To sustain static and dynamic loads on lunar lander.

Special attention shall be paid to the landing impact load for design of lunar lander structure in addition.

The mechanisms on different spacecraft have different requirements of specific function. The specific functions of different mechanisms might be different even on the same spacecraft. It is difficult to fully describe or summarize all the functions of the spacecraft mechanisms in a few words [1]. According to the characteristics of the lunar lander mechanism, the main functions of the lunar lander mechanism are as follows:

(1) To provide support and withstanding loads for accessories during launch and landing.
(2) To complete predefined motions including deployment, separation, extension, docking, cushion and etc. after orbit injection or landing of lunar lander.

D.-Y. Yu et al., *Technology of Lunar Soft Lander*, Space Technology Library 38,
https://doi.org/10.1007/978-981-15-6580-9_4

(3) To ensure the attachment and connection stiffness with lunar lander, to maintain the relative position of the attachment and the primary structure, to withstand the temperature alternation load, to ensure pointing accuracy of the attachment as required during operation.
(4) To provide landing cushion to ensure safety of the instrument on the lunar lander during soft-landing on lunar surface.

Special attention shall be paid to the inertial load on the deployed mechanism during orbit maneuver, the impact load upon landing, gravity of the lunar surface, and the relatively harsh temperature environment for design of lunar lander mechanisms in addition.

In this chapter, design and verification of the structure and mechanism of the lunar lander will be illustrated for the structure, solar panel mechanism, connection/unlocking and deployment mechanism of the lunar rover of the Chang'E-3 lunar lander. The landing cushion will be described in Chap. 11.

4.2 Structural and Mechanism Features of Lunar Lander

4.2.1 Structural Features of Lunar Lander

The structure of the lunar lander varies greatly depending on its mission. The series of Surveyor lunar lander of United States configuration is shown in Fig. 4.1. The surveyor's primary bearing structure is shown in Fig. 4.2. It was mainly composed of thin-walled aluminum pipes, forming a hexagon around the main retro-rocket, which provided installation location and interface for the landing cushion mechanism, connection structure with launch vehicle, main retro-rocket, vernier engines and fuel tanks, thermal partitions, antenna, flight control systems, and scientific experiment equipment [2, 3].

The general Artemis lunar lander proposed by the NASA Johnson Space Center is shown in Fig. 4.3, whose primary structure is a radially symmetric truss structure consisting of carbon fiber tubes connected by metal fasteners. At the top there is a plain surface that provides interface for the installation of the payloads with fuel tank inside [4, 5].

The American Apollo lunar module is shown in Fig. 4.4. It consisted of upper and lower sections. The upper structure was an irregularly shaped sealed capsule with life support equipment. The lower section was an octagonal prism, including landing engine, propellant, water, oxygen, and scientific experimental equipment. The internal structure of the descent module was a cross box structure consisting of truss and stiffened plate, as shown in Fig. 4.5. The main retro rocket was located in the center of the structure with four fuel tanks in the four cubic boxes around. Its launch load and landing impact load were supported by 4 pyramidal frames [6, 7].

An exploded view of the lunar lander for the ESA Euro Moon 2000 project is shown in Fig. 4.6 [8]. The structure was composed of three major parts: an aluminum

Fig. 4.1 Configuration of Surveyor lunar lander

Fig. 4.2 Frame structure of Surveyor lunar lander

ring for interface ring with the Ariane 4 launch rocket; a primary bearing cylinder; a transition conic section for connection a small lunar spacecraft. Both the primary bearing cylinder and the transition conic segment were composed of aluminum skinned honeycomb sandwich shell or aluminum corrugated shell. There was a platform for installation of electronic equipment between the primary bearing cylinder and the transition conic segment, which was also a honeycomb sandwich panel structure. The lower parts of the fuel tanks were fixed on the strut and the upper parts are connected to the primary bearing cylinder. Eight thrusters were mounted on the aluminum struts that held the fuel tank.

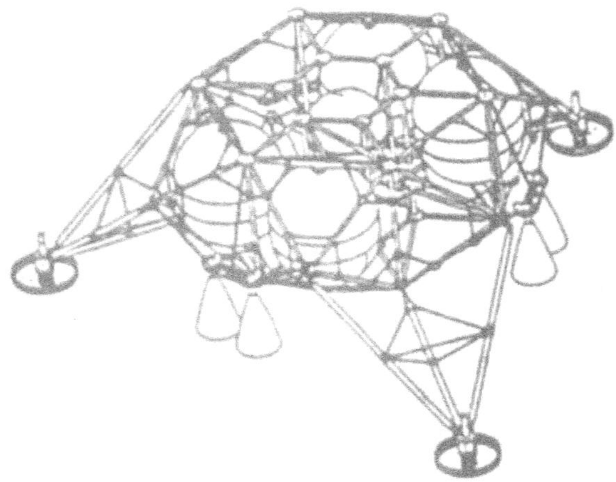

Fig. 4.3 General Artemis lunar lander

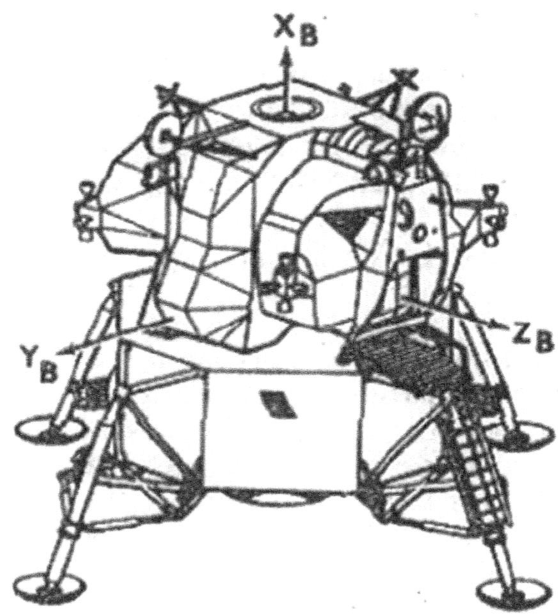

Fig. 4.4 Apollo lunar module

It can be seen that the primary structure bearing the launch load is quite different while the primary structure bearing the landing impact load is mostly trussed structure by the lunar lander launched and to be launched in other countries. The most important feature of the trussed structure is its capability to withstand concentrated loads. Landing impact loads are usually acted to the structure in the form of concentrated loads through the landing cushion mechanism [9, 10].

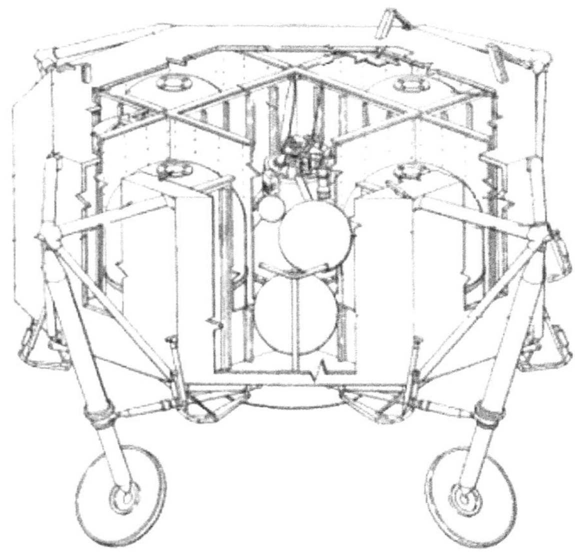

Fig. 4.5 Internal structure of the Apollo descent module

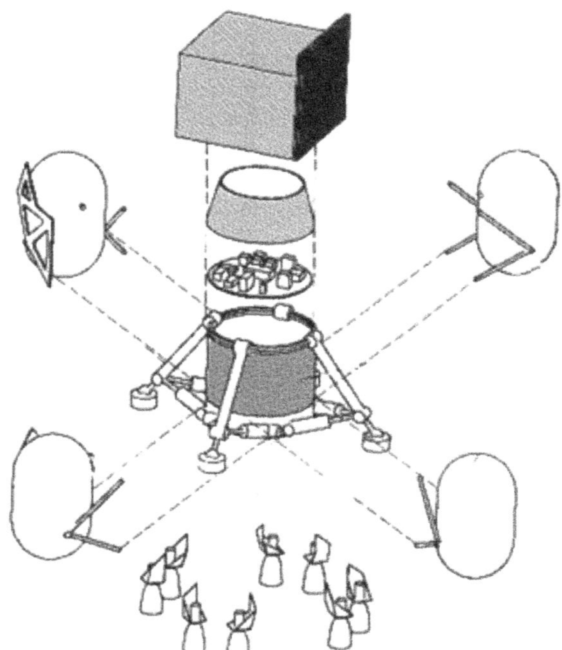

Fig. 4.6 Exploded view of the ESA lunar lander

In the design of lunar lander structure, the following aspects should be considered:

(1) Lightweight structure. Lightweight is the eternal theme of spacecraft design, especially for deep space probes such as lunar lander. Due to the constraints of the launch vehicle, the mass of the lunar lander is strictly limited, while the structural mass accounts for a large proportion of the dry mass of lunar lander.
(2) The design of structural configuration of lunar lander must focus on the installation of large fuel tanks, large engines, and the lower center of gravity requirements of soft-landing. The lunar lander needs to carry a large amount of propellant in order to change orbit and descend. The propellant mass of lunar lander is a large proportion of the mass and usually is about 2/3 of the launch mass. Therefore, large fuel tanks need to be installed. A large engine is necessary for powered descent. In order to ensure stability after soft-landing, it is more important to lower the center of gravity of lunar lander. Therefore, a flat structure shall be used for lunar lander.
(3) The lunar lander structure is subjected to a landing impact load. The lunar lander has a certain speed (kinetic energy) before landing. The energy shall be absorbed instantly when the landing cushion mechanism collides with lunar surface, which will cause a large landing impact load upon lunar lander. The landing impact load is transferred to the structure by the landing cushion mechanism. For a typical landing gear type of landing cushion mechanism, the landing impact load is a concentrated load, and the force transfer path of the landing impact load may be completely different from the transfer path of the launch load (the launch load is transferred to the structure by the docking surface of the launch vehicle, but there was also the same design for the force transfer path, such as the Apollo series spacecraft). It requires that the structure of the lunar lander is totally different from traditional spacecraft and spaceship.
(4) The structure is subjected to a landing impact load after experiencing high temperature and low temperature environment. While landing, the lunar lander structure has experienced a harsh temperature environment during flight, and the temperature level of the structure at landing is higher. Therefore, special attention shall be paid to bearing capacity on a special temperature environment.

4.2.2 Mechanism Features of Lunar Lander

Different mechanisms are often needed to achieve different specific functional requirements on lunar lander. The types of mechanisms are up to specific mission requirements of the lunar lander. The typical lunar lander mechanisms include landing cushion mechanism, solar panel mechanism, antenna pointing mechanism, and mechanism for separation and deployment of the lunar rover. The type and approach of lunar lander mechanism is not only up to the mission characteristics of lunar lander, but also closely related to the general configuration of lunar lander, technical specification, environment on lunar surface and other factors.

Fig. 4.7 Deployment process of the Apollo the lunar rover by astronauts

The deployment of the lunar rover to lunar surface is an important means to implement lunar scientific exploration. Because it is difficult to launch the lunar rover directly to the Moon and land safely, usually the lunar rover is installed on lunar lander and deployed to lunar surface after safe landing. The American spacecraft Apollo deployed the lunar rovers to lunar surface in the presence of astronauts. When launched, the lunar rover was mounted on the side of the lunar module. After landing, the astronauts unlocked the lunar rover. Then, the lunar rover was controlled by ropes and pulleys to roll over to lunar surface, and finally the wheels were deployed. The deployment process of the Apollo the lunar rover by astronauts is shown in Fig. 4.7.

Lunar rovers were deployed in Soviet Union's robotic missions of Luna 17 and Luna 21. When launched, the lunar rovers were installed on the top of the lunar lander by means of tie rods. Folded ladders were installed on both sides of the lunar landers. After landing, the folded ladders on both sides of the lunar lander were deployed and touched on lunar surface for the lunar rover to move to lunar surface on a ramp, as shown in Fig. 4.8. The foldable ladder-type mechanism for the lunar rover deployment is suitable when height of the lunar rover to lunar surface was not too high after landing.

The United States Mars Pathfinder lander used airbag to achieve soft-landing on the surface of Mars [11]. Its lander was a tetrahedron structure with one basic plate and three side plates. The Sojourner Mars rover was installed inside the tetrahedron. After landing, the tetrahedral structure was unfolded and two ramps were deployed in front and rear of the Sojourner Mars rover. Since the Sojourner rover was only about 0.6 m from the surface of Mars after landing, two ramps could directly touch the surface of Mars, providing a passageway for deployment of Sojourner's rover to Mars, as shown in Fig. 4.9. The "Spirit" and "Opportunity" Mars rovers of the United States MER, due to its lower height to Mars surface after landing, could be deployed through the unfolded side plates instead of a special ladder as shown in Fig. 4.10.

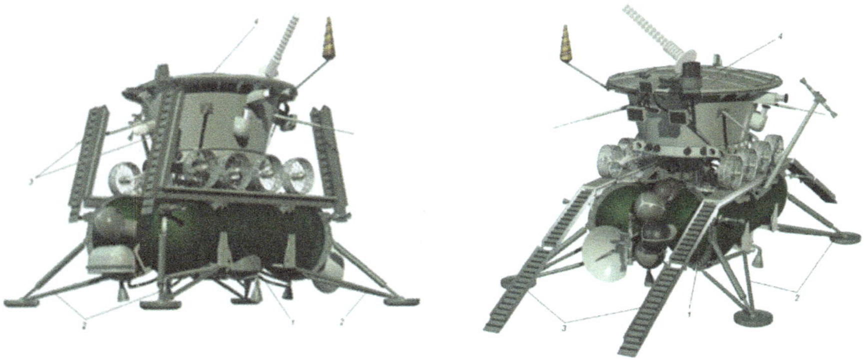

(a) Before landing (b) After landing

Fig. 4.8 Luna 21 of Soviet Union

Fig. 4.9 Deployment of Sojourner

It shows that the types and functions of mechanisms are significantly different due to significant differences of the characteristics of the mission and the general configuration for lunar and Mars landers that has been launched. For the lunar lander, the design of the mechanism shall meet the specific mission requirements of the lunar lander. Special attention shall be paid in design of lunar lander mechanisms as follows [12]:

(1) Lightweight. The requirement is the same as the structure.
(2) For the mechanism unlocked and deployed during orbital flight, it is difficult to endure landing impact load if it is not folded before landing. The load during orbit maneuver and powered descent shall not be ignored.
(3) Lunar dust environment. Before landing, the plume of engine will activate lunar dust. When landing, the landing cushion mechanism will also activate lunar dust. It will affect the normal movement of mechanisms if lunar dust enters into the motion join.

Fig. 4.10 Deployment of MER rover "Spirit" and "Opportunity"

(4) The harsh temperature environment. Mechanisms are usually installed on surface of lunar lander. Before soft-landing, lunar lander must go through various orbital and attitude maneuver such as LTO, circumlunar orbit and powered descent, which means that mechanisms will be faced with sophisticate high and low temperature environment for operation and survival. After landing, lunar lander will go through solar elevation variation, lunar daytime and night, lunar eclipse, and etc. At the same time, the attitude of lunar lander on lunar surface is uncertain to some extent. All above means that the mechanisms operating on lunar surface must endure harsher high and low temperature environment than ordinary spacecraft.

(5) Lunar terrain. Lunar terrain shall be considered for mechanisms operating on lunar surface. For example, slope, pits, and rock size on lunar surface should be considered in design of deployment mechanism of the lunar rover.

(6) Lunar gravity environment. When mechanisms move on lunar surface, effect of lunar gravity environment shall be considered. For the deployment or folding, lunar gravity is a drag, sometimes lunar gravity is driving force, which depends on the position and attitude of mechanisms. The lunar gravity environment must be considered when determining the driving force of the mechanism. Generally, lunar gravity is much higher than friction resistance in magnitude.

4.3 Design and Verification of Lunar Lander Structure

The design process of lunar lander structure was from general design to part design, from stiffness design to strength design and from function specification to function implementation. The major tasks in different phases included determining structural design and completion of preliminary design in phase B, detailed design and design verification in phase C, design, manufacture, acceptance, test of flight hardware and operation after launch in phase D and E.

The major function of lunar lander structure was to provide platform for equipment installation and bearing load, i.e. to provide surface and room for equipment installation onboard lunar lander, as well as to ensure accuracy of equipment installation surface and integrity of configuration during ground operation, launch, landing and operation on lunar surface.

4.3.1 Structural Design

1. Determination of Requirements for Structural Design

Before structural design, it was necessary to define requirements. Trade-off and comparison of different options was the key to optimal design that meets the requirements.

For the design of the lander structure, although a number of specific requirements existed, it could be summarized as "The structure shall support operation of the instruments on board lunar lander with sufficient strength, stiffness, and accuracy."

The structural design usually began with the primary structure, taking into account the strength and stiffness of the structure. These requirements came from configuration of lunar lander, mass property, launch loads, landing impact and fundamental frequency limitations.

The primary structure was the framework of lunar lander which was usually made up of a primary body, interface ring with launch vehicle, and support structure for landing legs. Typical requirements for primary structure design are listed in Table 4.1 [15].

For the special impact load to lunar lander structure, due to cushion effect of landing cushion mechanism, the load on support structure for landing legs could be treated as static load, except that the load transfer path of landing impact load on the structure was different from that of launch load. The response of onboard instruments due to landing impact could be treated as a sinusoidal vibration response.

2. Structural Configuration Design

The configuration design of lunar lander was the basis of the structural design. The structural design could only be initiated after completion of the configuration design. The system scheme, especially the configuration scheme, had a great influence on the structural scheme. The structural design requirements should be derived from the system configuration. Therefore, joining the system configuration design was a good starting point for the structural design.

In order to make the structural design more reasonable, the force transfer path was an important issue in the configuration design. The principle of force flow continuity and the shortest direct transfer path were the basic principles of structural design. A rationally designed force transfer path could provide higher structural rigidity with less structural material and make strength design easier.

Table 4.1 Typical requirements for primary structure design

Requirements	Definition	Instructions	Required information
Strength	The components should not produce damage that might jeopardize mission goal under specified load such broken, instability, excessive deformation, etc.	For primary structure, the load highest usually occurred at lifting, launching, and landing. Loads were derived from launch and landing impact environment, mass property of lunar lander	1) Load coefficient and dynamic environment of launch vehicle 2) landing impact load 3) mass property
Natural frequency (or stiffness)	The natural frequency (usually the fundamental frequency, i.e. the lowest natural frequency, with directionality) is greater than the specified value or beyond a certain frequency range	Should meet natural frequency requirements of the launch vehicle. Stiffness requirements could be assigned to the parts of primary structure to ensure that the spacecraft was always within the dynamic envelope	1) Natural frequency and dynamic environment envelope of launch vehicle 2) mass property
Mechanical interface	Satisfaction requirements for mechanical match and connection with launch vehicle and interface with components and subsystems	Requirements for interface ring with launch vehicle include connection form, flatness and load distribution; Mechanical interface requirements with each sub-system include flatness of installation surface, connection form, and connection position; Interface requirements between components including the connection form between the non-detachable components, and the connection form and accuracy between the removable components	Mechanical interface requirements

(continued)

Table 4.1 (continued)

Requirements	Definition	Instructions	Required information
Position stability	Maintaining the specified position or pointing accuracy in a certain part or direction of the structure shall include errors caused by temperature deformation, material yield, mechanical connection offset, etc.	Primary structure should maintain the relative position accuracy between sensors, antenna, engine, payload, and etc.	1) Assigned accuracy index to primary structure 2) Temperature range
Structure mass	Total mass of the structural sub-system and primary components were less than the specified value	The mass was usually assigned to the structural sub-system, so it was often the goal of preliminary design, not as a fixed requirement	Assigned mass
Operability	All parts should be easy to install, remove, adjust, transport, store and ground test, and adapt to assembly operation of the probe	The construction and installation design of the components should be taken into account that all ground operations could be performed	Operation requirements
Material processability	The materials and manufacturing processes used for the components could be realized	The materials used could be purchased or produced, and the manufacturing processes used are mature	

There were two issues that should be paid special attention in configuration of lunar lander structure:

(1) Configuration of soft-landing mechanisms. It would directly determine the force transfer path of the landing impact load, and thus affect the primary structure configuration. Usually transfer paths of the landing impact and the launch load were different.
(2) Configuration of large fuel tanks and orbit maneuver engines. The inertial load of fuel tanks with propellant was the maximum load of the structure during launch; on the other hand, the room for large fuel tanks and orbit maneuver engines was big. For lunar lander with large device, the location and room for installation were also the focus.

The general feature of lunar lander configuration was that the fuel tanks were arranged in parallel (usually 4 or more), and the lower part of fuel tanks was directly mounted on the interface ring or short beam, in order to shorten the force transfer path and lower the center of gravity.

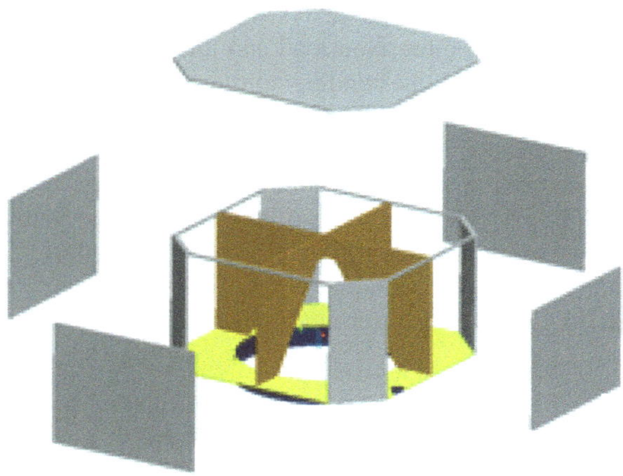

Fig. 4.11 Cross-support structure configuration of the Chang'E-3 lunar lander

For example, the Chang'E-3 lunar lander structure was a cross-support structure configuration made up of a cross-support structure at the center of the structure, interface ring, side plates, bottom plates and roof plates, as shown in Fig. 4.11. The cross-support structure was composed of four baffles and a center joint. Four fuel tanks were directly installed on the interface ring at the lower part. The upper part was connected to the upper corner of the cross-support structure with rods. The large room in the middle was for orbit maneuver engine. In order to adapt to the cross-support configuration, the baffles was designed as a beam-plate composite structure, where the beam was embedded in the inner side of the baffles for major support and the plates could provide auxiliary support role and installation for devices, cables and pipelines. In term of mechanics, the structure was a box structure with truss, and four beams were connected to form a pyramid truss. the lunar rover was mounted on the top by a bracket connected to a cross baffles, which could be treated as a big device for lunar lander.

3. Analysis of Force Transfer Path

The analysis of force transfer path was a qualitative analysis of the rationality of structural configuration design from the perspective of mechanics, which was a part of structural configuration design. After preliminary configuration design, quantitative mechanical analysis should be conducted to confirm the rationality of structural configuration design further.

The general structural components in spacecraft structure included rods, beams, plates, and shells. The capability of structural components such as rods and beams to withstand tension and compression was strong, while the capability to withstand bending and torque was relatively poor. The capability of structural components such as plates and shells to withstand in-plane tension and shear loads was strong, while the capability to withstand out-of-plane bending, torque, and shear loads was relatively poor. Therefore, in the analysis of force transfer path, it was considered that only tension and compression load and the in-plane load were transferred by structural components, and the bending, torque and the out-of-plane load were ignored (unless the structural component was specially designed to transfer such loads).

The analysis of force transfer path could be conducted after the general configuration, mass property, and configuration of large equipment was determined together with bearing characteristics of structural components.

For example, the primary structural payload of the Chang'E-3 lunar lander was launching and the landing impact load. The force transfer path of launching and landing impact load was analyzed as follows.

1) Transfer of launch load
(1) Lateral load: The transverse load caused by the fuel tanks consisted of upper and lower parts. The upper load was transferred to the cross support structure through the rods of fuel tanks and then to the interface ring. The rods of fuel tanks bore tension and compression, and the cross baffles bear the in-plane tension, compression and shear load. Pre-embedded beams bore tension and compression. The lower load was transferred directly to the interface ring. The load of the lunar rover was transferred to the cross baffle through the bracket. Most of other equipment was installed on the side and top plates. The load on the side plates was transferred to the interface ring through the cross support structure, the top plate and the bottom plate. The cross-support structure, the top plate and the bottom plate bore in-plane tension and shear loads, while the pre-embedded beams bore tension and compression. The load on the top plate was transferred to the interface ring through the top plate and the cross-support structure. The load on the interface ring was transferred to the launch vehicle.
(2) Longitudinal load: The longitudinal load of the fuel tanks was directly transferred to the interface ring. The longitudinal load of other equipment was transferred to the interface ring through the side plates and the cross support structure. The side plates and the cross support structure bore the in-plane tension and shear loads. The pre-embedded beams bore tension and compression. The load on the interface ring was transmitted to the launch vehicle.

From above analysis, it implied that the cross baffles and interface ring were the major bearing components of the structure for launch load. Special attention should be paid to the mechanics design of such structure.

2) Transfer of landing impact load
(1) Lateral load: A part of the lateral load was transferred to the cross-support structure, then to main pillars of landing cushion mechanisms through the support of landing gears, and to lunar surface through main pillars of landing gears at last. Another part of the lateral load was transferred to the bottom plates, then to auxiliary cushion of landing gears, then to main pillars and finally to lunar surface.

(2) Longitudinal load: Longitudinal load of the fuel tanks (in this case, the propellant was almost exhausted and the load was small) was transferred to the cross-support structure through the interface ring, and the longitudinal load of other equipment was directly transferred to the cross-support structure. The load on the cross-support structure was transferred to main pillars of landing cushion mechanisms through the support of landing gears, and finally to lunar surface.

From above analysis, it implied that the cross baffles and the support of landing gears was the primary bearing components of the structure for landing impact load. Special attention should be paid to the mechanics design of such structure.

4. Structural Component Design

The structural components design could be started after configuration design was finished. The factors in structural components design included: strength, stiffness, mechanical interface (including accuracy), assembly requirements, functional requirements (such as heat dissipation, conductive insulation, etc.), and manufacturing processes.

After the material and basic shape of the structural components was determined, the cross section size and thickness should be determined through mechanical analysis. For the design of the component connection point, especially the primary force transfer point, the load was generally given through mechanical analysis. The specific form of the connection point sometimes should be determined after local tests verification (the local structure static test was generally required due to the boundary).

A brief description of the structural design method is given as following for the cross support structure of the Chang'E-3 lunar lander as an example.

The cross support structure of the Chang'E-3 lunar lander was the key bearing component of the lander structure and made up of four baffles, as shown in Fig. 4.12. The main body of the baffles was a honeycomb sandwich structure. In order to construct a stable truss structure and strengthen the force transfer path, it was

Fig. 4.12 Cross-support structure configuration

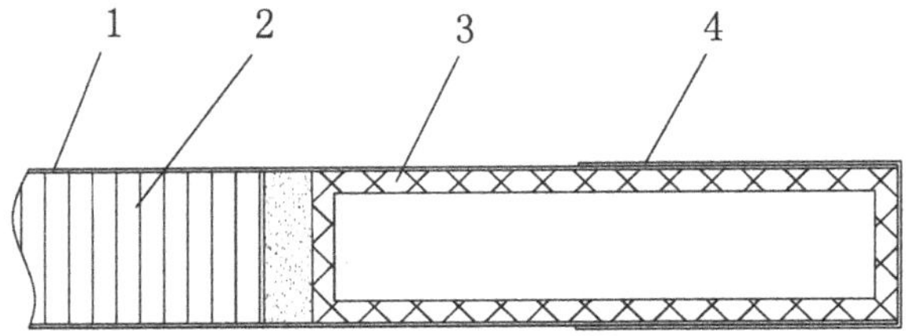

Fig. 4.13 Beam-plate composite structure. 1-Plate; 2-Honeycomb; 3-Rectangular beam; 4-Woven fabric

designed to be a beam-plate composite structure. The cross section thereof is shown in Fig. 4.13. The beams in the baffles were embedded inside the baffles and assembled to construct a quadrangular pyramid, which guarantees the rigidity of the platform; the honeycomb plates could distribute the load and provide the equipment installation surface.

Because of the embedded beams in the plates, the force conditions were complex, and the plates material was made of high-strength carbon fiber composite material, which could well coordinate the stiffness gradient between the plates and the embedded beams, as well as bearing landing impact load. The beams in the plates were pre-embedded rectangular hollow beams, and the material was high modular carbon fiber composite material to improve the rigidity of the entire lander.

The force near the lower joint connected with the baffles and the interface ring was bigger, and the reinforcing measures were taken near the lower joint of the baffles. In order to prevent the edges of the beam-plate joints from peeling, a carbon fiber woven cloth was wrapped around the inner edge of the baffles to increase the local thickness. In addition, the connection between the baffles with the landing cushioning support and the lunar rover support was locally reinforced.

5. Structural Connection Design

The connection between the structural components should be considered simultaneously when the types of structural components were determined. The major connection methods included mechanical fastening (such as screwed or riveted), welding and glued. Three types of connection methods are compared in Table 4.2. For critical connections (such as big load or high precision requirements), a combination of screw and glued connections was often used.

4.3.2 Structural Analysis of Lunar Lander

Structural analysis was an important part of structural design throughout the development of structures. At the beginning, conceptual design was determined based on the design requirements, previous design experience, and simple preliminary analysis. After conceptual design was determined, detailed analysis of the designed structure was performed to verify whether the design met the requirements mainly including

Table 4.2 Comparison of connection methods

Method	Advantages	Shortcomings	Application
Mechanical fastener	1) General, various types 2) Easy to get 3) Standard size and screw thread shape 4) Could be installed in almost any device 5) Connection for different materials 6) Easy to disassemble 7) Increased structural damping	1) Heavy 2) Difficult to avoid the loss of stiffness 3) Connection may drift 4) Often necessary to use end fittings to transfer loads from structure to fasteners 5) Introduction of stress concentration that may cause premature destruction of brittle materials	Suitable for most structures made from ductile materials
Welding	Keep rigid	1) Couldn't disassemble parts 2) Difficult to maintain the size of parts (warpage due to residual stress) 3) Reduced strength of aluminum and some other materials 4) Mass varies with process and product, must develop and demonstrate process	Dedicated to connecting weldable metals
Glued	1) Light weight 2) Increased the structural damping 3) Might be the easiest and most economical connection method 4) The load was fairly evenly distributed in the joints 5) Produced a smooth surface 6) Effective for different materials	1) Usually couldn't be disassembled 2) Mass varied greatly with process, material and products, process must be developed and demonstrated 3) Low tensile (peeling) strength 4) Limited shelf life of glue 5) Some glue was brittle 6) Temperature limit	Adapted to assembly of brittle materials Connecting composites to metal end fittings

strength and stiffness requirements. The lander structure was usually complex structures. It was difficult to design a rational and optimized structure with simple qualitative analysis. Structural analysis could effectively select a reasonable structural design. In structural design, especially in the initial stage of design, in order to obtain a high-quality design, it was usually necessary to make various design proposals and choose a reasonable one among various alternatives. There were many factors that need to be considered in the selection of a solution, but the requirements of structural strength, stiffness, and the adaptation to the mechanical environment were the most fundamental and must be considered, and these problems were exactly what structural analysis could solve. A reasonable design could be determined with structural analysis of different design, comparison of analysis results and comprehensive consideration of other factors.

Structural design should meet a variety of design requirements or design constraints. The most fundamental requirement was to meet the requirements of static characteristics (static strength) and dynamic characteristics of the structure. Therefore, after the design was completed, a very important issue was how to verify that the design had met these requirements, i.e. the verification of the structural design. Verification of structural design was also an integral part of structural design. Structural analysis was one of the common methods for verification. Compared with structural tests, structural analysis could be performed without structural hardware, therefore saved development time and costs. For some issues that were difficult to conduct ground tests, it was more necessary to solve them through analysis. The analysis could also guide the test and predict the test results.

The central task of structural analysis was to quantitatively evaluate the mechanical properties of the structure. The most widely used analysis method at present was the finite element method. The mechanical basis of the finite element method was the variational principle of elastic-plastic mechanics.

The analysis model was the basis of structural analysis. In structural analysis, it was crucial to establish a valid model. The model should commonly contain all the mechanical properties of the structure. In general, the main issues that should be considered in establishing a structural analysis model (referred to as modeling) included: idealization of structures, selection of finite element types and division of cell grids, preparation of cell characteristics, determination of boundary (and initial) conditions and load conditions, etc., as well as the specific content and accuracy requirements of analysis. All above issues were not isolated and they were organically related with each other. Therefore, they must be considered in a unified manner.

There was not much difference between the structural analysis of lunar lander and general spacecraft. The attention should be paid to additional landing load conditions (including the corresponding boundary conditions)and the accuracy of the fuel tanks model.

1. Modal Analysis

The lander modality was an inherent property of the lander instead of the lander structure. The lander modality was determined by the stiffness of lander structure, mass distribution and boundary conditions.

The requirements of the lander modality were the basic requirements for structural design of the lander. Because the lander modality was determined by the stiffness of lander structure, the modality requirements of the lander were sometimes referred to as the stiffness requirements of structure. When the lander structure was designed, at first it was necessary to consider the requirements of the modality. Therefore, the modal analysis of the lander was one of the important aspects of the structural analysis as well as basic aspect of structural analysis.

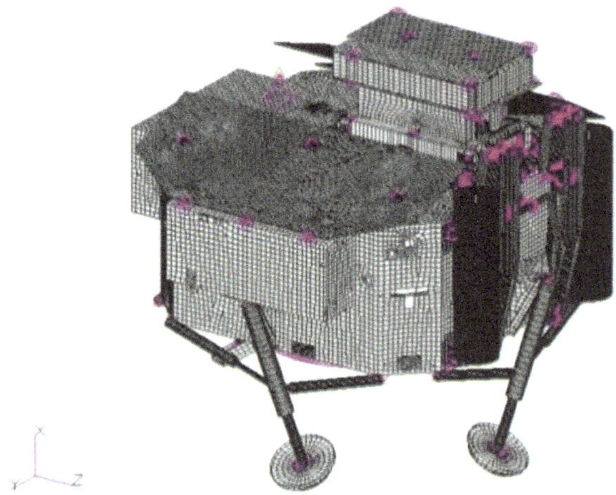

Fig. 4.14 Cross-support structure configuration

In the practical analysis process, mass distribution of the entire lander and model of large-scale accessories, etc. were continuously changing along with the progress of the engineering development, so the analysis model was also updated accordingly. The structure itself was also subjected to multiple modal analyses according to the continuous optimization of the section size, ply, local reinforcement, and functional requirements of structural components. Then a reasonable and optimized structural design was determined.

The finite element model of the Chang'E-3 lunar lander is shown in Fig. 4.14. According to analysis results, the effective mass of the longitudinal local frequency of the fuel tanks of the Chang'E-3 lunar lander was relatively large (theoretically it could also be considered as the general longitudinal frequency of entire lander), which caused the modal effective mass of the longitudinal frequency of the entire lander relatively small. Therefore, these frequencies must be paid special attention in the longitudinal vibration test of the lander, which was also one of the characteristics of the lunar lander different from other spacecraft.

2. Static Analysis

The mechanical behavior of the aircraft structure under static or quasi-static conditions was studied in structural static analysis in order to solve the issues of static strength, stiffness and stability of the structure.

Structural static analysis included the analysis of stress, deformation and stability. The purpose of static analysis was to guide and verify the structural design, parameter design, strength and stability of the structure by obtaining the stress and deformation characteristics of the structure under the load of static load (static load or quasi-static load).Then the structural mass could be reduced. In terms of structural load, the design of primary bearing structure was focused on static load.

For the lunar lander, the quasi-static load in launching process and landing impact load were usually key inputs for primary structural design, as well as important inputs for static analysis. The determination of load should be taken into account all critical load cases.

Prediction analysis of static test as also an important part of structural static analysis. The prediction analysis of static test could verify the rationality of test plan on the one hand and guide the test on the other hand.

The force transfer path and bearing characteristics of the structure should be analyzed in the static analysis modeling, and the structural parts of the detailed stress analysis should be determined, and the element mesh was refined in the stress concentration area.

The landing load in landing state used as structural design was simplified to quasi-static load, which was given by system design output. The load was acted on the support of landing cushion mechanisms of the lander structure. The analysis of the force transfer path showed that the structural components such as the support of landing cushion mechanisms and its adjacent structural components should be specially considered. Compared with launch state, the static analysis model of landing state should only adjust the fuel tanks mass (part of propellant used during landing), and the rest was the same as the model of the launch state. It was necessary to explain that the element mesh should be refined in the stress concentration area (such as the support of landing cushion mechanisms) according to analysis results of the load transfer path and bearing characteristics of the lander structure in landing state.

The finite element model of the support of landing cushion mechanisms and related structure is shown in Fig. 4.15 for static analysis of the Chang'E-3 lunar lander. Analysis model was simplified as follows:

(1) Composite plate element (PCOMP) simulation for horizontal triangulations and vertical plates.
(2) Plate unit common is used to simulate the plate of rectangular cross stiffening beam of vertical plate, and the beam element was used to simulate the wing plate.

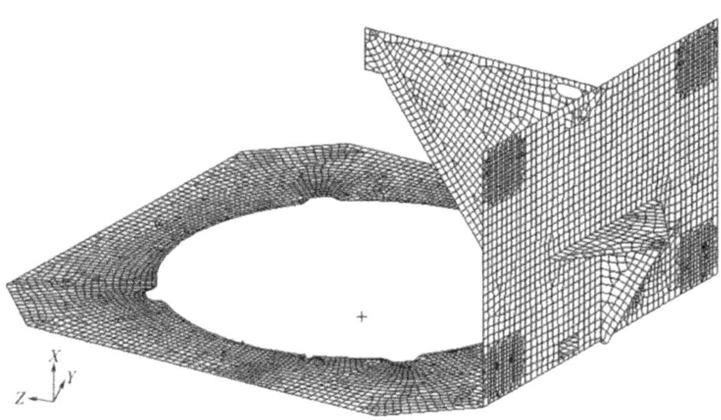

Fig. 4.15 Static model of the support of landing cushion mechanisms of the Chang'E-3 lunar lander

(3) Angle bar for structure connection was simulated by plate element.
(4) Bond between angle bar and structural panels was simulated by element common nodes, and the Multi Point Constraint (MPC) method was used to simulate the screw connection.
(5) Boundary condition is simple support by the far boundary, and the load was applied at the connection point of main pillar of landing cushion mechanisms.

Under a certain landing condition, the analysis results of maximum principal stress for the Chang'E-3 lunar lander are shown in Fig. 4.16.

In the structure design and analysis, the yield strength margin of the analysis was generally greater than 0 for the metal structure, and the first layer failure strength of the composite structure was generally greater than 0.25.

3. Stability Analysis

Although stability was a typical eigenvalue issue, it was consistent with the principle of stress analysis. In the expression of the finite element equation for the stability issue, the geometric stiffness matrix (differential stiffness matrix) must be obtained according to the deformation model determined by the stress analysis. In the analysis process, the boundary conditions and the load conditions were the same for both stability analysis and stress analysis. Therefore, the same finite element model could be used.

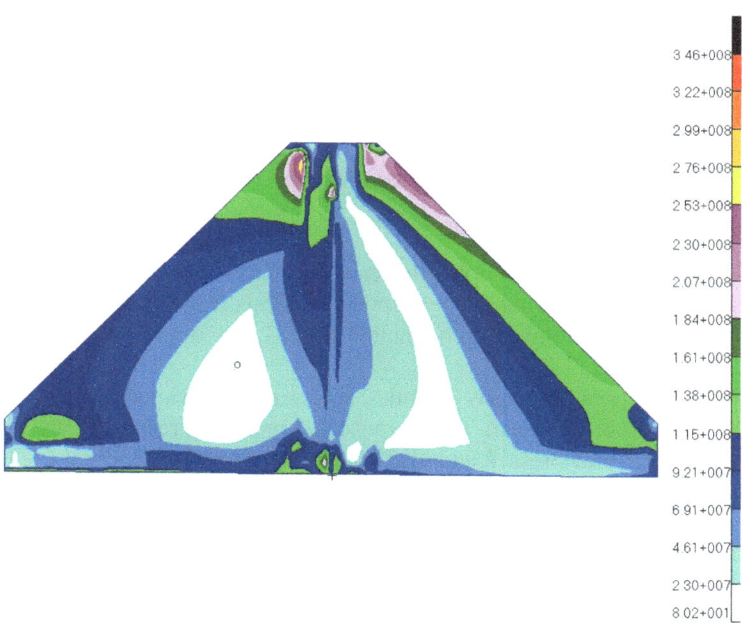

Fig. 4.16 Static analysis results of the support of landing cushion mechanisms of the Chang'E-3 lunar lander

Different from stress analysis, the establishment of the stability analysis model was not very concerned with the stress concentration area. And it was concerned with the areas where compressive stress and shear stress was relatively serious. Because the model of stress analysis had been able to get the deformation mode relatively accurately, there was no need to consider mesh refinement as a whole. However, in the local area where compression stress and shear stress were serious, it was still possible to refine the mesh further more so as to ensure that there were enough units to accurately represent the instability waveform. Otherwise, the unit mesh was too sparse, causing the structure to stiffen relative to other regions, resulting in inaccurate calculation results.

In the stress analysis model of complex structure, spring element was usually used to simulate the local connection. When the stability analysis as carried out, the spring element couldn't provide the differential stiffness, so other suitable units were used in the local connection where the instability waveform was greatly affected.

The results of stability analysis were greatly influenced by boundary conditions. Although the boundary condition was the same as stress analysis, it was necessary to draw lessons from past analysis experience and test results. The determination of boundary conditions was mainly based on engineering experience. In addition, the subjective judgment of the rationality of the instability waveform could check whether the boundary conditions were accurate.

It was worth noting that the stability analysis results obtained from the finite element method as a linear eigenvalue might be more deviations from test results, especially for plate and shell structure. This was mainly due to the geometric defects and material defects in the structure manufacture process and the uneven load.

The stability analysis of the Chang'E-3 lunar lander during the development process was carried out. Because the stability coefficient of the analysis was greater than 3 and it was local instability, the model which was exactly the same as static analysis (including load and boundary condition) was used in the stability analysis.

4. Frequency Response Analysis

The frequency response analysis was the response analysis of the spacecraft under the sinusoidal excitation condition. The response analysis results under different sinusoidal frequency excitation could be plotted as the curve of the frequency response curve.

Frequency response analysis includes analysis of acceleration, displacement, stressed, load (force) and etc. The purpose of frequency response analysis was: (1) to obtain structural stressed and load at connection pointed under the sinusoidal load, to guide and verify the structure design and parameter design, to verify the dynamic strength of the structure; (2) to obtain the acceleration response of each part under the sinusoidal load, and to judge the secondary structure, large parts and so on. The frequency coupling of spacecraft with secondary structure and large parts was used to evaluate the stiffness design rationality of secondary structure and large components, which provided the basis for the establishment of the sine vibration environment conditions of equipment onboard the lunar lander.

The modal superposition method was usually used in frequency response analysis. The selection of frequency range should be determined according to the frequency range of forced excitation, at least all modes in the range of forced excitation frequency should be included. In ordered to improve the accuracy of analysis, the frequency range of modal truncation shall be broadened furthermore.

The selection of modal damping was generally determined based on past test data or empirical data, which had an important impact on the analysis results. Due to the complex mechanism of damping, there was a large error in the measurement of damping in the test, and there was no other analysis method to make up the shortage of the test. Therefore, the modal damping issue had been puzzling the improvement of the analysis accuracy. The modal damping of structures with different magnitudes of sinusoidal vibration was also different, and generally increases with the increase of excitation leveled.

In the analysis of frequency response, for acceleration analysis and load analysis, the model was the same as that used in modal analysis. For stressed analysis, issues that should be considered in the analysis model were similar to that in static analysis, and the element mesh was refined in the area with the stressed concentration. Therefore, the frequency response analysis model integrated the characteristics of modal analysis and static analysis.

4.3.3 Structural Tests of Lunar Lander

Structural tests are generally used to verify that structural products meet design requirements during and after the relative environment, and are the primary methods for structural design validation, and are also the most effective and direct methods. There were four types of tests for verification of structural design requirements: development test, qualification test, acceptance test and validation test.

The purpose of the development test is to demonstrate the design scheme and obtain the necessary information for design. For structural design, the strength test of local structure is usually carried out. Development tests are usually carried out at the early stage of phase A to phase C.

The purpose of the qualification test is to verify whether the design is reasonable and meet the requirements. The experimental magnitude is 1.3~1.5 times of the maximum load, and the qualification test is usually carried out at phase C.

The purpose of the acceptance test is to check the defects of materials, processes and so on, so as to verify the adaptability of the hardware. The acceptance test is mainly to verify the tested hardware, not to verify the design, and the operation load is generally used. Normally, the acceptance test of the structure is not carried out alone, but with the acceptance test of the whole probe.

The purpose of validation test is to confirm the validity of the analysis and provide test data to confirm the key analysis, or to modify the analysis model. The validity confirmation test has no damage to the tested hardware. Stiffness test and modal test are typical validation tests.

1. Static Test

The static test is a test method to study the mechanical behavior of the spacecraft structure under static or quasi-static conditions, and to solve the static strength, stiffness and stability of the structure. As the content of static analysis, it's justa different research method. For the lunar lander, the quasi-static load in launch process and landing impact load were still static test loads.

The static test of the lunar lander structure mainly included static tests of primary structure in launch state and landing state.

1) Static test of the primary structure in launching state

The static load of launching state was inertia load. For practical loading, it was necessary to simplify the actual load in static test. In addition to ensuring the effective verification of the structure, the simplification of the load should also be taken into account the feasibility and operability of the test plan, as well as the actual conditions of the test site and equipment. The simplified load and the actual load should be equivalent in macro sense, and the simplification of load shall meet the purpose of the experiment.

As an example, the static load of the Chang'E-3 lunar lander was simplified as follows: 4 tanks loading points, 1 the lunar rover loading points, 4 lateral plates loading points, 4 loading points in +Y module and 4 loading points in −Y module. The diagram of the lever loading system for X to Z joint loading test of the Chang'E-3 lunar lander is shown in Fig. 4.17.

In order to verify the rationality of the simplified load test, i.e. that the simplified load could effectively simulate the actual load of the structure and the test could effectively simulate the actual stress situation of the structure to make the verification effective, generally the prediction analysis should be done for the structural static

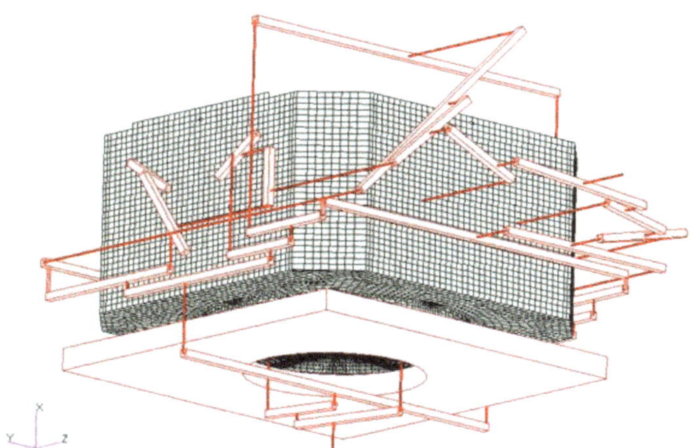

Fig. 4.17 Lever loading system for X to Z joint loading test of the Chang'E-3 lunar lander

test. The object of prediction analysis was test condition and simplified test load and boundary condition were introduced exactly. In addition, according to the results of prediction analysis, the layout of test points could be targeted reasonably.

2) Static test of the primary structure in landing state

From the previous static analysis, it could be found that the support of landing cushion mechanisms could be extracted from the whole lander. By design of the local structure test parts with applying the corresponding load and certain boundary constraints, the test could be greatly simplified. The static test of the support of landing cushion mechanisms and the surrounding structure should accurately reflect the stress status of the landing gears and the lander structure at installation points under landing load condition, in order to verify whether the design of the support of landing gears and the adjacent structure of the lander could meet the requirements. Therefore, the design of the test parts should include support of landing gears and the adjacent structure, which were subject to large load under the landing load.

The diagram of the static test parts of landing gears and its support is shown in Fig. 4.18 for the Chang'E-3 lunar lander.

The loading diagram of structural static test in landing state is shown in Fig. 4.19 for the Chang'E-3 lunar lander.

2. Dynamic Test

The dynamic test is the experimental method to study the mechanical behavior of the spacecraft structure under dynamic environmental conditions and solve the dynamic strength issue of the structure. For the lunar lander structure, the landing impact load environment had been covered by the static state of the landing state. Therefore, only the dynamic environment in launching state was studied for lunar lander just the same as common spacecraft. Though due to the great differences in configuration, especially the fuel tanks in parallel installation and propellant in tanks occupied about 2/3 mass of whole lander, its characteristics was different from common spacecraft.

Usually, the dynamic test of spacecraft structure is not carried out in phase B, because the design of primary bearing structure is centered on the static load for structural load. In addition, there might be some uncertainties in system design of spacecraft and large parts which have a great influence on the dynamic characteristics of whole spacecraft in phase B. Once the design is changed, it will directly affect the verification of dynamic characteristics of spacecraft structure.

Usually, the spacecraft primary structure is insensitive to the acoustic environment. However, acoustic environment will cause the spacecraft to generate a random vibration response, which have a great impact on equipment, cable, pipeline, and equipment connections on spacecraft. So, the major purpose of the acoustic test is to verify the ability of equipment, cable, pipeline, and equipment connection on the spacecraft to withstand the random vibration environment, and to provide inputs for the random vibration environment conditions of the equipment onboard the spacecraft.

The major purpose of the sinusoidal vibration test is to verify the rationality of the spacecraft structure design, to evaluate the ability of the spacecraft structure to

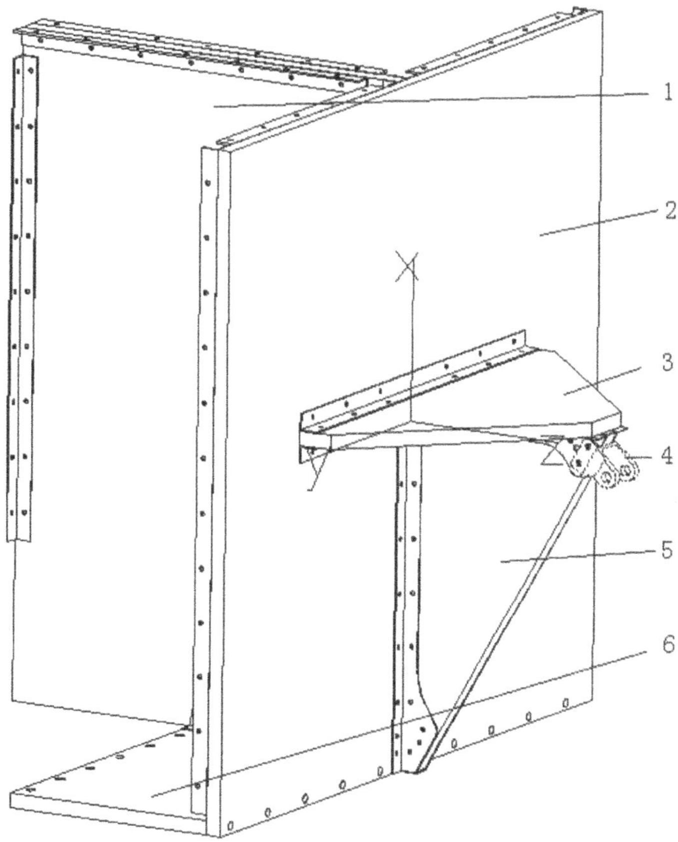

Fig. 4.18 Static test parts for the Chang'E-3 lunar lander. 1-Clapboard; 2-Lateral plate; 3-Horizontal triangle plate; 4-Upper joint; 5-Vertical triangle plate; 6-Base plate

withstand sinusoidal vibration environmental conditions, i.e., to verify the dynamic strength of the structure. Secondly, the components (such as pipeline systems, cable bindings, etc.) can be verified in the sinusoidal vibration test of whole lander, which are not likely to be verified in tests at component level.

In the sinusoidal vibration test, in order to avoid damage to the structure or the onboard equipment due to excessive response, the test conditions are reduced in certain frequency bands (i.e., the test conditions can be concave-down). The reason for the test conditions concave-down is that the conditions specified in test specifications are not the actual environmental conditions that the spacecraft is subjected to, rather than the straight line envelope of the maximum value of all different environment. The actual environmental conditions experienced by a spacecraft are only high in certain frequency bands and relatively low in other frequency bands. The magnitude of different frequency bands shall be obtained through analysis of coupled load

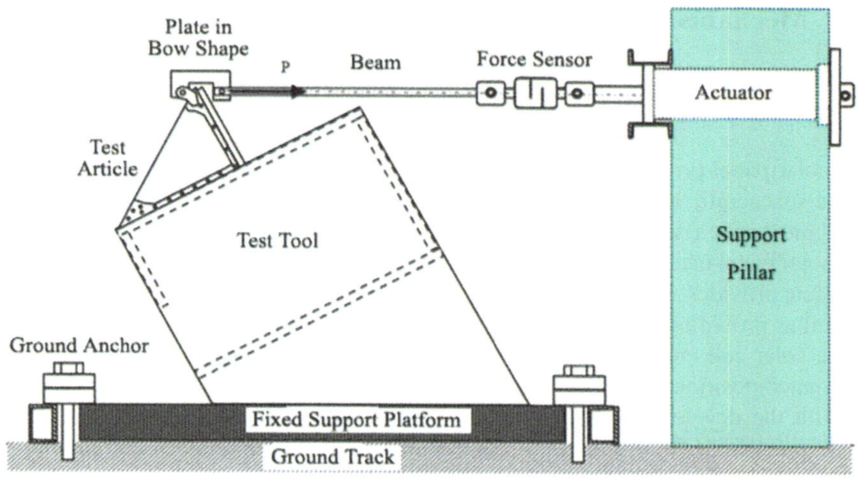

Fig. 4.19 Loading diagram of structural static test in landing state for the Chang'E-3 lunar lander

with launch vehicle. The principles of determining the magnitude of test condition concave-down is: the stress of the primary structure in the test is not greater than the stress of the primary structure under static load conditions; the vibration level is not lower than the coupled analysis results with launch vehicle. Generally, there shall be no contradiction between the two conditions. If there is a contradiction, it indicates that the matching of the dynamic stiffness of spacecraft and launch vehicle is unreasonable. The upper and lower limits of test conditions are given by both of these conditions. The connection of key equipment and the response of key equipment shall be paid attention during tests to determine reasonable magnitude between upper and lower limits. As an example, special attention had been paid to the response of fuel tanks (with propellant filled) during the sinusoidal vibration test. In the longitudinal vibration test, concave-down control was not necessary for the structural load, but concave-down at two local frequencies (each for the nitrogen tetroxide tanks and the monomethyl hydrazine tanks) for fuel tanks. The minimum level was 40% of the specification level, which was still higher than the coupled analysis of the lunar lander and launch vehicle, while there was no concave-down at other frequencies. In the transverse vibration test, the concave-down was conducted at the first-order frequency and at the local frequency of fuel tanks. The concave-down magnitude at the first-order frequency as derived from the magnitude of enlargement of the strain of interface ring with launch vehicle at the resonant frequency combined with the force at quasi-static load at interface ring, and the concave-down magnitude at the local frequency of fuel tanks was determined concerning fuel tanks response in qualification test.

4.4 Mechanism Design and Verification of Lunar Lander

4.4.1 Solar Panel Mechanism

The solar panel provide electric energy for the spacecraft and are key components of the spacecraft. The solar panel is generally divided into mechanical parts (solar panel mechanism) and electrical parts according to the functions of the components. The solar panel mechanism serves as a carrier for the electrical part, and its structural substrate provides an installation surface of a specified area for the solar cell circuit, as well as providing sufficient strength and rigidity to support the normal operation of the solar cell circuit; And the mechanism driving part deploys solar panel to a designated position as well as keeping solar cell's orientation to the sun.

With the development of aerospace technology, solar panel has evolved from body-mounted to deployable solar panel to meet the greater power requirements of spacecraft. To date, the solar panel used on Chinese satellites is all multi-panel and deployable type of solar panel. The solar cell circuit is laid on a rigid substrate. The solar panel is folded and held down on the lateral plates of space craft in launch vehicle fairing and deployed into designated position to form a complete plane after orbit injection. The sun orientation of solar panel is realized by Solar Array Drive Assembly (SADA) controlled by the GNC subsystem according to the characteristics of orbit. For the solar panel mechanism of the lunar lander, in addition to the functions including holddown and release on orbit as usual spacecraft, it was generally necessary to be capable of folding and unfolding repeatedly, i.e., the solar panels could be folded(and sometimes be held auxiliary if necessary) before orbit maneuver and soft-landing to withstand load, as well as unfolded after orbit change and soft-landing.

1. Determination of Design Requirements

As an important component of the entire spacecraft, the basic specification of solar panel is determined by spacecraft configuration, orbit and power energy requirement, such as the area, folding/unfolding position, orientation and loading conditions, etc. in phase B.

Generally, there were following requirements for the solar panels of the lunar lander:

(1) Capable of folding, hold down and release, unfolding. In order to meet the requirements of fairing envelope volume and load of the launch vehicle, the solar panels were often folded and held down, while released and deployed after orbit injection.
(2) Supporting the solar cell circuit. As part of the solar panel mechanism, the solar panel substrate provided effective support for the solar cell circuit to ensure the normal operation.
(3) Capable of folding and unfolding repeatedly. During orbiting maneuver, there would be overload when the engine as ignited and shut off. To prevent the solar

panels from being damaged due to overload in the process of orbit maneuver, the solar panels should be folded before orbit maneuver and unfolded after orbit maneuver. In addition, there was a large landing impact load during soft-landing. Therefore, the solar panels should be folded before landing and, if necessary, auxiliary held was also required. After successful landing, the solar panels were deployed again and oriented to the Sun.

(4) Oriented to the sun. In order to meet power supply requirement of the lunar lander, the solar panel mechanism should drive solar panels to be oriented to the Sun and be able to maintain its position. In addition, the orientation to the Sun should meet certain requirements of angle accuracy.
(5) Fundamental frequency requirement. The solar panels should meet certain fundamental frequency requirements in the folded and unfolded state.
(6) Life requirement. The solar panels of the lunar lander should be folded and unfolded repeatedly during the mission. Therefore, there are certain requirements for the operation times of the solar panel mechanism, i.e., the working life requirements.
(7) Reliability requirement. The reliability of solar panel mechanism had a direct impact on the power supply of solar panels. Therefore, the solar panel mechanism should be highly reliable.
(8) Other functional requirements. After the solar panel mechanism was unfolded or folded to designated position, an accurate signal or angle information should be given to judge whether the solar panel mechanism is in position.

The requirements of the solar panel mechanism are listed in Table 4.3.

2. Task Analysis

The task analysis of flight hardware on spacecraft should be done according to different mission profiles. All flight events, environmental conditions, and operating conditions during flight operation should be defined to provide comprehensive and correct inputs for design.

All matters during flight operation should be analyzed in task analysis to identify critical environmental factors that affected the function, performance and reliability of the flight hardware and made them quantization. The main contents of task analysis included:

(1) Analysis of operation conditions combined with space environment and operation mode of flight hardware itself to identify all operation conditions and determine the worst case of all operation conditions.
(2) Analysis of key failure modes that affected the function implementation of flight hardware and might cause mission failure to identify key failure modes and acceptable degradation modes, as well as establish criteria for task success.

Compared with solar panels of other spacecraft, the solar panels of the lunar lander had a complex mission profile and tough working environment. Therefore, it was particularly important to analyze tasks of solar panels of the lunar lander for all mission profiles. As an example, the task analysis of the solar panels is given as follows for the Chang'E-3 lunar lander:

Table 4.3 Requirements of the solar panel mechanism

Requirement	Definition	Description	Information needed
Strength	The components should not produce catastrophic damage (such as fractures, instability, excessive deformation, etc.) to the lunar lander under specified load	1) Usually the largest load on the solar panel was encountered during launch and landing, which was subjected to the solar panel substrate 2) The load was derived from launch environment, landing impact environment and mass property of the lunar lander 3) The requirement was same in all stages including launching, flight and landing	1) General mechanical environmental conditions 2) Landing impact load 3) Mass property 4) Temperature
Natural frequency (or stiffness)	The natural frequency was beyond a specific value or a certain frequency. For mechanism products, there were requirements of natural frequencies for both unfolded and folded state	1) The fundamental frequency requirement given by system design must be met 2) In launch state, the stiffness of the solar panels was the major concern, and it was necessary to ensure that the lunar lander was always within the dynamic envelope 3) During flight, the dynamic analysis of the entire lunar lander was performed considering the stiffness of unfolded solar panels	1) Fundamental frequency given by system design 2) Mass property
Mechanical interface	Satisfaction with the requirements of mechanical mating, connection and motion interference with structure, components, and all subsystems	Requirements for interfaces with the structure included accuracy of installation surface and position of connection Requirements for interfaces between components included connection between components, accuracy and the envelope during folding, unfolding and motion	Mechanical interface requirements

(continued)

Table 4.3 (continued)

Requirement	Definition	Description	Information needed
Mass	The total mass was less than the specified value	Preliminary allocation in phase B, adjustment after detailed design	Mass allocated
Orientation requirements	Requirement of relative position between solar panels and the lunar lander coordinate system	Requirements of the solar panel position during flight and operation on lunar surface	Given position requirements to spacecraft coordinate system in all flight phases
Drive capability	Output torque of drive	1) The motion requirements of the solar panel were different in flight and operation on lunar surface 2) The maintenance capability requirements of the solar panel at stop position were also different in flight and operation on lunar surface	1) Mass property 2) Load 3) Resistance moment 4) Operation temperature

(1) Launch phase: The solar panels should withstand the load during launch.
(2) Flight phase: After orbit injection, the solar panels would be deployed at 180° according to power supply requirements of the lunar lander. At this phase, the solar panels should provide sufficient driving capacity to carry out folding and unfolding under vacuum and radiation conditions by command and provide correct position signal. At the same time, the solar panels also had to withstand the load of 7500 N engine ignition and stop during orbit maneuver.
(3) Landing phase: Solar panels should withstand landing impact loads.
(4) Operation phase on lunar surface: On the lunar daytime, solar panel mechanism should perform folding/unfolding and change angle under environmental conditions such as high temperature, radiation and lunar dust on lunar surface to provide sufficient power energy. The drive assembly should have sufficient capability of driving and position maintenance under different attitude of the lunar lander and 1/6 lunar gravity. The position switch should provide correct in-position signal. On the lunar night, the solar panel mechanism should survive in extreme low-temperature environment and recover to normal operation after lunar night.

From above analysis, it could be seen that the key design of solar panels of the Chang'E-3 lunar lander is as follows:

(1) The solar panels could withstand mechanical loads caused by launch, landing, high and low temperature environments, and the structural strength can meet requirements.

(2) The solar panel mechanism had sufficient capability of driving and position maintenance with load under mechanical and thermal environmental conditions during flight and operation on lunar surface to ensure normal unfolding and folding.

3. Configuration, Components and Operation Mode of Solar Panels.

The installation position and basic configuration of the solar panels on the lunar lander should be determined according to the factors such as orbit, plume, and occultation. The configuration of the solar panel mechanism was designed to meet the requirements of strength, stiffness, and deployed position in all mission profile. The main design contents included: strength of solar panels substrate, stiffness, hold down and release mechanism (including the number and location of hold down points), driving mechanism of solar panels deployment, and etc.

In initial stage of the configuration design of solar panel mechanism, multiple concepts should be compared to determine final configuration through comparison of functions and performances.

The configuration and components of solar panel mechanism are shown in Fig. 4.20 (the lunar rover is removed in the figure) for of the Chang'E-3 lunar lander.

Two sets of solar panels of the Chang'E-3 lunar lander were symmetrically arranged on both sides of the lander. The solar panel mechanism was made up of solar panels substrate, Hold Down and Release Mechanism, rotating hinge, drive assembly, safety limiters, angle sensors and etc. The functions of each component of solar panel mechanism are listed as follows:

(1) Substrate: To attach solar cells and provide support for solar cell circuits.

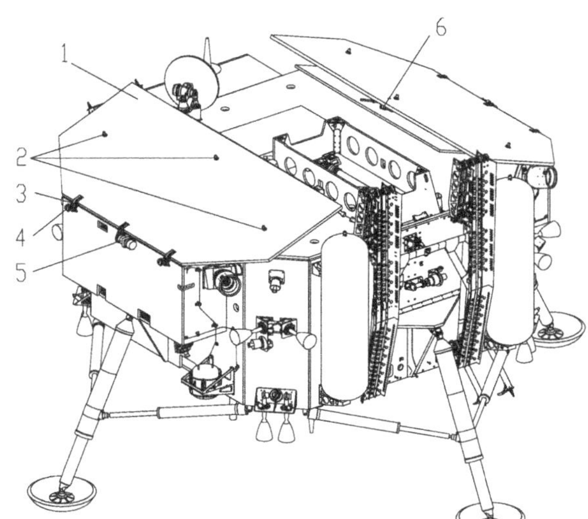

Fig. 4.20 Configuration and components of solar panel mechanism in hold down state for the Chang'E-3 lunar lander. 1-Substrate; 2-Hold Down and Release Mechanism; 3-Hinge; 4-Angle sensor; 5-Drive Assembly; 6-Safety limiter

(2) Hold Down and Release Mechanism: To fold sun panels and hold down to increase strength and rigidity of solar panels to prevent solar panels from being damaged by the load during launch.
(3) Rotating hinge: For connection of solar panels substrate and lander structure, and to provide support for the single-degree-of-freedom rotation of the substrate relative to lander structure.
(4) Drive assembly: To provide driving torque for solar panels deployment and folding, and provide position holding torque for solar panels in specific position.
(5) Safety limiter. It was used to limit the rotation of solar panels substrate in landing phase, and improve structural strength and rigidity of solar panels substrate to avoid damage of solar panels substrate and other related components (such as rotating hinge) under the landing impact load.

The operation modes of solar panel mechanism are listed as follows for the Chang'E-3 lunar lander:

(1) The solar panels were held down on the top plate of lander structure by Hold Down and Release Mechanism to withstand load during launch.
(2) After orbit injection, the Hold Down and Release Mechanism was unlocked and the solar panels were released.
(3) In orbital flight, the sun panels were deployed and oriented toward the Sun. In order to withstand load at the ignition and stop of the 7500 N engine of the lunar lander and avoid the influence of engine plume on solar panels, the solar panels were folded before the main engine ignition and deployed again after the engine stop. According to the flight procedure, the solar panels should be folded and unfolded more than once in orbit.
(4) Before powered descent and soft-landing, solar panels were folded to ensure that it could withstand landing impact load. Then the safety limiter was powered on to constrain solar panels to improve the strength and stiffness of solar panels structure.
(5) After landing, the safety limiter was power off and the constraint of solar panels was released. Then solar panels were unfolded to a specified angle and able to be changed as required to ensure enough power supply. Before the lunar night, sun panels were folded and powers off to maintain this position until the end of the lunar night. After the lunar night, solar panels were unfolded by command and oriented to the Sun.

4. Solar Panel Mechanism Analysis

The solar panel mechanism was also a mechanical component. During launch, the solar panel mechanism was folded and held down and there was no mechanism to move, which meant that the solar panel mechanism could be analyzed as a static structural product. The elemental requirement was to satisfy the static (static strength) and dynamic requirements of the structure. Analysis included modal analysis, static analysis, sinusoidal vibration response analysis, etc. The analysis method was basically the same as the spacecraft structural analysis. After orbit injection, the solar panel mechanism was deployed. The deployment of the solar panel mechanism was a process of the mechanism movement. Special analysis and calculations were required according to the characteristics of solar panel mechanism in order to meet design requirements of the solar panel mechanism. For example, in addition to structural mechanics analysis of folded solar panel mechanism, following analysis was done to the solar panel.

1) Analysis of inertia overload impact of engine ignition

The 7500 N throttling engine on Chang'E-3 lunar lander was used for orbit maneuver and powered descent. When engine was ignited or shut off, there would be large inertial overload. The pattern of load when engine ignition and shutdown is shown in Fig. 4.21. As shown in Fig. 4.21, the instantaneous load at ignition reached twice as nominal output load of the engine, thus acceleration would be generated along the normal direction of solar panel substrate and there would be larger torque and shear at the root of connection between solar panel substrate and lander structure. So, it was necessary to conduct strength analysis for the bearing part of solar panel mechanism root. In addition, during orbit maneuver of the Chang'E-3 lunar lander, the solar panel substrate was held by the self-positioning moment of solar panel driving mechanism. Therefore, it was also necessary to analyze the static torque margin of position maintenance for the solar panel driving mechanism when there was inertia overload of engine ignition.

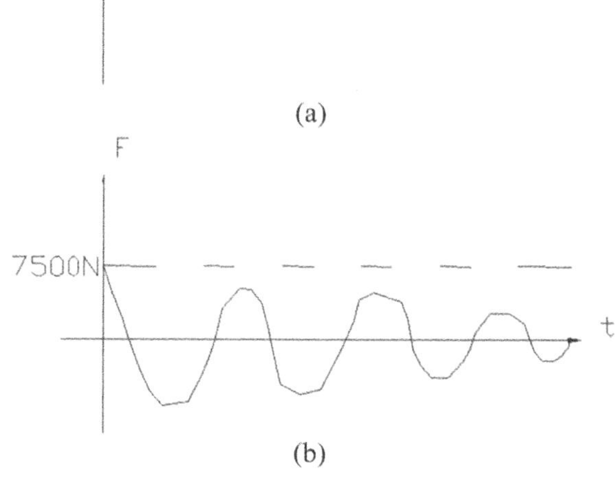

Fig. 4.21 Pattern of load when engine is ignited and shut off for the Chang'E-3 lunar lander. (**a**) Ignited; (**b**) shut off

2) Analysis of static moment margin for folding/unfolding on lunar surface

Usually, mechanisms of spacecraft on orbit operate under zero gravity. The inertial load generated by spacecraft orbital maneuvers is the major load that affects the movement of mechanisms whose magnitude is generally small. Nevertheless, the solar panel mechanism of the Chang'E-3 lunar lander should implement unfolding and folding on lunar surface under lunar gravity. So, the lunar gravity on lunar surface had a greater influence on movement of solar panel mechanism. It was an important part of the design and analysis of solar panel mechanism of the Chang'E-3 lunar lander to analyze the static moment margin for folding/unfolding of solar panel mechanism on lunar surface.

During the deployment of solar panel mechanism of the Chang'E-3 lunar lander, it should be overcome including lunar surface gravity moment, frictional resistance moment of hinge and additional resistance moment on output shaft of electric motor during the cable unrolling process, while the static moment margin was minimal. Among the moment, lunar surface gravity moment could be accurately obtained through analysis and calculation, and its uncertainty was far below friction resistance moment of hinge and the cable deployment resistance moment. Therefore, when the static moment margin during development of solar panel mechanism was analyzed, only uncertainty of the friction resistance moment of hinge and the cable deployment resistance moment was analyzed, while lunar surface gravity moment was removed with a certain coefficient. The formula is as follows [15]:

$$\eta = \frac{Ts - kTg}{T\gamma} - 1 \tag{4.1}$$

where T_s is drive torque; T_g is gravity moment; k is gravity moment coefficient, generally 1.2~1.5; T_γ is the total resistance moment.

In the early design, the driving torque in the calculation of static moment margin of the solar panel mechanism was derived from design value. The resistance moment was obtained based on empirical data. Before the end of the development phase, the static moment was corrected according to test results of driving torque and resistance moment.

5. Solar Panel Mechanism Tests

For spacecraft flight hardware, generally the function is verified through function tests and the adaptability to flight environment is verified through environmental tests. The verification contents and methods of function and environmental tests are briefly introduced as follows for solar panel mechanism of the Chang'E-3 lunar lander

1) Functional tests

The basic function of solar panel mechanism was the ability of unfolding and folding. Generally, the solar panel mechanism of spacecraft is deployed once on orbit and needs not to be folded under zero gravity conditions. The solar panel mechanism of the Chang'E-3 lunar lander should not only be unfolded and folded under zero gravity on orbit, but also be unfolded, folded and maintain under gravity environment of 1/6 g on lunar surface. Therefore, for the functional verification of the solar panel mechanism of the Chang'E-3 lunar lander, it was not only necessary to simulate unfolding and folding under zero gravity on orbit, but also to simulate unfolding, folding and maintaining functions under 1/6 g gravity and different landing attitude conditions.

The functional tests of solar panel mechanism in simulated zero-gravity environment could be carried out on the hanging-type simulated zero-gravity deployment equipment. After the weight of solar panel substrate was balanced by spring, it could be verified for unfolding and folding.

The function tests of unfolding, folding and maintaining for solar panel mechanism under simulated 1/6 g gravity could be carried out by adjusting vehicle attitude angle to simulate possible landing attitude of the lander and balancing 5/6 ground gravity of solar panel substrate by balloon.

2) Environmental tests

The Chang'E-3 lunar lander solar wing should adapt to environment of launch, orbit flight, powered descent, soft-landing and lunar surface. Therefore, environmental tests were conducted for solar panel mechanism of the Chang'E-3 lunar lander including:

(1) Mechanical environmental tests were conducted for the loads during launch and landing including sinusoidal vibration tests, random vibration tests, acoustic environment and static test tests for launch, and static tests for landing.
(2) Thermal environment tests were conducted for thermal environment during flight. The temperature conditions for the solar panel mechanism of the Chang'E-3 lunar lander's on orbit and lunar surface were worse than solar panel mechanism of general spacecraft. Therefore, thermal environment tests were conducted during development of solar panel mechanism including extreme high and low temperature vacuum environment tests, extreme low temperature environment tests on lunar surface, and life tests under extreme high and low temperature environment tests on lunar surface.
(3) Lunar dust environmental adaptability tests were conducted in order to verify the adaptability of motion parts of solar panel mechanism to lunar dust environment.
(4) Low temperature function tests were conducted to verify the function of unfolding and folding of solar panel mechanism under low-temperature environment.

4.4.2 *Separation Mechanism for Lunar Rover*

1. Determination of Design Requirements

Before primitive design of the separation mechanism for the lunar rover, it was necessary to have an accurate understanding of configuration of the lunar rover besides general configuration, envelope and system design requirements of the lunar rover. The general function requirements of the connection and release mechanism for the lunar rover included: reliable connect of the lunar rover to the lander and capability to withstand the load conditions during launch, earth-moon transfer, circumlunar, powered descent and soft-landing, release of the lunar rover after landing on lunar surface, other auxiliary functions including lifting and guiding for deployment after release.

(1) To provide support for the lunar rover.
(2) To connect the lunar rover reliably with the lander from launch to landing, in order to increase structural stiffness and strength of the lunar rover so as to avoid damage due to launch and landing loads. When the lunar rover was transferred to lunar surface, it could release the lunar rover.
(3) Other necessary auxiliary functions including release, lifting and moving guidance.
(4) High reliability. the lunar rover separation mechanism was a single point of failure, which was crucial to the safe deployment of the lunar rover to lunar surface. Therefore, the lunar rover separation mechanism should be high reliable.

Typical requirements for the lunar rover separation mechanism are listed in Table 4.4.

2. Task Analysis

Compared with separation mechanism of general spacecraft with launch vehicle, the separation mechanism of the lunar lander should withstand landing impact load besides launch load. For example, task analysis of the lunar rover separation mechanism of the Chang'E-3 lunar lander is as follows:

(1) Launch phase: the lunar rover separation mechanism should connect the lunar rover to the lander reliably and withstand the load during launch.
(2) Orbital flight phase: the lunar rover separation mechanism should withstand the space environment such as temperature alternation, radiation and vacuum.
(3) Landing: the lunar rover separation mechanism should connect the lunar rover to the lander reliably and withstand the landing impact load.
(4) Operation on lunar surface: the lunar rover separation mechanism should release the lunar rover and raises the lunar rover to a certain height by the lifting device, provides guidance for the lunar rover to move on the lander for deployment. During operation on lunar surface, the lunar rover separation mechanism should withstand high temperature and the lunar gravity environment.

Table 4.4 Typical requirements for the lunar rover separation mechanism

Requirements	Definition	Description	Information required
strength	The components did not produce damage (such as fractures, instability, excessive deformation, etc.) that could imperil mission of the spacecraft under the specified load	1) the lunar rover was connected to the separation mechanism and usually had the highest load during launch and landing. The load of the lunar rover separation mechanism was withstood by the lunar rover bracket and separation mechanism 2) Load was derived from launch environment, landing impact and mass properties of the lunar lander and the lunar rover 3) Same requirements should be satisfied in all phase including launch, flight, powered descent and landing	1) General mechanical environmental conditions 2) landing impact load 3) mass properties of the lunar rover
Natural frequency (or stiffness)	The natural frequency (usually the fundamental frequency, i.e. the lowest natural frequency, with directionality) should be greater than specific value or beyond a certain frequency range	1) The fundamental frequency of system requirement should be met 2) During launch, the stiffness of separation mechanism of the lunar rover should be enough to keep the space craft within the dynamic envelope	1) Fundamental frequency specified by system design 2) mass properties

(continued)

Table 4.4 (continued)

Requirements	Definition	Description	Information required
Mechanical interface	Satisfaction with the structure, the lunar rover, various parts, and the mechanical mating, connection and movement envelope with other sub-systems	1) Requirements for interfaces with structures included: mounting surface accuracy and connection point position 2) Requirements for the interface with the lunar rover included: the accuracy of connection points, the configuration and size of the mating surface 3) Requirements for interfaces between components included: connection form between the components, the accuracy, and the enveloping during unfolding, folding and movement	Requirements of mechanical interface
Mass	The total mass should be less than specific value	The initial allocation might be adjusted according to the detail design results	Mass allocated
Bearing requirements	Under the impact of the lunar rover load, the structure of the lunar rover separation mechanism should not be damaged	For the lunar rover of allocated mass, the separation mechanism under external dynamic load and landing impact load conditions should ensure a reliable connection between the lunar rover and the lander	1) Quasi-static qualification load for launch and landing 2) Mass of the lunar rover

Through above analysis, it showed that the key points for design of the lunar rover separation mechanism of the Chang'E-3 lunar lander were:

(1) The structure of the lunar rover separation mechanism should withstand the launch and landing mechanical loads, and the structural strength should meet the requirements for use;

(2) The lunar rover separation mechanism was connected with the unlocking device and the lifting device can work reliably under the high and low temperature environment on lunar surface to ensure the normal separation and release of the lunar rover.

3. Configuration, Composition and Operation Mode off the Lunar Rover Separation Mechanism

For the Chang'E-3 lunar lander, the lunar rover was on the top of the lander. According to the configuration characteristics of the lunar rover, special attention was paid to the design of configuration of hold-down point, separation (release) mechanism, support structure, lift and guidance mechanism for the lunar rover. There were main components including the lunar rover bracket, separation mechanism for rover body, hold-down and release mechanism for wheels, lift and guidance mechanism as shown in Fig. 4.22.

The component function of separation mechanism for the Chang'E-3 lunar rover was as follows:

(1) Lunar rover bracket: to provide support for the connection between the lunar rover and lander, and installation interface for the separation mechanism for rover body and the hold-down and release mechanism for wheels.

(2) Separation mechanism for rover body: to connect and release the lunar rover body with the lunar rover bracket and withstand most of the loads during launch and landing of the lunar rover as the major connection part of separation mechanism for the lunar rover.

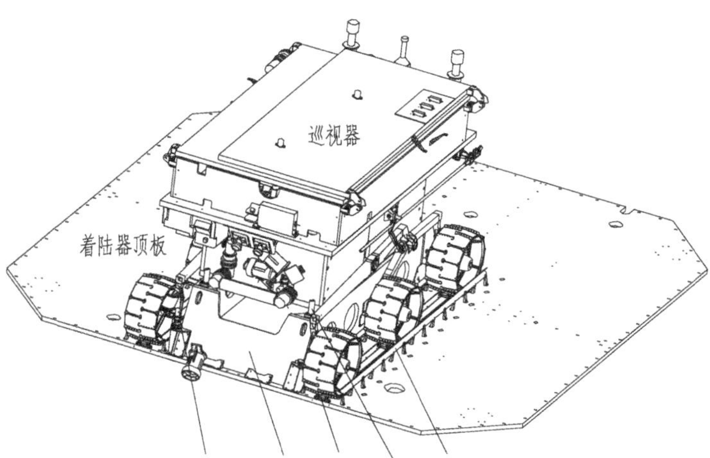

Fig. 4.22 Connection between the lunar lander and the lunar rover. 1-Lunar rover bracket; 2-Separation mechanism for rover body; 3-Hold-down and release mechanism for wheels; 4-Lift mechanism; 5-Guidance mechanism

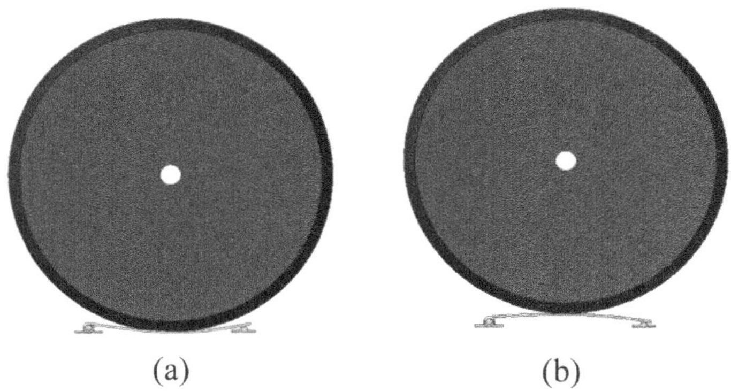

Fig. 4.23 Status of lift mechanism. (a) Pre-deformed when hold-down; (b) lifting after release

(3) Hold-down and release mechanism for wheels: to hold down and release the locomotion subsystem of the lunar rover, and withstand part of the loads during launch and landing of the lunar rover.
(4) Lift mechanism: to be installed on the top plate of the lander structure at the position of the lunar rover wheels. The lift spring undergoes pre-deformation under Earth gravity of the lunar rover when the lunar rover is connected with the lunar lander. On lunar surface, after the lunar rover was released, the lift spring deformed and recovered to provide lift and support for the lunar rover during release. The principle of life mechanism is shown in Fig. 4.23.
(5) Guidance mechanism: to provide guidance to move to the deployment mechanism after release of the lunar rover.

The operation modes of separation mechanism for the Chang'E-3 lunar rover were as follows:

(1) From launch phase to landing phase, the lunar rover was connected reliably with the lunar lander by the separation mechanism for rover body and the hold-down and release mechanism for wheels, and the lift spring undergoes a pre-deformation under the lunar rover connection state;
(2) After landing, the hold-down and release mechanism for wheels was released and then the separation mechanism for rover body was released by ground command to release the lunar rover by separation of the connection between the lunar rover and the lunar rover bracket; the lift spring pre-deformed recovered to lift the lunar rover to a certain height; with the aid of the track and guidance mechanism, the lunar rover moved on the top of the lander structure to reach the deployment mechanism and was deployed to lunar surface.

4. Static Analysis for the Lunar Rover Separation Mechanism

When the lunar lander was launched, the lunar rover was connected with the lunar lander reliably by the separation mechanism and withstood the loads during launch and landing. Therefore, the lunar rover separation mechanism should have sufficient stiffness and strength. the lunar rover bracket was the primary component that withstood and transferred the load. The basic requirement was that it should satisfy the static characteristics (static strength) requirements of the structure. The mechanics analysis for the lunar rover separation mechanism was mainly for the static analysis of the lunar rover bracket.

In the static analysis, the finite element modeling of the lunar rover and the lunar rover bracket was carried out respectively. According to the specific load conditions of three coordinate axes, the complex model of the lunar rover and its bracket was statically analyzed. The finite element model is shown in Fig. 4.24.

5. Test of the Lunar Rover Separation Mechanism

1) Function tests

The basic function of the lunar rover separation mechanism was to connect the lunar rover with the lander and release by command. After the lunar rover was released, the lunar rover would be lifted to a certain height and provided guidance for the lunar rover by the separation mechanism. There would be the connection and separation/lift function test in function tests of the separation mechanism.

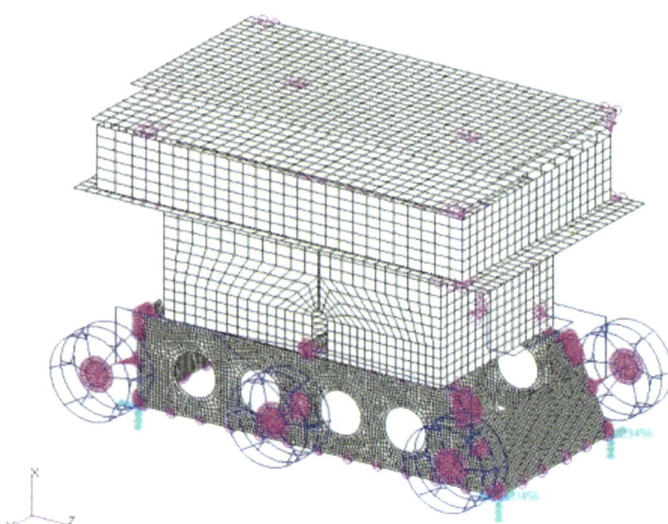

Fig. 4.24 Finite element model of the complex of the lunar rover and its bracket

The major purpose of connection function test was to examine the connection function between the lunar rover and the lander under the launch load and landing impact load, which was generally verified in the mechanical environmental test and landing impact test of the lander.

The major purpose of separation/lift function test was to verify the release function of separation mechanism after experiencing the launch and landing impact environment, and to verify the lift height after the lunar rover was released. It was important that the 1/6 g gravity environment on lunar surface shall be simulated in the test.

For the Chang'E-3 lunar rover separation mechanism, its auxiliary function such as the lunar rover guidance was verified in the release and deployment test for the lunar rover.

2) Environmental tests

There were mechanical and thermal test in the environmental tests for the lunar rover separation mechanism. The specific test items should be determined according to the mission environment profile of the lunar rover separation mechanism. The Chang'E-3 lunar rover separation mechanism experienced complex mechanical environments including launch and landing, as well as extremely high temperature on lunar surface. During development of the Chang'E-3 lunar lander, the lunar rover separation mechanism was qualified by mechanical environment and landing impact environment with the lunar lander. In addition, critical extreme high (low) temperature environmental tests were conducted for key components such as pyrotechnics separation nuts, pyrotechnics cutters and lift mechanism.

(1) Mechanical environmental test for the lunar lander

During development of the Chang'E-3, the lunar rover separation mechanism was qualified by qualification mechanical tests including sinusoidal vibration, random vibration, noise and landing impact. After mechanical environmental tests, the function of release, lift and guidance of the lunar separation mechanism was tested and it passed.

(2) Environmental test verification of extreme high temperature on lunar surface

The environmental test verification of extremely high temperature of the Chang'E-3 lunar rover separation mechanism was conducted only for the key components which were sensitive to temperature environment, such as pyrotechnics and lift mechanism. For example, the performance and reliability verification of pyrotechnics under extremely high temperature environment are briefly introduced here.

The most sensitive parts of the pyrotechnics to high temperature environment were the pyrotechnic composition. Under the high temperature environment, the physical stability and chemical stability of the pyrotechnic composition wouldbe instable to some degree, which would affect the work of the pyrotechnics reliability. In order to verify the effect of extreme high temperature environment on the reliability of a pyrotechnic device on lunar surface, the following test verification was conducted on the pyrotechnic device:

(1) Pyrotechnics pharmacy stability assessment. The key to the reliable operation of pyrotechnics in extremely high temperature environments was whether the pyrotechnics is stable. Through the analysis of vacuum stability, DSC, TGA and purity, and flame sensitivity test, the pyrotechnics pharmacy stability after experiencing extreme high temperature environment could be fully evaluated.
(2) Pyrotechnics performance under high temperature and reliability test. In order to further verify the reliability of pyrotechnics after experiencing high temperature environment, a pyrophoric performance test was conducted on the pyrotechnics and the gas pressure after firing was tested. The reliability of the pyrotechnics could be evaluated through the gas pressure test results.
(3) Pyro technic spyrophoric test under high temperature. Based on the evaluation of the pharmacy stability of pyrotechnics, the pyrophoric performance and reliability test, the pyrotechnics was conducted pyrophoric test after high temperature storage to verify the function of pyrotechnics.

4.4.3 Transfer Mechanism for Lunar Rover

The lunar rover was an important approach to implement scientific lunar exploration. It was often necessary to design a special transfer mechanism for the release of the lunar rover to satisfy the mission requirements for the deployment of the lunar lander. The design of transfer mechanism for the lunar rover should be based on characteristics of the mission, configuration and function of the lunar rover, and lunar surface environment. The design and verification of the transfer mechanism for the Chang'E-3 lunar lander and the lunar rover is briefly introduced in this section.

1. Determination of Design Requirements

As an important part of the Chang'E-3 Lunar Lander, the major requirements for the transfer mechanism are as follows:

(1) Hold down and release. In order to meet the requirements of the launch vehicle fairing envelop and load, the transfer mechanism should be held during launch and released after soft-landing of the lander.
(2) Support for transfer of the lunar rover. When the lunar rover was transferred to lunar surface, the structure of the transfer mechanism should have sufficient strength and stiffness to provide support for the lunar rover and keep the lunar rover stable.
(3) Transfer of the lunar rover. After the lunar rover was released, the transfer mechanism should move the lunar rover to the lunar surface and ensure that the tilt angle of the lunar rover did not exceed the required value during transfer. At the same time, the transfer mechanism should be able to adapt to certain landing attitude and lunar terrain.
(4) Reliability. Because the reliability of transfer mechanism was crucial to the mission goal of the lunar rover, the transfer mechanism should be as simple and reliable as possible.

(5) Fundamental frequency. The transfer mechanism should satisfy certain fundamental frequency requirements in hold down status.
(6) Others. The transfer mechanism should provide the position information during transfer of the lunar rover, etc.

All above requirements are listed in Table 4.5.

2. Task Analysis

For the Chang'E-3 lunar lander, landing legs were designed for soft-landing. the lunar rover was mounted on the top of the lander. The transfer mechanism for the lunar rover was on the lateral side of the lander and should satisfy requirements of the launch vehicle's fairing envelope. During launch, the transfer mechanism should be unfolded and held down. After landing, it should be deployed and transfer the lunar rover to lunar surface. The task of the transfer mechanism is analyzed as follows:

(1) Launch phase: The transfer mechanism was held down on the lateral side of the lander to withstand the launch load.
(2) In-orbit flight phase: The transfer mechanism should withstand space environment including temperature alternation, radiation and vacuum.
(3) Landing phase: The transfer mechanism should withstand the landing impact load.
(4) Operation phase on lunar surface: The transfer mechanism was released and deployed, while the lunar gravity should be overcome and the movement interference with the landing gears should be avoided. During the process of transfer and release of the lunar rover by the transfer mechanism, the transfer mechanism should withstand the lunar gravity load of the lunar rover, transfer the lunar rover with high altitude and large displacement, avoid the movement interference with the landing gears and maintain the stability of the lunar rover. During operation phase on lunar surface, the transfer mechanism should adapt to complicate environment including high-low temperature, lunar dust and terrain.

From above analysis, it could be seen that there were two key points in design of the transfer mechanism for the Chang'E-3 lunar rover as follows:

(1) The transfer mechanism should be held down and deployed. Due to the constraints of the size of structure, the transfer mechanism should be able to be held and deployed, while the impact of lunar gravity should be overcome and the movement interference with the landing gears should be avoided during deployed.
(2) the lunar rover should be transferred with high altitude and large displacement. The transfer mechanism should be designed to transfer the lunar rover from the top of the lunar lander to lunar surface (maximum height of 2.2 m) and avoid movement interference with the landing legs, as well as maintain the stability of the lunar rover during transfer and adapt to lunar terrain.
(3) Configuration, composition and operation mode of transfer mechanism

Table 4.5 Typical requirements for the lunar rover transfer mechanism

Requirements	Definition	Description	Information required
Strength	The components did not produce damage (such as fractures, instability, excessive deformation, etc.) that could imperil mission of the lunar lander under the specified load	1) The transfer mechanism usually had the highest load during launch and landing. The load of transfer mechanism was withstood by its structure and hold down mechanism 2) Load was derived from launch environment, landing impact and mass properties of the lander and the lunar rover 3) Same requirements should be satisfied in all phase including launch, flight, powered descent and landing	1) General mechanical environmental conditions 2) landing impact load 3) mass properties of the lunar rover
Natural frequency (or stiffness)	The natural frequency (usually the fundamental frequency, i.e. the lowest natural frequency, with directionality) should be greater than specific value or beyond a certain frequency range	1) The fundamental frequency of system requirement should be met 2) During launch, the stiffness of transfer mechanism in hold down status should be enough to keep the lunar lander within the dynamic envelope	1) Fundamental frequency specified by system design 2) mass properties

(continued)

Table 4.5 (continued)

Requirements	Definition	Description	Information required
Mechanical interface	Satisfaction with the structure, the lunar rover, various parts, and the mechanical mating, connection and movement envelope with other sub-systems	1) Requirements for interfaces with structures included: mounting surface accuracy and connection point position 2) Requirements for the interface with the lunar rover included: the accuracy of connection points, the configuration and size of the mating surface 3) Requirements for interfaces between components included: connection form between the components, the accuracy, and the enveloping during unfolding, folding and movement	Requirements of mechanical interface
Mass	The total mass should be less than specific value	The initial allocation might be adjusted according to the detail design results	Mass allocated
Bearing requirements	Under the impact of the lunar rover load, the structure of the lunar rover transfer mechanism should not be damaged	For the lunar rover of allocated mass, the transfer mechanism should with stand static load of the lunar rover gravity on lunar surface	1) Quasi-static qualification load for launch and landing 2) Mass of the lunar rover

The transfer mechanism was designed according to the mission characteristics of the Chang'E-3 lunar lander as well as the functional requirements and task analysis of the transfer mechanism. Through a four-bar linkage transfer mechanism was determined after comparison of different design. The four-bar linkage transfer mechanism was composed of four-bar linkage assembly, ladders assembly, hold down and release mechanism, deployment and hold down mechanism for ladders, and slow release mechanism as shown in Fig. 4.25.

The operation mode of the transfer mechanism was as follows:

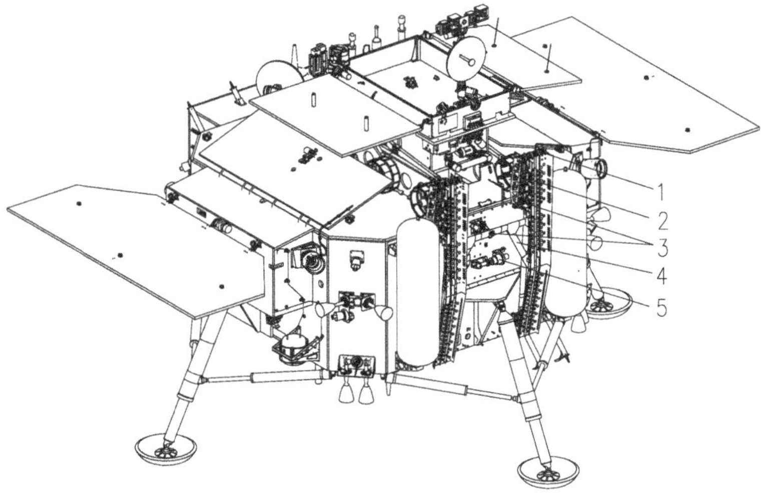

Fig. 4.25 Composition of the lunar rover transfer mechanism of the Chang'E-3 lunar lander. 1-Deployment and hold down mechanism for ladders; 2-Ladders assembly; 3-Hold down and release mechanism; 4-Four-bar linkage assembly; 5-Slow release mechanism

(1) From launch to soft-landing: The transfer mechanism was folded and held down on the lateral side of the lunar lander to withstand the launch and landing loads.
(2) Operation on lunar surface:

1. After the lunar lander was landed, the ladders of transfer mechanism was released, deployed and held down.
2. The lunar rover drove to a predetermined position on the ladders of transfer mechanism, then stopped.
3. The four-bar linkage assembly of transfer mechanism was released. Then the four-bar linkage assembly was deployed by the slow release mechanism under the effect of lunar gravity to touch lunar surface with the lunar rover on the ladders.
4. After the transfer mechanism touched lunar surface, the transfer mechanism stopped and the lunar rover drove from the ladders of transfer mechanism to lunar surface.

The transfer process is shown in Fig. 4.26.

4. Transfer Mechanism Analysis

The analysis of the transfer mechanism included structural modal analysis, static analysis, frequency response analysis, and mechanism dynamics analysis at hold down status. For structural analysis such as modal analysis, static analysis and frequency response analysis, the analysis method was the same as the analysis method of general spacecraft structure. Therefore, the structural analysis of the transfer mechanism will not be introduced here. Because the movement of the transfer mechanism

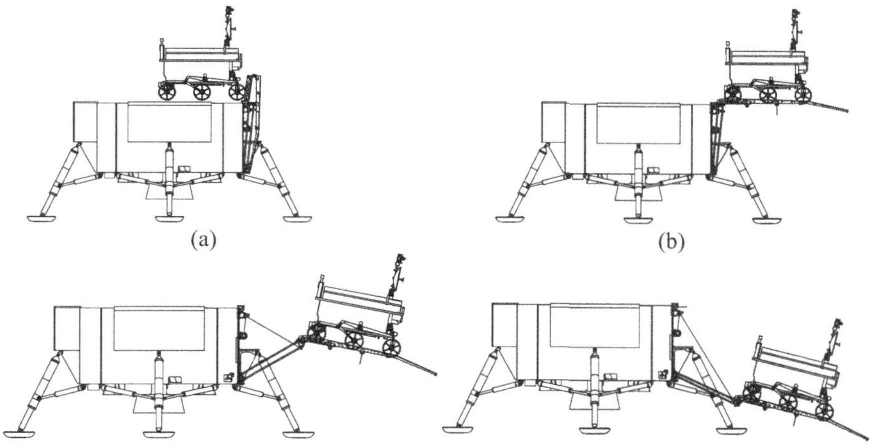

Fig. 4.26 Lunar rover transfer process. (**a**) Before deployment; (**b**) Deployment of ladders; (**c**) Deployment of four-bar linkage assembly; (**d**) Transfer of the lunar rover to lunar surface

was complex and there were many factors during operation on lunar surface, the kinematics and dynamics of mechanism was conducted for the transfer mechanism including the dynamic simulation analysis of ladders deployment and transfer on lunar surface.

1) Dynamic simulation analysis of ladders deployment

The ladders assembly of the transfer mechanism was foldable and composed of a short ladder and a long ladder connected by hold down and release mechanism. During launch, the ladders were folded and held down on the lateral side of the lander. After the lander was landed the hold down and release mechanism was released, then the ladders were deployed and held down. During deployment of the ladders, the movement of the long ladder was constrained by the linkage between the long ladder and the short ladder in order to ensure that the movement path of the long ladder was certain before the linkage didn't work, so the movement interference between the transfer mechanism and landing legs could be avoided during the deployment of the ladders. The dynamic simulation of the ladders deployment of the transfer mechanism was conducted. The simulation analysis showed that the linkage could effectively restrain the trajectory of the long ladder and interference with landing legs could be avoided as shown in Fig. 4.27.

2. Dynamic simulation analysis of transfer process on lunar surface

The dynamic simulation analysis of transfer process on lunar surface included the analysis of the startup torque and adaptability to lunar terrain.

The four-bar linkage type of the lunar rover transfer mechanism was designed for the Chang'E-3 lunar lander. The active force was lunar gravity for motion of the transfer mechanism. When the maximum pitch was $+15°$ after the lander was landed,

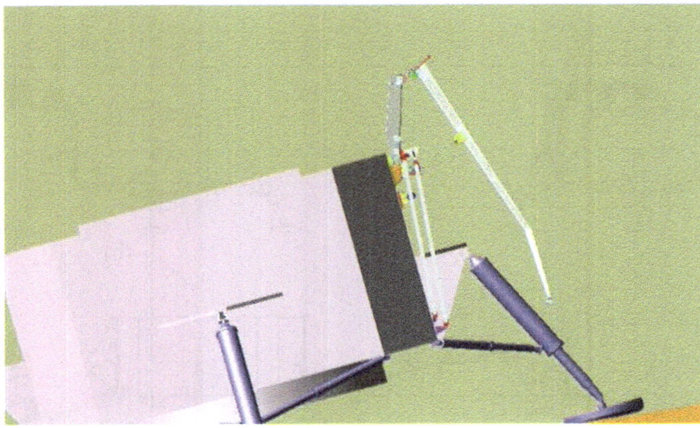

Fig. 4.27 Simulation of ladders deployment

Fig. 4.28 Simulation of driving torque of the transfer mechanism when maximum pitch of the lander is +15°

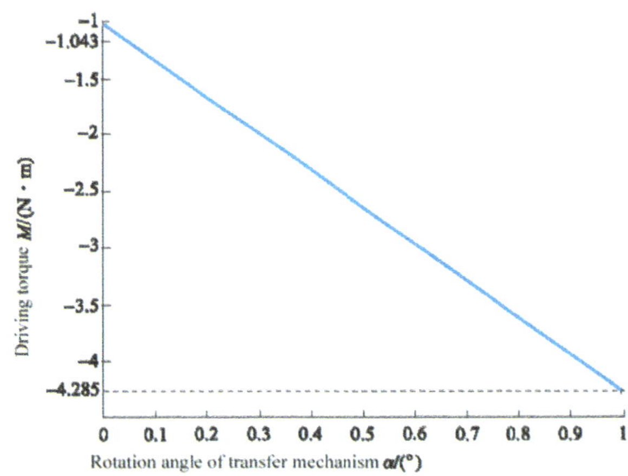

the moment of gravity could be a "negative moment", i.e. the lunar gravity was drag for motion of the transfer mechanism instead of driving force. In order to avoid the transfer mechanism's failure under the extreme landing attitude, a dynamic analysis of motion of the transfer mechanism under extreme landing attitude was conducted through dynamic simulation to determine the driving force required. The design of the transfer mechanism was optimized through analysis to make the required driving force of the transfer mechanism under extreme landing attitude as small as possible. The analysis result of the driving torque of the transfer mechanism when the lander was tilted by maximum +15° is shown in Fig. 4.28.

The two ladders of the transfer mechanism were deigned to move synchronously before touching lunar surface and adapt to lunar surface respectively after touching lunar surface. The simulation analysis of lunar surface adaptability of ladders of

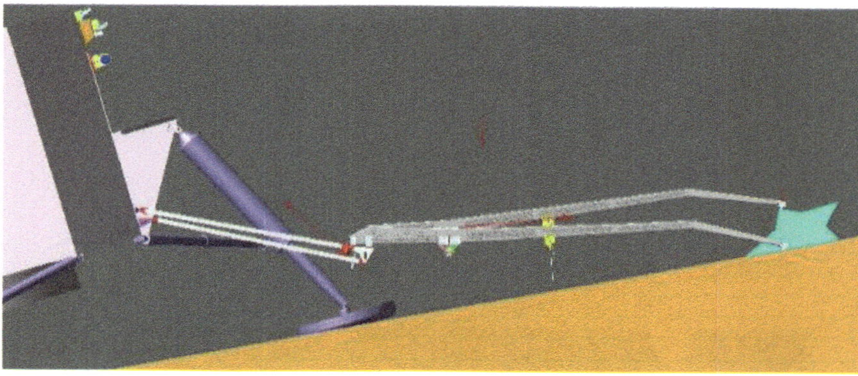

Fig. 4.29 Simulation analysis of lunar surface adaptability of the ladders

the transfer mechanism was conducted. The analysis condition was that one ladder touched a lunar rock and the other ladder directly touched lunar surface, i.e. there was height difference between two ladders. The result of simulation analysis is shown in Fig. 4.29. When one ladder touched lunar rock, the rope of slow release mechanism continued to release until the other ladder touched lunar surface. It was showed that the ladders had a good adaptability to lunar terrain.

5. Transfer Mechanism Tests

1) Functional test

The basic function of the transfer mechanism was to transfer the lunar rover from the lander to lunar surface with a series of actions such as the deployment of the ladder, transfer and release.

(1) Function test of ladder deployment

Since the two ladders of the transfer mechanism were independent in deployment, the ladder deployment test of the transfer mechanism could be conducted respectively. In order to simulate 1/6 g lunar gravity on lunar surface, the slope equivalent method was used to simulate lunar gravity environment. When the ladder was deployed, the Earth gravity component on the slope was equal to lunar gravity components in direction of the ladder deployment on lunar surface. The deployment function test of single ladder of the transfer mechanism is shown in Fig. 4.30.

(2) Transfer function test

When the ladders of transfer mechanism were deployed, the lunar rover on the ladder was a load. Therefore, in order to simulate the lunar 1/6 g gravity environment in transfer function test of the transfer mechanism, the lunar rover simulator was used to simulate gravity of the lunar rover on lunar surface. The weight of the lunar rover simulator should equal to 1/1 of the real the lunar rover weight. 6 (including 5/6 of the weight of the ladder), the test situation is shown in Fig. 4.31.

Fig. 4.30 Function test of single ladder deployment of transfer mechanism (Low temperature test)

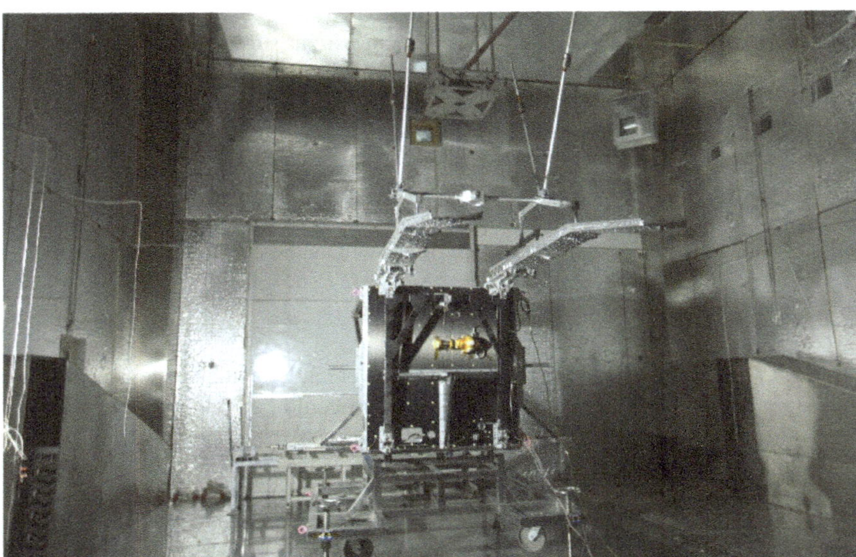

Fig. 4.31 Transfer function test of the transfer mechanism (low temperature test)

2) Environmental test

The transfer mechanism should withstand complex load environment including launch and landing. During the development of CE-5, the transfer mechanism should be verified through component environmental test, mechanics environment of test the lander and the landing impact environment.

(1) Mechanical environmental test

For the mechanical environment during launch phase of the lunar lander, the transfer mechanism was conducted a sinusoidal vibration test and a random vibration test at the qualification level. For the quasi-static load during the landing process, an acceleration test was conducted on the transfer mechanism. In addition, the transfer mechanism would conduct sinusoidal vibration, random vibration, and noise tests with the lander and passes a landing impact test.

(2) Verification test of lunar surface environmental adaptability.

The lunar surface environmental tests of the transfer mechanism included extreme high and low temperature environmental tests and lunar dust environmental tests, and the high and low temperature performance and dust resistance of the transfer mechanism and its main moving parts were verified.

4.5 Summary

In this chapter, the characteristics of the lunar lander structure and mechanism are briefly introduced. Based on this, the factors that should be considered in the lunar lander structure and mechanism design are analyzed. For example, the lunar lander structure and mechanism design, analysis and verification methods are described for the Chang'E-3 lunar probe that has been successfully launched. On December 2, 2012, the Chang'E-3 lunar probe was successfully launched. The lander structure and mechanism successfully completed the expected tasks. Its function and performance passed the flight test assessment. It was indicated that China had initially mastered the lunar lander key technologies for structural and mechanism design.

It shall be noted that in deep space exploration, the structure and mechanism design of the deep space probes are closely related to the task characteristics of the probe. Different deep space probes may not have the same structure and require different mechanisms to complete different functions. The structure and mechanism design of the probe should be based on a comprehensive analysis of mission of the probe to carry out corresponding design, analysis and verification work.

References

1. Yuan J, Yu D, Chen L et al (2004) Design and analysis of spacecraft structure. China Astronautic Publishing House, Beijing
2. Milton B (1964) Surveyor lander mission and capability. JPL, 8
3. Hughes (1964) Surveyor spacecraft A-21A model description. JPL, 9
4. Jim R et al. (1991) The final report on the design of the Common Lunar Lander.
5. Nikola B, et al. (1993) Extended Duration Lunar Lander.
6. Williamson M (2002) Aiming for the moon: the engineering challenge of Apollo. Eng Sci Educ J 11(5):164–172
7. Floyd VB (1970) Apollo Lunar Descent and ascent trajectories. Aerospace Science Meeting, New York
8. Wubbo JO (1997) EuroMoon 2000 a plan for a European lunar south Pole expedition. Acta Astronaut 41(4–10):579–583
9. Marien GJ (2004) Engineering Design Challenges of the Lunar Lander. AIAA
10. Siegfried WH (1999) Lunar base development missions. IAF Congress 44:755–767
11. Cadogan D, Sandy C, Grahne M (2002) Development and evaluation of the mars pathfinder inflatable airbag landing system. Mars Pathfinder Landing System 50:633–640
12. Chu G, Zhang H (2007) Technology of lunar probes. China Science and Technology Press, Beijing
13. Sun Z, Zhang H, Wu X et al (2010) Mission analysis of a lunar soft lander. Spacecraft Eng 19(5):12–16
14. C Wang, Deng Z, Gao H et al (2006) Development status of lunar landers. Missiles and Space Vehicles 284(4):31–36
15. L Chen (2005) Spacecraft structures and mechanisms. China Science and Technology Press, Beijing

Chapter 5
Thermal Control Technology of Lunar Lander

5.1 Introduction

Compared with Earth orbit and circumlunar spacecraft, lunar soft-landing probes will go through complicated phases including launching, LTO, lunar orbit, powered descent, landing and operation on lunar surface. The external environment and operation modes of instruments in different phases are different. It's necessary to take proper thermal control measures to make sure appropriate temperature range of structure and instruments to achieve mission objectives. The major function of thermal control subsystem is to maintain the structure and instruments of the lunar lander within acceptable temperature range throughout lifetime by controlling the heat exchange between interior of the lunar lander and external environment with thermal material and devices. The requirements of mass and power consumption of thermal control subsystem shall be as low as possible. The temperature range is different, for example, general electronic equipment is from −20 °C to 55 °C and inner structure of the lunar lander is from −100 °C to 80 °C before landing. Therefore, there will different thermal measures for different temperature range.

The environment is harsh during powered descent and operation on lunar surface for the lunar lander. The structure of the lunar lander shall be far away from the descent engine, whose nozzle temperature would be as higher as 1300 °C during powered descent, in order to avoid higher temperature of structure and instruments near the engine. It is important to take measures to insulate the structure and instruments from the high temperature of descent engine. During the lunar daytime, the highest temperature might be as high as 120 °C on lunar equator, while it might be as low as −180 °C in lunar night for about 14 earth days [1]. The thermal control subsystem shall adapt to the extreme high and low temperature alternation between lunar daytime and night. Compare with the microgravity in LTO and lunar orbit, there will be 1/6 g gravity on lunar surface, which will affect operation of heat pipes and two-phase liquid loop. The simulation of the lunar 1/6 g gravity on earth is a challenge as well. On the other side, the gravity on lunar surface makes it possible to transfer heat in new method for thermal control.

D.-Y. Yu et al., *Technology of Lunar Soft Lander*, Space Technology Library 38,
https://doi.org/10.1007/978-981-15-6580-9_5

The lunar dust is a special issue for thermal control of the lunar lander, which will degrade heat rejection system performance seriously even only little amount of dust accumulated on thermal radiators [2–4]. There is no effective cleaning technique now to remove lunar dust from radiator surface because of its fine grain size and static electric effect by sunlight on lunar surface [5]. It is difficult to precisely predict the orientation of the lunar lander before landing because of lunar terrain at landing site. The tilt and rotation of the lunar lander will affect the input of external heat flux and the performance of thermal control devices such as heat pipes, so the adaptability of thermal control subsystem shall be high.

Because the mission profile is complicated with harsh environment and lot of uncertainties, the thermal control subsystem should take different philosophy, solutions and ground test methods from other kinds of spacecraft on the basis of mature thermal control devices and technology to satisfy the requirements. In this chapter, design constraints, philosophy and solutions, typical technologies, and ground test methods of thermal control are described briefly.

5.2 Current Status of Thermal Control Technology

The Apollo program of the United States and Lunokhod of the Soviet Union had successfully implemented lunar soft-landing and operation on lunar surface, while lots of thermal control technologies were developed. Compared with passive thermal control technologies, active thermal control technologies such as variable conductance device, fluid circulation loop, water sublimator heat sink, lunar dust prevention were paid more attention recently because of extreme thermal environment in lunar daytime and night for the lunar lander in different countries [6–9].

In the design of thermal control of the lunar landers in different countries, the philosophy was the same. In order to minimize the influence of external environment change on the temperature of structure and instruments, the internal of lunar probe were insulated from external by insulator between main body and outer components and multilayer insulation blankets on surface [6, 9]. Thermal radiators should be oriented toward proper direction to dissipate excessive heat of working instruments to space, and minimize the lunar surface radiation flux and lunar albedo flux by conducting or transferring through heat pipes or fluid loops [8].

In order to solve the thermal control challenges to survive on the Moon for long-term the lunar lander or lunar base, many efforts were taken in different countries to developing "heat switch" techniques such as variable conductance devices and fluid loop devices [9]. Variable conductance devices include VCHP, Loop Heat Pipes (LHPs), Capillary Pumped two-phase fluid Loops, mechanically Pumped two-phase fluid Loops, mechanically Pumped single-phase fluid Loops and etc., while the mechanically Pumped single-phase fluid Loops are suitable for short-term manned the lunar landers. Considering for the temperature differences between the internal and external parts, as well as crews' safety, the thermal design of the lunar lander is usually divided into internal systems and external systems and there is the heat

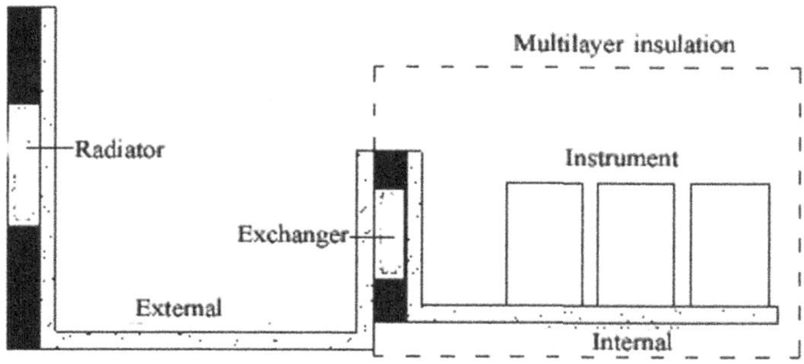

Fig. 5.1 Design of thermal control subsystem for the lunar lander

exchanger between them as shown in Fig. 5.1. The major function of the internal systems with narrow temperature limits will collect the heat of working instruments and transfer to the exchanger during the lunar daytime, and cut off the heat connection with the external systems to keep the inner instruments warm by minimizing the heat leak during lunar night. The major components of external system with wide temperature limit to adapt to extreme temperature alternation during day and night, which are usually thermal radiators, will transfer excessive heat from heat exchange to the radiators to radiate to space.

The United States manned Apollo program and the Soviet Union robotic lunar exploration program were implemented in 1970s, when the long-term scientific experiment platforms and the probes used the Radioisotope Thermoelectric Generators (RTGs) to generate electrical power, or use Radioisotope Heat Units (RHUs) to provide thermal energy. The Apollo spacecrafts succeeded the short-term crewed missions landing on moon and returning to Earth, and the environmental control subsystem of the Lunar Module used mechanically pumped single-phase fluid loops and water sublimator for heat sink [10]. Between the descent engine and the spacecraft structure, there was the heat shield to isolate the spacecraft structure from the high temperature of the engine. The Apollo lunar surface experiments package, which was designed for long-term working on lunar surface to obtain physical and environmental properties of the Moon, was powered by a SNAP-27 RTG [11]. The mass of SNAP-27 with 3.8 kg ^{238}P was 31 kg, which was designed to provide 1480 W heat energy and generate 63.5 W electric power energy. The Soviet Union lunar rover Lunokhod 1 (landed on the moon in 1970) and Lunokhod 2 (landed on the moon in 1973) worked successfully on the lunar surface for about 10.5 lunar days and 5 lunar days respectively. The Soviet Union lunar rovers had pressurized vessel in which equipment and payloads were contained, filled with gaseous mixture. In lunar daytime the solar panel was deployed to expose thermal radiator to radiate excessive heat into space. In lunar night the solar panel was folded and covered the thermal radiator, and the gaseous mixture was diverted to circulate by fans to transfer the heat energy of isotopic materials to pressurized vessel to keep the temperature during lunar night [12]. The mass of ^{210}P in Lunokhod was 11 kg.

The lifetime of Chinese Chang'E-3 lunar lander was one year, when the probe would go through about 13 lunar day/night circles. The Chang'E-3 lunar lander took use of Variable Conductance Heat Pipes (VCHPs) to transfer the electronics waste heat to the radiators toward space on the top of the lunar lander during the lunar day, and VCHPs could be "shut off" during the lunar night. The lunar lander was equipped with Radioisotope Heat Units (RHUs) to produce heat energy, which was translated into the electronic cabins by Gravity Driven ammonia two-phase fluid loops, to maintain survival temperature during the lunar night. A new type of heat shield was used to protect the lander structure from the high temperature environment of the descent engine during landing.

5.3 Thermal Technical Characteristics

5.3.1 Design Constraints

In order to keep equipment working normally in orbit and on lunar surface, the lunar lander must adapt to and even take advantage of the different environmental factors to control the heat exchange between the internal equipment and the external environments. Compared with the earth orbit spacecraft and circumlunar spacecraft, besides similar environment such as the space vacuum, the background heat sink, the heat input and the microgravity, the lunar lander would go through unique environments such as the lunar surface heat radiation, the extreme low temperature at lunar night, the lunar surface gravity except and lunar dust.

Besides the natural environments, special constraints of lunar soft-landing including the high temperature and the plume impingement of the descent engine, and the uncertain orientation of the lunar lander before landing on moon should be adapted by design of thermal control subsystem.

1. Natural Environmental Constraints

1) Space vacuum

After the lunar lander was more than 100 km away from the Earth, the space pressure was gradually reduced to less than 10^{-2} Pa, about 10^{-6} Pa at 500 km, 10^{-10} Pa at 2500 km, and less than 10^{-11} Pa beyond 3000 km. The atmosphere on lunar surface is very thin where the atmospheric density is about 14 orders of magnitude smaller than that of the Earth, and the pressure is about 10^{-6} Pa to 10^{-10} Pa [13]. Therefore, the lunar lander usually worked under high vacuum conditions. The pressure on lunar surface is basically the same as the Earth orbit with a height of 500 km to 2500 km. The material deflation and the thermal insulation performance of multilayer insulations that should be considered in thermal design were the same as that of an Earth orbit spacecraft.

The only heat exchange between the lunar lander and external environment under vacuum conditions was radiation. For non-hermetic structures, the heat exchange between internal structure and instrument of the lunar lander was heat conduction and radiation, where there was no atmospheric convection heat transfer on the ground which would cause difficulty for heat exchange. But it was suitable to achieve excellent heat insulation performance of multilayer insulation components.

The adverse effects such as surface evaporation, cold welding, and non-uniform temperature on the surface brought by space high vacuum environment should be paid careful attention to in the solution and material selection.

2) Low temperature

Except Sun, Earth, Moon and other planets, the space faced by the lunar lander is equivalent to an absolute black body with a temperature of 4 K, and the space radiation is about 10^{-5} W/m^2 outside the Earth's atmosphere and on lunar surface [14, 15]. Because other stars are too far away, the heat fluxes received by the lunar lander were from solar radiant heat flow, Earth radiation and its reflection heat flow, lunar surface radiation and its reflection heat flow.

During operation on lunar surface, the energy reflection between the lunar lander and lunar surface should be considered. In other phases, due to the relatively small size of the lunar lander, it was generally believed that the heat radiated from the lunar lander surface was totally absorbed by the space.

3) Solar radiation

Solar radiation is the major heat source for the lunar lander. Due to the fact that the distance between Earth and Moon is small relative to the distance between Sun and Earth, the distance between Sun and Earth together with the distance between Sun and Moon is equal to one Astronomical Unit. Therefore, the solar radiation intensity for the lunar lander was the same as that of the Earth orbit spacecraft during all flight phases, i.e., one solar constant (S, the mean value is 1367 W/m^2) at average distance between Sun and Earth. The solar constant was corrected according to the change of distance between Sun and Earth in different seasons, the maximum was about 1414 W/m^2 for the winter solstice (Perihelion) and the minimum was 1322 W/m^2 for the summer solstice (Aphelion).

The spectral distribution of solar radiation is shown in Fig. 5.2, which is basically consistent with the absolute blackbody radiation spectrum of 5760 K [14].

4) Lunar Albedo

Part of the solar radiation received by the Moon is absorbed, and the other part is reflected. The part that is reflected off is called the reflection heat flow. According to observations on the Earth, the reflectivity of solar radiation on lunar surface is related to local chemical and mineral composition, particle size, and material density, which could be referred to Chap. 2 for specific values.

For the lunar lander who orbits the Earth sometime, the Earth's reflection heat flow shall be considered during thermal analysis. Earth's reflection heat flux varies with the earth's surface morphology, seasons, geographic latitudes, and orbital inclination. The factors are complex. However, due to high altitude and fast velocity of the lunar

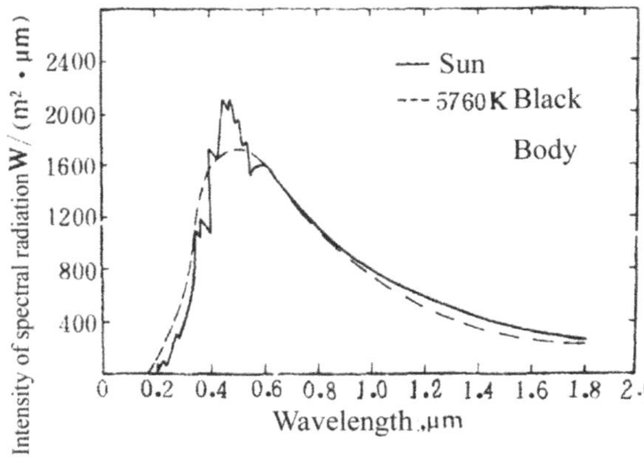

Fig. 5.2 Spectral distribution of solar radiation

lander, the Earth's albedo energy is smaller than that of the direct solar radiation. Therefore, average value of 0.35 is usually taken in thermal analysis.

The spectral properties and spatial distribution of the Earth and Moon albedo are relatively complex. Different reflectors have obvious selectivity for the absorption or reflection of solar radiation, resulting in a significant change in the reflectance spectrum with time and location. Generally, the reflection spectrum is still approximately considered to be the same as solar spectrum. In addition, it is generally assumed that the reflection on the surface of Earth and Moon is diffuse reflection.

5) Lunar radiation

The Moon absorbs solar radiation while radiating energy continuously. Lunar infrared radiation heat flux varies with the geomorphology, latitude, etc. In thermal analysis, it is generally assumed that lunar surface is a gray body. The temperature of lunar surface is determined by the solar radiation absorbed and the heat from its interior. The lunar surface temperature at sub-solar point is about 120 °C, and the temperature at the moon night and moon shadow is as low as −180 °C. The distribution of lunar surface temperature relative to sub-solar point is shown in Fig. 5.3. Ignoring difference of lunar terrain, lunar surface temperature is related to local distance and position to sub-solar point. Because the angle between the Moon's equator and ecliptic is 1.5°, i.e., the sub-solar point is almost on lunar equator, the landing latitude of lunar latitude greatly affects lander's temperature.

A lunar day includes lunar daytime and lunar night, both of which are 14 Earth days. During the lunar daytime, the Sun arises from east to west. At the 'noon' of lunar daytime, lunar surface temperature reaches the highest. While during the lunar night, there is no solar radiation at all.

Ignoring difference of lunar terrain, assuming that lunar surface material has the same thermal property, lunar surface temperature only depends on the thermal inertia γ ($cm^2 \cdot s^{1/2} \cdot K/J$) of the lunar soil defined as follows:

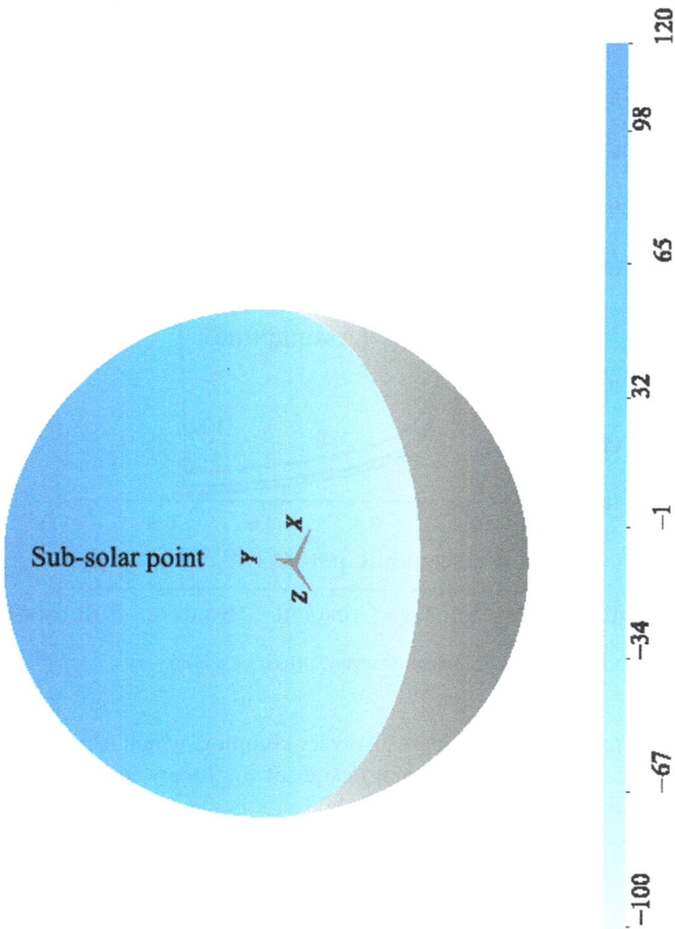

Fig. 5.3 Model of temperature distribution of lunar surface at sub-solar point

$$\gamma = \frac{1}{\sqrt{k\rho c}}$$

where: k is heat conductivity (W/(cm · K)); ρ is density (g/cm^3); c is specific heat capacity (J/(g · K)).

The temperature variation curve of lunar equatorial surface on a lunar day is shown in Fig. 5.4 [9]. The maximum temperature difference between lunar day and lunar night in the figure exceeds 300 °C, thus the thermal control subsystem shall have a strong environmental adaptability.

For thermal analysis of the lunar lander, the average infrared heat flux of Earth surface is approximately 220 W/m^2. The spatial distribution and spectral distribution

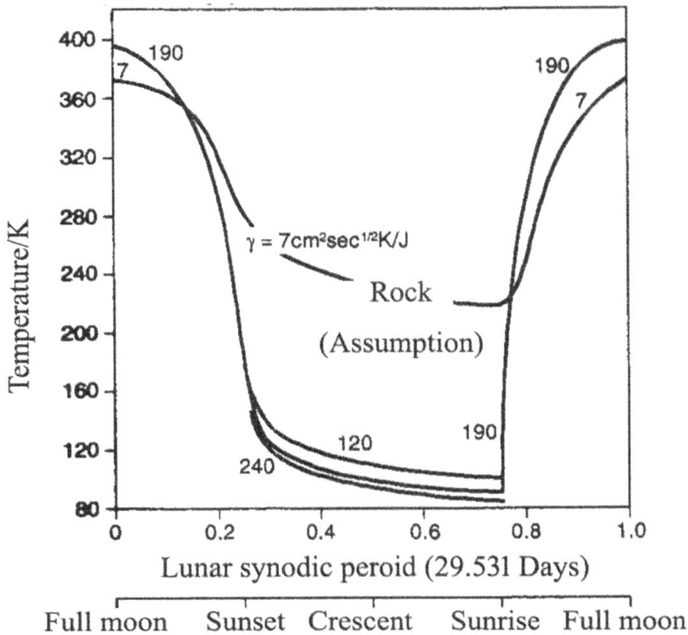

Fig. 5.4 Temperature changes of lunar equatorial surface on a lunar day

of the infrared radiation of Earth are also very complex, which is generally assumed to be diffuse radiation in the thermal design of the lunar lander. And its spectral distribution is treated as absolute black body radiation of 250 K. More about the introduction of the Earth's radiation heat flow can be found in related works.

6) Gravity

The lunar lander usually experienced three types of gravity conditions:

(1) Before launch: 1 g of ground gravity;
(2) LTO and circumlunar phase: microgravity;
(3) Lunar surface: (average) 1/6 g gravity.

There are gravity anomalies due to mass distribution and other reasons on some lunar mares (as shown in Fig. 5.5). There are differences between those regions and average lunar gravity of 1/6 g.

Gravity conditions will affect the working conditions of thermal products that have working fluids such as heat pipes and fluid loops. It is necessary to consider the difference in the gravity conditions at all phases and make full use of the gravity conditions in thermal design.

In addition to above gravity conditions, the acceleration caused by main engine of the lunar lander during launch, LTO, orbit maneuver and landing will generate similar effects of gravity. These short-term processes have less impact on performance of

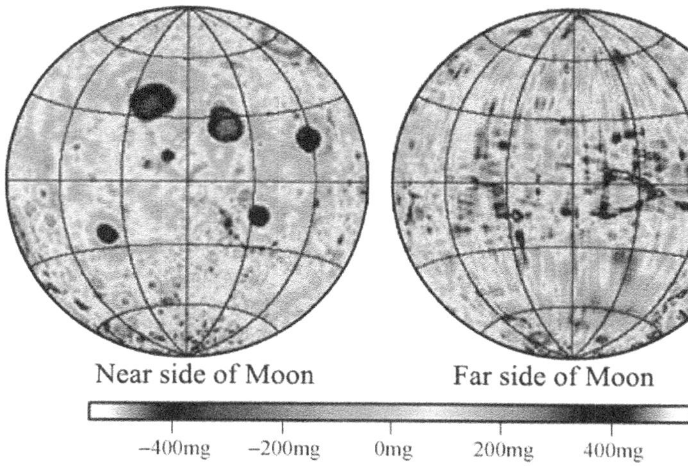

Fig. 5.5 Gravity anomaly regions on lunar surface

thermal control products, but products such as fluid loops shall be able to recover after experiencing above phases.

7) Lunar dust

The lunar surface is covered with a layer of lunar soil, which can produce very fine dust particles under long-term environmental conditions. Due to some reasons such as impact of micrometeorite, impact of engine plume, interaction between locomotion subsystem and lunar soil, operation of drill, lunar dust of fine dust particles may be activated. A small amount of lunar dust deposition will degrade heat dissipation performance of thermal control subsystem [3]. Due to fine particle size and surface charging during lunar daytime, cleaning of lunar dust is very difficult [4, 5]. Compared to the Earth orbit spacecraft and circumlunar spacecraft, lunar dust prevention is an important issue to be focused on for thermal control subsystem of the lunar lander.

The causes of lunar dust include natural factors and artificial factors. Two natural factors known are the meteorite impact and electrostatic suspension effect at terminator of morning and evening [5]. According to analysis, lunar dust activated by natural factors is few, but reasonable management of heat dissipation surface (for example, as far as possible from lunar surface) will help to ensure the long-term life of the thermal control subsystem.

Artificial factors include engine plume impact during landing and take-off of the lunar lander, drilling, movement of the lunar rover, crew extravehicular activity and etc., where movement range of lunar dust caused by impact of engine plume during landing and take-off of the lunar lander is the largest [5]. Because there is no obvious atmosphere on lunar surface, lunar dust particles follow the Newton's law with parabolic motion. Therefore, lunar dust caused by impact of engine plume mainly affects area around the lunar lander, and lunar dust is few on the top surface, which can be utilized for configuration of heat dissipation surface.

2. Design Constraints

There were design constraints for thermal control subsystem of the lunar lander including environmental conditions during launch, nominal lifetime, thermal boundary of engine, and the uncertainty of the landing attitude and operation modes of instruments.

1) Environmental conditions during launch

Thermal control subsystem shall adapt to the mechanical conditions and thermal environmental conditions during launch. Based on experience of other spacecraft, general thermal control products can withstand mechanical conditions during launch, but effects of acceleration on the working conditions and performance of heat pipes, fluid loops and other products shall be considered.

As there was heat insulation in launch vehicle fairing, although the temperature of the outside surface of the fairing was high due to the aerodynamic heating, it had less effect on instruments of the lunar lander. Due to the short duration of launch phase, the temperature of instruments on outer surface of the lunar lander would not exceed allowed temperature level. After jettison of fairing, the temperature of internal instruments would generally not exceed allowed value because of thermal inertia of the lunar lander and onboard instruments.

2) Lifetime

If the lifetime of the lunar lander on lunar surface was less than one lunar daytime after landing, there was no need to consider the issue of survival in lunar night. Then the thermal design was relatively simple, while single-phase fluid loop and heat pipe could be used for heat transfer, or an evaporator (or sublimation) consumption heat sinks for heat dissipation. Through setting a reasonable landing time, landing points for such the lunar lander could even be selected near lunar equator.

If the lifetime of the lunar lander was longer than one lunar daytime after landing (for example, the lunar lander with lifetime of one year will experience about 13 cycles of lunar day and night), it was necessary to consider the issue of survival in lunar night as well as the long-term high temperature in lunar daytime. In order to provide heat required for survival during lunar night, isotope materials were usually used instead of large-capacity battery because of large mass requirement. The isotope materials could provide heat for the lunar lander by utilizing the heat generated by radioactive decay. Because of uncertainties such as impact of lunar dust, high temperature issue in lunar daytime must also be paid attention to in thermal design. There would be requirement of heat dissipation for thermal control subsystem if the landing point is chosen to be near lunar equator.

3) High temperature boundary of engine

A large engine was used by the lunar lander to perform lunar orbit insertion and soft-landing on lunar surface, while small thrusters (such as 150 N and 10 N thrusters) were used to realize attitude control and assist landing. Due to the long-term working and high temperatures (throat temperature exceeding 1300 °C, as shown in Fig. 5.6), the

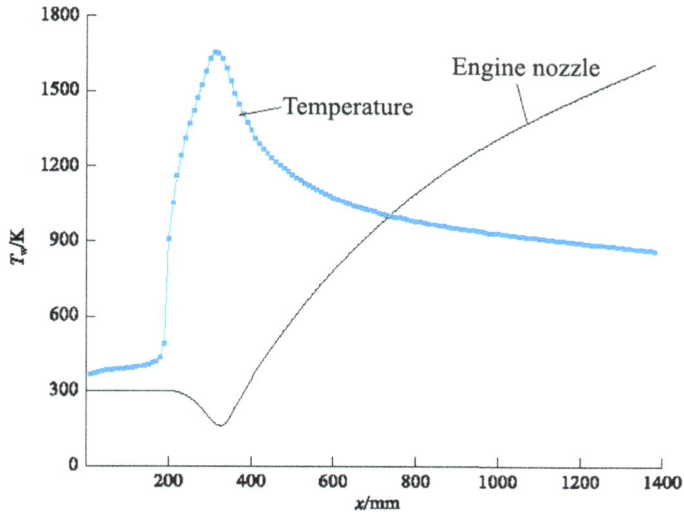

Fig. 5.6 Nozzle temperature distribution of 7500 N engine of the Chang'E-3 lunar lander

high temperature of the high-thrust engine nozzle might cause the probe structure and equipment temperature to exceed the standard and affect the safety of probe. Due to configuration limitations, this issue was even more serious if the landing engine was embedded inside the probe structure. The solution was to adopt thermal protection measures to thermally isolate the probe body from the high temperature boundary of the engine.

4) Uncertainty of landing attitude

The lunar terrain is complex and there may be slopes, rocks and pits in landing site. It is difficult to predict actual lunar terrain or avoid unfavorable lunar terrain before landing. Due to the accuracy of attitude control performance of landing legs of the lunar lander, it might cause uncertainty of the lunar lander attitude to tilt, twist on lunar surface after landing, which would cause a large deviation between the external heat flow of thermal control surface and the expected situation to affect implementation of thermal control function and performance, even safety of the lunar lander. In design of thermal control subsystems, the combination of the worst-case tilting and twist conditions of the lunar lander attitude should be considered in order to determine the maximum temperature envelope for typical conditions and satisfy temperature requirement of the lunar lander.

5) Operation modes of instruments

There were sub-systems including propulsion, GNC (Guidance, Navigation and Control), landing cushion, power supply and distribution, thermal control, On-Board Data Handling (OBDH), TT&C (telemetry, tracking and control) in the lunar lander. Some of them including propulsion, GNC, and some mechanisms would not work after landing, while some of them such as power supply and distribution, thermal control, and OBDH would work throughout the lifetime of the lunar lander. Only a few of instruments were working during lunar night. The operation modes of the lunar lander indifferent phase were different, which also affected the determination of thermal control design.

In the system design and thermal control design of the lunar lander, the differences of operation modes of instruments at all phases should be considered. The differences in temperature requirement in different phases should be treated carefully to minimize the demand for mass and power through optimization of instruments configuration and thermal control design.

5.3.2 Technical Characteristics

The mission profile of the lunar lander was complex and the environmental was harsh. The thermal control subsystem should satisfy special requirements such as great changes in the external heat flux during LTO, circumlunar orbit, landing and lunar surface, different operation modes, and survival in lunar night. At the same time, it should solve the technical difficulties from harsh high-temperature boundary of landing engine, uncertain landing attitude, and deposition of lunar dust.

According to mission requirements of the lunar lander, the technical characteristics of thermal control subsystem were as follows:

(1) Strong adaptability to changes in external heat flow.
(2) Strong adaptability to changes in operation modes of instruments.
(3) Strong adaptability to changes in gravity conditions by taking full advantage of lunar gravity to improve capability of heat pipes and fluid loops.
(4) Adaptability to uncertain landing attitude by reasonable configuration of heat dissipation surface and system prevention of lunar dust to minimize its influence.
(5) Application of RHU (Radioisotope Heat Unit) for heat source during lunar night together with active thermal control measures to achieve reasonable control of RHU.
(6) Use of high temperature insulation material to prevent the high temperature effects of engine nozzle during landing.

5.4 Design Methodology

5.4.1 Principles of Thermal Design

The thermal design of the lunar lander was based on the characteristics and requirements of the mission, as well as the internal and external heat load and typical flight events experienced by the lunar lander in different phases. Various thermal control measures were used to organize the heat exchange process inside and outside the lunar lander. And it could ensure that the temperature of all instruments and structures satisfies the requirements during the entire lifetime of the lunar lander. Besides, it was also necessary to minimize the need for resources such as mass and power.

The main design principles of thermal control subsystem for the lunar lander were as follows:

(1) Insulation among different modules to minimize mutual influence. The concept of "modular design" was applied where thermal insulation was designed for each module. As for the instruments that were no longer used after landing, the temperature should satisfy the requirements only before landing. While for the instruments working on lunar surface, it should satisfy the requirements of high temperature during the lunar daytime. The introduction of heat from RHU could make sure that the storage requirement during lunar night are satisfied.

(2) Favorable surface orientation for heat dissipation. After landing, the heat dissipation surface of the lunar lander was oriented to zenith in order to minimize the effect of the high temperature of lunar surface on the heat dissipation capability. In addition, the configuration of heat dissipation surface, heat pipe and two-phase fluid loops should be able to adapt to the possible tilt of the lunar lander after landing. They could operate well even if the maximum tilt attitude occurs.

(3) Insulation of thermal coupling between exposed parts and main body. Except for the necessary radiation surface, the outer surface of the lunar lander was covered with multilayer insulation component. And the exposed components such as TT&C antenna and mechanisms were connected to main body with insulation material. Such design could minimize the effect of environmental changes on temperature of internal instruments.

(4) New approach for harsh environmental issues. The use of variable conductivity heat pipes (VCHP), two-phase fluid loops and isotope heat sources (RHU) could satisfy the heat dissipation requirement for internal instruments during LTO, circumlunar and lunar daytime on lunar surface. In the lunar night, VCHP was shut off and two-phase fluid loop transfer the heat of RHU into relative modules to keep warm for survival requirement. In design of VCHP and two-phase fluid loop, the ground gravity (1 g), lunar gravity (1/6 g) on lunar surface and effect of micro-gravity during LTO and circumlunar phases should be considered to make sure normal operation in microgravity conditions and make use of lunar gravity to improve performance.

(5) Minimizing effects of lunar dust deposition on performance of thermal control products. Thermal performance of OSR (Optical Sun Reflector) heat dissipation surfaces, thermal coatings such as white paint was greatly affected by lunar dust. However, there was no simple and effective method to eliminate lunar dust on lunar surface. The system prevention could help to prevent adverse effects of lunar dust on thermal control products during landing by shut off of landing engine before landing and shielding. There should be enough margins for temperature of instruments in thermal design. In addition, the effect of lunar dust on thermal control products such as OSR heat dissipation surface and white paint should be verified by simulated lunar dust if possible.

(6) Consideration of the effect of engine's high temperature on the lunar lander. The thrust generated by engines was used by the lunar lander to perform lunar orbit insertion, soft-landing and attitude control. Usually there were number of engines in the lunar lander. When an engine was firing, the temperature of engine nozzle was very high, while the velocity and heat flux of its plume was also high, which had a great influence on the temperature of structure of the lunar lander and instruments. When the configuration of the lunar lander was designed, the engine should be exposed as much as possible to reduce its influence on the lunar lander. In addition, high temperature thermal shields and other products should be designed to protect the lunar lander by thermal insulation of high-temperature engine.

5.4.2 Design of Thermal Control

1. Thermal Control Measures

There are passive and active measures in thermal control design of spacecraft. Passive thermal control measures are used when the change of internal heat source and external heat flux is small, or when the allowable temperature range of structure and instruments is relatively wide. By selecting materials with different thermophysical properties and controlling the heat exchange process between internal and external reasonably, the temperature of structure and instruments can be controlled within the allowable range. Passive thermal control measures are simple, highly reliable and long life because there are no moving parts, which are generally used as the preferred measures and basic measures for thermal control design of spacecraft. There are thermal control coatings, multilayer thermal insulation materials, heat pipes, thermal conductive fillers and thermal insulation pads in passive thermal control products.

When the internal heat source or external heat flow of spacecraft varies greatly, or the temperature requirements of structure and instruments are within a narrow range, passive thermal control measures cannot meet the requirements alone. Then the active thermal control measures shall be added. Active thermal control measures can adjust working conditions according to changes in the spacecraft environment and keep the temperature of structure and instruments within the required range. Such thermal control system has strong adaptability, but the system is more complex, while more mass is required and reliability will be lower. There are electric heaters, thermal switches, variable conductivity heat pipes, single-phase fluid loops, two-phase fluid loops and loop heat pipes in active thermal control products.

In design of thermal control, at first the passive thermal control measures are preferred according to changes of external heat flow, internal heat source and temperature requirement of instruments. If the environment is harsh, or the heat source in the module or instrument changes so greatly that the passive thermal control measures cannot meet the requirements, the appropriate active thermal control measures can be selected. The change of the environment faced by the lunar lander was great, and high temperature in lunar day and low temperature in lunar night was harsh, therefore only passive thermal control measures couldn't meet the requirements. So, it was necessary to take more active thermal control measures to control the temperature range of structure and instruments.

2. Selection of Heat Source and Method of Heat Transfer during Lunar Night

It is up to the lifetime of the lunar lander whether special heat source such as RHU is necessary. For example, if the lifetime was more than a lunar daytime, the heat source was necessary to provide heat for survival of the lunar lander in lunar night.

During lunar night, the heat from the heat source should be introduced to the lunar lander by fluid loops driven by bump, thermal switches, and loop heat pipes. Because the pump that drove the fluid loop requires electric power, a high-power isotope thermoelectric generator was necessary to meet the power supply requirement, which meant a large cost of the system. And there were also active components such as pump that may affect long-term operation reliability. The solution of thermal switch with loop heat pipe was feasible, but the thermal switch with moving parts might cause low reliability and is prone to malfunctions resulting in poor system reliability.

For example, the lifetime of the Chang'E-3 lunar lander on lunar surface was 1 year. It was necessary to use RHU to provide heat. According to configuration of the lunar lander, two RHUs were used to provide heat for the +Y and −Y modules of the lunar lander. The isotope material was ^{238}Pu with 87.7 years half-life period. Each RHU could provide heat power of 120 W. A two-phase fluid loop driven by lunar gravity was developed for thermal control subsystem, which could transfer heat of RHU into instrument modules during lunar night. The two-phase fluid loop used control valves to realize the on-off control of the fluid loop system as shown in Fig. 5.7. There was no moving part except control valves, therefore no electric power or other resource requirement was needed during operation in lunar night, which was highly reliable.

3. Modular Design

The operation modes of the lunar lander in different phases might be different. The idea of modular design could isolate thermal coupling among different modules to reduce thermal impact of the changes in operation modes of instrument. In each module, the thermal control was independent. Intensive heat exchange was designed to increase the uniformity of instruments temperature in the module. According to operation modes of instruments, heat consumption distribution and total amount

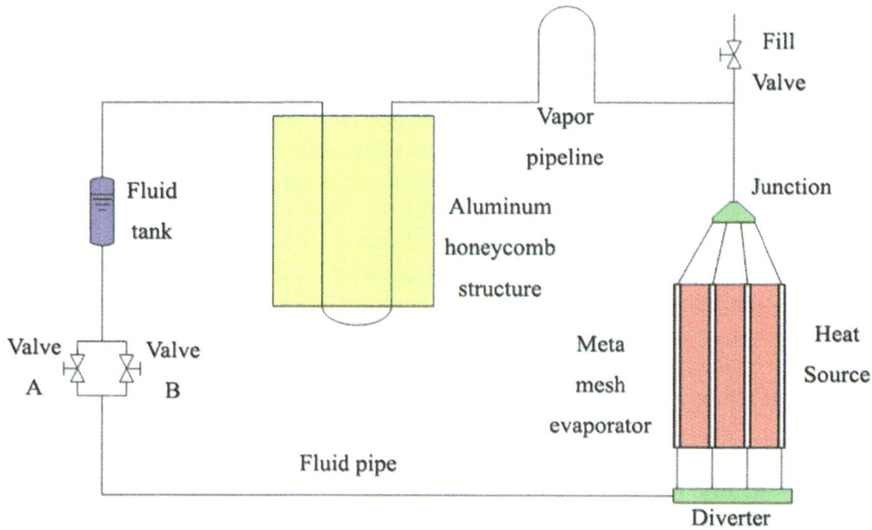

Fig. 5.7 Diagram of two-phase fluid loop

of heat in the module, configuration of heat dissipation surface, active and passive thermal control measures could be determined.

For example, there were four modules including center module, +Y module, −Y module, and −Z module in the Chang'E-3 lunar lander. Each module was thermally insulated with others. The instruments of OBDH, TT&C, power supply and distribution and scientific payloads subsystems were in +Y module and −Y module. Those instruments would work in all phases of mission profile and have large heat consumption and need to be ensured at storage temperature in lunar night. In the design of thermal control, there was one radiation plate for each module to dissipate heat of instruments, while the heat of instruments during operation was transferred to the radiation plate by the VCHP. The variable conduction heat pipe transfers heat to the thermal radiation plate for heat dissipation during operation of the device. During lunar night, the thermal coupling between instruments and heat dissipation surface was insulated by filling the condensation section (the radiation plate) with control gas to block instruments from dissipating heat to external environment. Together with the heat of the RHU introduced by the two-phase fluid loop, keeping warm in the module could be achieved in lunar night.

4. Selection of Thermal Control Coating

The thermal control coating is a surface material used to adjust the thermal radiation properties of the solid surface for thermal control. The method of heat exchange between the lunar lander and external environment was only radiation, while the methods of heat exchange between the internal structure of the probe and the equipment were radiation and heat conduction. Therefore, the selection of thermal control coating was significant to heat exchange in modules and between internal and external of the lunar lander. The two important parameters of thermal control coating were the solar absorptance α_S and the infrared hemispherical emissivity ε_H, which respectively represented the absorption ratio of solar radiation energy and the ability to radiate energy.

There are many kinds of thermal control coatings, which can be classified into external thermal control coatings and internal thermal control coatings. For the lunar lander, external thermal control coating should generally have the characteristics of low solar absorption ratio and high infrared emissivity. In addition, the effects of lunar dust deposition on performance of thermal control coatings were different. It was generally believed that the effect on the type of vacuum deposition thermal coating was smaller than that of thermal paint. Based on above factors above, thermal control coating of vacuum deposition was selected for external instruments. Available products included glass second surface mirror of vacuum deposition, or flexible thin-film second surface mirror of vacuum deposition which could be easily installed.

The spray-type thermal control coating was generally used for thermal control coating in the module including the surface of structural plates and instruments, whose infrared hemisphere emissivity was high to improve the temperature uniformity in the module.

When thermal control coating was determined, beside the α_S and ε_H, other factors such as adaptability to temperature and environment should also be considered. For example, for RHU, its temperature was relatively higher under high temperature conditions than the normal operating temperature of some coatings. Therefore, thermal control coating with high temperature should be selected.

5. Selection of Multilayer Insulation

The multilayer insulation is composed of a high-reflectance screen and spacers of low thermal conductivity laminated alternately. The layers reflect the radiation heat flow to form a very high thermal resistance, which is a good insulation under vacuum conditions. Because of its light weight, it is ideal for spacecraft.

According to the difference in application environment, multilayer insulation is classified into low-temperature, medium-temperature and high-temperature multilayer thermal insulation. Different types of multilayer thermal insulation differ in the highest temperature that the reflective screen and spacer materials can endure. The reflectors and spacers of typical low-temperature multi-layer insulation are double-sided aluminum-plated polyester film and mesh respectively, which can be used in environment temperature below 150 °C. The reflective screen and spacers of medium-temperature multi-layer insulation are polyimide aluminum film and glass fiber cloth, which can be used in environment temperature below 350 °C. The reflective screen and spacers of high-temperature multilayer insulation assembly are nickel foil or other metal foil and glass fiber cloth, which can be used in environment temperatures below 900 °C.

Low-temperature multi-layer insulation was generally used for insulation on surfaces and instruments in the modules of the lunar lander. Medium-temperature and high-temperature multi-layer insulation were generally used for high-temperature protection of attitude and orbit control engines. For medium-temperature and high-temperature multilayer thermal insulation, they were commonly used together with low-temperature multi-layer thermal insulation.

The high-temperature insulation shield was a special multi-layer insulation including high-temperature, medium-temperature and low-temperature multi-layer insulation. The reflective screens and spacers of the high-temperature multilayer insulation components were replaced to meet the high temperature conditions of nozzles with 1300 °C of the landing engine.

6. Thermal Control for Propulsion System

The propulsion system of the lunar lander performed attitude adjustment, orbit change, and landing by commands. It was crucial to mission of the lunar lander. In addition, the thermal control of propulsion was complex due to due to complexity of propulsion system including gas, liquid and valves.

There were in propellant tanks, gas tanks, pipeline and engines in propulsion system.

(1) In order to ensure that the temperature of propulsion system components meet the requirement of temperature at low external heat flux, electric heaters were used for temperature control of propellant and gas tanks, liquid pipeline, gas pipeline and other components. And a backup was designed for thermal control loops to improve reliability.
(2) According to distribution of environment temperature, the pipeline was divided into different sections for thermal control to ensure the effectiveness of temperature control. The pipeline was heated by electric heating tapes, while film-type electric heating film was used for other components.
(3) Because of large deformation of the gas tanks after filled, the heating device on surface of tanks should be able to adapt to the deformation.

7. Design of Heating Loops

For instruments such as pyrotechnics, batteries, mechanisms and propulsion, it was necessary to use electric heating for thermal control. This method had the advantages of simple structure, convenience and high controlling accuracy, but it should consume a certain amount of electric power.

In accordance with the form of electric heating devices, they can be classified into film type electric heaters and electric heating tapes. The electric heating material is copper-nickel-manganese alloy, and the insulating material is a polyimide film, where the electric heater is mainly used for propulsion system pipelines. For thermal control of engine's head, armored electric heaters could be selected to withstand higher temperatures.

The heating loops could be controlled by an independent heating controller according to a preset threshold or ground commands. It was also possible to centralize the control of the heating circuit to OBDH subsystem or an integrated electronic subsystem to increase the degree of system integration, whose functions and control methods were the same as for the individual heating controllers.

8. Thermal Control Solution for Exposed Parts

The exposed parts of the lunar lander ere mechanisms, antennas, sensors and optical devices such as cameras. Generally, the solar panel mechanism and some antennas should work normally again after experiencing harsh environment in lunar night.

Different exposed parts had different temperature requirement and various environmental conditions, therefore, independent thermal design as required. In order to reduce the impact on main body of the lunar lander, the exposed components were generally installed with insulation to the lunar lander. For exposed parts with special requirements for temperature during lunar night, a small RHU could also be used to provide heat to ensure that the temperature requirement was met.

Compared with Earth orbit and circumlunar spacecraft, due to the harsh environment during the lunar daytime and nights and large cost, it was impossible to take measures to ensure that each exposed component was in a similar temperature environment to other spacecraft. Generally, it was necessary for exposed parts of the lunar lander to withstand more severe temperature ranges and be verified experimentally.

5.4.3 Thermal Analysis and Calculation

The thermal analysis and calculation was an important means to verify the thermal design of the lunar lander. Its objective was to determine the temperature change pattern of each part in the lunar lander according to the thermal conditions inside or outside the lunar lander and the thermal control measures. It was also to check whether the thermal design had kept the various parts controlled within the required temperature range and to predict the actual operating temperature of the lunar lander. Due to the complex flight process of the lunar lander and uncertainty of landing attitudes, the thermal balance test couldn't cover all flight events. The thermal analysis model could only be modified based on limited thermal test conditions, and then give the temperature prediction for various flight phases and task. The thermal analysis generally ran through the entire development process of the lunar lander. After launch and orbit injection of the lunar lander, according to flight control requirement, the results of thermal analysis were required to provide the basis for dealing with flight events. After landing, it was necessary to predict the subsequent temperature conditions based on actual attitude. Therefore, the thermal analysis and calculation was an important task for development of thermal control subsystem in the lunar lander.

Like other spacecrafts, the thermal network analysis method was the major approach for thermal analysis of the lunar lander. Currently, many mature commercial software of thermal analysis was available for modeling and calculation. However, secondary development on the basis of existing commercial software might be necessary for calculation of lunar surface temperature field in order to conduct the thermal analysis calculation in circumlunar and lunar surface phases.

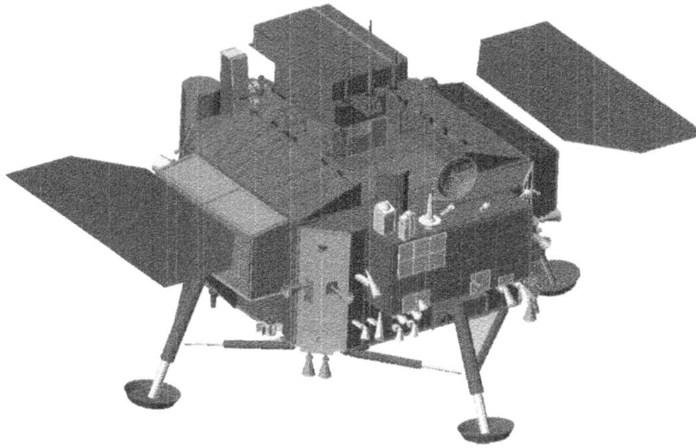

Fig. 5.8 Thermal analysis model of the Chang'E-3 lunar lander

1. Thermal Analysis Modeling and Simplification

The lunar lander had a complex structure, a large number of instruments, various shapes and assembly relationships of instruments. The workload was very large if the thermal analysis model was completely based on the actual conditions of the lunar lander and its components. It might exceed the capabilities of the software. Thus, it was not necessary in practice. Therefore, according to the physical model and assumptions in thermal analysis, a variety of simplified methods were used to simplify the local structure that did not affect the temperature of instruments and the structure to improve the computational specificity. For VHCP, two-phase fluid loop and other products, they were simplified reasonably to reduce the computational complexity and improve the computational efficiency without affecting the simulation of heat transfer process. The thermal analysis model of the Chang'E-3 lunar lander is shown in Fig. 5.8.

An important part of thermal analysis modeling of the lunar lander was the processing of lunar infrared heat flow. In LTO, the lunar lander was far away from Moon, and it was not necessary to consider effects of lunar infrared radiation. In circumlunar phase and operation on lunar surface phase, lunar infrared radiation heat flow had a great influence on the lunar lander. The detailed model of lunar temperature field and lunar surface should be established to calculate temperature of the lunar lander under lunar infrared radiation heat flux and different situation. The lunar dust caused by engine during landing and natural factors might deposit on surface of thermal control coating to degrade coating performance. In thermal analysis and calculation, the influence of lunar dust pollution on thermal coating should be considered by setting reasonable parameters for thermal control coating.

Another major concern in thermal analysis of the lunar lander was the impact of engines. There were two effects of engine in thermal analysis modeling including engine's radiation heat flow and plume heat flow. The radiation heat flow of engines

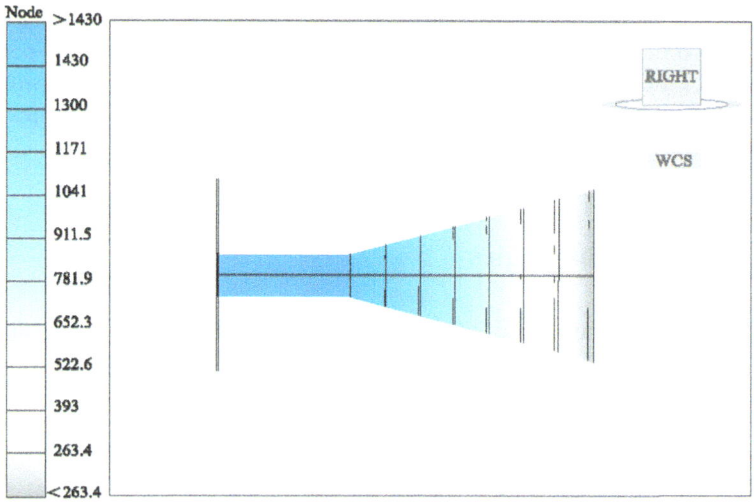

Fig. 5.9 Thermal analysis model of the 150 N engine of the Chang'E-3 lunar lander

could be directly calculated by thermal analysis software. The heat flow of engine plume should be calculated by professional software, and the result could be used as an input of heat flow. The thermal analysis model of the 150 N engine of the Chang'E-3 lunar lander is shown in Fig. 5.9.

2. Correction of Thermal Analysis Model

In initial design of the lunar lander, a detailed thermal analysis model should be established for calculation to determine the actual performance of thermal control. The thermal control measures should be modified and improved according to the results. The modified thermal control measures must pass through thermal analysis and calculation to be verified. Therefore, thermal control design and thermal analysis were iterative.

The mission profile of the lunar lander was complex with complex environment and uncertainty in landing attitude. It was not possible to verify all conditions by thermal balance tests directly. The thermal analysis was necessary to verify thermal control design for many flight phases. Improving the accuracy of thermal analysis model as very important for thermal control subsystems and the lunar lander. In order to correct the errors such as the error of conductivity from model simplification and calculation error from node separation, temperature prediction and model corrections should be conducted before and after thermal balance tests. After the lunar lander was launched, the thermal analysis model should be corrected again by using the actual temperature in orbit to predict and correct temperature in subsequent phases. According to telemetry data on orbit, the thermal analysis model was revised to make calculation results closer to actual situation. The high confidence of temperature prediction on orbit could be used as an important basis for flight control and emergencies handling. If it was necessary, the flight procedure of the lunar lander could be modified to meet temperature requirements. According to past experience, correction of thermal analysis model should focus on heat capacity of instruments and structure, heat conduction among different components, heat consumption and operation modes of instruments.

5.5 Typical Technologies

The environment experienced by the lunar lander such as lunar orbit insertion, landing, lunar surface after landing was quite different from other spacecrafts. In order to meet the mission requirements, the thermal control subsystem should adopt several typical technologies suitable for the lunar lander including variable thermal-conductive technology, high-temperature protection technology for engines, RHU application technology and two-phase fluid loop technology.

5.5.1 Variable Thermal Conductive Technology

According to the task analysis, the external heat flow in different phase and the operating modes of instruments varied greatly. For conventional heat transfer components of constant thermal conductivity such as conventional heat pipes, there would be relatively large fluctuation of instrument temperature. If a variable thermal conductivity product such as a variable conductive heat pipe is applied, the length of its working period could be automatically adjusted according to the operating mode of instrument and the external heat flow to reduce temperature fluctuation of instrument.

The lunar lander should dissipate heat in lunar day time and keep the instrument module warm in lunar night. Such requirements could only be achieved by variable thermal conductivity technology at the same time. The relation between the lunar lander structure and heat transfer system of variable thermal conductivity is shown in Fig. 5.10.

There were available variable thermal conductivity technologies including VCHP, LHP, pump-driving two-phase fluid loop and heat pipes of special fluids, where VCHP and LHP technologies were more mature. For heat pipes of special fluids, the fluids of different freezing points were selected to work at the required temperature in gas and liquid phase. When the temperature was below freezing point, the fluid is frozen to solid and the heat pipe stops working to block the heat transfer. For specific applications, two or more variable thermal conductivity products could be combined and used at the same time. It could be evaluated by power requirement, mass, lifetime, technology readiness, inheritance, complexity, and experimental feasibility to choose the most suitable product for mission requirements.

The rationale of the VCHP with gas tank on cold side is shown in Fig. 5.11. Compared with the conventional heat pipe, there was an additional gas tank with

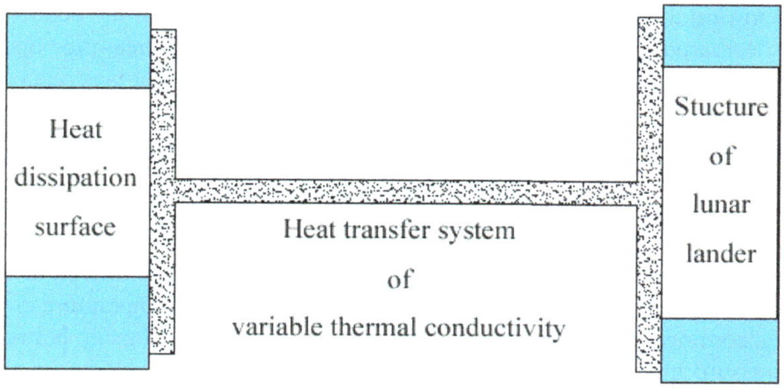

Fig. 5.10 Relation between the lunar lander structure and heat transfer system of variable thermal conductivity

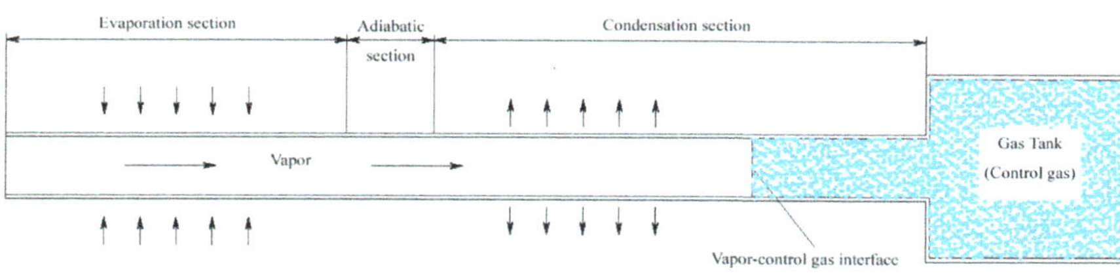

Fig. 5.11 Rationale of VCHP

control gas besides working fluid vapor in the VCHP [16]. According to the calculation results of high temperature and low temperature working condition, a certain amount of control gas (non-condensable gas, generally helium, nitrogen, etc.) and working fluid were filled into the heat pipe. The working fluid and control gas were in a more uniform mixed state when the heat pipe is not working. While the heat pipe was working, the vapor of working fluid flowed to condensation section with the control gas, where the vapor condenses into liquid and the liquid flowed back to the evaporation section through the capillary core and participates in evaporation again, with the control gas retaining in the gas tank and partial condensation section near the gas tank. When the input heat to the evaporation section increases, the vapor temperature increased and the pressure increased, the control gas was compressed, causing the vapor-control gas interface moving toward the gas tank. Then the effective length of the condensation section increases with heat radiation of the heat pipe increasing and the operating temperature of the heat pipe could be prevented to rise further. When the input heat was increased to a certain extent, the control gas is completely compressed into the gas tank and the condensation section was fully involved for heat dissipation. Similarly, if the heat input to the evaporation section decreases, the vapor temperature decreased and the pressure decreased. Due to the pressure difference, the control gas expanded causing the vapor-control gas interface moving toward the adiabatic section. Then the effective length of the condensation section decreased with the heat dissipation decreasing, and the heat pipe operating temperature would no longer reduce further. When the input heat was reduced to a certain extent, the vapor-control gas interface entered the adiabatic section or even the evaporating section, and the heat pipe no longer transferred heat to the condensation section by phase change.

According to rationale of VCHP, the position of the vapor-control gas interface was also affected by the average temperature of the control gas, i.e., the temperature change of the gas storage chamber also affected the working temperature of heat pipe. The relationship between the control gas pressure and temperature could be treated according to the ideal gas state equation, while the relationship between the vapor pressure and the temperature of working fluid was exponential, so the position of the vapor-control gas interface was more sensitive to vapor temperature.

Actually, there was no obvious vapor-control interface in VCHP, but there was a transition section in the condensation section where the concentrations of two gases in the transition section gradually change. The length of the vapor-gas transition section could be calculated based on the gas mass diffusion theory. The actual application could be verified based on experimental data. The existence of the transition section resulted in the maximum heat transfer capacity of VCHP smaller than the theoretical value, and the minimum heat transfer capacity larger than the theoretical value. In order to minimize the length of transition section and the heat leakage during lunar night, measures should be taken to reduce heat conduction of heat adiabatic section and condenser section of VCHP along the length direction.

VCHP was used as heat transfer between instruments and heat dissipation surface in the Chang'E-3 lunar lander to achieve heat transfer during equipment operation and block during lunar night. Eight and six variable conductance heat pipes were settled on +Y and −Y modules respectively, where the evaporation section was connected with instruments in modules and the condensation section was embedded in the heat dissipation surface. The configuration of VCHP shall consider gravity assistance under various landing attitude. The volume of gas tank and control gas should satisfy heat transfer requirement during instruments operation and block during lunar night. The shape of a VCHP is show in Fig. 5.12.

When instruments of the Chang'E-3 lunar lander were working in LTO, circum-lunar and operation on lunar surface in lunar daytime, the heat of instruments was transferred to heat dissipation surface and radiated to space to ensure temperature requirement of instruments. During lunar night, the control gas of VCHP would "choke" the condensation section and heat in module would no longer be transferred to heat dissipation surface through the condensation section by phase change, so the temperature of instruments in modules would meet storage requirement in lunar night.

The moon-based observatory of the Chang'E-3 lunar lander used heat pipe with special fluid to achieve heat transfer during operation and insulation during lunar night. The freezing point of the fluid was −40.19 °C. The temperature in lunar daytime was higher than the freezing point, and the fluid was in the two-phase state. The heat pipe transferred the heat from instruments to heat dissipation surface when

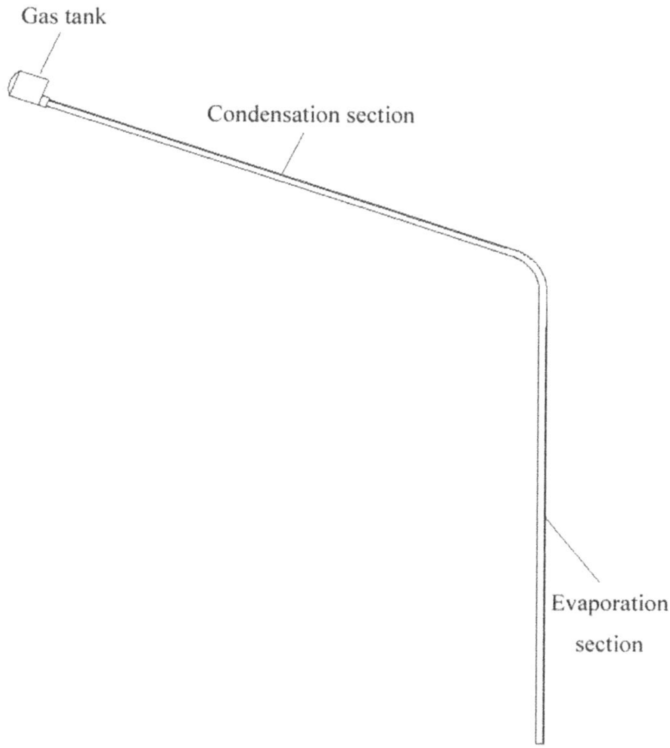

Fig. 5.12 Shape of a VCHP

operation. When instruments stopped work during hibernation in lunar night, the temperature of instruments and fluid in heat pipe gradually decreased with external heat flow decreasing. When the temperature was lower than the freezing point, the fluid was frozen and stopped flowing. The heat pipe no longer transferred heat from instruments to heat dissipating surface in phase change to maintain instrument temperature for storage requirement in lunar night. When the fluid was selected for special heat pipe, freezing point and temperature range should meet requirements at first. Besides, compatibility between fluid and material of heat pipe shell should be verified as well as heat transfer characteristics.

5.5.2 Protection Technology for High-Temperature Engine

During LOI and soft-landing, large engine was used by the lunar lander to decelerate with small thrusts (such as 150 N and 10 N thrusters) for attitude control and landing assistance. During LOI and soft-landing, due to long duration and high temperature of engine nozzle that could be higher than 1300 °C, high temperature insulation shield could be effective to insulate main body of the lunar lander with engine high temperature for protection up to date.

The high-temperature insulation shield was a multilayer heat insulation material that could withstand the high temperature of engines. In order to reduce the weight, the material of reflective screen and the spacer layer at different positions was optimized. The performance and characteristics of high-temperature insulation shield factors was the same as conventional multilayer insulation. Compared with conventional multilayer thermal insulation, in design of high temperature insulation shields for engines in structure, the configuration of heat shield should minimize blockage of the engine, so that the engine surface could radiate heat as much space as possible. The thermal shield material should be able to withstand high temperatures during fire of engines, and the volatilization of metal at high temperature should meet requirement. In addition, the high temperature thermal shield was generally fixed on bracket. The method of fix should minimize the heat leakage and meet the temperature limit of fixed bracket.

The performance of high-temperature insulation shield is related to the number of layer units. For thermal shield of engines, the lag of temperature response caused by heat transfer lag in the thickness direction is an important feature of the high-temperature insulation shield besides heat leakage. The thermal analysis method is generally used to predict temperature response and verified by vacuum thermal test when thermal shield is developed.

Due to configuration limitations, major part of the 7500 N landing engine was embedded inside the Chang'E-3 lunar lander resulting in protection issues of high temperature. The relative position of the high-temperature insulation shield to the engine is shown in Fig. 5.13. When the engine was firing, the surface temperature of high-temperature insulation shield opposite the engine throat was higher than 1000 °C.

High-temperature insulation shields were one of key components for safe landing of the Chang'E-3 lunar lander, whose thermal insulation properties were crucial to the landing process. High-temperature insulation shield was composed of high-resistance insulating materials. Heat insulation effects and mass reduction were considered, and the mass was reduced as much as possible in the premise of heat insulation performance requirement satisfied when high-temperature insulation shield was designed. When the 7500 N engine was firing, the high temperature of engine nozzle was insulated from main body of the lunar lander. Only a small amount of heat was introduced into the lunar lander through the high-temperature insulation shield to ensure that the temperature of structure and instruments in modules would not be higher than requirement.

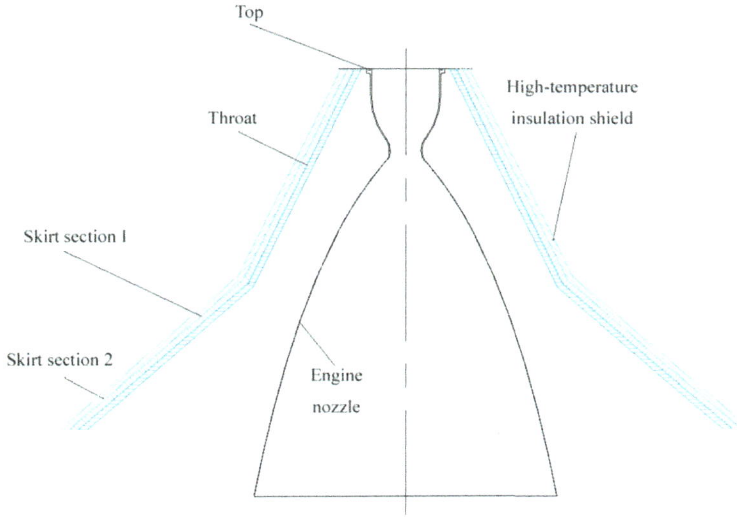

Fig. 5.13 Relative position of high-temperature insulation shield to engine

The size of 7500 N was big and a great many high-temperature gas was generated during firing. The performance of high-temperature insulation shield could not be verified directly through firing of actual engine. The high-temperature boundary was obtained through the 7500 N engine firing test when the engine was developed. The thermal insulation performance of shield was verified in vacuum test by simulating temperature boundary with an experiment product.

5.5.3 RHU Application Technology

In order to make the lunar lander survive through lunar night, usually RHU was used to provide necessary heat energy. The quantity of heat from the RHU decays with exponential relationship which is not affected by external environments. Therefore, RHU was preferable for long-term operation of the lunar lander under harsh environment on lunar surface. The output of heat power of RHU is given as follows:

$$P(t) = P_0 e^{-\lambda t} \tag{5.1}$$

where: $P(t)$ is the heat power (W) at time t, P_0 is the heat power (W) at the initial time, t is the time, λ is the decay constant, $\lambda = 0.693/T_{1/2}$, $T_{1/2}$ is the radioactive half-life period of the radio isotope.

There are few of radioisotopes suitable for application in space. When the radioisotope is selected, such factors shall be considered including:

(1) It shall be α nuclear source which is easy for radiation protection.
(2) Its mass specific power shall be high to reduce the mass as much as possible.
(3) Its radioactive half-life shall be longer enough for long-term application.

According to above requirements, the ^{238}Pu with long radioactive half-life (87.7 years), high mass specific power (0.55 W/g), α nuclear source, heat power decay rate (0.8% per year), is preferred for space application among over 3000 isotopes available. Because ^{238}Pu is radioactive and highly toxic, the most important issue was radiation protection when RHU is developed to ensure personnel health and environment safety by isolation protection measures in AIT (Assembly, Integration and Test) according to international treaty for space application.

The RHU of the Chang'E-3 lunar lander was connected with heat transfer devices, which were turned on to transfer the heat into instrument modules to maintain the temperature within allowable range during hibernation in lunar night. After the lunar lander was awakened, the heat transfer devices were shut off by ground commands to stop transferring heat into the module to avoid that instruments in the module was overheated resulting temperature arising or even higher than requirement. The temperature of RHU would arise after the heat transfer devices stopped to transfer heat in lunar daytime. Therefore, it was necessary to take thermal control measures such as radiator fins to meet requirement in lunar daytime and night. The diagram of RHU composition and energy transfer is shown in Fig. 5.14.

Besides RHU, the RTG can be also applied to produce electrical power for scientific instruments operation such as environment investigation in lunar night along with heat energy. The RTG can convert the heat power of radioisotope decay directly to electric power by Seebeck effect of thermoelectric material. The diagram of RTG composition and energy transfer is shown in Fig. 5.15. Compare to the RHU, the RTG increase the thermoelectric material, which needs to focus on the temperature limits of the thermoelectric material in thermal designing.

RHU in the Chang'E-3 lunar lander was the very first application of RHU in Chinese spacecraft. The RHU provided heat energy required during lunar night by radioactive decay heat of ^{238}Pu. The configuration of RHU is shown in Fig. 5.16.

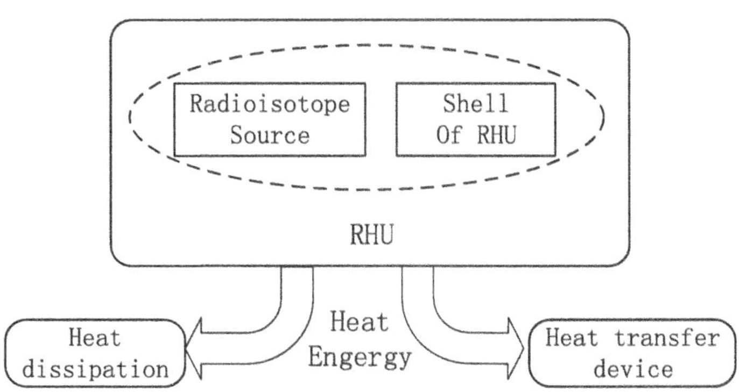

Fig. 5.14 Diagram of RHU composition and energy transfer

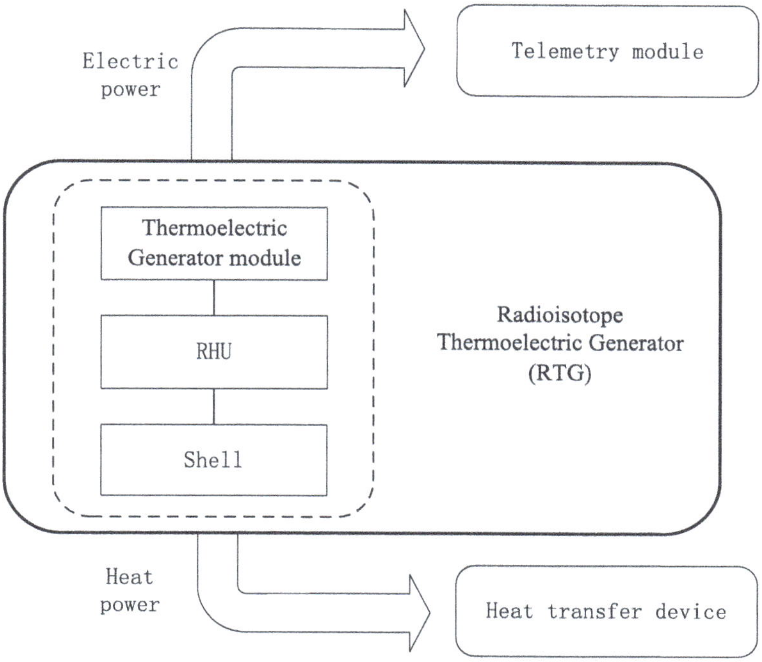

Fig. 5.15 Diagram of RTG composition and energy transfer

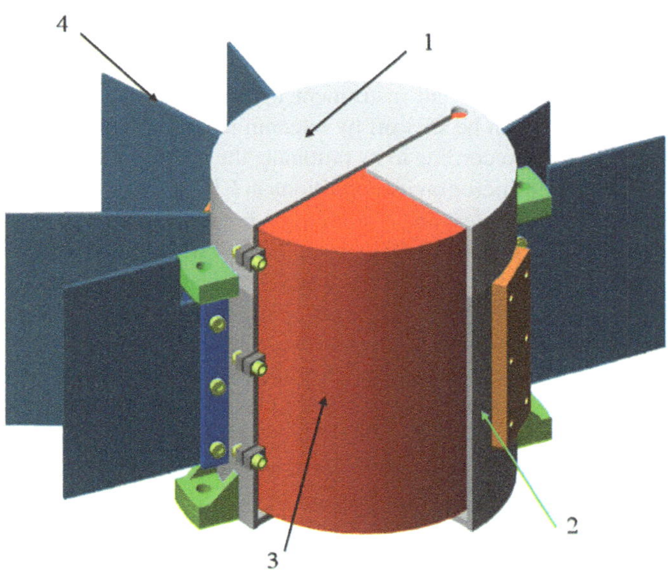

Fig. 5.16 Configuration of the RHU of the Chang'E-3 lunar lander. **1** RHU cover; **2** Housing; **3** Radioisotope source; **4** Fin

There were two RHUs installed outside instrument modules in thermal control subsystem of the Chang'E-3 lunar lander, each of which could produce 120 W heat power.

5.5.4 Two-Phase Fluid Loop Technology

The RHUs of the Chang'E-3 lunar lander was installed outside instrument modules. The heat of RHUs should be transferred into instrument modules to maintain instrument temperature. For the Chang'E-3 lunar lander, the heat transfer devices should not consume any electric power and be highly reliable during continuous operation of 14 days in lunar night.

The technical challenge of thermal control under low temperature in lunar night was heat transfer on basis of RHU application. The requirement was to transfer heat without any electric power consumption. Theoretically the heat of RHU could be transferred directly into instrument module by heat conduction, or by heat loop pipe, heat pipe with thermal switch, or by gravity-driven two-phase fluid loop. The thermal switch was necessary to control heat transfer for design of heat conduction, heat loop pipe or heat pipe, which should be highly reliable (If the thermal switch failed on lunar surface, it could result in mission failure).

The gravity-driven two-phase fluid loop is driven autonomously by the density difference between the vapor in ascending pipeline and the liquid in the return pipeline under the natural lunar gravity on lunar surface. No electric power or other resources are necessary during operation and there is no mechanism in the gravity-driven two-phase fluid loop, therefore it is highly reliable, which is suitable for heat transfer between heat source and instrument modules during lunar night. During lunar daytime, the loop can be shut off by telecommand to block heat transfer from heat source to modules. According to its rationale, the gravity driven two-phase fluid loop cannot work under microgravity conditions in LTO and circumlunar phase, i.e., it can only work on ground test and operation on lunar surface.

The gravity-driven two-phase fluid loop technology was applied in thermal control system of the Chang'E-3 lunar lander, which transferred heat from RHUs into modules during lunar night. When the two-phase fluid loop was designed, it was necessary to consider the possible effects of uncertainty of landing attitude on operation of the loop. And the normal operation should be ensured even in the worst case. The two-phase fluid loop should be designed to be lightweight with no mechanism and simple structure and highly reliable. There were evaporator, condenser, fluid tank and pipelines in the two-phase fluid loop with ammonia fluid. There were two sets of two-phase fluid loops for heat transfer of two RHUs in the Chang'E-3 lunar lander. The configuration of the two-phase fluid loop of the Chang'E-3 lunar lander is shown in Fig. 5.17.

When the two-phase fluid loop was working, the loop was in connected state. Due to a certain height difference between the fluid tank and the evaporator, the fluid flowed through the evaporator and absorbed heat of the RHU under lunar gravity.

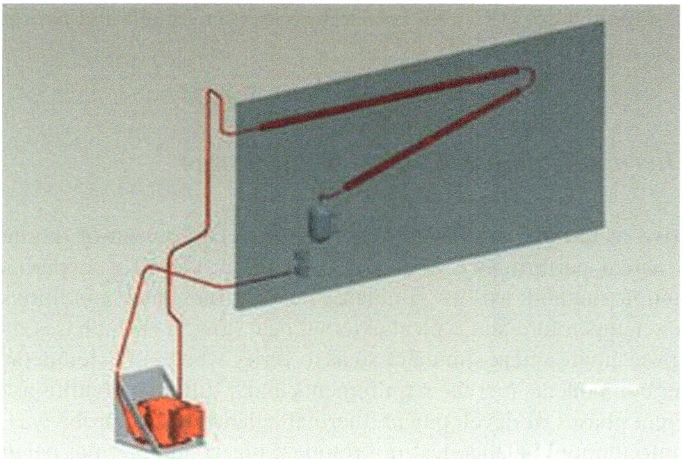

Fig. 5.17 Configuration of two-phase fluid loop of the Chang'E-3 lunar lander

Then the fluid became gaseous or gas-liquid two-phase after phase change and flowed into the vapor junction, and then flowed along the vapor pipeline to the condenser pipeline for condensation. The cooled fluid flowed into the fluid tank and finally returned to separate evaporators through the fluid pipeline. A circulation loop was formed to complete heat transfer.

When the two-phase fluid loop stopped, the loop was in disconnected state. Most of the heat leakage from the RHU to the condenser was heat conduction of stainless steel pipeline and vapor, which were little so as to "block" the heat transfer.

For long-term operation of the Chang'E-3 lunar lander, the two-phase fluid loop did not operate in launch, LTO, circumlunar, landing and lunar daytime. During hibernation in lunar night, the two-phase fluid loop started up to transfer heat of the RHUs into modules.

Gravity conditions affected the driving ability of the two-phase fluid loop and the flow pattern of the fluid. A technical challenge in development of the two-phase fluid loop was that the heat transfer performance under 1/6 g lunar gravity could not be verified directly on Earth gravity. It was necessary to verify through simulation analysis and equivalent simulation.

5.6 Testing and Verification

There were two phases in design and verification of thermal control subsystem of the Chang'E-3 lunar lander including system verification and key technology verification. The focus of system verification phase was thermal balance test of the lunar lander. The focus of key technology verification phase was verification of new technologies such as VCHP, high-temperature thermal shield and two-phase fluid loop.

5.6.1 Thermal Balance Test of Lunar Lander

The objective of the thermal balance test was to verify design of thermal control subsystem actual performance of thermal control products by experiments. The implementation method was to simulate the environmental conditions such as vacuum, low temperature and typical external heat flow in each phases of the lunar lander in space environment simulator so as to verify whether the temperature of the lunar lander components met the requirements under different conditions.

In different phases of development, thermal balance test of probe was generally classified into thermal balance test in prototype phase and thermal balance test in flight model phase, where the difference was the state of test articles. In thermal balance test in prototype phase, the test articles were generally analog articles with same external dimensions, surface conditions, heat capacity, heat consumption and operation mode as instruments prototypes. In thermal balance tests of flight model, the test articles are generally flight models. Considering the capabilities of flight models such as pressure and heat sink temperature, safety and other special requirements, analog articles with same thermal parameters could be used to simulate individual flight models. But the analog articles should adapt to test conditions such as temperature, pressure, and duration in thermal balance test.

1. Test Objectives

The major objectives of thermal balance test were as follows:

(1) To obtain temperature distribution data of the lunar lander, and verify the ability of thermal control subsystem to maintain temperature of instruments and structure within specification.
(2) To provide ground test data for correction and improvement of thermal analysis model.
(3) To obtain temperature distribution of instruments in special phases such as landing, hibernation and awakening, and support for formulation of flight control strategies.

For the thermal balance test of flight model, the obtained temperature data could also provide basis for determination of temperature in thermal vacuum test of the lunar lander.

2. Simulation of Space Thermal Environment

The space environment simulator is used to simulate the vacuum and low background temperature environment of spacecraft in orbit. Since the effect of heat convection and conduction of gas is almost negligible when the pressure is less than 10^{-2} Pa, the pressure shall not be higher than 1×10^{-3} Pa in the space environment simulator during thermal balance test.

Space background heat sink (black body radiation of about 4 K) can be simulated by liquid nitrogen heat sink in a space environment simulator. The surface of heat sink is usually black painted with infrared emissivity with ε_H greater than 0.9. The heat sink temperature shall not be higher than 100 K during thermal balance test. For a 100 K heat sink, the radiation energy intensity during the test is approximately 5 W/m^2, equivalent to 2.3% of the Earth's infrared radiation heat flux (220 W/m^2). The error will not significantly affect test result. Liquid nitrogen heat sink is acceptable in engineering implementation as other spacecraft. Compared with liquid helium and other heat sink, the liquid nitrogen heat sink can simplify the simulation and reduce the cost greatly.

After the lunar lander was landed on lunar surface, the temperature of components such as landing legs would be affected by thermal conductivity of lunar surface and the temperature of the lunar lander would be affected by lunar surface infrared radiation and reflected of solar heat flow. In the thermal balance test, the radiation of lunar surface and the reflected heat flow were generally equivalent to external heat flow, and the thermal effect of lunar surface heat conduction was simulated by the surface temperature contacted with lunar surface. Thus, it was possible to simulate lunar surface environment with same method as for space thermal environment in orbit.

3. Simulation of External Heat Flow

The external heat flow of the lunar lander included solar radiation, lunar infrared radiation and lunar albedo. The simulation accuracy of external heat flow directly affected thermal balance temperature level of the lunar lander. The simulation method of external heat flow should be determined in design of thermal balance test.

The simulation methods of external heat flow included non-contact type and contact type up to date. There were solar simulator, infrared heater (such as infrared array, far-infrared cage, heating plate and etc.) in non-contact simulation methods of external heat flow to simulate incident heat flow onto surface of the lunar lander. The spectrum and irradiance of the solar simulator should be measured in advance to meet the requirements when it was used to simulate solar radiation heat flow directly. Generally, a heat flow measurement device with same state as surface of the lunar lander was necessary when other non-contact methods are used to simulate the total incident external heat flow. Except solar simulator, the response time of other non-contact simulation systems of external heat flow were generally longer and suitable for steady state test conditions. When the solar simulator was used, a motion simulator was necessary to simulate the change of relative position and attitude between solar radiation and the lunar lander in orbit, which might make the test system relatively complicated. In contact-type simulation of external heat flow, flexible film-type electric heater was installed on surface of the lunar lander to simulate heat absorption flow on surface of the lunar lander. Due to fast response of the electric heating method, the contact-type simulation method of external heat flow was suitable for transient and quasi-transient test conditions.

Generally, the attitude of the lunar lander was solar oriented during LTO and circumlunar phase. In LTO, the external heat flow was a stable solar radiation heat flow, while it was periodic alternation of solar radiation, lunar infrared radiation and reflected heat flow in circumlunar phase. After the lunar lander was landed on lunar surface, the solar radiation, lunar infrared radiation and reflected heat flow varied with solar elevation. However, due to long lunar daytime (approximate 14 days), the change rate of solar elevation was relatively small (average 0.5°/h on lunar equator and about 0.3°/h at 45° latitude). Therefore, it could be assumed that the temperature of the lunar lander could reach balance at each time. Because the time of LOI and landing process was shorter (from several minutes to about ten minutes), the external heat flow was transient.

The simulation method of external heat flow should be the methods suitable for transient conditions such as solar simulator and contact type method after comprehensive consideration of current status of simulation technology of external heat flow, variation of external heat flow of the lunar lander and determination of test conditions (focuses are verification in circumlunar phase, landing process and operation on lunar surface). According to actual external heat flow of different surfaces of the lunar lander, infrared heaters (infrared lamp array or far infrared cage) could be used to simulate the external heat flux on the surface of small change, but the simulation method of external heat flux should be optimized according to actual working conditions to minimize error.

The infrared radiation heat flow might also be the largest on the surfaces of the lunar lander toward Sun on lunar surface. The combination of two external heat flow caused a large external heat flow above 1200 W/m^2 on some surfaces of the lunar lander. When infrared heaters (infrared lamp array or far infrared cage) were used to simulate external heat flow on such surfaces, the proportion of those simulators' area was large. When the temperature was low, the temperature would be higher because external heat flow was large than actual condition. In order to solve this problem, simulators with different area ratios could be used to simulate external heat flow before and after landing respectively, or a mobile infrared heater could be used which was removed at low temperature conditions.

4. Test Conditions

The mission profile of the lunar lander was complex. The different launch windows would also cause great change of environment conditions. Uncertainty of landing attitude resulted in numerous theoretical combinations of landing conditions on lunar surface. Therefore, it was not feasible to simulate all of actual conditions of the lunar lander in ground tests and only a limited number of typical conditions could be selected for testing. The typical test conditions for thermal balance tests of the lunar lander are as follows:

(1) The high temperature condition and low temperature condition should cover the whole process of the mission, i.e., the highest temperature condition and the lowest temperature condition should cover all phases of the mission. Since lunar dust during landing and after landing might have great effect on thermal

physical properties of the lunar lander surfaces resulting in temperature raise, the effect of lunar dust on the test conditions should be considered. In addition, because the distribution of external heat flow in circumlunar phase and operation on lunar surface was different, the high temperature conditions should be considered separately. The major external heat flow was solar radiation heat flow in LTO and there were few of instruments on operation at cruise attitude, therefore the whole the lunar lander probe was generally at low temperature. In a circumlunar orbit period, there were generally conditions including in illumination and in lunar shadow during different time. For exposed components and some instruments in modules, high temperature conditions and low temperature conditions might occur in one orbital period. Thus, there should be high temperature conditions and low temperature conditions in circumlunar phase.

(2) The transient conditions should generally be verified in the test. For a short-time transient process such as landing, the operation mode of the lunar lander was quite different from other phases, while the external heat flow varied greatly, which caused the temperature of instruments affected by transient characteristics greatly. In order to obtain the temperature change of instruments accurately and provide more detailed data for correction of thermal analysis model, it generally should be verified directly.

(3) Special test conditions should be set for requirements of thermal analysis and calculation. Due to limited number of thermal test conditions, a large number of thermal analysis and calculations should be conducted to predict the temperature in all mission phases for the lunar lander. Therefore, it was very important to improve the accuracy of thermal analysis model. Generally, relatively simple external heat flow conditions or other on-orbit conditions easy to be simulated accurately should be selected for verification of thermal analysis model. Or it could also be included in other test conditions of high and low temperature. Test conditions such as LTO phase or noon on lunar surface were suitable for verification of thermal analysis model.

(4) Test conditions including lunar night and wakeup should be set to verify the lowest temperature of the lunar lander in lunar night and the temperature changes in wakeup process. The lunar night was the most typical environment condition for long-term lunar land which was harsh. The lunar night condition should be set to verify the temperature level and distribution of the lunar lander. Since the temperature changes in wakeup process was related to the temperature in lunar night, the wakeup condition was generally set after the lunar night condition. Since the time of lunar night was as long as 14 days and the temperature was generally dropping down in lunar night, the criterion of temperature balance in the lunar night condition should be stricter than usual.

5.6.2 VCHP Test

Generally, the function and performance of VCHP were verified in heat balance test and mechanical test of the lunar lander. However, due to factors such as limited operation modes of instruments, it was inconvenient to conduct quantitative assessment of heat transfer performance and operation characteristics in the test. Therefore, it was generally necessary to conduct special tests for validation during key technology validation phase.

1. Test Objectives

The major test objectives of vacuum performance of VCHP during key technology verification phase are as follows:

(1) To measure the heat transfer capacity of VCHP during operation and verify the heat transfer "blocking" capability at low temperature during lunar night.
(2) To study the effect of different heat source distribution in evaporation section on the work characteristics and start-up process of the heat pipe;
(3) To study the working characteristics of VCHP during the wake-up process and hibernation process.

2. Status of Test Article and Working Conditions

The shape of pipeline and the volume of gas tank should be considered for test article of VCHP. Since the bending of pipeline will increase the flow resistance of fluid, the heat transfer capacity could be reduced. It was impossible to test each type of heat pipe. Therefore, the product with the most unfavorable working performance should generally be selected for testing to ensure that the test result covered all products.

Since the condensing section of VCHP was generally embedded in the heat-dissipating surface, there should be a piece of corresponding heat-dissipating surface on test article to simulate the actual situation. The area of the simulated heat-dissipating surface should be the same as the actual product. Generally, the electric heater pasted on the body of heat pipe directly was selected for instrument on the evaporation section. The size and configuration of the heater should simulate the actual instrument. The heat consumption and operation modes of test article should also be the same as the actual situation.

The effect of landing attitude on performance of VCHP should be considered when the status of test article was determined. Generally, the test should be conducted on the most unfavorable landing attitude.

During the test, the temperature of the condensation section was used to determine the location of the vapor-control gas interface. The temperature measurement point distribution should consider the possible location of the interface. In order to obtain accurate temperature distribution, more temperature measurement points should be arranged in the condensation section, which could provide sufficient data for determination of vapor-control gas interface, calculation of heat dissipation capacity and correction of thermal analysis models. In addition, the heat leak test in lunar night was also an important objective of the test and the temperature measurement points should be managed easy to calculate the heat leakage. Due to the large number of temperature measuring points, thermocouples and other sensors with a small wire diameter should be used to reduce the effect of temperature leakage of the temperature measuring units.

The test conditions of VCHP should be determined according to the type of heat pipe and test objectives. In order to study the performance of the heat pipe under different conditions, special tests could be conducted to test the sensitivity of some parameters.

5.6.3 Vacuum Thermal Insulation Testing of High-Temperature Heat Shield

The high-temperature thermal shield was designed to protect structure and instruments of the lunar lander during when the landing engine was firing. Because of the large size of the engine and the large amount of plume generated during firing, the thermal insulation performance of the high-temperature heat shield could not be verified directly with the actual engine firing. A test article of the high-temperature heat shield with certain area was selected to conduct heat insulation performance test alone.

1. Test Objectives

The major test objectives of the vacuum thermal insulation performance test of the high temperature heat shield are as follows:

(1) To validate the design of high-temperature heat shield;
(2) To obtain the temperature distribution inside the heat shield when the engine was firing and to provide test data for correction of the thermal analysis model.

2. Simulation Method of High Temperature Boundary

Because the performance of high-temperature heat shield could not be directly tested by the engine firing and the boundary temperature was high (the highest temperature of the nozzle was higher than 1300 °C), the challenge in the performance test of high-temperature thermal shield was to accurately simulate the high temperature boundary of the engine when the pressure was lower than 10^{-2} Pa.

When the engine was ignited, the temperature of the nozzle would be stable after a few seconds. Then after several minutes to over ten minutes of stable firing of the engine, the temperature of the surface on the high-temperature thermal shield which was opposite to the engine nozzle was generally stable in 60 s to 100 s with the highest temperature 1000 °C. Two heating methods including high-power quartz lamp and wind tunnel could achieve such a high temperature in a short period of time up to date. It was longer (about tens of seconds) for the high-power quartz lamp to heat to the specific temperature. However, the system of high-power quartz lamp was relatively simple and easy to be integrated with available space environment simulator to achieve vacuum condition with lower cost. The wind tunnel heating method could simulate the temperature response of the actual operation of the engine, but the test system was complicated and it was difficult to maintain vacuum condition near the test article. Furthermore, the duration of test was limited and the cost was high. After various factors were considered comprehensively, it was more reasonable and feasible to use a high-power quartz lamp or other similar electric heating source to simulate the high temperature boundary of the engine.

The heat flux obtained on the surface of heat shield (second type boundary condition) could be simulated in the heat insulation performance test. Because the high temperature of the engine resulted in rapid temperature raise of the heat shield surface and the temperature was stable during the engine firing, as well as the main objective of the test was to examine the overall thermal insulation properties of heat shield, it was also possible to simulate the surface temperature of heat shield (first type boundary condition). For the quartz lamp heating method, because the response time was slower than the wind tunnel, it was suitable for simulating the temperature response process of heat shield surface. In addition, the temperature measurement was simpler and more accurate than measurement of high temperature heat flux. Based on the above analysis, it was more feasible to simulate the surface temperature condition of heat shield in the vacuum performance test with quartz lamp heating.

3. Status of Test Article and Working Conditions

The composition of test article for the high-temperature heat shield should be consistent with the design status. For the parts with different composition, special test pieces should be made separately. In order to examine the influence of joint and fastened methods on temperature, special test pieces should be manufactured for testing. The joint status and fastened method should be consistent with the design status. High-temperature heat shield was generally symmetrically configured to the engine. Therefore, the surface temperature conditions were uniform. At the same time, the lateral heat transfer of heat shield was poor and the edge effect was not obvious. So, the temperature distribution and temperature response in the thickness direction could be accurately obtained with a mall area. The specific size of the test articles should be determined according to the size of effective heating area of quartz lamp, so as to avoid the accuracy of test results affected by non-uniform heat flux.

In order to obtain temperature distribution in the thickness direction, a plurality of temperature measurement points should be arranged in the thickness direction, wherein a temperature measurement point should be arranged on the surface facing the heat source for controlling the test temperature. The temperature sensor should adapt to the highest temperature during the test, and in order to reduce the heat leakage, the temperature sensor with wire of small diameter should be chosen generally. After the quartz lamp works, the temperature of the test piece would rise rapidly, and the sampling frequency of temperature measurement system should be higher than 1 Hz.

In order to simulate the pressure on orbit, the test system should be placed in space environment simulator (as shown in Fig. 5.18). Before the test, in order to empty the

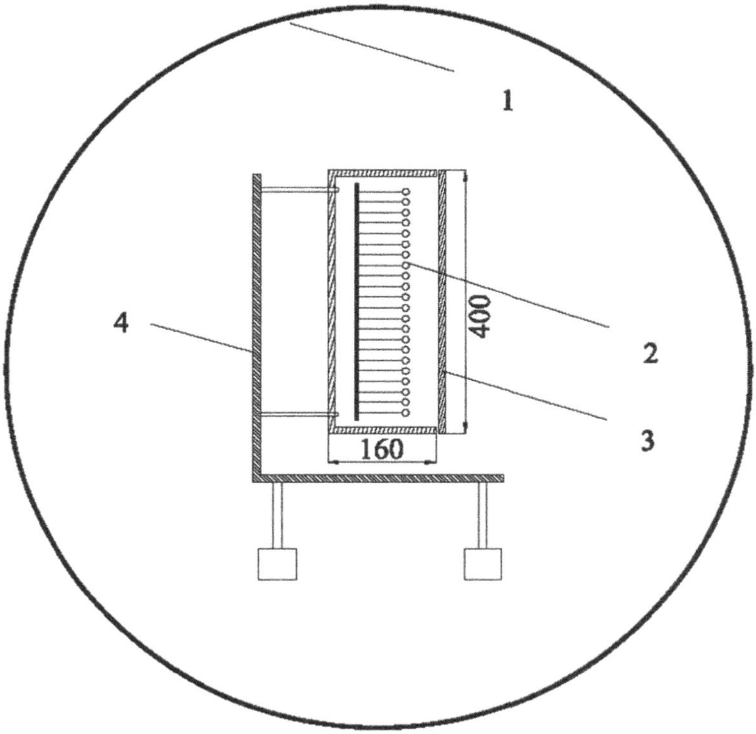

Fig. 5.18 Diagram of the test article of heat shield in space environment simulator. **1** Heat sink; **2** Infrared lamp; **3** Test article; **4** Support

residual gas inside the test article, the internal pressure of the vacuum system should be reduced to the order of 10^{-3} Pa for a certain period of time. Because of the high temperature outgassing of the heat shield material, the pressure of the space environment simulator would increase during the test, which was relevant to volume of space environment simulator and capacity of vacuum system.

Differences in the operating conditions of the engine might result in deviations in the nozzle temperature. In order to cover the possible temperature conditions on the orbit, the surface temperature of the heat shield should be biased in the test. In addition, the temperature changes in different parts of the heat shield were related to operation time of the engine. It should be considered that the operation time of the engine should be biased in the test. A set of typical temperature response curves for vacuum performance test of the high temperature heat shield of the Chang'E-3 lunar lander is shown in Fig. 5.19.

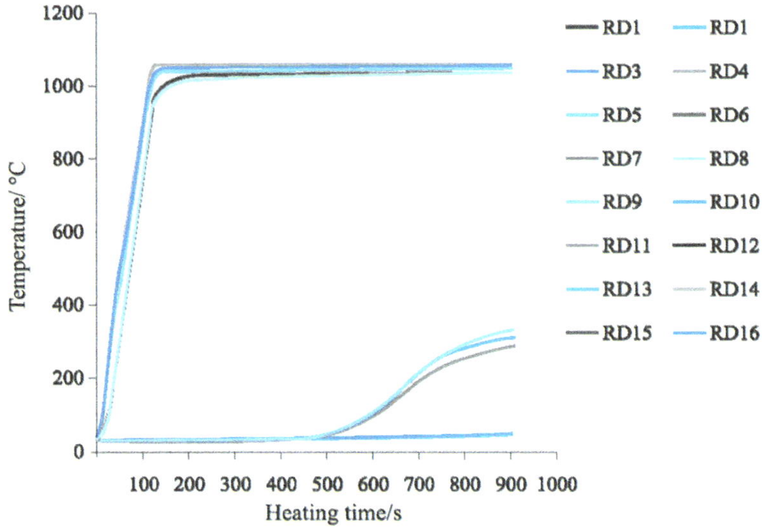

Fig. 5.19 Typical temperature response curves for vacuum performance test of high temperature heat shield of the Chang'E-3 lunar lander

5.6.4 *Two-Phase Fluid Loop Test*

The two-phase fluid loop introduced heat of RHUs into the lunar lander during lunar night, whose heat transfer capability was a key parameter. During development of the lunar lander, operation characteristics and mechanical environment adaptability of the two-phase fluid loop were verified in thermal balance test and mechanical test of the lunar lander. The focuses were the heat transfer capability test under different temperature conditions and the data obtained were also utilized in correction of the heat transfer model of the two phase fluid loop when the key technology was developed.

1. Test Objectives

The heat transfer performance tests of the two phase fluid loop were conducted when the key technology was developed and verified. The major objectives are as follows:

(1) To examine thermal resistance and heat transfer capacity of the two-phase fluid loop under different temperature levels;
(2) To provide experimental data for correction of theoretical model of the two-phase fluid loop.

2. Simulation Method of Driving Force

Heat transfer capacity of the two-phase fluid loop was up to the driving force and flow resistance. When the lunar lander was landed on the Moon, the driving force was the lunar surface gravity (average gravitational acceleration 1.62 m/s^2), while it was Earth gravitation (average gravitational acceleration 9.81 m/s^2) in ground tests. The lunar gravity was about 1/6 of the Earth. The flow resistance of the two-phase fluid loop was not only related to flow velocity, but also related to flow pattern of the fluid during operation of the two-phase fluid loop. Compared to 1 g Earth gravity, there were differences under 1/6 g lunar gravity in weakened or elimination of the asymmetrical and stratified flow caused by the density difference in two-phase flow, mechanism of flow pattern generation and transformation, phase distribution, pressure drop and heat transfer characteristics. It was difficult to simulate a stable 1/6 g gravity condition on Earth. The difficulty in ground tests of the two-phase fluid loop was to equivalently simulate the driving force and resistance under lunar gravity.

There were little flow characteristics and empirical parameters of two-phase flow under different gravity conditions for reference in China or other countries. When the two-phase fluid loop was developed, envelop processing methods were generally used. Ground tests conditions should consider various influencing factors and cover the worst combination of driving force and flow resistance that might occur on lunar surface, i.e., the most unfavorable conditions for heat transfer should be covered.

The rationale of ground simulation method for driving force was that the driving force in all operation conditions of ground tests should not be greater than the actual driving force on lunar surface so as to ensure that the ground verification was not under-tested. Nevertheless, the over-conservative simulation of driving force that might cause failure of ground tests should be avoided. The diagram of driving force simulation of the two-phase fluid loop with consideration of difference and change of the liquid level is shown in Fig. 5.20, where the cross-sectional area of fluid tank was 6 times of actual tank and reference height was 1/6 of actual reference height on lunar surface [17].

3. Status of Test Article

The actual resistance of the two-phase fluid loop was related to the length of each part of the pipeline, the number of joints, the bending form, and etc. According to the ground simulation method of driving force and resistance, the ground test article was inconsistent with the design status besides adjustment of fluid tank and reference height with increasing adaptive bend.

The start-up operation characteristics of the two-phase fluid loop were related to the heat capacity and contact method of the heat source. The heat capacity and installation method of the heat source in ground tests should be the same as the actual heat source. There might be obvious temperature oscillation in the two-phase heat transfer. Therefore, more temperature measurement points with high sampling frequency should be arranged at the possible interfaces such as the condenser. The heat transfer capability under different working temperatures should be examined in tests. The electric heating mode should be adopted to conveniently control the temperature of condensing plates. A set of test curves for heat transfer capability of the two-phase fluid loop of the Chang'E-3 lunar lander at −50 °C is shown in Fig. 5.21 [17].

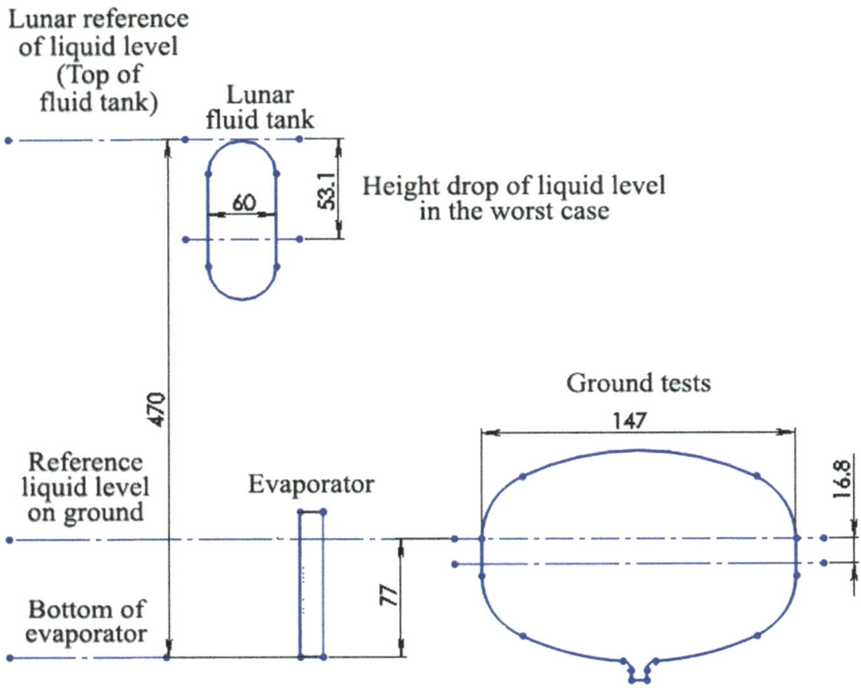

Fig. 5.20 Diagram of driving force simulation

5.6.5 Flight Testing on Orbit

On December 2, 2013, the Chang'E-3 lunar lander was launched in the Xichang Satellite Launch Center. After LTO and circumlunar flight, the probe was landed on lunar surface and the lunar lander was successfully separated with the lander. During the entire flight and separation process, the thermal control subsystem of the lunar lander worked normally. The characteristic of VCHP was consistent with the expectation (the temperature change curve in circumlunar phase is shown in Fig. 5.22). The high-temperature heat shield effectively protected the safety of structure and instruments of the lunar lander during LOI and landing process (the temperature change curve of the heat shield itself is shown in Fig. 5.23). The lunar lander had been working on lunar surface for several lunar days. It had passed multiple hibernation, awakening and lunar eclipses. After wakened up, the instruments of the lunar lander were operated normally, which proved that the two-phase fluid loop, VCHPs, RHUs and the thermal control subsystem worked normally. The temperature change of the two-phase fluid loop during hibernation is shown in Fig. 5.24. The temperature of instruments and structure in all phases met the requirements. During the lunar daytime, scientific payloads such as lunar observatory were operated normally

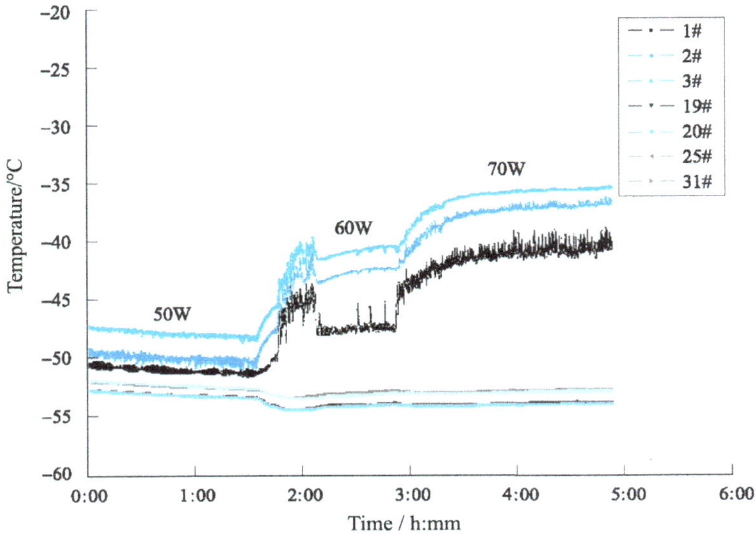

Fig. 5.21 A set of test curves for heat transfer capability of two-phase fluid loop of the Chang'E-3 lunar lander at −50 °C

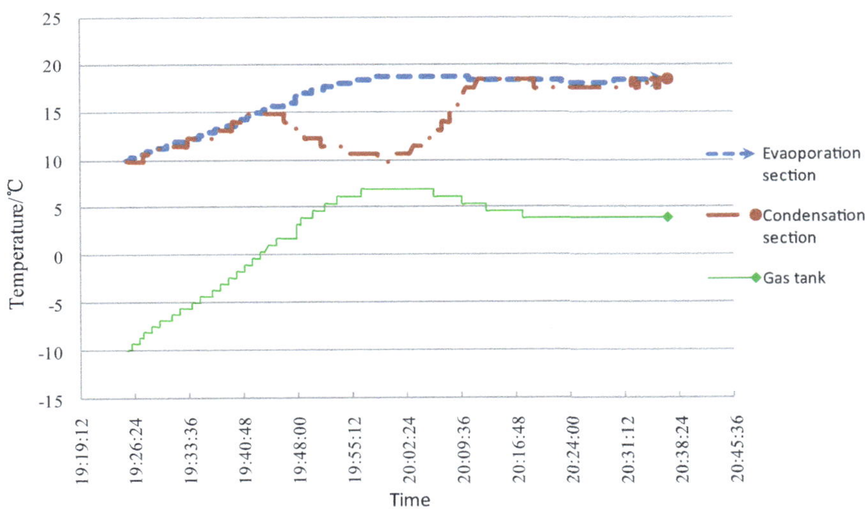

Fig. 5.22 Temperature change curve of VCHP of the lunar lander in circumlunar phase

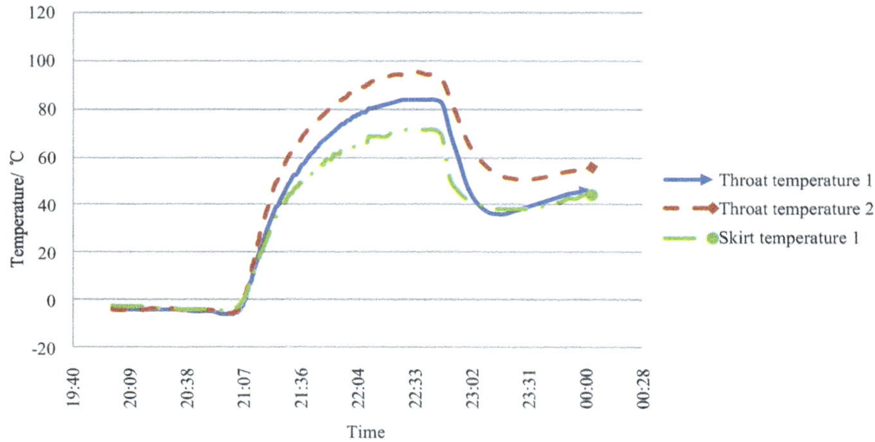

Fig. 5.23 Temperature of inner surface of heat shield during landing process

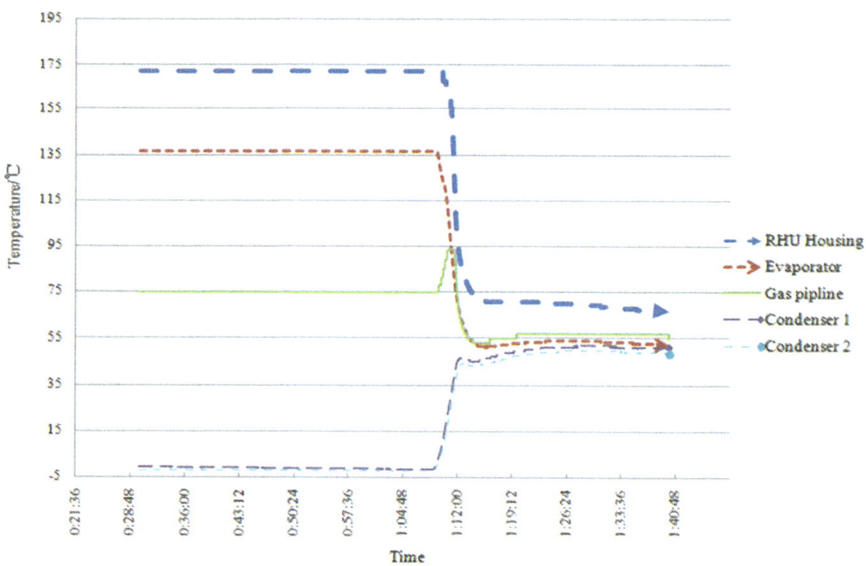

Fig. 5.24 Temperature of two-phase fluid loop in +Y module in start-up process during hibernation of the lunar lander

and a great many effective scientific data was obtained. On April 15, 2014, the lunar lander passed the first lunar eclipse. During the lunar eclipse, the temperature of instruments and two-phase fluid loops met the design requirements. The typical temperature changes of the working equipment and fluid loop are shown in Figs. 5.25 and 5.26, where the lowest temperature can be seen before exit of the umbra.

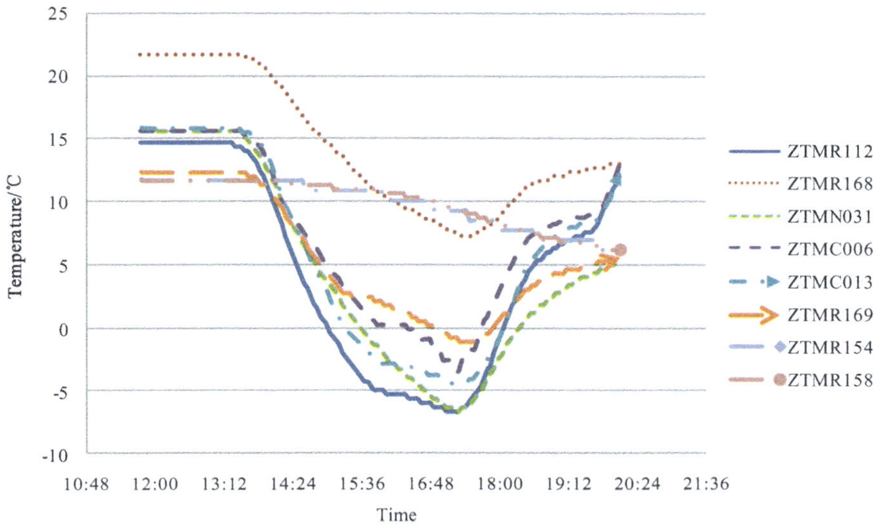

Fig. 5.25 Temperature of operation instruments during lunar eclipse

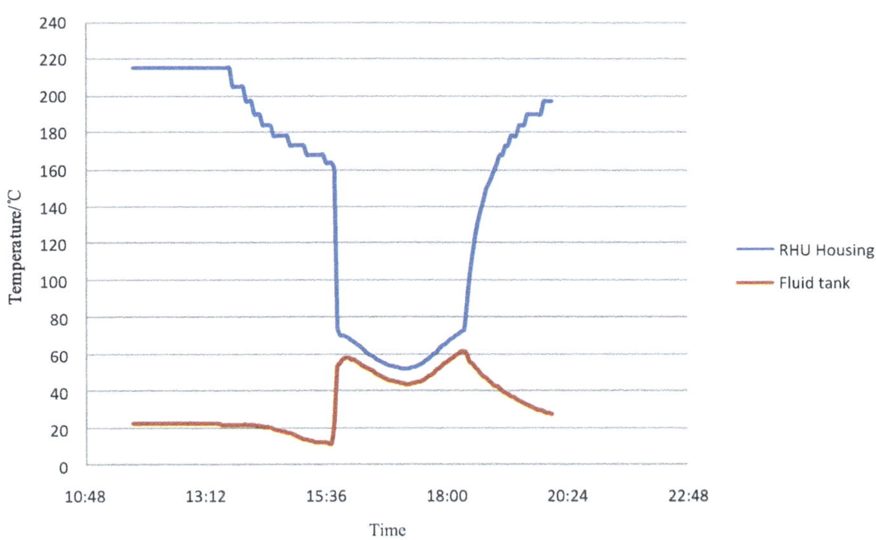

Fig. 5.26 Temperature of RHU and fluid tank of the fluid loop in +Y module during lunar eclipse

During landing process and operation on lunar surface, a great many temperature data and changes of key products such as VCHP, high-temperature heat shield, two-phase fluid loops, RHUs and different parts of the lunar lander were obtained. With these data, the rules and actual effects of environmental conditions such as engine plume, lunar dust deposition and 1/6 g of lunar gravity on lunar surface could be analyzed. These data and analysis results were important for further exploration of the Moon and other celestial bodies.

5.7 Summary

Compared with spacecraft on Earth and circumlunar orbit, the lunar lander should pass special environment on lunar surface. There were technical challenges in development of thermal control including design principles of thermal control, thermal analysis and calculation, thermal balance test and new thermal materials. On the basis of successful experiences of spacecraft on Earth and circumlunar orbit, it was necessary to study innovative technology and products such as VCHP, high temperature protection of the engine, application of RHU, controlled fluid loop to develop design and ground test method of thermal control. In this chapter, the design constraints, design principles and methods, typical thermal control technology, ground test methods and flight results on orbit of thermal control of the Chang'E-3 lunar lander are described. The thermal control technology was verified through the development and verification on orbit of the Chang'E-3 lunar lander.

According to the strategic plan of CLEP, robotic sample and return mission was under development and subsequent manned lunar missions and lunar base construction would be implemented. Further study shall be conducted including high-power and adaptable fluid loop, VCHP, loop heat pipe, new thermal radiator and lunar dust prevention and etc., which can provide necessary support for further lunar exploration missions and other space exploration missions.

References

1. Ouyang Z (2005) Introduction to lunar science. China Astronautic Publishing House, Beijing
2. Liu Z, Xiang Y, Si D, et al (2014) Design and verification of thermal control system for Chang'E-3 probe. Sci Sin Tech 44:353–360
3. Gaier JR, Siamidis J, Panko SR, et al (2008) The effect of simulated lunar dust on the absorptivity, emissivity, and operating temperature on AZ-93 and Ag/FEP thermal control surfaces. NASA/TM-2008-215492
4. Agui JH, Nakagawa M (2005) Dust on the Moon and Mars. AIAA 2005-539
5. Gaier JR, Jaworske DA (2007) Lunar dust on heat rejection system surfaces: problems and prospects. NASA/TM-2007-214814
6. Stephan RA (2011) Overview of the Altair lunar lander thermal control system design and the impacts of global access. AIAA 2011–5001

7. Hanford AJ, Ewert MK (1997) Advanced active thermal control systems architecture study. NASA TM 104822
8. McGhee JR (1993) Lunar base thermal management/power system analysis and design. NASA N93–27985
9. Peter E (1999) The lunar base handbook. The McGraw-Hill Companies Inc, New York
10. Gillen RJ, Brady JC, Collier F (1972) Apollo experience report-lunar module environmental control subsystem. NASA TN D-6724
11. Wayne Tobery E, Pantano DR (2006) Heat rejection concepts for radioisotope power systems on the lunar surface. AIAA-2006-4200
12. Harvey B (2006) Soviet and Russian lunar exploration. The Springer-Verlag, New York Inc., New York
13. Chu G, Zhang H (2007) Technology of lunar probes. China Science and Technology Press, Beijing
14. Ming G, Guo S (1998) Spacecraft thermal control, 2nd edn. Science Press, Beijing
15. Ming G (2005) Spacecraft thermal control technology. China Astronautic Publishing House, Beijing
16. He Z (1988) New thermal control material, device and its application. China Astronautic Publishing House, Beijing
17. Zhang H, Miao J, Wang L, et al (2014) Ground test method and results of closed two-phase thermosyphons for the moon exploration spacecraft Chang'E-3. Sci Sin Tech 44:589–596

Chapter 6
Power Technology of Lunar Lander

6.1 Introduction

The power system of the lunar lander was responsible for generation, storage and controlling of electrical power [1]. During launch, LTO, circumlunar and operation on lunar surface, the electric power was provided continuously and stably by the power system.

In the Sun light period, the power system generated electricity on photovoltaic effect of solar cells to provide electrical power for the lunar lander and charge battery pack. At the same time, the excess electrical power generated by solar cells was shunted and converted into thermal energy for heat dissipation. In the shadow period, the battery pack discharged to the main bus to power the lunar lander through regulation and control. The power system controlled and regulated power source, managed battery charge and discharge according to the predefined design, and provided stable, continuous and high quality power to the lunar lander during flight.

Lunar environment is extraordinary. How could the lunar lander survive through 14 days lunar night? How could the power system ensure the power supply requirement in lunar night? Could the power system withstand the -180 °C low temperature and $+120$ °C high temperature on lunar surface? How could the solar cells sustain high power output in case of a wide range of solar elevation on lunar surface? How could the reliability and safety of the power supply be ensured in flight process, especially in orbit changes and powered descent? How could lunar dust effects on power generation of solar cells be dealt with?All above issues should be considered and analyzed in design of the power system.

D.-Y. Yu et al., *Technology of Lunar Soft Lander*, Space Technology Library 38,
https://doi.org/10.1007/978-981-15-6580-9_6

6.2 Status of Power Supply Technology Development

The United States and the former the Soviet Union had implemented many lunar soft-landing exploration missions. Based on the analysis of lunar landing missions in different countries, the combination of "solar array + battery" was the option in the power system design for most of these missions except fuel battery in early Apollo missions. Solar array converted light energy into electricity to power the probes, while excess energy was converted into chemical energy for storage and the chemical energy was converted into electrical power for joint power supply if necessary.

In terms of space solar cells, there were several types such as silicon (Si) cells, single-junction gallium arsenide (GaAs/Ge) cells, and triple junction gallium arsenide (GaInP/GaAs/Ge) cells [2]. Up to date, silicon solar cell was the most popular one. Along with development of fundamental technologies, the photoelectric conversion efficiency of solar cell has been gradually increased from 14% (silicon solar cell) to 30% (triple junction gallium arsenide cells). With successful research & development and space applications of gallium arsenide solar cell, gallium arsenide solar cell with higher photoelectric conversion efficiency will be gradually applied in the future.

There were several types of battery such as nickel-cadmium (Ni/Cd) battery, nickel-hydroxide (Ni/H2) battery and lithium-ion (Li-ion) battery [3]. The specific energy of the battery had been gradually increased from 50 W•h/kg (nickel-cadmium battery) to 150 W•h/kg (lithium-ion battery), while the specific energy of hydrogen-nickel battery was in between. Lithium-ion battery had the advantages of high specific energy, low heat consumption, and low self-discharge [4]. Now lithium-ion battery had been gradually applied to deep space exploration.

The Surveyor series missions were robotic the lunar landers launched by the United States. There were total 7 missions launched, of which 5 were successful. The spacecraft buses of these 7 the lunar landers were exactly the same, whose power system was a combination of solar array and zinc-silver battery with one fully adjustable main bus and one unregulated main bus. The area of solar panels was 0.83 m^2, and power output was 81 W. The battery pack connected by 14 units could output power 3375Wh.

Euro2000 was a lunar landing exploration mission planned by Europe. It consisted of the lunar lander and a circumlunar spacecraft. It was originally planned to be landed on lunar south pole. The power system also used a combination of solar cell and battery. The solar array of the lunar lander consisted of a fixed plate and a deployable plate. The area of fixed plate was 2.5 m^2, and the area of deployable plate was 2.0 m^2. The solar cells were GaAs solar cells with power density of 207 W/m^2. In addition, there was also a group of 16 kg NiH_2 battery packs with a specific energy of 60 W•h/kg which could provide 300 W power supply.

Apollo manned spacecraft consisted of service module, command module and lunar module. After launched, the spacecraft was inserted into LTO. When the spacecraft approached the Moon, the lunar module was separated. The operation on lunar surface lasted for one day. After life-off on lunar surface, the ascent stage of lunar module rendezvoused and docked with service and command modules. Then service and command modules returned to Earth. When approaching Earth, the command module with astronauts was separated and landed on Earth. The lunar module had independent power supply requirement during LTO, landing, operation on lunar surface, lift-off on lunar surface and docking with command module. The command module had independent power supply requirement after separated with service module and during return to Earth. The lunar module and command module would not work independently for a long-term and requirement for electric power was not large. The flight time of Apollo spacecraft with service module, command module and lunar module was about 14 days. The average power consumption was 2 kW and the total energy consumption was 672 kW•h. There was no solar photovoltaic system in Apollo spacecraft. Instead, battery packs were used. During LTO, the spacecraft was not influenced by the sun light. At the same time, the air drag of very thin atmosphere could be avoided because there was no large area solar array on the spacecraft so as to save propellant for engine firing. There were two main buses in the spacecraft. The two main buses could be powered independently or jointly. The three fuel cells could be flexibly combined to supply power to two main buses. The secondary load on the spacecraft was powered by a secondary main bus. The secondary load could be shut off by disconnecting the secondary main bus from the main buses. Such power distribution method could effectively isolate fault battery and enhance the power distribution capability. The power system design of the Apollo spacecraft is determined by the electrical energy required for each mission segment. The type of battery and its capacity were different in different mission phases. The composition of power system is listed as follows:

(1) Service module. There were three 100 kg fuel cells to provide total power 570 kW•h and a cadmium-nickel battery pack could assist power supply for peak power.
(2) Command module. There were three rechargeable cadmium-nickel battery packs with a total capacity of 3.5 kW•h to power command module when it returned to Earth. If necessary, it could be connected with the service module battery for joint work and be charged by the fuel cells.
(3) Lunar Module: There were two zinc-silver storage batteries with a total power of 8.4 kW•h to power lunar module when it left the Moon. There were four zinc-silver batteries with a total energy of 46 kW•h for landing and operation on lunar surface.

The combination of silicon solar array and nickel-hydraulic battery were used in power systems of the Chang'E-1 [5] and the Chang'E-2 adopt. The peak output power was higher than 2 kW. The triple junction GaAs solar array and lithium ion battery were use in power system of the Chang'E-3 lunar lander, where the capacity of lithium ion battery was 60Ah and peak output was less than 2 kW.

6.3 Mission Requirements and Characteristics

Compared with conventional spacecraft, the orbit of the lunar lander was complex with harsh environment and stringent mass constraints [6]. After landed on the moon, the lunar lander would experience near 300 °C temperature difference between lunar daytime and night. In addition, the effect of lunar dust on solar cell performance should be considered. At the same time, it was a challenge to provide power during the lunar night.

6.3.1 Functional Requirements

The major functions of the lunar lander power system included power generation, energy storage, power supply (including battery charging and discharging), autonomous hibernation and awakening control.

1. Power Generation and Storage

During all phases of the lunar lander, when the solar cells were exposed to the sun, the solar cells converted solar energy into electrical energy to power the lunar lander and stored energy for the battery packs to meet power requirement of the lunar lander.

During all phases, when the solar cells did not work (in the shadow, solar panels unfolded, orbit change, powered descent and lunar night) or the energy was insufficient, the lunar lander was powered by the battery packs to meet power requirement.

2. Power Supply

Before launch, when the power system was switched to provide power by battery packs of the lunar lander, the power system would pass launch, LTO, circumlunar, powered descent and landing, operation on lunar surface and lunar night, when the power system performed battery charge and discharge control, and support power for all electric devices of the lunar lander. At the same time, the power system of the lunar lander would power the lunar rover and charge the battery of the lunar rover, as well as be powered by the battery of the lunar rover

3. Autonomous Hibernation and Awakening

During lunar night of 14 Earth days, the solar cells of the lunar lander could not output power without illumination. The energy requirement of the lunar lander in the minimum mode could not be satisfied by the battery packs. Therefore, the lunar lander should be able to be hibernated will all instruments shut off before lunar night and awakened autonomously to recover normal operation when the lunar daytime was coming.

6.3.2 *Performance Requirements*

In performance specifications of power system, the focuses were power output, main bus voltage, solar array area, photoelectric conversion efficiency, capacity and life of battery pack, and charging method.

1. Power Output

In all phases of the mission, the power output of power system of the lunar lander should be higher than power requirement of the load and there should be certain margin to deal with unexpected failures.

2. Main Bus Voltage

According to scale of spacecraft power, the main bus voltage could be divided into three categories as 29 V, 42 and 100 V. Because the power scale of the lunar lander was small, 29 V was suitable for main bus voltage.

3. Solar Array Area

The solar array area was closely related to configuration of the lunar lander. If solar array area was too large, there would be waste in mass and cost. If the solar array was too small, the power output of solar array would be insufficient, which might result in catastrophic consequences. The power generation efficiency of solar cell should also be considered when the solar array area was determined.

4. Photoelectric Conversion Efficiency

Photoelectric conversion efficiency was a key specification of solar cell, which directly determined power output of solar arrays.

5. Battery Pack Capacity

The capacity of the battery pack determined the lunar lander's ability to work in the shadow period. In crucial phases such as launch, orbit change, eclipse in circumlunar orbit and powered descent, the lunar lander was powered by the battery pack alone. The capacity of the battery pack must meet the power requirement during crucial phases and there should be certain margin to deal with accidents.

6. Battery Pack Life

The life of the battery pack was closely related to operation temperature, depth of discharge, and discharge rate. According to mission profile of the lunar lander, when the battery pack was operating at the maximum depth of discharge, the number of cycles should be greater than the requirement to ensure the reliability of power supply during nominal mission period.

7. Charging Method

The charging method should match the charging characteristics of the battery pack. Usually the charging method of constant current-constant voltage was used for lithium-ion battery packs, which could complete the battery charging in a short time and ensure a high charging efficiency. The shift point of constant current to constant voltage was very important.

6.3.3 Mission Characteristics and Analysis

There were three special features of power system of the lunar lander including flight profile and illumination, power supply requirement, lightweight and miniaturization requirements [7].

1. Flight Profile and Illumination

The flight profile had a great impact on the power system including the time from launch to deployment and orientation to the Sun of the solar array, time and attitude of orbit changes, operation modes of spacecraft platform and load in all flight phase, and etc. The flight orbit would affect the time of illumination and in shadow, which would determine the scale of the solar array and the battery pack. The flight attitude would affect incident angle of sunlight on the solar array and blocked area of the solar array by the spacecraft. All above factors would affect power output of the solar array.

Before landed on lunar surface, the lunar lander would pass flight phases such as launch, LTO, orbit changes on circumlunar orbit and powered descent. During such phases, the power load varied greatly. The long-term load was the platform load with low power. But the transient load was very large such as powered descent when the main engine was firing frequently. Therefore, the adjusting capability of the power system should be sufficient to ensure stable and reliable power supply of the lunar lander [6].

After the lunar lander was landed on lunar surface, the battery packs would be heavily involved in charging and discharging according to operation modes. During lunar night without illumination, the energy requirement of the lunar lander could not be supported by the battery packs. It was necessary to power off all instruments before lunar night and power on autonomously to recover to normal operation mode when the lunar day was coming.

The illumination of the lunar lander on lunar surface was very complex. For example, the illumination is repeated with a period of 29.53 days on the Sinus Iridum (45.6°N, 18.2°W). In one lunar daytime, the local solar elevation/azimuth angle is shown in Fig. 6.1.

2. Power Supply Requirements

(1) Power supply requirement on different status. Because the lunar lander was on different status in different flight phases, the power requirement for different status were different. For example, there was a need for charging and discharging between the lander and the rover of the Chang'E-3. The joint power supply control strategy and special control circuit should be designed to ensure the reliability and margin of power supply in different phases.
(2) Power requirement for different phases of the mission. The lunar lander would pass different flight phases such as launch, LTO, orbit change, circumlunar orbit, powered descent, initialization and operation on lunar surface with different operation modes. The power supply requirement for each phase were quite different. Therefore, specific design was necessary.

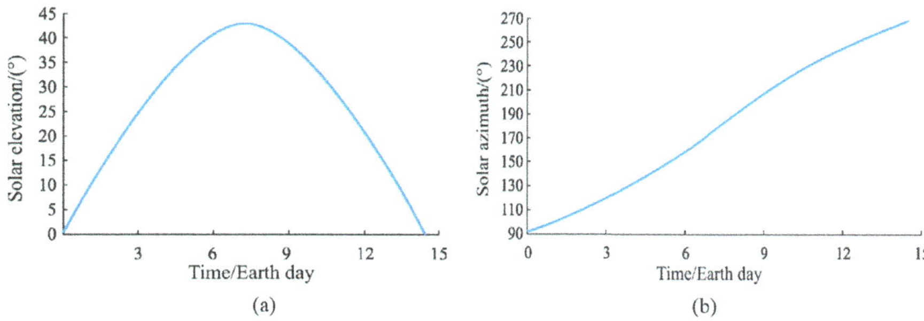

Fig. 6.1 Solar elevation/azimuth angle in one lunar daytime (**a**) Solar elevation angle (**b**) Solar azimuth angle

(3) Power supply interface requirement between multiple probes. The power supply interface of the lunar lander included the lander's internal power supply interface, interface between the lander and the rover, interface between the lander and launch vehicle, and the ground power supply interface. Due to power supply requirement, information transmission and ground test before launch, the design of the electrical interface including lander to rover, lander to launch vehicle and lander to ground was difficult under strict mass constraints.

3. Lightweight Requirement

For conventional spacecraft, the mass constraints are relatively loose. For example, for the LEO (Low Earth Orbit) spacecraft, the mass of propellant is only about 1/10 of the total launch mass of the spacecraft. The mass of propellant will be about half of the total launch mass of the MEO (Medium Earth Orbit) and GEO (Geosynchronous Earth Orbit) spacecraft. The propellant mass of the lunar lander might be 2/3 or more of total spacecraft mass due to constraint of launch mass. Therefore, the mass constraints of other sub-systems of the lunar lander should be strictly constrained. According to data from NASA, the mass of power supply and distribution subsystems in conventional spacecraft is about 22~27% of the total mass. For the lunar lander, the figure was only 10~15%.

6.4 Design Methodology

6.4.1 Design Principles

The power system technology of the lunar lander could be divided into three categories including power supply in ground test, on launch site and during flight. Because the power supply mode of the lunar lander was not same during ground test, pre-launch and flight, when the power supply system was designed, it was necessary to focus on the reuse of designed function, simple interface, convenient operation, and low resource consumption and to minimize the complexity of interface for tests and reduce mass of test interface as much as possible.

Following principles should be followed when the power system was designed.

(1) Miniaturization

The mass constraint of the lunar lander was very strict. On the basis of guaranteeing reliability, comprehensive measures should be taken including instrument function integration, components selection, housing material selection, housing design, selection of internal and external connectors, and wire selection to achieve miniaturization, so as to improve specific power of power system as much as possible.

(2) Integration of inheritance and advancement

The power system of the lunar lander should fully inherit power supply technology of previous spacecraft, especially mature experience of the Chang'E lunar probes and Shenzhou spacecraft. At the same time, it was necessary to research key technologies. The system requirements could be satisfied by three aspects including system optimization, high specific energy and high specific power products, high integration and lightweight design. For the research products of key technology, ground simulation analysis and test verification should be strengthened as well as risk analysis and control to ensure that the product design could satisfy the mission requirements reliably and reasonably.

6.4.2 Illumination Conditions

Analysis of illumination condition was the basis of power system design [8]. For the lunar lander, the analysis of illumination included analysis of LTO, initialization process after landing and operation on lunar surface.

1. Illumination During Flight

During flight phases before the lunar lander was landed, the solar array was always toward the Sun. The incident angle of solar array depended on the accuracy of attitude control of the lunar lander. During launch, orbit change and powered descent, the solar array was unfolded and did not generate electricity. In circumlunar orbit, the lunar lander would pass alternation process of light/shadow. For example, analysis of illumination conditions is described as follows.

(1) Launch

The solar array was unfolded before launch and deployed till the lunar lander was separated with launch vehicle, when there was no power generated and the lunar lander was fully powered by the battery packs. The battery packs of the lunar rover could power the lunar lander jointly.

(2) LTO

After the solar array was deployed, the battery packs of the lunar lander and rover were charged. The solar array was kept toward the Sun during LTO except period of trajectory correction maneuvers. During trajectory correction maneuvers, the solar array was unfolded and did not generate electricity, when the lunar lander was powered by battery packs. After trajectory correction maneuvers, the battery packs were charged.

(3) LOI

When the lunar lander approached the Moon, the LOI was performed to make the lunar lander captured by Moon gravity. The lunar lander was inserted into a 100 km altitude LLO. During LOI, the engine was firing to decelerate. The solar array would be unfolded in advance. After LOI, it would be deployed again. After deployed, the solar array would enter alternation process of light/shadow. The power would be provided alternately by the solar array and the battery packs.

(4) 100 km circular LLO

After the lunar lander was inserted into LLO, the light/shadows alternated in a 2 h cycle. The solar array and the battery pack would alternately power the lunar lander, when the battery packs were charged and discharged alternately. The time of the lunar lander in 100 km LLO was about 5 days.

(5) Orbit change on LLO

Five days after LOI, the lunar lander was inserted into a 100 km $\times$ 15 km elliptical lunar low orbit. Before the engine was firing, the solar array would be folded in advance. After orbit change, it would be deployed again. Then the solar array would enter alternation process of light/shadow and the power would be provided alternately by the solar array and the battery packs.

(6) 100 km $\times$ 15 km elliptical LLO

After the orbit was changed to 100 km $\times$ 15 km elliptical LLO, the light/shadows alternated in a 2 h cycle. The solar array and the battery pack would alternately power the lunar lander. On such orbit, the battery packs were charged and discharged repeatedly. The time on such orbit was approximately 4 days.

(7) Powered descent

During powered descent, the engine was firing and the solar array would be unfolded in advance. The lunar lander was powered by the battery packs. The battery packs of the lunar rover could power the lunar lander jointly.

2. Occlusion Analysis of Solar Array during Initialization on Lunar Surface

After the lunar lander was landed on lunar surface, the operation status should be setup as well as TT&C, thermal control and depletion of propellant. However, the discharge depth of battery packs was higher after powered descent and the power output of the solar array should be analyzed in order to manage operation procedure of the lunar lander reasonably.

For the Chang'E-3 lunar lander, after the lunar lander was landed on lunar surface, the lander rover was powered by the lunar lander through separation plug. Due to limitations of temperature and TT&C conditions, the separation of the lander and the rover should be accomplished as soon as possible. Before the separation, the camera of the lunar rover would take photograph of lunar terrain near the landing site by the navigation cameras and panorama cameras. The working time was long and the load power was relatively high. The solar array of the lander would supply power to the rover main bus and charges the battery packs of the lander and the rover. After the battery pack of the rover was fully charged, the rover could be separated. It was necessary to analyze of the power out of solar array of the lander accurately to ensure the safety and reliability of separation.

The rover was on the top of the lander before separation, part of the solar array of the lander was occluded by the rover before it was deployed. The power output of the solar array of the lander was up to not only the solar incident angle but also the occlusion before separation of the rover.

In order to accurately analyze the occlusion area of the solar array, the solar array was divided into grids. Generally, the smaller the grid of the solar array was, the higher the accuracy was in analysis of occlusion. But at the same time, the larger the calculation burden was. Therefore, it should be traded off in actual occlusion analysis. Due to the special configuration of the solar array of the lunar lander and the size of the solar cells, when the occlusion analysis was conducted, the mesh was divided as follows.

(1) For regular area of substrate structure of the solar array, the rectangle grid was used.
(2) For irregular part of the solar array, because the irregular part of the solar array was trapezoidal, it was divided into small trapezoids when the grid was divided and the bottom side of upper trapezoid coincided with the top side of lower trapezoid.

The grid of the solar array of the Chang'E-3 lunar lander is shown in Fig. 6.2.

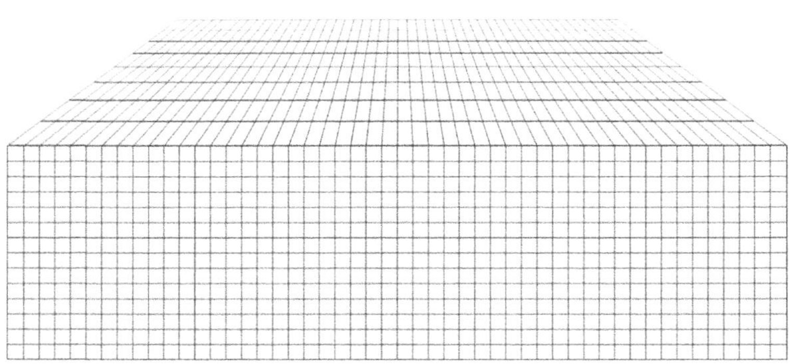

Fig. 6.2 Grid of the solar array of the Chang'E-3 lunar lander

In order to calculate the power output of the solar array accurately, the error of landing attitude should be considered. When the occlusion of the solar array was analyzed, the maximum occlusion area of the solar array and the maximum solar incident angle could be analyzed respectively to simplify the calculation. When the maximum occlusion area of the solar array was calculated, the occlusion area under different solar elevation and azimuth should be analyzed with bias of landing attitude. When the maximum solar incident angle of the solar array was analyzed, the incident angle under different solar elevation and azimuth should also be analyzed.

3. Analysis of Maximum Solar Incident Angle of Solar Array

When the lunar lander was working on the moon, it was necessary to adjust the angle of deployment of the solar array according to the solar elevation in order to achieve the maximum power output. Due to the limitation of the landing site, the local solar elevation angle was relatively low, and the solar incident angle of the solar array was relatively large, and the power output of the solar array at a large solar incident angle should be analyzed.

For example, the deployed angle of ± Y solar array of the Chang'E-3 lunar lander was from 20° to 245°. The angle was 180° when the plane of the solar array was parallel to the local horizontal plane. The maximum deployed angle of the solar array was 245°.

In the nominal landing attitude, the solar incident angle of ± Y solar array of the Chang'E-3 lunar lander varied from 0° to 48° in a lunar daytime as shown in Fig. 6.3.

From analysis of the solar incidence angle of the solar array, it can be seen that the maximum solar incident angle could be 50° during the lunar daytime, which was different from that of conventional spacecraft on orbit. It was critical to analyze the effects of such solar incident angle on power output of the solar array of the lander. It was necessary to conduct power output test at large incident angle for solar cells of the solar array. The test results showed that when the solar incident angle was less than 45°, the power output of the solar array was approximately cosine of the solar

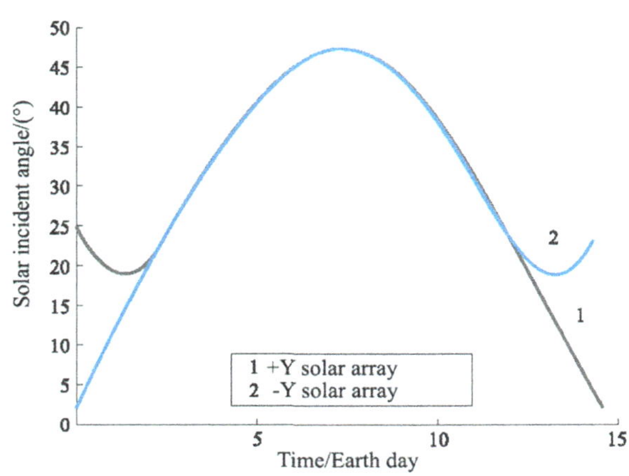

Fig. 6.3 Solar incident angle of ± Y solar array of the Chang'E-3 lunar lander

incident angle, while the solar incident angle was greater than 45°, the power output of the solar array was worse with increasing of the solar incident angle.

6.4.3 Lightweight Design

The lunar lander had limited mass resources. In order to achieve lightweight, the power system had been specifically designed from both system and instrument levels. For example, such design of the Chang'E-3 lunar lander is given as follows.

1. System-Level Optimization Design

The Chang'E-3 the lunar lander consisted of the lunar lander and the lunar rover. The interface between the lander and the rover was complex. The power supply requirement was different during all flight phases. To improve the efficiency of power supply and optimize the energy configuration of system, the system-level joint power supply design was necessary, which will be introduced in the following sections.

2. High Specific Energy/Power Product Design

The major contents of optimization design of the power supply controller included configuration, structural component optimization, and selection of components and parts. In configuration design, the modular design could reduce internal cables and volume. In structural optimization, the thermal characteristics of the separately installed components were analyzed to optimize their structure, with aluminum alloy for structure and titanium alloy for the rods with high strength and stiffness. In selection of component and part, lightweight and small units were selected for electrical connectors, relays, and magnetic cores, and reduce the number of filter capacitor arrays and the size of the filter inductor core by increasing the frequency of the discharge circuit.

The wires, strips, silicone rubber, and fasteners in lithium-ion battery packs were determined by electrical properties and mechanical properties, so it was not easy to reduce weight. Therefore, the weight could only be realized by reduction of structural parts such as lithium ion battery cells and its wall plates as follows.

(1) New material. The use of high-strength materials such as titanium alloys and carbon fiber-reinforced composite materials instead of traditional structural materials such as stainless steels and aluminum alloys could reduce the mass while structural strength was ensured.
(2) Reduction of material. The mass of material could be reduced by reduction of the thickness or shortening of the length to reduce the amount of the current collector material while the material properties met the requirements of the product performance.
(3) Reduction of redundant design. Because the lithium-ion battery technology was matured and stable, the redundancy of single battery design could be reduced appropriately.

6.4.4 Analysis of Space Environmental Effects

The power system of the lunar lander should pass various special environments including Earth's orbit, LTO, circumlunar orbit and lunar surface [9].

1. Analysis of Space Radiation Effect

For the power supply system, the effects of space radiation included two aspects. On the one hand, the solar array installed outside the module directly exposed to the space environment. On the other hand, the power supply control device and battery packs installed in the module were affect too.

(1) External equipment

There were two main effects on solar arrays. One was the total dose effect of ionization, and the other was displacement damage. The major result was the degradation of the electrical performance of solar cells, which was reflected in the drop of voltage and current. The germanium-doped cover glass on triple-junction gallium arsenide solar cell (GaInP2/GaAs/Ge) was used to reduce the impact of radiation effects. In design of solar cell circuit, the power loss should be considered to ensure that the lunar lander load requirement was met at the end of life.

(2) Internal equipment

The structural materials of the power controller and the battery pack were made of aluminum, and the non-plastic-sealed components such as resistors, capacitors, operational amplifiers, diodes, bipolar transistors, and relays were all insensitive to radiation. The major components sensitive to radiation were mono-stable integrated circuits. The major consideration was the design of total dose effect protection and single event latch-up protection in the design of radiation-resistance of the power system.

2. Analysis of Space Thermal Environment Impact

The solar array was installed outside the module, while the lithium-ion battery packs and power controller were in the module. The temperature outside the module was affected by lunar surface environment [10], and the temperature range might be relatively large as -180 °C ~+120 °C. The internal temperature of the module was managed by the thermal control system to ensure the normal operation temperature of the power controller and the battery.

(1) Solar array

Extremely high and low temperature environments might increase the risk of damaging the material of the solar array including single cell, cover glass, backsheet glue, wires, and solar cell array solder joints. The components and parts of the solar arrays such as wire-to-connectors, wires, isolation diodes, and grounding resistor should be verified by pertinent specific tests. The forward conduction performance and reverse cut-off performance of the diode, and the electrical performance of the triple-junction gallium arsenide solar cell should be verified by high-low temperature cycling test and low-temperature storage test in air.

(2) Lithium-ion battery pack

The electrical performance of lithium-ion battery pack was closely related to temperature. Especially under low temperature and high temperature environments, the electrical performance of lithium-ion battery pack would degrade rapidly and might even fail. At low temperature, when the charging current was large, many lithium ions would migrate to the negative electrode surface. Due to the slow diffusion rate of lithium ions and low ion permeability of the Solid Electrolyte Interphase (SEI) film (as a negative electrode protection) at low temperature, part of lithium ions did not diffuse through the SEI film but were deposited directly on the negative electrode surface, resulting in the production of metallic lithium. Lithium might interact with the electrolyte easily to generate gas or form lithium dendrites which would penetrate the separator to cause a short circuit in the battery.

At high temperatures, the performance of SEI film is unstable and prone to decomposition. The newly exposed carbon negative electrode will consume a portion of lithium and form new SEI film on the surface, resulting in irreversible capacity loss of the cell. At the same time, the formation of the SEI film is usually accompanied by gas generation, resulting in increase of internal pressure of the battery to increase internal resistance of the battery.

3. Analysis of Lunar Dust Effect

After the lunar lander was landed on lunar surface, lunar dust was mainly activated by engine plume on lunar surface during landing. Before landing, the solar array was folded. After landing, there would be a certain period of time before the solar array was deployed again, which could effectively prevent lunar dust from covering surface of the solar array to pollute. In the phase of normal operation in lunar daytime, the impact of lunar on solar cell power generation was compensated by retaining certain margin of power output and adjusting the operation mode.

4. Analysis of Other Effects

Similar to nickel-cadmium battery in space application, lithium-ion battery uses lean liquid and the electrolyte inside the battery cell is completely absorbed in the separator and the pole pieces. There is no free electrolyte inside. It was necessary to analyze whether 1/6 g gravity had an effect on the normal operation of lithium-ion battery pack. The lunar eclipse will occur during the lunar daytime. During such period, the lunar lander load would be completely provided by the battery pack. Therefore, the discharge capacity of the battery pack should be evaluated to design a suitable operation mode in lunar eclipse to ensure the power supply safety.

If the nominal lifetime of the lunar lander on lunar surface was one year, it would pass 12 lunar nights. Every lunar night lasts for 14 earth days. During lunar night, there is no sunlight and the solar array cannot generate electricity. If there was only lithium-ion battery pack, it was impossible to provide continuous energy for such long period of time. Therefore, all electric instruments should be shut down before the lunar lander entered into lunar night. The battery discharge switch was disconnected to make the lunar lander go into hibernation. When the lunar night was over, the lunar lander would be awakened again. For detailed design, see Sect. 6.5.4.

6.4.5 Energy Balance Analysis

Energy balance analysis could determine whether the lunar lander energy was balanced during a specific period of time. When the power balance of the lunar lander power was calculated, there were many factors to be considered including illumination condition, load requirement, solar array and battery of power system, charging and discharging efficiency and etc. The analysis process and calculation method were basically the same as that of Earth orbit spacecraft.

In LTO, the lunar lander was powered mainly by the solar array. During orbit changes and phases in Moon shadow, the lunar lander was powered mainly by the battery pack. At the same time, the long-term power load of the lunar lander was small and short-term high power loads occurred mainly in orbit changes. In addition, the interval between orbit changes was usually several orbital periods during circumlunar phase. For the lunar lander, a multi-turn energy balance method was suitable, which allowed the battery of the lunar lander to be discharged at a greater depth and fully charged in subsequent several orbital periods to achieve energy balance. Such design could reduce the total power output of the solar array so as to reduce the area and mass of the solar array, and make it possible to select small-capacity battery pack. If the energy balance was achieved in one orbital period, the power output of the solar array and the capacity of the battery should be greatly increased. At the same time, in cruise process of circumlunar orbit, the power output of the solar array was excessive and a great many shunting was required, which was not optimized design.

The power load of the lunar lander changed dramatically during the mission profile and the flight process was complex including the launch [11], LTO, 100 km circular lunar orbit, 100 km × 15 km circumlunar orbit, powered descent, initialization on lunar surface, operation on lunar surface and lunar eclipse on orbit. The change of external heat flux was complex during the mission profile and the temperature range of the battery pack varied greatly, as well as occlusion of the solar array of the lunar lander was complex during operation on lunar surface.

When energy balance of the lunar lander was calculated, comprehensive consideration should be conducted including the system power load conditions, illumination conditions, occlusion of solar array and battery support capabilities under a wide temperature range for each flight phases, as well as various unexpected conditions.

Due to wide range of battery operation temperature on orbit, in order to verify the discharge capacity of the lithium-ion battery pack under different working conditions, the same batch of single cells of the lander lithium-ion battery was verified by experiments for different discharge conditions including LOI, circumlunar orbit, powered descent and lunar eclipse. At the same time, in order to examine the discharge capacity of the battery pack, the flight hardware of the battery pack was tested according to the orbital power and temperature conditions to ensure that the battery pack would work properly on orbit.

6.4.6 Analysis of Power System Architecture

1. Distribution and Grounding System

The power system of the Chang'E-3 lunar lander provided a fully regulated main bus, which was used by the power distribution unit to provide primary power for the instruments. Secondary power supply could be provided according to requirements by distributor for secondary distribution.

Compared with conventional spacecraft, the interface design of the Chang'E-3 lunar lander was more complex in grounding [12]. The probe consisted of the lunar lander and the lunar rover, while the complex worked together sometime and the lunar lander or the lunar rover worked separately sometime. In grounding design of the lunar lander, it was necessary to meet the single-point grounding requirement during flight alone, as well as the single-point grounding requirement during flight in the complex status. The lunar lander and the lunar rover respectively set a single point of grounding point within the respective power distribution unit, and the two probes was connected electrically with a high resistor. Before the lunar lander was landed of lunar surface, the rover was considered as an electrical load of the lunar lander. Single grounding was achieved respectively for each probe after landing.

2. Power Supply System

According to task characteristics and requirements of the power system, the power system consisted of solar arrays, power controllers and battery packs.

Compared to silicon solar cells and single-junction gallium arsenide solar cells, the triple-junction gallium arsenide solar cell has the advantages of high photoelectric conversion efficiency and high area specific power, which becomes the preferred main power source for various types of spacecraft in the world. Considering power requirement and mass constraint, the triple-junction gallium arsenide solar cell was used as a power generation unit in power system.

Compared to nickel-cadmium and nickel-hydrogen batteries, lithium ion battery has the advantages of high specific energy, high cell voltage, wide operating temperature range, low heat generation, and low self-discharge rate. Due to the constraints of mass and volume, lithium-ion battery pack was used as energy storage after the power density and low self-discharge rate requirements during lunar night was considered.

The S4R power regulation technology was used the power controller. The S4R power regulation technology combined the advantages of S3R-type and hybrid-type power regulation technologies. It implemented two-domain control, which not only overcame the disadvantages of the three-domain complicated control of the S3R power regulation technology, and the large power consumption and mass of the charging controller, but also avoided the disadvantages of low energy utilization of the solar array induced by the separate design of the charging array in the hybrid power adjustment technology.

For the task requirements of autonomous hibernation and wake-up of the lunar lander, a wake-up control circuit was separately designed in the power controller, which was responsible for the transfer of operation modes in lunar daytime and lunar

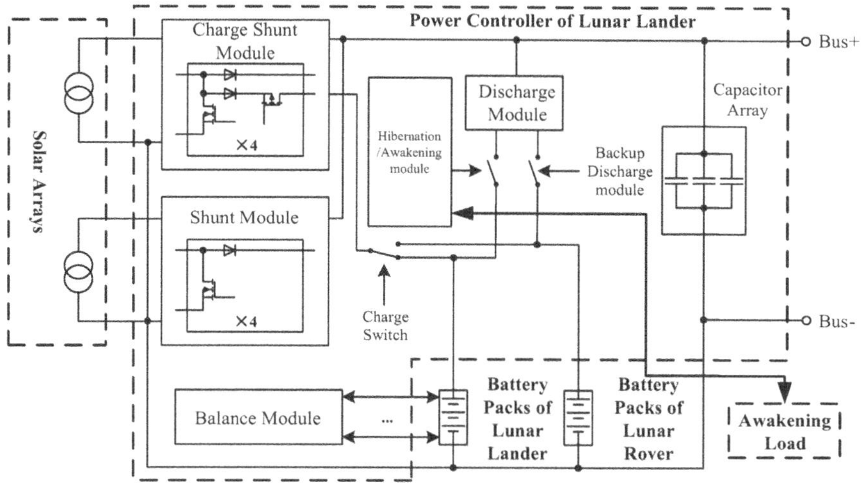

Fig. 6.4 Block diagram of the power system of the Chang'E-3 lunar lander

night to realize autonomous wake-up. The block diagram of the power system of the lunar lander is shown in Fig. 6.4.

6.5 Typical Technology

6.5.1 Design of Power Controller

For the solar array-battery pack power system, the power controller is designed for the control of the solar array and the battery pack. In specific design of the power supply controller, it should be considered including main bus regulation mode, discharge control mode of battery pack, charge control mode of battery pack, over-current and over-discharge protection. The framework of the power supply system for the lunar lander was different for different spacecraft because it was different for their scale, life and operation orbits. However, with the continuous development of spacecraft power technology, spacecraft power system is being developed in the direction of full regulation, switch regulation and modularization.

The long-term load of the lunar lander was low, so the 29 V main bus could be used. The power controller could adopt the fully-regulated main bus. Due to requirement for lightweight and miniaturization of the lunar lander, the shunt/charge shunt circuits and the discharge regulation circuits in the power controller could adopt modular design. The modular structure of the power controller was designed for each group of circuit functional modules to design a corresponding independent frame-type internal structure. The external structure of each frame was basically the same. The advantage of such design method was that a more reasonable installation and configuration method could be achieved for each group of printed circuit boards or circuit functional modules, which facilitated the design of a single module in reliable internal cable, lightweight and convenient maintenance. In addition, the internal structure of each functional module in the modular structure was independent, which facilitated further optimization in its thermal design and electromagnetic compatibility design. Moreover, the same external shape and flexible combination of modular structures created favorable advantages for development of new power transmission methods.

Due to the special lunar surface environment and lunar night of 14 days, the battery pack could not meet the power load requirement during lunar night. Before it was entering lunar night, the power load must be disconnected and the discharge switch of the battery pack should be disconnected to make the system go into hibernation. When the lunar daytime was coming, the discharge switch of the battery pack should be closed and the OBDH computer of the lunar lander was awakened. The hibernation and awakening control circuits should be designed for the power controller.

Due to strict mass constraint, the lunar lander and the lunar rover were designed with energy reuse circuits to save mass. In design of the power controller, it was necessary to consider the method of sharing the energy of the two probes. The detail of common design method of energy share is given in Sect. 6.5.5.

In order to reduce the mass, the aluminum alloy structure was usually used in the power controller. For further reduction of the mass, the magnesium alloy mechanism was used in the power controller. Each structural module was formed by a whole piece of magnesium plate and the chassis was composed of left and right wall plates with titanium alloy connecting rods.

6.5.2 Design of Battery Pack

The energy storage device used in the power system of the Chang'E-3 lunar lander was lithium-ion battery packs. It was composed of several and even dozens of rectangular battery cells connected in series and parallel.

Because of the strict mass constraint, the equalization control of lithium-ion battery packs was not used for the Chang'E-3 lunar lander. To ensure that the single cell's performance of the battery packs was not dispersed during the one-year life on orbit, single cells were screened to ensure consistency. After the monomer formation efficiency, charge and discharge efficiency, internal resistance, self-discharge efficiency and monomer capacity were compared, the monomers consistent with each specification were selected to form the battery packs.

The drawbar type structure was adopted for the battery packs to achieve a tight assembly between the individual battery cells for preventing the battery cells from being deformed due to internal pressure and facilitating heat dissipation at the same time. The space between the single storage battery and the structural member was an insulating polyimide tape to ensure good insulation performance between the single storage battery and the structural member. Lithium-ion battery cells were electrically connected with the guide bars. The battery and the wires were connected through the terminals. All the current wires and voltage wires were integrated into different electrical connectors according to different functions, so the signal transmission was isolated from the power transmission.

With increasing number of charge and discharge of lithium-ion battery packs, the difference in the charging capacity and self-discharge rate of each cell in battery packs gradually accumulated, so the voltage gap between battery cells within battery packs became larger and larger with a divergent trend. In order to ensure safety of lithium-ion battery packs, each single cell in battery packs must be charged with voltage-balanced charge control. In charge design of the power controller, the battery pack voltage and the cell voltage were limited to ensure the charging safety of lithium-ion battery packs.

6.5.3 Design of Solar Cell Circuits

The solar array circuit was composed of many solar cells in series and parallel. The design of the solar array circuit was design of series, parallel and combination of solar cells. The method for calculating the number of solar cell series and parallel plates was basically the same as the calculation method for Earth orbit spacecraft. For the lunar lander, when the solar cell array was designed, it was also necessary to be designed for the ultra-wide operation temperature range and occlusion prevention of solar array.

During LTO and operation on lunar surface, operation temperature of the solar array working temperature might be as higher as +120 °C. The output voltage of solar array at the highest temperature was low. In order to ensure satisfaction of the minimum main bus voltage requirement, the worst case temperature should be considered during design of series number of solar cells. The design of solar cell array configuration was based on the theoretical calculation results of solar cells series and parallel design, along with the substrate area of carbon-fiber aluminum honeycomb, direction and coefficient of distribution. For the Chang'E-3 lunar lander, there was energy reuse for the lunar lander and the lunar rover. After the lunar lander was landed on lunar surface, there were two sub-arrays in solar panels of the lunar lander to charge the battery packs of the lunar rover. Therefore, in configuration design of the solar array, the two sub-arrays were designed to prevent occlusion as much as possible.

In design of the solar arrays, some measures should be taken to ensure the reliability, safety, maintainability and interchangeability of solar cell array as well as reliability of diode welding and circuit output.

There are two methods in installation of spacecraft solar array including body-mounted and deployable solar array. The solar arrays of the Chang'E-3 lunar lander were composed of four panels including two deployable solar arrays and two body-mounted solar arrays of carbon-fiber aluminum honeycomb.

6.5.4 Design of Hibernation and Awakening Control

1. Technical Difficulties

Technical difficulties of hibernation and awakening control included two main aspects. When the lunar night came, all equipment must be shut down completely. The normal sequence for power-off implementation of the power controller of main bus and the OBDH computer for command driving power were coupled. Therefore, the power-off sequence for power controller and OBDH computer was cross coupled. So, it was a technical difficulty to shut off the last instrument. When the lunar daytime came, the lunar lander should be transferred from hibernation to awakening. The awakening of the first equipment should work autonomously by the external input conditions, and then other instruments were powered on according to predetermined procedure to recover telecommand and telemetry of the lunar lander.

2. Implementation Method

After the telecommunication coverage, temperature adaptability, angle range of solar array and power balance were considered for the lunar lander, hibernation and autonomous awakening method were designed for solution. The instruments of the lunar lander would be shut down subsequently by ground commands before lunar night to hibernate. The awakening process was triggered by sunlight when the solar array power output met the requirement for minimum operation mode load. The awakening control circuit would send instruction autonomously to power on battery packs, OBDH computer and power supply controller, and then power on other instrument to implement transfer to the normal operation mode in lunar daytime.

3. Key Control Switch

The hibernation and awakening control circuit consisted of many key switches for control of the power supply. The center of hibernation and awakening method was to design the power on/off sequence of the key control switches.

4. Process of Hibernation and Awakening

The workflow diagram of hibernation and awakening is shown in Fig. 6.5.

6.5.5 Power Supply Reuse Between Probes

Due to strict mass constraint, additional mass would be necessary for the lunar lander and the lunar rover if they used their own power system to provide energy.

For the Chang' E-3 lunar lander, no additional mass was required by making full use of original power supply controller modules of the lunar lander and the lunar rover. The energy share circuit and logic control command and was designed to implement energy share between the lunar lander and the lunar rover. The interface of energy share is shown in Fig. 6.6. After the lunar lander was landed on lunar surface, during

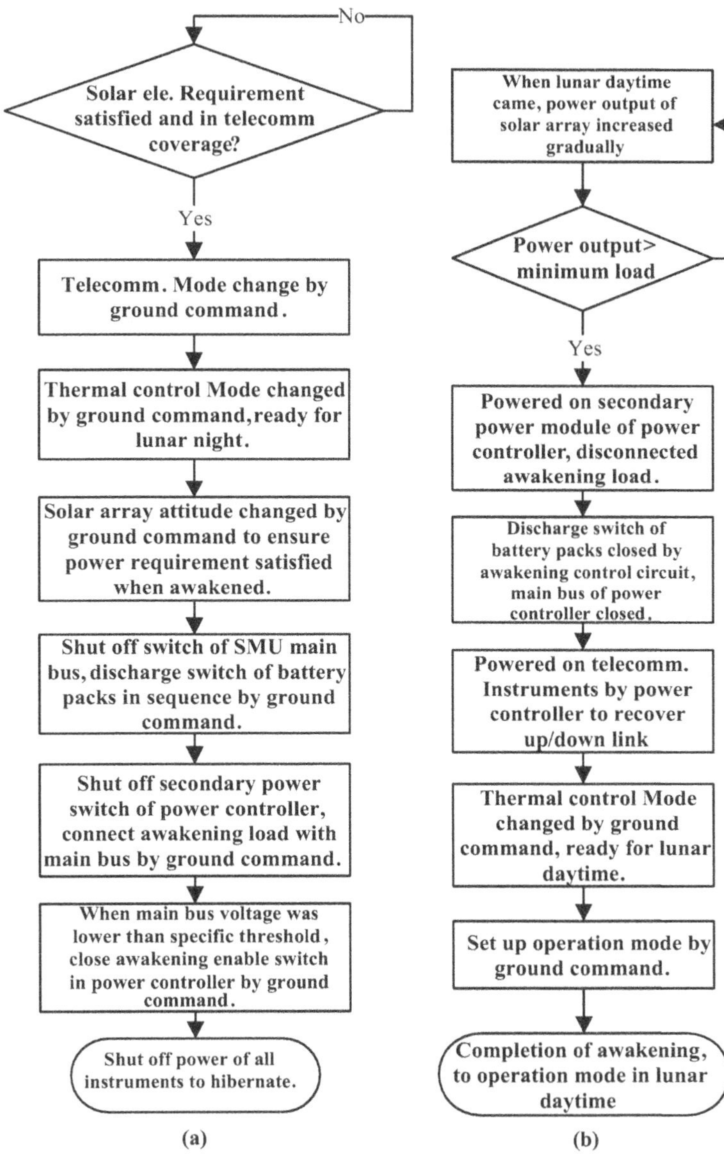

Fig. 6.5 Workflow diagram of hibernation and awakening (**a**) Hibernation process; (**b**) Awakening process

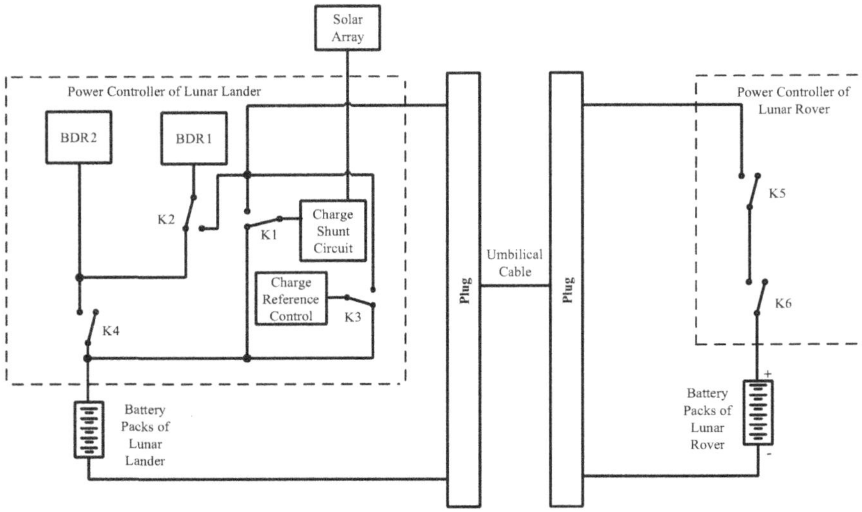

Fig. 6.6 Interfaces of energy share between the lunar lander and the lunar rover

initialization before separation of the lunar lander and the lunar rover, the solar array of the lunar rover was not deployed yet and the load was not powered on, while the battery pack of the lunar rover was charged by the solar array of the lunar lander and the main bus of the lunar lander could be powered by the battery pack of the lunar rover.

Compared to the lunar lander without energy share, the major change in power system of the Chang'E-3 lunar lander was reflected in the power controller with additional control circuit for power share switch.

During launch and powered descent, the battery pack of the lunar rover could supply power to the main bus of the lunar lander. During launch and illumination on lunar surface, the battery pack of the lunar rover was charged by the solar array of the lunar lander.

1. The Battery Pack of the Lunar Rover Supplied Power to the Lunar Lander

When the battery pack of the lunar rover supplied power to the power main bus of the lunar lander, because the power controller of the lunar rover was not powered on, when the discharge switch was closed, the battery pack would charge high current to main bus capacitors. In order to avoid the electric current impact, the charging circuit in the power controller of the lunar lander was used to charge the main bus capacitor ahead by the electric current limited characteristics.

2. The Battery Pack of the Lunar Rover Was Charged by the Lunar Lander

After the lunar lander and the lunar rover were powered jointly, the battery pack of the lunar rover should be charged by the lunar lander. Timing and the control process were similar to the process of the lunar lander powered by the battery pack of the lunar rover. The only difference was that the battery pack of the lunar rover was not connected to BDR module of the lunar lander.

6.6 Testing and Verification

6.6.1 Contents of Verification

Compared with conventional spacecraft, in addition to the conventional performance tests of mechanical, electrical and thermal, special tests should be conducted for the power system of the lunar lander including hibernation and awakening function verification, verification of large incident angle for solar cells and performance test of solar cell circuit under extremes temperature.

6.6.2 Verification of Hibernation and Awakening

The hibernation and awakening circuit was in the power controller to control the process of hibernation and awakening of the lunar lander. The solar array simulator was used to simulate the current generated by sun light in the special test, while all other status was consistent with condition on orbit. The output current of the solar array simulator was adjusted to simulate the change of light intensity and reproduce the transfer process of lunar daytime and night.

In the test of hibernation, the ground test station sent commands one by one to shut off instruments of the lunar lander in sequence, while the output current of the solar array simulator was reduced gradually to simulate the gradual weakening of sun light. When the output current of the solar array simulator was less than the load current, the secondary power supply of the power controller was shut off and the lunar lander entered hibernation, while all telemetry remained in the last frame.

In the test of awakening, the output current of the solar array simulator was gradually increased to simulate the gradual strengthening of sun light. When the output current of the solar array simulator reached the load current, the secondary power supply relay of the power supply controller was closed, and the secondary power supply was powered on, and the pulse trigger circuit was triggered to generate high level pulse to connect the battery packs and OBDH computer. The lunar lander was awakened autonomously.

For the hibernation and awakening function, it was necessary to verify in normal temperature and pressure, thermal cycle, thermal vacuum, system-level electrical performance test, and system-level thermal test to ensure that the test items and test conditions were comprehensively covered. The function must be tested at the highest working temperature and lowest working temperature, and compared the output current value of the solar array simulator at awakening time under different temperature conditions to ensure the consistency of the operation status at awakening time.

6.6.3 Verification of Large Incident Angle for Solar Cells

Similar to condition on Earth, the intensity of sun light is changed during on lunar daytime. For the Chang'E-3 lunar lander, landing site was about 44 degrees north latitude, and the local solar elevation angle was always less than 45 degrees. The solar array incident angle was greater than 60 degrees during the alternation of lunar daytime and night. The incident angle has a great influence on the power generation of the solar array, so it was necessary to verify the power generation of the solar array at a large incident angle.

An aluminum honeycomb substrate was used and three-junction gallium arsenide solar cells were stuck on the surface as a test article. A pulse solar simulator was used to simulate solar illumination under AM0 condition. An angular rotation tool was used to adjust the light incident angle of the test article to simulate changes in the light angle of the solar array on lunar surface. The test article is shown in Fig. 6.7.

As the incident angle changed, the output current of the test article also changed. The output current of the solar array began to gradually deviate from the theoretical curve. Compared to silicon solar cells, the triple-junction gallium arsenide solar cells had higher power generation at high incident angles as shown in Table 6.1.

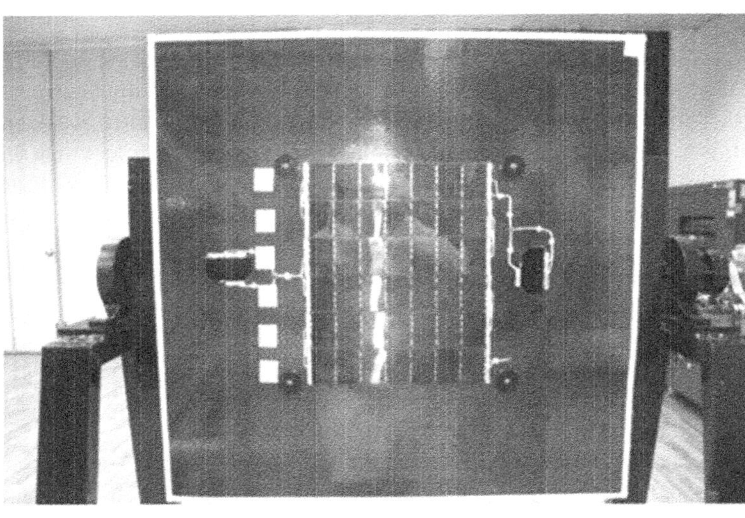

Fig. 6.7 Test article of solar array

Table 6.1 Comparison of test values of three-junction GaAs and Silicon solar cells with large Light incident angle

Angle /(°)	Test value of three-junction GaAs solar cell (ratio of direct light)	Test value of three-junction Silicon solar cell (ratio of direct light)
60	0.440	0.4
75	0.191	0.158
90	0	0

6.6.4 Extreme Temperature Verification of Solar Cells Performance

Due to the high temperature during lunar daytime and the low temperature during lunar night, the storage temperature range of the solar cells of the lunar lander was from −180 to +120 °C. In order to ensure the reliability of solar array design and procedure, special high-low temperature environmental test was necessary for the solar array.

For extreme temperature environments, two tests were conducted for the solar arrays of the Chang'E-3 lunar lander including high and low temperature test and low temperature storage test. Both high and low temperature test and low temperature storage test were under atmospheric pressure. The temperature of low temperature storage test was as low as –210 °C. The appearance of the three-junction gallium arsenide solar array did not change before and after test. The solar array did not suffer from cracking, chipping, and etc. The solar array changed little in electrical properties before and after test. The performance of the isolation diode in forward conduction and reversed cut-off did not change.

6.7 Summary

Based on the mission characteristics of the lunar lander, such as harsh environment, requirements of power supply load and strict mass constraint, a set of analysis and design method was come up for the power system. The task requirements and characteristics analysis, design methods, typical technologies and verification is described. Emphasis was on the analysis of design, autonomous hibernation and awakening, energy share, and lunar environmental adaptability.

China's deep space exploration mission has just begun. After the lunar soft-landing was achieved, more deep space missions such as Mars landing, lunar base construction and interplanetary exploration will be implemented [13]. Such missions will be more complex and requirements for power systems are getting higher and higher. Nuclear power such as nuclear reactors and isotope thermoelectric generators, and large space energy systems such as fuel cells and space solar power stations will become future options. Along with development of power technology, the space power system will also be updated and improved. Therefore, more resources are necessary in research of new energy sources.

References

1. Li G (2008) Spacecraft power system technology[M]. China Astronautic Publishing House, Beijing
2. Ma S (2001) Satellite power system technology[M]. China Astronautic Publishing House, Beijing
3. Peng C (2011) Spacecraft system design[M]. China Science and Technology Press, Beijing
4. Han B, Chen Q (2010) Spacecraft power system [M]. China Astronautic Publishing House, Beijing
5. Ye P, Sun Z, Rao W (2007) Research and development of Chang'e-1[J]. Spacecraft Eng 16(6):9–15
6. Ye P, Peng J (2006) Retrospect of human deep space exploration history and its prospect in China[J]. Eng Sci 8(10): 13–18
7. Sun Z, Zhang H, Wu X, et al. (2010) Mission analysis of a lunar soft lander[J]. Spacecraft Eng 19(5):12–16
8. Ye P, Xiao F (2006, Feb) Issue about lunar environment in lunar exploration project[J]. Spacecraft Environ Eng 23(1):12–16
9. Chen L, Li F, Ren D (2010) Lunar surface and near lunar space environments and their effects[J]. Spacecraft Eng 19(5):76–81
10. Wang C, Deng Z, Gao H et al (2006) Development status of the lunar landers[J]. Missiles and Space Vehicles 4:31–36
11. Yang M, Jia Y, Chen J (2008) Research on system design of lunar rover[J]. Aerospace Control and Appl 34(3):3–7
12. Peng J, Liu Z, Zhang He (2008) Conceptual design of a lunar lander[J]. Spacecraft Eng 17(1):18–23
13. Wu W, Liu W, Jiang Y (2011) The revelation from the development of foreign deep-space exploration beyond the Moon[J]. Aerospace China (7):9–12

Chapter 7
Guidance, Navigation and Control Technology of Lunar Lander

7.1 Introduction

Guidance, Navigation, and Control (GNC) subsystem was one of key subsystems for the lunar lander [1]. Major functions of GNC subsystem were: trajectory corrections maneuvers during LTO, LOI at perilune, orbit phase adjustment and de-orbit burn, soft-landing GNC during powered descent.

Compared with GNC subsystem of conventional spacecraft, the main challenge for GNC subsystem of the lunar lander was the powered descent phase, which was also the key and the most difficult process of lunar soft-landing mission. Due to short duration of the powered descent phase, large velocity change and uncertainty of lunar terrain, it was impossible to depend on ground control. Therefore, it was necessary to design a fully autonomous landing control strategy based on range measurement and terrain recognition.

The guidance system autonomously generated commands for the main engine (including magnitude and direction) based on task and the lander status given by the navigation system. The navigation system determined the lunar lander status (including orbit and attitude information) based on the information given by relative navigation sensors and Inertial Measurement Units (IMUs). The control system implemented the desired attitude tracking or attitude maneuver through the actuators according to the current attitude determined by the navigation system and the target attitude given by the guidance system [2, 3].

The major technologies of GNC subsystem of the lunar lander are described in this chapter, including the general architecture design of the subsystem, hardware and software design techniques, and typical methods of guidance, navigation, and control, and etc.

D.-Y. Yu et al., *Technology of Lunar Soft Lander*, Space Technology Library 38,
https://doi.org/10.1007/978-981-15-6580-9_7

7.2 Development of GNC Technology

In the 1960s, the United States and the Soviet Union launched a number of lunar probes including the Luna missions of the Soviet Union, the Surveyor and Apollo missions of the United States, which achieved hard landing, soft-landing, lunar sample and return, and manned lunar landing.

For landing navigation, microwave range and velocity sensors were used to achieve high-precision autonomous navigation for most of the landers such as US Surveyor and Apollo missions, the Soviet Union's Luna missions, Viking and Mars Polar Lander mission for landing on Mars. With development of microwave imaging technology, in addition to microwave range and velocity measurement, the Mars Science Laboratory of the United States had also added millimeter wave imaging capability to identify slopes and obstacles on surface of Mars to ensure safe landing of the lander.

It can be seen in Table 7.1 that the radar ranging and speedometers were used for navigation sensors in the lunar landers of the 1960s and 1970s, such as Apollo, Surveyor and Luna missions. But due to the technical limitations at the time, no imaging sensors for obstacle identification were used, while the Apollo 11 lander avoided the obstacle to the lander through the translational maneuver made by the

Table 7.1 Navigation sensors used in foreign the lunar landers

Lander name	Navigation sensors
Apollo Lunar Module	Powered descent phase: 15 ~ 12 km, IMU; 12 ~ 7 km, IMU and Radar Altimeter (1-D); 7 ~ 2.3 km, IMU, Radar Altimeter and Doppler Velocimeter (3-D) Approaching phase: 2.3 km ~ 156 m, IMU, Radar Altimeter and Doppler Velocimeter (3-D) Landing phase: 156 ~ 1.7 m, IMU, Radar Altimeter and Doppler Velocimeter (3-D) Touchdown: 7 ~ 0 m, probe Note: The radar antenna had a driving mechanism that could switch two fixed angles, equivalent to two sets of range and velocity sensors
Surveyor	~7.8 km: IMU 7.8 km ~ h*: IMU (h* is the maximum range of Radar Altimeter and Doppler Velocimeter) h* ~ 4 m: IMU, Radar Altimeter, Doppler Velocimeter
Luna Lander	15 ~ 2 km: IMU 2 km ~ 700 m: Attitude Adjustment, free fall, IMU 700 ~ 20 m: IMU, Radar Altimeter, Doppler Velocimeter 20 ~ 3.5 m: IMU, Radar Altimeter, Doppler Velocimeter
SELENE-2	Radar Altimeter: 50 km ~ 50 m Laser Altimeter: 5 km ~ 10 m Laser Velocimeter: 50 m/s ~ 0 Navigation Camera: for navigation Laser Scanning Sensor and Camera System (for obstacle detection): 500 m ~ 1 m

astronauts after finding obstacles before landing. Lunar soft-landing mission in the twenty-first century, such as SELENE-2 mission of Japan, is not only equipped with advanced microwave range and velocity sensors to improve range and velocity accuracies, but also equipped with a Laser Scanning Sensor and optical sensors specifically for lunar surface recognition to improve landing safety, and also Camera System for navigation, and etc.

For landing guidance, the gravity-turn guidance law was applied in the Surveyor missions, while the nominal trajectory guidance law was applied in the Luna missions and the fourth-order polynomial guidance law was adopted in the Apollo missions.

In strategic plan of CLEP, there were three phases including orbiting, landing, and sample return, which would be implemented step by step. In 2007 and 2010, the Chang'E-1 and the Chang'E-2 probes achieved the target of the "orbiting" phase and verified the guidance navigation and control technology in LTO and circumlunar orbit. The Chang'E-3 the lunar lander achieved the target of the "landing" phase and verified the guidance navigation and control technology of lunar soft-landing [4].

7.3 Analysis of Technical Characteristics

During lunar soft-landing, the task and work conditions of the lunar lander GNC subsystem were very different from those of satellites and spacecraft developed in the past. The duration of lunar soft-landing process was short (about 10 min). At the same time, the Moon was a non-cooperative target. Therefore, it was necessary to design a highly autonomous and real-time GNC system. The soft-landing process was achieved by active deceleration and the propellant consumption was large (about 2/3 of the lunar lander mass at beginning of powered descent phase). The optimization of the propellant consumption was necessary when the guidance law was designed. The high safety should be ensured for soft-landing mission, which needed rigorous requirements for navigation, hazard detection, and obstacle avoidance guidance for the landing process. High-precision tracking of the guided target attitude should be achieved, while the engine disturbance torque should be compensated at the same time in attitude control. In addition, the effect of lunar surface dust and the engine plume on performance of the sensors and the system during the landing process should be considered [5].

7.4 Design Methodology

7.4.1 System Configuration

Generally, there are three parts in GNC system including controller, sensors and actuators [6]. According to the task requirements, the configuration of GNC subsystem of the Chang'E-3 lunar lander was determined as follows (as Fig. 7.1) after the redundant design of the subsystem and the product status was considered.

1. Controller

Central Control Unit, including control computer, image processing board, secondary power supply, subsystem power supply and distribution.

2. Sensors

Including Star Tracker Sensor, Digital Sun Sensor, Analog Sun Sensor, IMU (including gyro assembly, accelerometer assembly, IMU processing circuits), Laser Rangefinder, MRVS, 3-D Laser Image Sensor, Optical Image Sensor, Touch-down Switches (affiliated to landing cushion subsystem) and Gamma Shutdown Sensor.

3. Actuators

Propulsion subsystem composed of a 1500 N ~ 7500 N throttling engine, 16 × 150 N and 12 × 10 N thrusters.

7.4.2 Operational Modes

Only one part of the GNC subsystem should run during a certain flight phase. Different function systems were formed as required. One operation mode of the GNC subsystem was a generalization of operation modes of the GNC subsystem that performed the same or similar tasks and functions at certain phases of the lunar lander mission.

There was no definite standard for the operation modes. The operation modes of the GNC subsystem of the Chang'E-3 lunar lander were given in Table 7.2.

7.4.3 Component Design

The GNC subsystem usually consisted of three major parts: sensors, controllers and actuators as described in Sect. 7.4.1.

The sensors could be classified into attitude sensors and navigation sensors. There were typical attitude sensors including infrared Earth sensors, sun sensors and star sensors, inertial sensors including gyros and accelerometers, radio sensors including

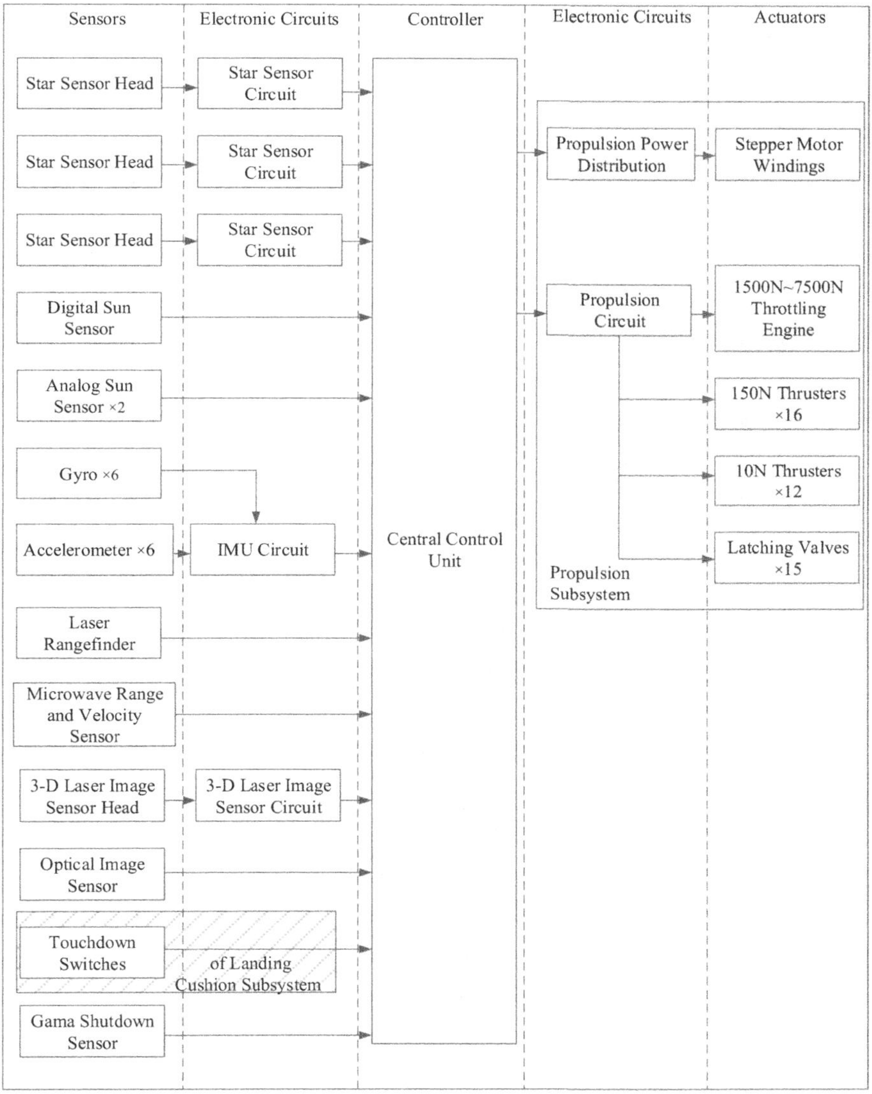

Fig. 7.1 Schematic diagram of GNC subsystem of the Chang'E-3 lunar lander

RF sensors and magnetometers. There were typical navigation sensors including relative range and velocity sensors, imaging navigation sensors, etc.

The controller could be implemented by analog circuits, digital circuits or electric computers.

Generally, there are three types of actuators including jet actuator, electromechanical actuator and environmental actuator. The jet actuator discharges the gas (work

Table 7.2 Operation modes of GNC subsystem

Operation modes	Attitude determination (Sensors)	Attitude control (actuators)	Orbit control	Phase
Launch	/	/	/	Before separation of lander and launch rocket
Sun orientation	Sun capturing and orientating (sun sensors for pitch and yaw angle, gyros for tri-axial angular rate and integration)	Phase plane (10 N thrusters)	/	After separation, LTO cruise, circumlunar orbit, failure safe mode for non-orbit control phase
Star orientation	Star orientating (star trackers and gyros)	Phase plane (10 N thrusters)	/	Before and after establishment of ignition attitude for orbit maneuver
Inertial attitude adjustment	Gyro prediction (gyros)	Phase plane (10 N thrusters)	/	Attitude adjustment for ignition attitude before orbit maneuvers
Orbit control orientation	Star orientating (star trackers and gyros)	PID + PWM (10 N and 150 N thrusters)	Main engine or 150 N thrusters on-off control	Orbit control during LTO and circumlunar orbit
Ignition preparation	Star orientating (star trackers and gyros)	PID + PWM (10 N and 150 N thrusters)	/	Establishing ignition attitude before powered descent
Main deceleration	Main by gyro prediction (gyros), star trackers information introduced if necessary		Orbit determination (IMU, MRVS and Laser Rangefinder), Orbit control (7500 N)	Main deceleration during powered descent
Quick attitude adjustment			Orbit determination (IMU, MRVS and Laser Rangefinder), Orbit control (1500 N ~ 5000 N throttling thrust and 150 N thrusters)	Quick attitude adjustment during powered descent
Approaching				Approaching during powered descent
Hover				Suspending during powered descent
Hazard avoidance				Hazard avoidance during powered descent

(continued)

Table 7.2 (continued)

Operation modes	Attitude determination (Sensors)	Attitude control (actuators)	Orbit control	Phase
Slow descending				Slow descending during powered descent
Rate damping	Gyro prediction (gyros)	Phase plane (10 N thrusters)	/	Failure recovery during orbit control
Non-control	Gyro prediction (gyros)	/	/	Fault

medium) generated by propellant at a high speed to generate reaction thrust, which is also called mass ejection actuator. The electromechanical actuator is generally composed of a drive circuit, a motor, bearing, transmission, and rotational inertia, such as inertial flywheel, control moment gyro for attitude control of space station, de-spinning component for dual-spinning spacecraft, directional drive assembly for solar panel, gimbal drive assembly for antenna pointing control, and etc. The environmental actuator uses the devices that generate moments through the interaction between space environment and the spacecraft, such as magnetic torque, gravity gradient rod, solar radiation pressure adjustment wing, and aerodynamic torque wing, and etc.

According to requirements of the lunar lander mission, there were following components in the GNC subsystem of the Chang'E-3 lunar lander.

1. Sun Sensor

The cruising attitude (sun orientation attitude) was maintained to ensure the power supply of the lunar lander in most time of LTO and circumlunar orbit.

The Sun sensor was used to determine the direction of the Sun in the body coordinate system of the lunar lander and send solar orientation information to the Central Control Unit for sun capture and orientation control, which was a key component of the lunar lander during LTO and circumlunar orbit phases. It consisted of two parts: sensor head and sensor circuit. The sensor head was used to sensitize Sun light and convert it into electrical signals such as digital sun sensor and analog sun sensor. The sensor circuit (integrated in the Image Processing Board of the Central Control Unit) could process the signal output from the head to obtain the information about the sun orientation and send it to the control computer.

The Sun sensor circuits were integrated into the Image Processing Board of the Central Control Unit and there was a cold backup of Image Processing Board. The output information of the Sun sensor was collected by its circuits and transferred to the control computer through the internal serial port of the Central Control Unit. The block diagram of the Sun sensor is shown in Fig. 7.2.

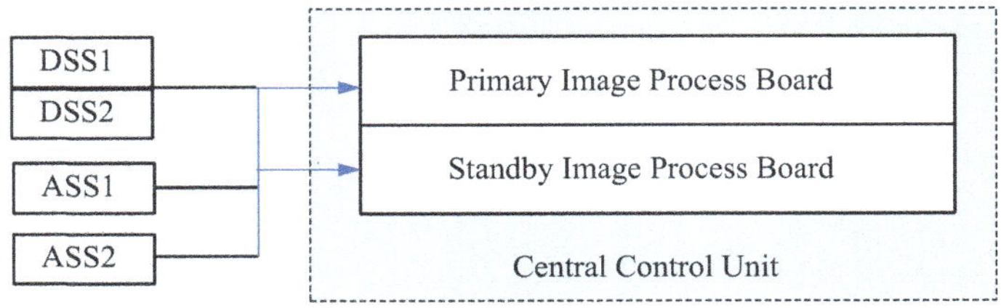

Fig. 7.2 Block diagram of Sun sensor

2. Star Tracker Sensor

The principle of the Star Tracker Sensor was to determine the orientation of the sensor in the inertial frame by the distribution of different star maps in a specific region of the celestial sphere and then the attitude of the spacecraft in the inertial frame could be determined, which had a full-sky capture and tracking capability. The Star Tracker Sensor consisted of a star sensor head and a star sensor circuit, which were connected by cables.

The Star Tracker Sensor head consisted of lens hood, optical system, refrigerator, APS timing/drive and A/D conversion processing circuit.

The Star Tracker Sensor circuit consisted of image processing hardware circuits related to attitude determination, refrigerator control circuits, and secondary power supply, and etc.

The lens hood was used to block the light directly from the sun into the lens, while blocking the reflected sunlight form the Earth, the Moon or the surface of the spacecraft and other components.

The lens was used to photograph the star. It should ensure the performance specification of angle of field of view, focal length, relative aperture, and other imaging quality requirements including chromatic difference of magnification, color distortion, distortion, and diffuse spots, and etc.

The mechanical structure was used to support and fix the optically sensitive head and circuit board, as well as the light tightness of the optical system. At the same time, the thickness was increased to resist radiation.

3. IMU

The IMU included gyros and accelerometers. The gyroscope was used to measure the angular rate of the lunar lander in the inertial frame. The accelerometer was used to measure the specific force of the lander. The IMU consisted of four components: two single-degree-of-freedom rate integrating gyro assemblies (GA1, GA2), an accelerometer assembly (AM) and an Inertial Measurement Unit Electronics (IMUE).

The gyro assemblies consisted of 6 single-freedom rate integrating gyros.

The accelerometer assembly consisted of quartz flexible accelerometers, secondary DC power supply, I/F conversion circuit, structure and shock absorber.

4. MRVS

The MRVS was used to measure the distance and velocity of the lunar lander relative to lunar surface so as to correct the navigation results of the IMUs. The MRVS consisted of two types of sensors: range sensors and velocity sensors, wherein the range sensor used pulsed wave Doppler at long distance and uses quasi-continuous wave for close range with separate receiving and transmitting antennas. The velocity sensors used continuous wave with separate transmitting and receiving antennas were standalone. The MRVS consisted of antennas, velocity measurement diaphragms, transmitting/receiving channels, signal process unit, frequency synthesizer, crystal oscillators, power distribution controller, and matching waveguides and cables.

The 10 antennas of the MRVS formed 5 measuring beams (a receiving antenna and a transmitting antenna for each beam): R1 (T1), R2 (T2) and R3 (T3) were three standalone velocity measuring beams, R4 (T4) was a long-distance ranging beam, and R5 (T5) was a short-range ranging beam. The transmitting/receiving channels included ranging transmitting/receiving channels and velocity transmitting/receiving channels (three velocity measurement beams shared the transmitting channel, but the receiving channel was mutually independent). The signal process unit included a range DSP, a primary velocity DSP, and their backup. The crystal oscillator, frequency synthesizer and power distribution controller were backed up, in which the crystal oscillator was hot backup, frequency synthesizer and power distribution controller were cold backup. In addition, three waveguide switches were configured to complete the primary-backup switching for speed measurement and beam switching for long and close range measurement.

Configuration of the MRVS is shown in Fig. 7.3.

5. Laser Rangefinder

The Laser Rangefinder was used to measure the distance between the lunar lander and lunar surface. The Laser Rangefinder used the pulsed laser to measure the distance. It adopted an integrated design and had two orthogonal directions to meet the requirements for long distance and near distance ranging respectively.

The block diagram of the Laser Rangefinder is shown in Fig. 7.4. The laser unit was an active Q-switched high-power laser. From a color separation sheet on a beam expander of the laser unit, the beam was divided into two orthogonal directions of –Z axis direction and –X axis direction.

The Laser Rangefinder consisted of laser and laser drive unit, laser receiving unit, signal process unit, secondary power unit and thermal control unit.

6. 3-D Laser Image Sensor

Because the scientifically valuable areas are usually cliffs or craters, where the obstacles are widely distributed and lunar terrain is complex. Therefore, in order to ensure that the lunar lander was landed accurately in a safe area, it was necessary to configure imaging sensors to image the lunar surface in order to achieve hazard avoidance and ensure landing safety. The 3-D Laser Image Sensor was one of the landing navigation sensors for 3-D imaging of the landing area when the distance was about 100 meters above lunar surface. The obtained 3-D elevation map data of lunar surface was used for hazard detection and obstacle avoidance.

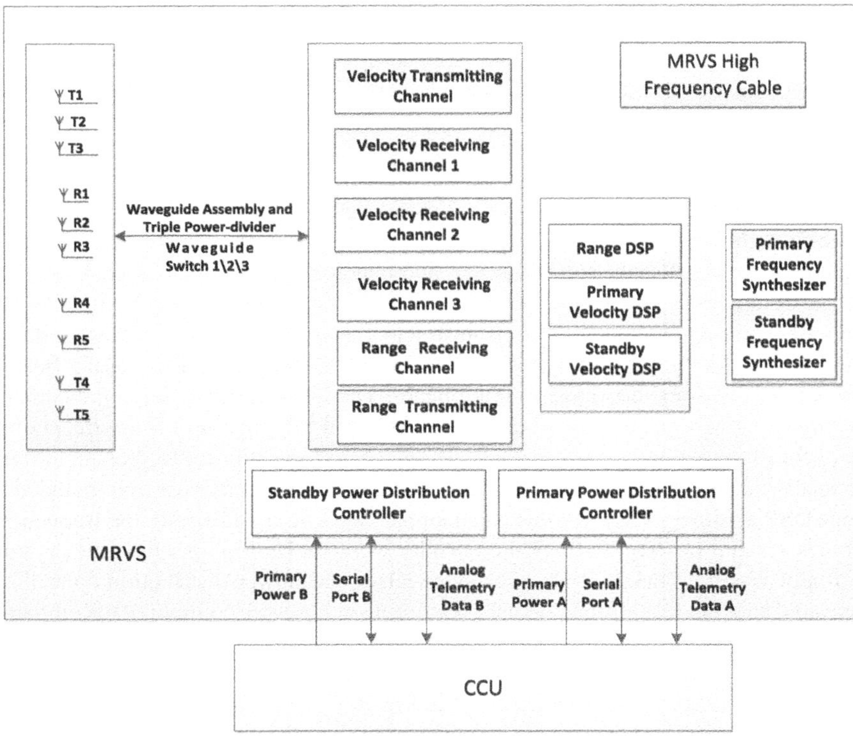

Fig. 7.3 Block diagram of the MRVS

The sensor consisted of two parts, the head and the electric control box. The sensor head was mainly composed of a fiber laser and emission optics, a two-dimensional scanning mechanism and a drive circuit, and a laser echo receiving and amplifying unit. The main wave and echo of the laser were amplified and transmitted to the electric control box for processing through cables. The electric control box consisted of a main control circuit, a motor control circuit, a time interval measurement circuit and a secondary power supply.

7. Optical Image Sensor

The Optical Image Sensor was used to image the landing area, select the landing point to ensure the safety of landing, and achieve long-distance obstacle avoidance. The sensor was a split-type structure with an optical head and an image processing board.

The Optical Imaging Sensor included an optical system, a detector circuit, a mechanical structure, a Digital Processing Unit (DPU), and software. The digital processing unit was on the image processing board and connected to the detector circuit with cables. The software included a hardware initialization module, an information processing module, an image transmission module, a communication module,

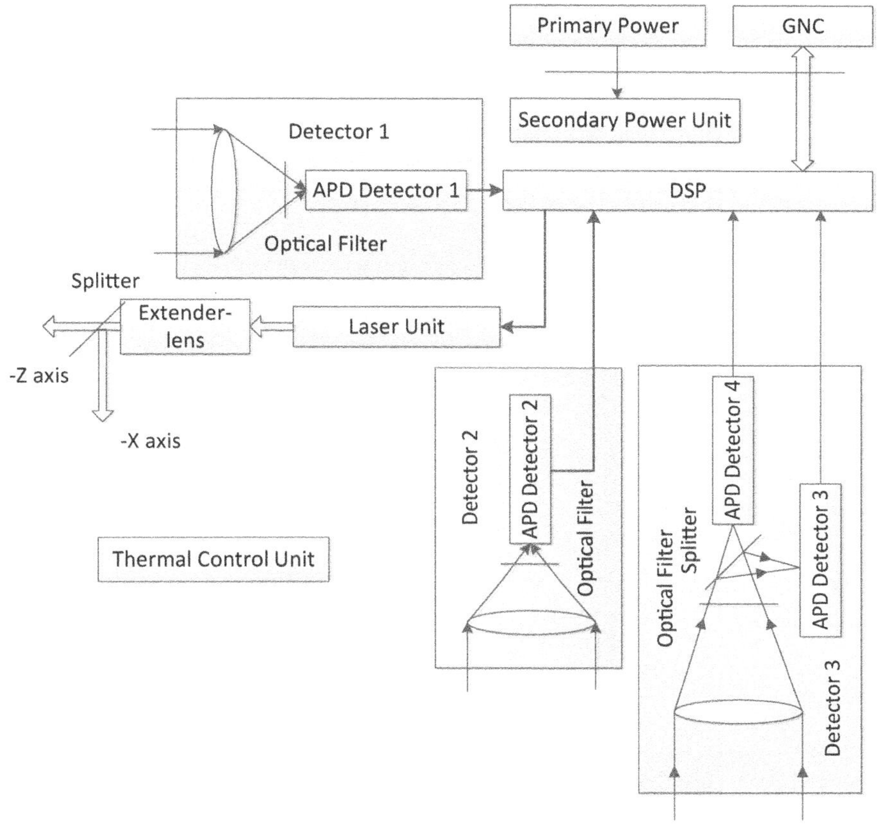

Fig. 7.4 Block diagram of Laser Rangefinder

and etc. burned in the digital processing unit. The block diagram of the Optical Image Sensor is shown in Fig. 7.5.

8. Gamma Shutdown Sensor

The Gamma Shutdown Sensor was used to accurately measure the height of the lunar lander from lunar surface when it was close to lunar surface, and gave the engine shutdown signal at a specific height from lunar surface. The Gamma Shutdown Sensor consisted of a transmitter and a receiver. The transmitter was a gamma source and was a passive component of the subsystem. The receiver included three separate gamma ray receiving and processing channels, and provided three altitude data along each gamma ray as shown in Fig. 7.6.

9. Central Control Unit

The Central Control Unit was the controller of GNC subsystem. It was the information processing center of the entire GNC subsystem. The measurement signals of all

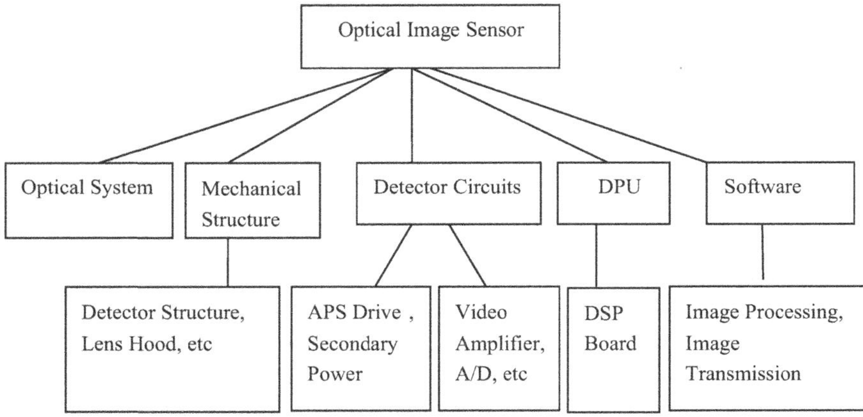

Fig. 7.5 Block diagram of Optical Image Sensor

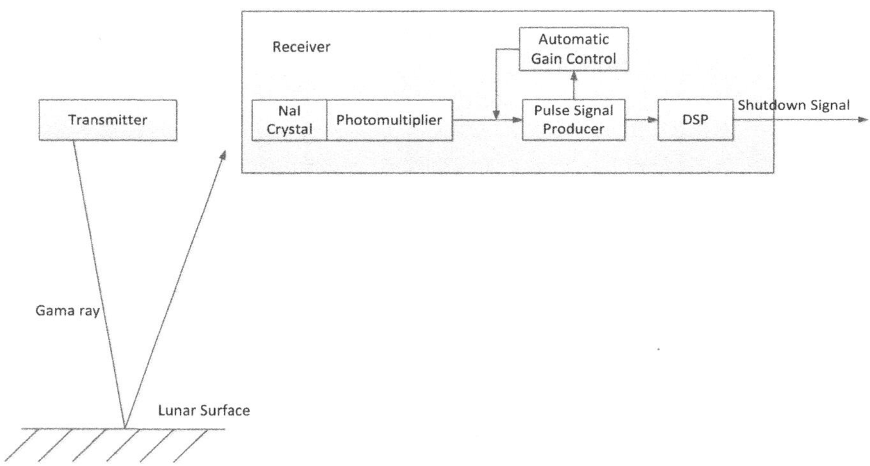

Fig. 7.6 Block diagram of Gamma Shutdown Sensor

sensors were output to the Central Control Unit. After the navigation, guidance and control laws were calculated, the command signals were output to the actuators to accomplish the function of GNC subsystem. The Central Control Unit consisted of on board computers, a secondary power supply module, a power distribution board, public I/O, and image processing boards as shown in Fig. 7.7.

The image processing boards received image data of the 3-D Laser Imaging Sensor and the Optical Imaging Sensor; collected current signal of the digital sun sensors and analog sun sensors; and output the safe landing point information and the sun orientation information obtained by image processing.

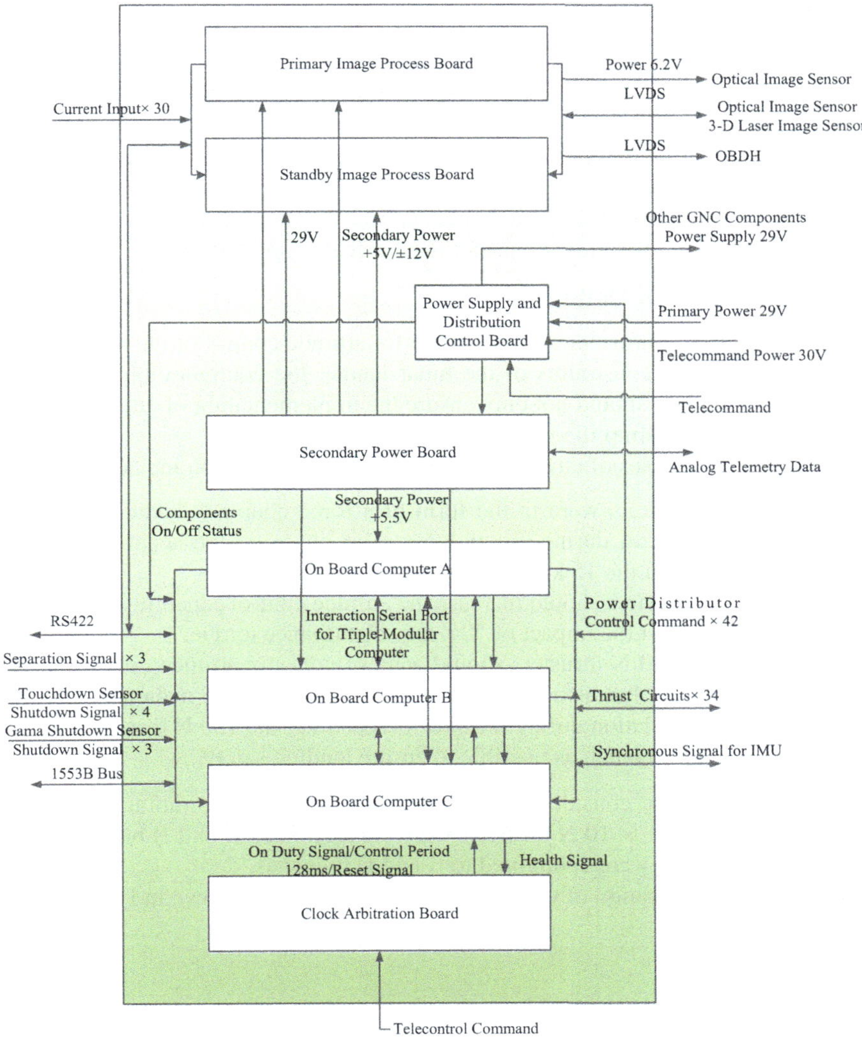

Fig. 7.7 Block diagram of Central Control Unit

The secondary power module provided secondary power for the computer module and the image processing module.

Power distribution module distributed primary power for each component of the GNC subsystem, computers and image processing board.

The computers collected and processed the information of each sensor, controlled the throttling engine and the attitude control thrusters, received and responded the ground remote control information to GNC subsystem, and transmit the GNC subsystem telemetry data to the ground.

10. Orbit Control Engine

The Orbit Control Engine was actuator of the lunar lander for orbit maneuver and powered descent. The engine was a 7500 N throttling engine, which could provide continuously throttling thrust through flow adjustment, so that the engine had less influence on the tremble of the lunar lander and it was much easier to design the guidance law when the engine thrust was throttling. The Orbit Control Engine was installed on the −X side of the lunar lander, and the direction of thrust vector pointed to the +X direction of the lunar lander's mechanical coordinate system.

11. Attitude Control Thrusters

The attitude control thrusters were actuators for attitude control of the lunar lander. Due to the resource constraints of the lunar lander, the configuration of the attitude control thrusters should not only ensure the implementation of attitude control function, but also be simplified as much as possible.

The design principles of attitude control thruster configuration included:

(1) The 10 N thrusters were in the form of a force couple to reduce the effect of attitude control during circumlunar orbit phase on the accuracy of orbit determination at the 15 km perilune.
(2) The 150 N thrusters should increase the attitude control capability as much as possible to resist the impact of Y, Z axis disturbance torque.
(3) Because the 150 N thrusters should achieve not only attitude control during orbit maneuver, but also the orbit control of the obstacle avoidance and the horizontal translation during the slow descending, the 150 N thrusters should be backed up as much as possible to ensure landing safety.

Based on above three considerations, the final thruster configuration was finally determined, where 12 × 10 N thrusters were configured to A and B branches, and 16 × 150 N thrusters were configured to C and D branches.

Thrusters Configuration of Chang'E-3 the lunar lander is shown in Fig. 7.8.

7.4.4 Software Design [7]

Recently the application of computer systems in spacecraft had become more and more widespread along with development of aerospace industry and computer technology in China. Due to complexity of navigation, guidance and control algorithms, GNC subsystem was one of the subsystems that prefer application of computer systems.

With the continuous improvement of computer hardware performance and quality, the requirement for computer software was getting higher and higher. The size of GNC subsystem software was getting larger and larger and the complexity was getting higher and higher. The code of the GNC subsystem application software had exceeded 20,000 lines.

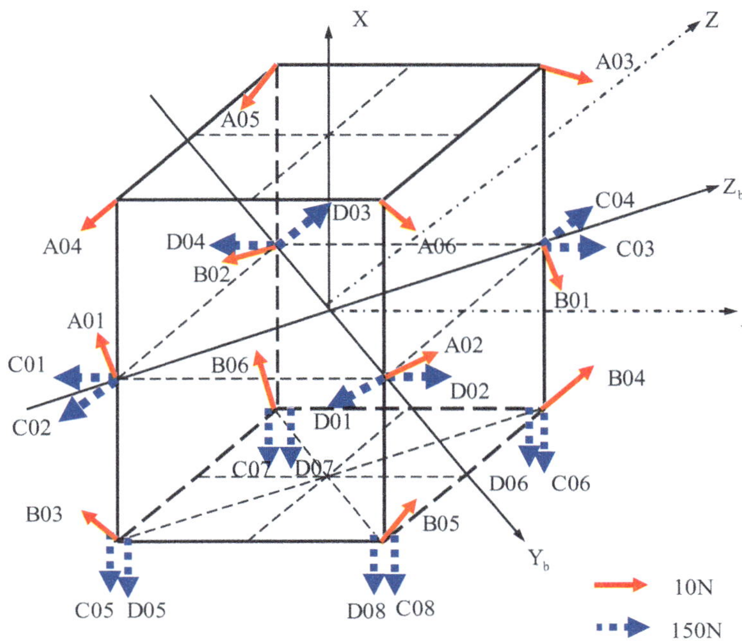

Fig. 7.8 Thrusters configuration of Chang'E-3 the lunar lander

1. GNC Software Characteristics

(1) Embedded real-time software. All subsystem software was an embedded system that worked together with the host computer to embed the entire subsystem. There were number of interfaces and real-time data transmissions within and outside the subsystem software.

The GNC subsystem was a system with high real-time requirements. The software played a key role in command and control within the subsystem. It undoubtedly appended high real-time requirements for software processing and data transmissions.

(2) High reliability and high security. The lunar lander had enormous economic value, as well as high reliability and safety requirements, and the requirements for the GNC subsystem were even more so. As an integral part of the subsystem, it was undoubtedly that the subsystem software should have high reliability and security requirements.

(3) Large scale and complex architecture. Because of the high quality, high reliability, and high security requirements of the lunar lander software, the size of the GNC subsystem software was enormous compared to similarly software. At the same time, the architecture of the subsystem software was very complex when it came to complicated data processing and accurate command & control calculating.

2. Software Configuration

Based on the design of subsystem components and above software characteristics, there were 10 types of software in GNC subsystem as shown in Table 7.3. The 10 types of software were embedded in the corresponding GNC subsystem components (host computers) to implement various functions of the GNC subsystem.

3. Software Function

Functions of the 10 types of GNC subsystem software are shown in Table 7.4.

4. Design of CCU Software

The control computer software was the "nerve center" of the entire subsystem and played the role of comprehensive dispatching and commanding. T design methodology and ideas of the CCU software are described in this section.

The CCU software was a multi-task real-time operating system, and consisted of system software and application software, both of which were burned in the PROM.

The software tasks could be classified into control, triple computer comparison, telecommand and telemetry, system monitoring and spare as shown in Table 7.5. Each task was uniformly scheduled and managed by the system software, according to its priorities and criticalities. Except the spare task, each task should be completed within a specified period of time.

The GNC subsystem application software of the Chang'E-3 lunar lander was designed to inherit the mature design concept of the Chang'E-1. The main operation modes included launch, Sun orientation, star orientation 2, inertial attitude adjustment, star orientation 4, orbit control orientation, non-control, and rate damping.

The control modes of powered descent phase included main deceleration, quick attitude adjustment, approaching, hovering, hazard avoidance, and slow descending. Under each mode, data of star tracker sensors, Laser Rangefinder, MRVS, and IMU are obtained for attitude and orbit determination, as well as to complete the GNC

Table 7.3 Software configuration of GNC subsystem

No.	Name	Configuration
1	Computer system software	1
2	Computer application software	1
3	Image Process Board DSP software	1
4	APS Star Tracker system software	1
5	APS Star Tracker application software	1
6	APS Star Tracker star catalog software	3
7	MRVS range DSP software	1
8	MRVS velocity DSP software	1
9	MRVS Power Distribution Controller software	1
10	IMU software	1

Table 7.4 Function of GNC software

No.	Name	Function
1	Computer system software	Managing the system resources based on real-time operating system, providing various system service functions related to the application software, and the application software ran in the environment of the system software
2	Computer application software	Lunar lander attitude data acquisition, real-time data processing and calculation, navigation and guidance command calculation, obstacle avoidance algorithm calculation, output of various control signals, and communication with OBDH subsystem to receive remote control command and output telemetry data, so that the attitude and orbit control of the lunar lander was achieved
3	Image Process Board DSP software	Completing image processing for Optical Imaging Sensor and Laser 3-D Imaging Sensor, handling digital sun sensors and analog sun sensors measurement data, obtaining self-check information and telemetry data, receiving CCU control command and sending safe landing point and self-check status to CCU, sending the image data of Optical Imaging Sensor and Laser 3-D Imaging Sensor to OBDH subsystem through LVDS interface
4	APS Star Tracker system software	Completing system self-test and error correction, achieving application software loading and operation mode transition, receiving CCU command data packet
5	APS Star Tracker application software	Achieving star identification and attitude determination, obtain the analog telemetry data
6	APS Star Tracker star catalog software	Including calibration parameters table, navigation star catalog, navigation star index table, navigation star template index table, parameters configuration table
7	MRVS range DSP software	FPGA signal acquisition and decoding, so the range information was obtained
8	MRVS velocity DSP software	FPGA signal acquisition and decoding, so the velocity information was obtained

(continued)

Table 7.4 (continued)

No.	Name	Function
9	MRVS Power Distribution Controller software	Completion of power supply and distribution of MRVS, receiving CCU command data and controlling the work status of MRVS, collecting the telemetry data of MRVS, obtaining measurement data of the ranging signal processor and velocity signal processor, and sending it to CCU via the serial port after the measurement data was packed, and providing range and velocity information of the lander relative to lunar surface to GNC subsystem during landing phase
10	IMU software	Obtaining signals of 6 gyros and 6 accelerometers, calculating the information about the angle and acceleration change rate, performing self-check on the status of IMU and collect telemetry data, receiving CCU control commands, send measurement and telemetry information

Table 7.5 Tasks of GNC software

Name	Priority	Task Function
1	1 (Highest)	Data acquisition, work mode identification, attitude computing, attitude control, navigation algorithm, guidance law, hazard avoidance algorithm, autonomous power management, system error detection and selection, gyro error detection and selection, accelerometer error detection and selection, LTU (Local Terminal Unit) self-check and LTU error detection, triple computer comparison data refreshing, etc.
2	2	Decoding the uplink data, responding the telecommand command and packing the telemetry data, etc.
3	3	Realizing vital data transmitting, receiving, comparison and 2 out of 3 voting of the triple computer
4	4	System monitoring (including task operation time, stack), updating self-check variables, assigning system software telemetry data, setting up computer health flag and clear software dongle
5	5 (Lowest)	Memory real-time refreshing

task by output commands to thrusters and main engine according to the control law requirements.

The control period was used to synchronize the internal of the subsystem. The synchronization signal inside the subsystem was initiated by CCU and responded by other components, i.e. the CCU was responsible for initiating communication signals with each component. Each component passively responds to CCU commands. Under normal circumstances, CCU initiated a communication signal each control period.

Real-time operating system time slices were used for real-time scheduling of control computer tasks. The synchronization between tasks of the CCU system software and the application software used time slices of the real-time operating system. Each task was assigned one or several time slices. If the task did not exit normally when it was timeout, the task was suspended and the CPU performs other tasks.

7.4.5 Autonomous Failure Diagnosis and Handling Logic Design

1. Autonomous Failure Diagnosis and Handling Principles

Under normal circumstances, the purpose of autonomous fault diagnosis and handling of GNC subsystem is to ensure that the critical tasks are completed on schedule in the event of a single onboard failure and the spacecraft is safe (to ensure power supply and prevent excessive fuel consumption). In combination with the characteristics of the lunar lander, the fault handling of the GNC subsystem was classified into two levels:

(1) Low level: component failure, including gyro failure, IMU circuit failure, MRVS failure, Star Tracker Sensor failure, accelerometer failure, and LTU communication failure;
(2) High level: Subsystem level faults, including unfinished subsystem functions within a specified time, continuous overproof attitude within a specified time period and continuous thrusters on within the specified time in the work mode.

In critical orbit maneuver phase, the principle of fault handling was to maintain the status as much as possible. In the powered descent phase, due to the phase was irreversible, the principle of fault handling was to ensure that the lunar lander could be landed safely.

The fault diagnosis logic should be as simple and clear as possible to prevent wrong judgment.

2. Components Autonomous Failure Diagnosis and Handling Methods

Gyro and accelerometer fault diagnosis principle was based on the number of work gyros and accelerometers that appointed by telecommand to determine whether to perform autonomous diagnosis.

Gyros and accelerometers that participated in the attitude determination were selected according to the scores. Gyro and accelerometer scoring included: scoring according to the consistency check results of six gyroscopes and accelerometers. The three gyros and accelerometers with the highest scores participated in the attitude determination and set the corresponding on-board gyro and accelerometer work status words.

The principle of IMU circuit fault diagnosis was to eliminate gyro and accelerometer outliers output data caused by single event upset. Set the upper limit of the output data of gyro and accelerometer according to the mass and inertia characteristics of the lander. If the output data exceeded the upper limit, the output data of the dynamic model was used instead of the output of the sensor.

According to the diagnosis allowable sign of telecommand, it was determined whether to perform fault diagnosis on the MRVS. The diagnosis method was: under the normal communication condition, if all the beam data of the microwave velocity measurement were continuously invalid, the MRVS was considered to be faulty. The diagnostic handling method was: switch to the backup MRVS when the main MRVS was fault and the status word showed that the MRVS could be chosen autonomously.

According to the star tracker status word of telecommand and communication error counter, the corresponding star tracker sensor was selected to work. When a star tracker sensor's validity status word was valid and there was no communication error, the star tracker's output was considered to be valid. When any of the star tracker outputs assigned by telecommand was valid, the star sensor was used. When a plurality of star sensors was normal, one output was sequentially selected as a measurement data for attitude determination. Otherwise, the gyro data was used to determine the attitude.

LTU communication fault diagnosis occurred during each control period. A single communication failure was a communication error, and consecutive control period errors were communication failure. If the LTU with active and standby cold backup failed to communicate, it was switched to the backup LTU. If the LTU was not backed up, the application software would not handle it autonomously after the communication failure, and if more than one LTU on duty was found to be in communication failure within one control period, then the control computer was switched.

3. GNC System Autonomous Failure Diagnosis and Handling Methods

(1) Diagnosis and handling methods of the main deceleration: If the attitude tolerance of the main deceleration caused by the disturbance torque was greater than the control torque continuously, a strategy of reducing the disturbance torque by using the thrust force of 5000 N was required. Based on the consideration of reducing the consumption of propellant, the target state of 3 km was changed to the target state of 100 m for the lunar lander above 4.5 km and the thrust of the main engine was changed from 7500 N to 5000 N, and the structure of the PEG law was unchanged, i.e., only the guidance target should be replaced.
    If altitude of the lunar lander was between 4.5 km and 3 km, the 7500 N main engine was switched to 5000–1500 N throttling thrust, and target status was changed to 100 m hovering status by using the designed safe mode guidance law.

(2) Diagnosis and handling methods of the approaching: at the beginning of the approaching, the nominal trajectory of the approaching was calculated based on the navigation data at current time. During approaching, if the deviation between the actual trajectory and the nominal trajectory was large, the safe mode guidance law was used.

(3) Diagnosis and handling methods of the hovering and below: If the speed of the lunar lander body axis Y_B or Z_B was large, the horizontal position control was used to reach the safe landing point.
(4) Propellant shortage: If the 7500 N throttling thrust engine had a lower actual thrust than the nominal value, it would cause the main deceleration to take longer and consume more propellant, causing inadequate propellant for follow-up phases, in this case, the strategy of cancel the approaching was used to reduce propellant consumption.
(5) Safe mode guidance law design: The principles of safe mode guidance law were: prefer to ensure vertical safety and propellant consumption as less as possible. The vertical optimization guidance law was adopted to meet the vertical height and speed constraints, and guide the lunar lander to reach the target altitude at a predefined velocity, and ensure vertical safety. To simplify calculations and satisfy vertical constraints, only vertical channel used polynomial guidance laws. The horizontal guidance law, in order to reduce the consumption of propellant as much as possible, controlled the velocity and acceleration only in the horizontal direction without controlling the position, and also eliminated the horizontal velocity as quickly as possible to reduce unsafe landing factors.

7.5 Typical Technology

7.5.1 Procedure of GNC Powered Descent Phase [8]

GNC subsystem hardware during the powered descent phase included sensors, actuators, and controllers. The close loop of GNC subsystem of the Chang'E-3 lunar lander during the powered descent phase is shown in Fig. 7.9. The controller was divided into a control computer and an image processing unit. The navigation, guidance and attitude control algorithms were run on the control computer; the lunar surface hazard identification and safe landing zone selection algorithms were run on the image processing unit. The control computer and the image processing unit could exchange information and data.

The general work flow of guidance, navigation and control of powered descent phase was: the navigation sensor measured and sent the measurement data to the control computer; the navigation algorithm processed the navigation measurement data and sent the navigation results to the guidance and attitude control algorithm; the guidance algorithm was based on navigation results and guidance parameters to form the main engine command and attitude command, and the attitude command was sent to the attitude control algorithm; the attitude control algorithm formed the thrusters command according to the navigation results and the attitude command; the control computer sent the main engine thrust command and the thrusters command to the main engine and thrusters, then the main engine and thrusters executed the commands.

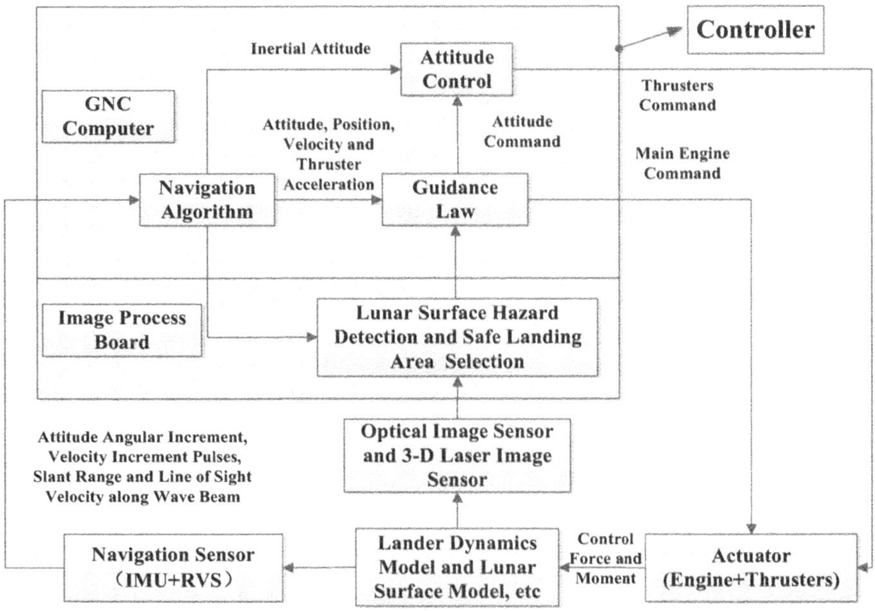

Fig. 7.9 Block diagram of the GNC loop for powered descent phase

During hazard detection and avoidance, the Optical Imaging Sensor sent the lunar surface image of the landing zone to the image processing unit, and the 3-D Image Sensors sent the range data corresponding to lunar terrain of landing zone to the image processing unit; the lunar surface hazard identification and safe landing zone selection algorithm used the navigation information sent by the control computer to process the lunar surface image and the lunar surface terrain data of the landing area, then the safe landing point information (the direction of the line of sight of the safe landing point with respect to the lander) was determined and sent to the control computer, and introduced into the guidance algorithm to complete the hazard avoidance task.

7.5.2 *Powered Descent Trajectory Design*

The trajectory of powered descent phase generally included main deceleration, hazard detection and avoidance, and final descending. The powered descent process of Chang'e-3 lunar lander consisted of 6 phases (trajectory and flight profile of powered descent phase can be seen in Fig. 7.10).

(1) Main deceleration: altitude to lunar reference surface (landing reference surface of the landing area selected before starting the landing) about 15 km to 3 km.

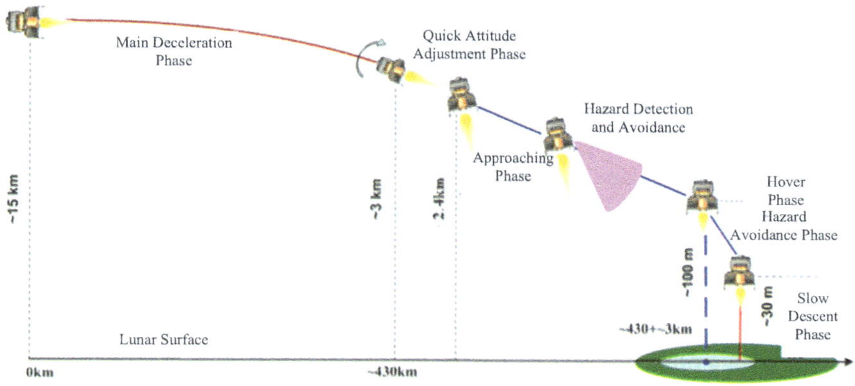

Fig. 7.10 Trajectory and flight profile of powered descent phase

The main task of this phase was soft-landing deceleration to reduce the velocity (about 1.7 km/s) to predefined value (about 70 m/s) and the altitude to about 3 km for of the lunar lander. According to requirements of the landing mission, the trajectory should be a gently descending trajectory formed by the deceleration of the main engine. The altitude was reduced by approximately 12 km and the downrange was approximately 430 km.

(2) Quick attitude adjustment: altitude to actual lunar surface was about 3 km to about 2.4 km. The main task of this phase was to quickly connect the main deceleration and the approaching phase. The fast attitude adjustment maneuver was executed to change the attitude to the entry attitude of the approaching phase. And the main engine thrust was synchronously reduced to a low thrust level. According to requirements of the landing mission, the trajectory should be a gently descending trajectory formed by the deceleration of the main engine. The altitude was reduced by approximately 300 m and the downrange was approximately 1 km.

(3) Approaching: altitude to actual lunar surface was from about 2.4 km to about 100 m. The main task of this phase was rough hazard avoidance. According to requirements of rough hazard avoidance, in order to ensure that the Optical Image Sensor's field of view was aligned with the landing zone, an approaching trajectory to the target landing zone that met requirements of the specific attitude and descent trajectory was designed. The lander straightly approached the landing zone gradually with an inclined angle of 45°. The Optical Image Sensor detected large obstacles (rocks or craters larger than 1 m in diameter), determined a safe landing zone and avoided obstacles, and finally reached a height of about 100 m above the landing zone. The relative velocity approaches zero and the downrange was about 3 km.

(4) Hovering: altitude to actual lunar surface was about 100 m. The main task of this phase was detection of fine obstacles in the landing area, keeping the lander in a hovering status and using the 3-D Image Sensor to image the landing area and selecting a safe landing site.

(5) Hazard avoidance: altitude to actual lunar surface was from about 100 m to 30 m. The main task of this phase was fine hazard avoidance and descending. According to the selected safe landing point, the lunar lander descended to 30 m above the landing point, the relative descent velocity was predefined (not more than 2 m/s), the horizontal velocity was close to zero, and the trajectory was slant to the landing site.

(6) Slow descending: altitude to actual moon surface was about 30 m to the shutdown signal to take effect. The main task of this phase was to ensure that the lunar lander descended steadily and slowly to lunar surface, and the control accuracy of velocity and attitude met the requirements. The lander slowed down at a preset relative velocity to lunar surface, eliminated the horizontal velocity with an attitude vertically to the lunar surface. Until receiving the shutdown sensor signal, the control computer provided command to shut off the main engine and the thrusters, free falling to lunar surface, and the trajectory was almost vertical.

7.5.3 Navigation Method for Powered Descent Phase

The conventional autonomous navigation for powered descent process is inertial navigation with correction by information of range and velocity sensors. In order to further improve the navigation accuracy, image matching navigation could be used. Navigation algorithm process of the Chang'E-3 lunar lander for powered descent process is shown in Fig. 7.11. In initial phase of powered descent, pure inertial navigation was used; in order to overcome the influence of terrain uncertainty and

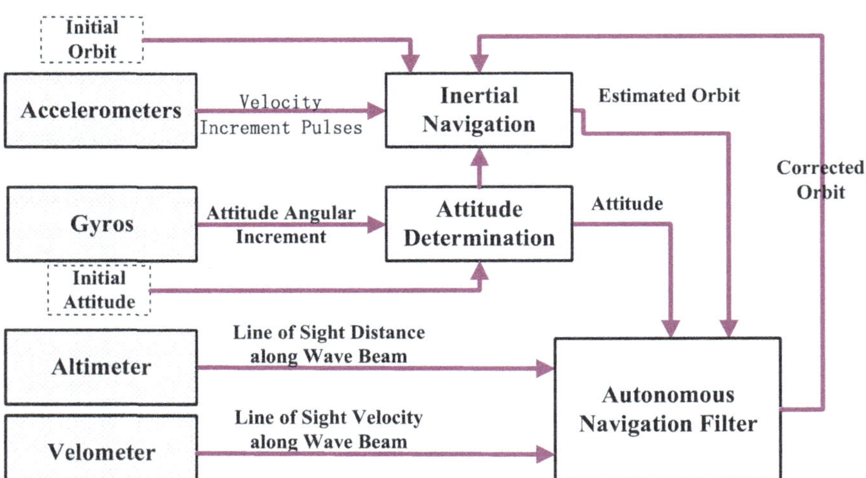

Fig. 7.11 Navigation algorithm process of the Chang'E-3 lunar lander for powered descent process

ensure accurate altitude relative to lunar surface after the end of main deceleration, ranging correction was used in the later stage of main deceleration; after the quick attitude adjustment phase, the lander remained relatively stable attitude to the Moon. Velocity accuracy requirements were getting higher and higher. At this time, the velocity correction was introduced. In the slow descending, due to the low height of the lunar lander to lunar surface, the engine plume would cause lunar dust to adversely affect the range and velocity sensors. Therefore, pure inertial navigation was restored in this phase.

1. Inertial Navigation

The basic navigation method used in the powered descent process was inertial navigation. The attitude, position, and velocity of the lunar lander could be extrapolated based on the angular velocity and non-gravity acceleration measurement information provided by IMU on the lunar lander (including gyros and accelerometers) and integration of translational and rotational kinematic equations of the lunar lander. The major error sources in the inertial navigation system included:

(1) Initial reference error: The initial position and velocity were provided by ground tracking system, the position accuracy was about 1 km, and the velocity accuracy was about 1 m/s. The initial attitude was given by the star sensor and the error was about tens of arc-second.
(2) IMU error. There were IMU errors including IMU installation errors and measurement errors. During LTO, on-orbit calibration of various IMU errors was performed.
(3) Gravitational field model error. Gravitational field model used a two-body model. For the brief 12 min of powered descent phase, the altitude calculation error caused by the error of two-body gravitational field model did not exceed 60 m, which was within an acceptable range.
(4) The error caused by the integration step. From the perspective of improving the accuracy of inertial navigation, it was desirable that the sampling period of the IMU should be as short as possible. Due to the ability of the lunar lander components, it was impossible to achieve a very high sampling frequency, so a four-sample algorithm was used for the attitude updating.

2. Range Correction and Multi-Beam Combination

The Chang'E-3 lunar lander was equipped with a Laser Rangefinder and a MRVS. Both sensors could measure the distance from the lunar lander to lunar surface to correct the altitude information of the inertial navigation system.

For a certain beam, the kinematics equation of the vertical channel of the lunar lander and the ranging sensor measurement equation were established, and the Kalman filter method was used to calculate the gain matrix. The preset gain matrix function on the control computer could be obtained by calculating the steady-state gain matrix for different altitudes based on inertial navigation performance, ranging sensor performance, and lunar surface fluctuations, as well as fitting the trend with curve function of altitude.

When both the Laser Rangefinder and the MRVS provided valid signal outputs, a fusion weighed correction method was used: the respective altitude corrections calculated by the two sensors were weighed and averaged, and then used for position correction. The weighing coefficients were calculated based on the traces of the Kalman filter variance matrix of the two modification system.

3. Velocity Correction and Multi-Beam Combination

The velocity of the Chang'E-3 lunar lander was corrected by the data of the velocity measuring beams in the MRVS. The sensor could provide three independent velocity measuring beams and a beam with both range and velocity measurement functions. In addition, the altitude estimation algorithm could also provide an estimate of the vertical velocity, which could also be introduced as a "beam" in one direction into the velocity correction. Due to the existence of multiple beams, the number of beams measurement information obtained at each time was not necessarily the same. Therefore, in the velocity correction strategy, a single beam calculation error and multi-beam combination correction method was used.

1) Calculation of single beam error

The kinematics equation of the lunar lander and the microwave velocity measurement equation were established. The Kalman filter method was used to calculate the gain matrix. Similar to the range correction, the steady-state gains of different velocity measuring beams were firstly obtained, and then they were fitted with the height-dependent curve function, and the preset parameters of the gain matrix was finally obtained.

2) Multi-beam combination correction.

According to different flight conditions, the number of available effective beams in each measurement period is different. Each error is calculated separately for each effective beam, and they were combined to calculate the velocity correction error vector, then the velocity of the lander was corrected. Obviously, if there were 3 or more effective beams in one measurement period, a complete three-dimensional velocity error correction could be obtained. When the number of effective beams was reduced to less than 3, only the dimensionality reduction velocity correction could be obtained. When two beams were effective, two-dimensional (in-plane) velocity correction could be obtained; when one beam was effective, only one-dimensional (single direction) velocity correction could be obtained.

7.5.4 Guidance Law Design of Powered Descent Process

In each phase of powered descent, different guidance laws should be designed according to the requirements. The main task of the main deceleration was deceleration; The major concern of propellant consumption optimization should be considered when the guidance law was designed. The main task of the hazard avoidance was to identify and avoid obstacles; The formation of a well downward trajectory should be considered so that the landing area could be observed when the guidance law was designed for the phase. The guidance law for final descent phase was designed to ensure that the lunar lander was landed on lunar surface smoothly. The design of the guidance laws for the powered descent process of the Chang'E-3 lunar lander are as follows.

1. Guidance Law for Main Deceleration

The main deceleration was the mission segment when the soft-landing process took the longest and the propellant was consumed the most. The main task of this phase was to eliminate the large initial horizontal velocity of the powered descent process (about 1.7 km/s). Therefore, the optimization of propellant consumption was the main design goal for guidance law of this phase. In addition, the autonomy and engineering feasibility should be taken into consideration. The linear tangent guidance law was close to fuel optimal, and the resulting attitude command was nearly linear, and the change was more gradual, which was conducive to the realization of attitude control. Based on the linear tangent guidance law, an adaptive powered explicit guidance was presented in the main deceleration, as shown in Fig. 7.12. According to the on-orbit acceleration measurement information, a recursive least squares method was used to estimate parameters such as engine specific impulse and guidance time in orbit, and through the terminal state and quick adjustment prediction, the guidance target was adaptively modified to meet the initial constraints of the approach phase. This method not only achieved relatively less propellant consumption, but also improved the system's adaptability to uncertainties such as mass, thrust and specific impulse.

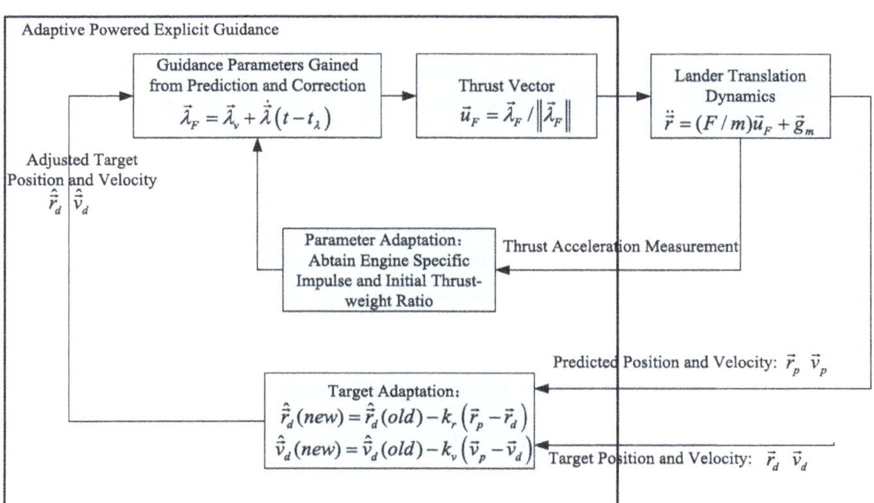

Fig. 7.12 Diagram of adaptive powered explicit guidance for main deceleration

In Fig. 7.12, the parameter adaptation section was to ensure that the engine thrust specific impulse and the initial thrust ratio were estimated to converge to the true value as the accelerometer data accumulated; the target adaptation section was to adjust the target position and velocity, and after the quick adjustment was embedded at the end of deceleration, the predicted terminal position and velocity was continuously approaching the target position and velocity; the two parts were combined to ensure that the position and velocity after quick adjustment converged to the target value gradually.

2. Guidance Law for Quick Attitude Adjustment Phase

At the end of the main deceleration, the attitude of the lunar lander was still close to the horizontal, the main engine was still working in the maximum thrust mode, and the thrust acceleration also reached the maximum; while the subsequent approaching required the attitude of the lunar lander to be close to vertical, the main engine worked at a low thrust level, and the altitude, velocity and acceleration should meet a certain relationship. It could be seen that it was difficult to directly connect the end status of the main deceleration and the initial status of the approach phase. In order to smoothly transition from the main deceleration to the approach phase, it was necessary to quickly adjust the attitude and thrust of the lunar lander to meet the status requirements at the entrance of the approach phase. According to the uniform transition requirements of the thrust of the main engine and the attitude of the lunar lander, a guidance law with the linear variation of the thrust magnitude and direction was proposed. The guidance parameters were determined by the actual state constraints of the final deceleration and the initial state of the approach phase. The same guidance law was also used to predict the quick deceleration process during the main deceleration to ensure that the actual state change of the quick adjustment was basically consistent with the main deceleration, thereby satisfying the constraints of the entrance status of the approach phase.

3. Guidance Law for Approaching

The main task of the approaching was to image the landing area and perform rough obstacle avoidance. The terminal relative velocity to lunar surface was close to 0 m/s. It was needed to ensure that the Optical Imaging Sensor could image the landing area and complete the rough obstacle avoidance during the approach phase. Therefore, the approach guidance law must be able to meet constraints of the position, velocity, attitude, and initial altitude and initial velocity of the guidance target. In order to satisfy the above constraints, based on the fourth-order polynomial guidance law, an improved polynomial guidance algorithm was proposed for the approach phase. When multiple constraints were satisfied, the time to go could be analytically calculated without iterative solution, so that the calculation was simplified and the stability of the algorithm was improved. At the same time, in order to ensure the high reliability of the guidance law for approach phase, a protection method for time to go, altitude and velocity out of tolerance were proposed. By design of the constraints in each direction, it was ensured that the Optical Imaging Sensor could always observe the landing area. After processing the image of the lunar surface, the safe landing site could be reset and landing maneuvers could be achieved within a certain range to complete rough obstacle avoidance.

4. Guidance Law for Hovering

The main purpose of the hovering was to use the 3-D Imaging Sensor to perform fine hazard detection on the landing area, and to provide the position information of the safe landing point relative to the lunar lander bottom point at the imaging time. The initial navigation altitude of the hovering was used as the guidance target of the suspending altitude, and the gravity of the lunar lander was compensated by the throttling engine to ensure that the lunar lander was in a suspending status with zero speed and stable attitude. The guidance goal of the hovering was: the altitude was the initial navigation altitude of the hovering, the target velocity in the three directions was 0 m/s, and the acceleration in the direction of the altitude was 0 m/s^2. In order to ensure the adaptability and robustness of the guidance logic for the hovering and subsequent mission phase, the guidance law of hovering and subsequent mission phase was based on the combination of outer loop guidance (PID) and inner loop guidance (phase plane control using position and velocity). Outer loop control was achieved by the main engine and attitude maneuvers; inner loop control was achieved by horizontal thrusters.

5. Guidance Law for Hazard Avoidance

The main task of the hazard avoidance was to accurately avoid obstacles and descent. According to the relative position information of the safe landing point given by the hovering, the lunar lander would horizontally maneuver above the selected safe landing point and the horizontal velocity at terminal of the horizontal maneuver was around 0 m/s; then the lunar lander descended to about 30 m above the landing point, and the terminal descent velocity relative to lunar surface was −1.5 m/s. The obstacle avoidance guidance law was similar to the guidance law of the hovering. The horizontal channel controlled the velocity and position, and the vertical channel controlled the altitude, velocity and acceleration. At the beginning of the obstacle avoidance, the time-based altitude, velocity, and acceleration to track target trajectory was designed in the vertical direction, and the target trajectory was tracked in real time in the obstacle avoidance descent phase, so that the terminal status of the obstacle avoidance could be better controlled.

6. Guidance Law for Slow Descending

The safety of landing was considered in the slow descending. To ensure the accuracy of the control of the speed and attitude for landing to lunar surface, the vertical descent should be made at a uniform speed with a low preset speed to eliminate the horizontal speed and acceleration until the shutdown sensor signals were received. The descent speed of −2 m/s was chosen when the propellant consumption and navigation position drift were considered. The guidance law during slow descending was the same as the guidance law during the hovering, besides that the guidance parameters were different. The target velocity in the horizontal direction was 0 m/s, and the position control target was the position of the lunar lander at the beginning of the slow descending. In the vertical direction, the velocity and acceleration was controlled when the altitude was above 20 m; when the altitude was less than 20 m, only the acceleration was controlled and the commanded acceleration was slightly less than the local lunar gravity acceleration, in order to improve the reliability of the lunar lander during descent.

7.5.5 Hazard Detection and Safe Landing Area Selection Method [9]

The hazard detection method was based on the 2-D or 3-D image data obtained by the sensor to identify the size of the obstacle and the slope gradient. The safe landing point selection method was to determine the safety radius of the safe landing site based on the identified obstacle characteristics, and select the landing site with the largest safety radius (defined as the safest) as the target landing site. The hazard detection and the safe landing point selection logic for the Chang'E-3 the lunar lander during powered descent process are as follows.

1. Large Hazard Detection and Safe Landing Area Selection Based on Optical Image

The rough hazard detection based on optical images was to identify large obstacles using image features of lunar rocks and craters to determine safe landing area.

The ratio of the shortest dimension to the longest dimension of the lunar rock is 1 ~ 0.2. The shape of the lunar rock is generally circular, rectangular pit shape or corroded bubble shape. The height is generally 1/2 of the diameter, and the standard albedo is 14 ~ 22%. Thus, the characteristics of the lunar surface rock image are: (1) The surface of the rock has a distinct bright target feature, the brightness is 2 times greater than the background; (2) there is a clear shadow area next to the bright target; (3) there is strong contrast between light area and shadow area, shown as edges; (4) The normal direction of the edge is consistent with the direction of the solar vector projection.

The slope of the moon crater is generally 25° ~ 50°, with an average of 35°. The outside of the crater is generally 3° ~ 8° with an average of 5°. The image features of the crater can be summarized as follows: (1) bright areas appear on the side towards the Sun; (2) shadow areas appear on the non-lighted dark areas; (3) arcs appear on the outer edges of dark areas.

Based on these characteristics of lunar rocks and craters, rough hazard identification and safe landing zone selection algorithms were designed: (1) image histogram analysis; (2) K-means clustering; (3) over-highlighted hazard identification; (4) over-dark hazard identification; (5) texture hazard identification; (6) using helix search algorithm to determine the safety radius of each cell (as shown in Fig. 7.13); (7) according to the safety radius, select a candidate safe landing point; (8) assess the

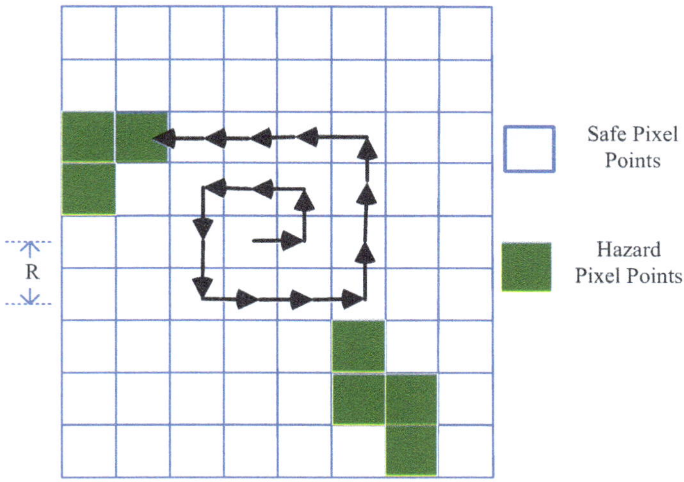

Fig. 7.13 Helix search algorithm for safe landing radium

velocity increment needed for reaching the candidate safe landing site; (9) comprehensively determine the safety landing point based on the safety radius and speed increment evaluation values.

2. Fine Hazard Detection and Safe Landing Area Selection Based on 3-D Laser Image

In the suspending status, the 3-D Imaging Sensor performed 3-D imaging on the landing area in its field of view to obtain high-resolution slant range data of the lunar lander relative to landing area. These data were processed by the fine hazard identification and safe landing zone selection algorithms, so that accurate hazard recognition and safe landing zone selection for lunar surface landing were achieved.

The design of the fine hazard recognition and safe landing zone selection algorithm was as follows: (1) Data preprocessing, which mainly included compensation for the lunar lander attitude and the translational velocity, and converting the slant range information of each "footprint" into a vertical distance to construct 3-D elevation map of the terrain in measurement coordinate system; (2) average slope construction, using the least squares method to fit the average slope of a certain area; (3) average slope calculation, calculate the average slope of the area according to the average slope surface; (4) calculation of obstacle altitude, calculate the obstacle height of each cell in the area based on the average slope surface; (5) Select the safe landing zone, using the method of spiral forward search starting from the center of the lander (as shown in Fig. 7.14) until a safe landing area that meet the landing requirements was found and the safe landing point was determined. If it was difficult to find a safe landing zone that satisfies the requirements completely within the field of view, the optimal zone was selected as a safe landing zone according to the weighing of the slope and the safety radius, and a safe landing point was determined.

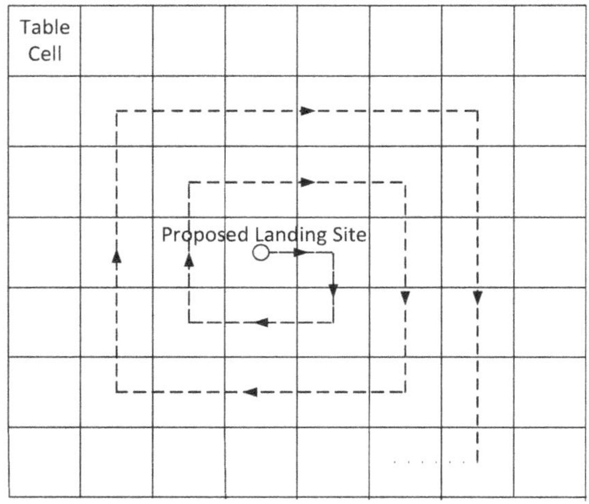

Fig. 7.14 Central helix search algorithm for safe landing area

7.5.6 Attitude Control Method for Powered Descent Process [10]

The attitude control methods in the powered descent process included phase plane control method and pulse width modulation (PWM) method. Lunar landers such as Apollo lander had a gimbaled main engine, so that attitude control could be performed by adjusting the thrust vector of the engine. The PID + PWM attitude control method for the powered descent process of the Chang'E-3 the lunar lander is as follows.

1. Partition Quaternion Control

When the current attitude quaternion of the lunar lander was known, and the guidance law gave the target attitude quaternion, the error quaternion could be calculated and the angle control deviation of the main body axes could be obtained. The phase plane of the attitude control was divided into 2 zones: nominal angular rate control zone, angle and angular rate control zone, as shown in Fig. 7.15. When the attitude angle deviation was greater than a certain threshold θ_{m}, an angular rate control with a target of $\mathrm{d}\theta_{\mathrm{r}}$ was performed, and the PI control of the angular rate deviation was adopted; otherwise, the PID control law of the attitude angle deviation was adopted.

2. Consideration of Integration Term

For the PID control system, the integral term was mainly used to cancel the low-frequency gradual change disturbance torque and improve the system performance under low-frequency disturbance. Before the powered descent process, according to the retention value of the integral term at the end of the previous orbit control, the initial value of the integral term was adjusted, so that the steady-state transient

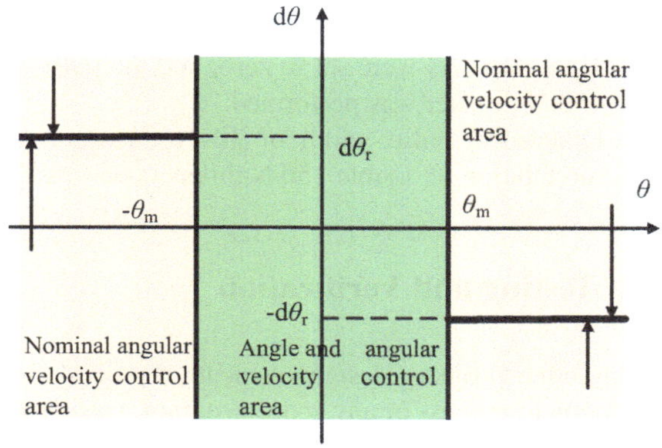

Fig. 7.15 Partition control strategy

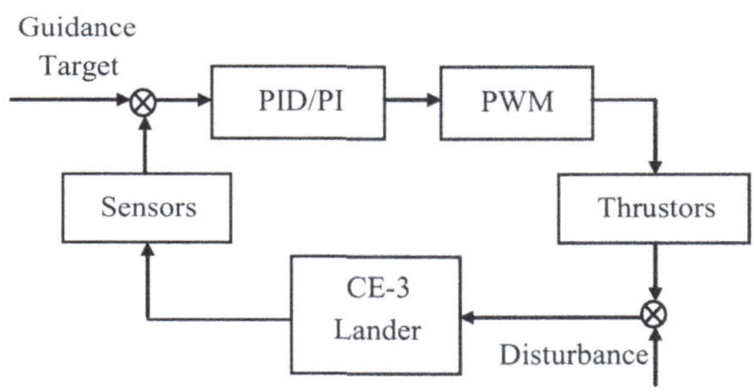

Fig. 7.16 Block diagram of PID/PI + PWM attitude control system of the Chang'E-3 lunar lander

time could be effectively reduced and the accuracy of the attitude control could be improved. When the phase point changed from the angular rate control zone into the angle control zone, the integral value of the angular rate deviation was used as the initial value of the angle deviation integral term, which could shorten the time for the system to enter the steady state.

3. Realization of Reaction Control System

A Pulse Width Modulation (PWM) method was proposed during orbit control of the Chang'E-1 lunar probe, which could be used as a substitute for a pseudo-rate modulator (PRM). The diagram of the Chang'E-3 lunar lander PID/PI + PWM attitude control system is shown in Fig. 7.16.

4. Combined Control of 10 N and 150 N Thrusters

According to the control torque required by the PID/PI control, pulse width modulation was performed in 3 cases.

(1) If the calculated control torque was less than the attitude control torque that could be provided by 10 N thrusters, then pulse width modulation of the 10 N thrusters was performed and the pulse width of the 150 N thrusters was set to zero.

(2) Otherwise, if the calculated control torque was less than the attitude control torque that could be provided by single 150 N thruster, the pulse width of the 10 N thrusters were set to zero, and the pulse width modulation of the single 150 N thruster was performed.
(3) Otherwise, pulse width of 10 N thrusters were set to zero, pulse width modulation of double 150 N thrusters was performed.

7.6 Testing and Verification

As any other projects, design verification process could not be separated from the development process of any space program. Design verification is an important part of the spacecraft development process and is integrated with the design, manufacture and assembly process during modern spacecraft production phase. The success and stereotypes of any space theories requires the support of advanced design verification technology.

The lunar surface environment faced by the lunar lander was a brand-new challenge. It WAs necessary to conduct complete and comprehensive design verification on the ground. According to the components features, navigation sensors calibration, subsystem flight test by airplane and semi-physical simulation were carried on during the development of GNC subsystem.

7.6.1 Key Navigation Sensors Calibration

1. Calibration of 3-D Laser Imaging Sensor
1) Test objective

In the calibration test, the laser ranging data of the 3-D Laser Image Sensor at multiple angles with no scanning of the motor was obtained, and the system error was compensated by the distance calibration device so that the slant range accuracy of the sensor satisfied the design requirements.

2) Test method

Ten test boards were placed at a depth of approximately 20 m in the test field, 12 or 13 test points were uniformly placed on each test board, and 4 test points were placed on the head of the sensor. The coordinates of all test points were measured with a total station. The measurement results of the 3-D Laser Imaging Sensor were compared with the total station results. The Laser 3-D Image Sensor outdoor calibration field is shown in Fig. 7.17.

2. Calibration of Optical Image Sensor
1) Test objective

Fig. 7.17 Outdoor calibration field of 3-D Laser Image Sensor

The purpose was to obtain the internal and external parameters of the Optical Imaging Sensor by calibrating it. The cube mirror which was mounted on the sensor was collimated and the relationship between the Optical Imaging Sensor and the cube mirror was obtained.

2) Test method

The Optical Image Sensor was calibrated by theodolite and the cube mirror was collimated in specialized test field. The calibration field for Optical Image Sensor is shown in Fig. 7.18.

3. Flight Testing by Airplane of MRVS
1) Test objective

A large transport airplane was used as a flight test platform, and the MRVS and the surrounding components were all installed in accordance with the actual configuration of the lunar lander to verify the influence of the surrounding components on the MRVS. The second purpose was to verify the performance of the MRVS when the distance was greater than 10 km by the airplane with high flight altitude. The third purpose was to confirm the reliability of the mathematical model for semi-physical simulation by flight test data and measurement results of various terrains in the desert zone.

2) Test method

According to the test requirements, the verification test of the flight test could be classified into two parts. One of which was qualitative verification. Its main purpose was to verify the effect of the surrounding components of the MRVS and the terrain relief on the performance of range and speed measurement. The second was to verify the performance of the sensor bellow 3 km after attitude adjustment of the lunar lander. The flight test route and local terrain are shown in Fig. 7.19.

Fig. 7.18 Indoor calibration field of Optical Image Sensor

Fig. 7.19 Flight test routine and local terrain

7.6.2 GNC Subsystem Flight Testing

1) Test objective

The purpose of the subsystem flight test was to use the components outdoor field test as a basis to specify the special operating dynamics environment of the subsystem and to verify the performance of the navigation sensor under specific dynamics between 2 km ~ 100 m during lunar landing.

2) Test method

The 3-D Laser Image Sensor and the Optical Image Sensor were fixed on the floor of the helicopter cabin through the adapter plate and protrude from the right side out of the cabin. As shown in Fig. 7.20, the cement floor and lawn were selected as the test field, and the cement floor was used as the large plane which was to be detected, and the target was set on the lawn to provide the simulation data for the subsequent accuracy calculation. The 3-D Laser Image Sensor and the Optical Image Sensor performed the distance measurement and imaging of the target and comparison with the preset target data when the helicopter was hovering, as shown in Figs. 7.21 and 7.22.

Fig. 7.20 Location of 3-D Laser Image Sensor and Optical Imaging Sensor during flight test

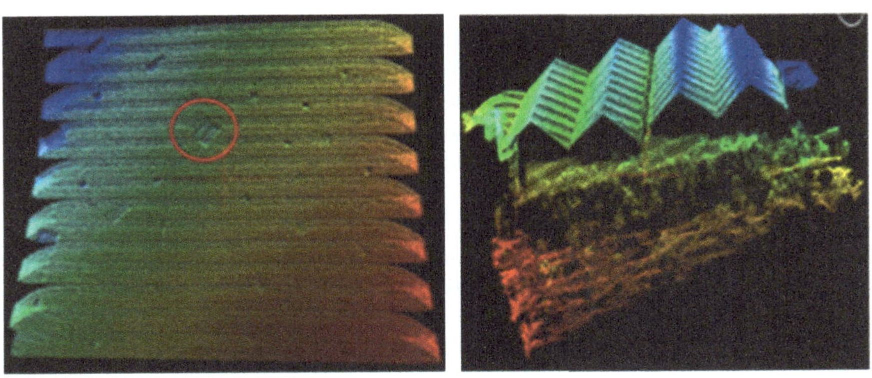

Fig. 7.21 Flight test result image of Laser 3-D Image Sensor

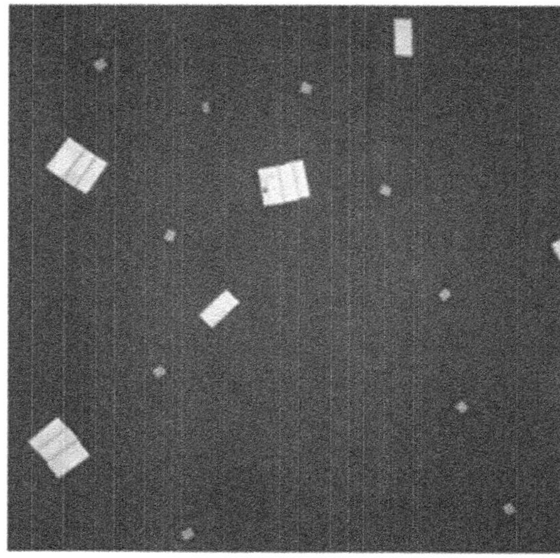

Fig. 7.22 Flight test result image of Optical Image Sensor

7.6.3 Hardware-in-the-Loop Test [11]

1. Test Objective

(1) To conduct ground test verification of GNC for hover, obstacle avoidance and slow descending during powered descent process, in particular verification of the feasibility and effectiveness of the precise hazard detection and obstacle avoidance with the Laser Rangefinder and 3-D Laser Imaging Sensor in the loop.

(2) To test the function and performance of some newly developed navigation sensors, such as Optical Image Sensor, Laser Rangefinder, 3-D Laser Image Sensor, and etc.

Because the closed-loop test system involved the landing navigation sensor, the Central Control Unit, the GNC software, and the test equipment, the system was relatively complicated, so the closed-loop test could be divided into two phases.

In the first phase of the closed-loop test, only the CCU, GNC application software, and interface box were used for the simulation verification of the test system, while navigation sensors such as Laser Rangefinder and 3-D Laser Image Sensor were not used. The main purpose of this phase was to set up the basic experimental framework to form a closed loop test on the software level, so as to achieve software level GNC algorithm verification, and remote control verification of the six-degree-of-freedom soft-landing motion simulation platform to confirm if it satisfied the dynamic environment requirements for testing.

On the basis of the first phase, in the second phase closed-loop test, the onboard equipment such as the Laser Rangefinder and the Laser 3-D Image Sensor was connected to the test system to form a closed-loop loop. The purposes of this phase are:

(1) To provide a true dynamic environment for the Laser 3-D Imaging Sensor and verify the dynamic performance of the Laser 3-D Imaging Sensor.
(2) To verify the effectiveness of the hazard detection and safe landing area selection algorithm for the hovering.
(3) To verify the feasibility of the navigation in the case of measurement corrections gained by real components;
(4) To verify the feasibility and effectiveness of GNC algorithm for obstacle avoidance and slow descending.

The GNC semi-physical simulation test system for the Chang'E-3 lunar lander was a large-scale system for soft-landing test on the ground (as shown in Fig. 7.23). The test system was complex, including a 6-DOF soft-landing motion simulation platform (as shown in Fig. 7.24), a lunar surface simulation model, as shown in, a dynamic computer, ground test main control computer, sensor and actuator interface box, digital simulation demonstration platform, optical terminal and optical fiber communication network, lander onboard components and other ground support equipment.

2. Test Method

The six-degree-of-freedom movement of the lunar lander was simulated by a soft-landing motion simulation system (including 3-DOF translation platform and flight rotation simulation platform). The onboard landing navigation sensor components (including Laser Rangefinder, 3-D Laser Image Sensor and Optical Image Sensor, etc.) were installed on the flight rotation simulation platform and connected into the control loop, the lunar surface simulation model was located on the ground to provide a relative measurement target for the landing navigation sensors (as shown in Fig. 7.25). The dynamics of the lunar lander and the actuators were replaced by mathematical models. Through the measurement of navigation sensors, GNC application software in CCU, and calculations of dynamics and kinematics in the dynamic computer, a closed-loop GNC semi-physical simulation test system was constructed to experimentally verify the hover, obstacle avoidance, and slow descending of GNC design and algorithms.

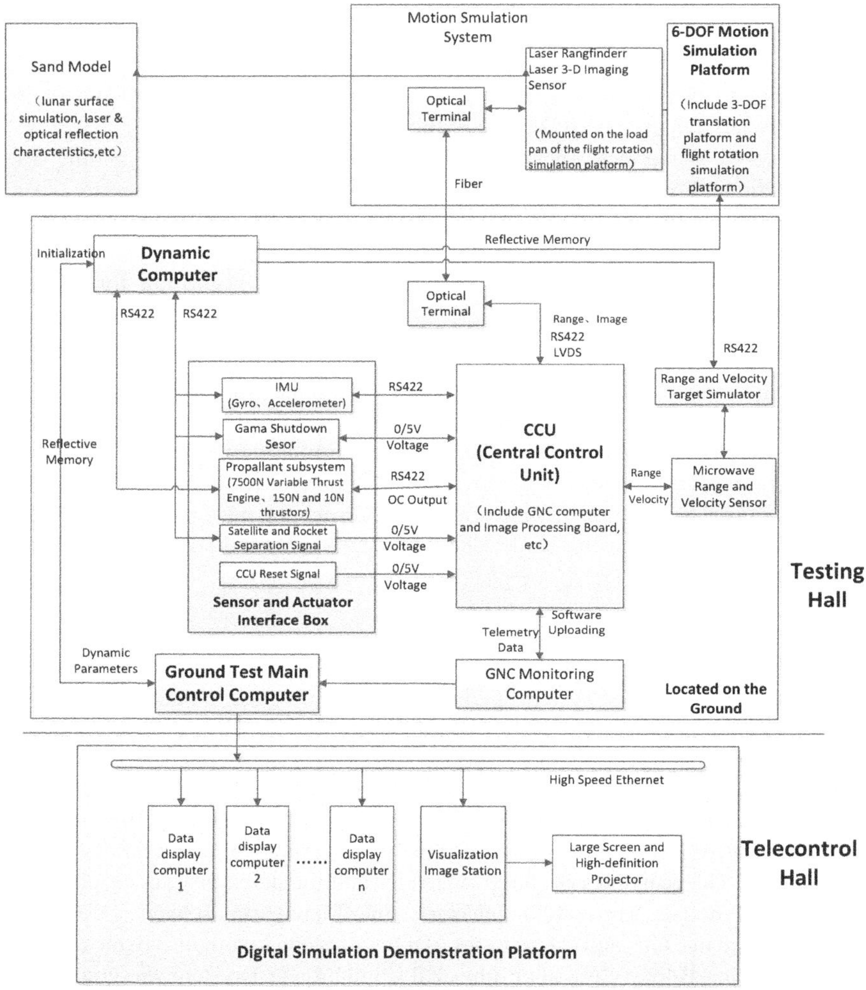

Fig. 7.23 Diagram of GNC Semi-physical simulation test of the Chang'E-3 lunar lander

7.7 Summary

Navigation, guidance, hazard detection and avoidance, and control methods, components selection, software and interface design, design verification of the lunar lander GNC subsystem are summarized in this chapter. Onboard robotic vision data was firstly implemented for autonomous hazard detection and avoidance during soft-landing process of the Chang'E-3 lunar lander. The landing accuracy of the Chang'E-3 reached the international advanced level, which proved the effectiveness of the GNC subsystem design and verification methods.

Fig. 7.24 Six degrees of freedom motion simulation platform for soft-landing

Fig. 7.25 Sand model for lunar surface simulation

It should be noted that lunar environment, especially the lunar terrain of the landing area deeply affected GNC subsystem design and landing accuracy for lunar landing. In order to accommodate more complicated lunar surface, to realize higher landing accuracy for prospective lunar missions, to establish a firm foundation for exploitation of lunar resources, future works should be focus on terrain matching algorithm, beacon based navigation technology and other potential methods to realize landing anywhere accurately on the Moon.

References

1. Qian X (1963) An introduction to interstellar voyage. Science Press, Beijing
2. Shancheng TU (2001) Spacecraft attitude dynamics and control. China Astronautic Publishing House, Beijing
3. Yang J (2002) Spacecraft orbit dynamics and control. China Astronautic Publishing House, Beijing
4. Zhang H (2007) Soaring the empire: from artificial satellite to lunar probe. Shanghai Science and Technology Education Press, Shanghai
5. Xu F (2001) Spacecraft engineering. China Astronautic Publishing House, Beijing
6. Zhang R (1998) Satellite orbit and attitude dynamics and control. Beihang University Press, Beijing
7. Yan J, Yan J (1990) An introduction to computer real time control software design. Tsinghua University Press, Beijing
8. Zhang H, Guan Y, Huang X, et al. (2014) Guidance navigation and control for Chang'e-3 powered descent. Sci China: Technol Sci 44(5):377–384
9. Zhang H et al (2014) Autonomous hazard avoidance control for Chang'E-3 soft-landing. Sci China: Technol Sci 44(6):559–568
10. Kaplan MH (1976) Modern Spacecraft dynamics and control. Wiley, London
11. Liu L (2001) Technology of satellite control system simulation. China Astronautic Publishing House, Beijing

Chapter 8
Propulsion Technology of Lunar Lander

8.1 Introduction

The propulsion subsystem of the lunar lander was one of key subsystems, which undertook attitude and orbit control tasks such as attitude control, stabilization, adjustment, orbit maneuver, orbit insertion and de-orbit, hover and translation and etc.

There are different types of propulsion subsystem such as solid, liquid, cold gas, gas generator, electric propulsion, nuclear propulsion and etc. Since the middle of twentieth century when mankind conducted lunar exploration, the propulsion subsystem of hypergolic and storable bi-propellant was the most popular one in various the lunar landers, which had advantages such as high specific impulse, high reliability, long life-time, multiple/pulse ignition, wide range of variable thrust and etc. [1].

Due to constraints of the lunar lander mass, configuration, landing condition and etc., usually the propulsion subsystem should have advantages such as large total impulse, small residual propellant, high specific impulse, large throttleable thrust, small structural mass, low redundancy, harsh load bearing, high reliability and etc. Compared to that of satellite and near-earth orbit spacecraft, the requirements for the lunar lander propulsion subsystem were stringent and special including adaptability with large variation of operation regime, continuous and rapid thrust modulation of orbit control engine, high accuracy thrust and mixture ratio control, equilibrium vent of parallel tank, propellant management in large flux and complicated overload conditions, light weight under high load bearing condition, long term autonomous fault management on orbit without ground support and etc.

In this chapter, based on basic information of typical propulsion subsystem of the lunar landers outside China, it is described including general design features, constraint, design, test and verification methods with development of propulsion subsystem of the Chang'E-3 lunar lander.

D.-Y. Yu et al., *Technology of Lunar Soft Lander*, Space Technology Library 38,
https://doi.org/10.1007/978-981-15-6580-9_8

8.2 Development Status [2, 3]

Since the 1960s, the world's leading space powers in succession conducted lunar exploration and developed various the lunar landers. Till now, only Luna series of the Soviet Union, Surveyor and Apollo series of the United States and Chinese the Chang'E-3 lunar lander successfully implemented lunar soft-landing exploration.

N_2O_4/Ethylamine bi-propellant propulsion subsystem was used in Luna 9 lander of the Soviet Union with a 45.5 kN pump-feed constant thrust main engine for orbit correction in LTO, braking and decelerating in descent and landing with second reigniting capability and 3 sets of 0.245 kN turbine gas nozzles for attitude control.

The propulsion for attitude and orbit control was independent in the descent stage of Luna 9 lander of the Soviet Union. A 7.35~18.92 kN throttling pump-feed engine and two sets of 2.06~3.43 kN throttling pressurized engines with N_2O_4/UDMH bi-propellant were used for orbit control. Gas pressurized monopropellant engines were used for attitude control. An 18.8 kN pump-feed constant thrust engine with N_2O_4/UDMH bi-propellant was used in the ascent stage for orbit control.

The orbit control engines of the descent and ascent stage of Luna series lander are shown in Fig. 8.1. Technical specifications for main engines of Luna series landers are given in Table 8.1.

The propulsion subsystem of Surveyor the lunar lander consisted of cold gas (Nitrogen) thrusters for attitude control, one solid rocket engine for deceleration in powered descent, and auxiliary liquid propulsion subsystem consisted of 3 sets of 133~463 N throttling engines for attitude control while the solid rocket engine was working.

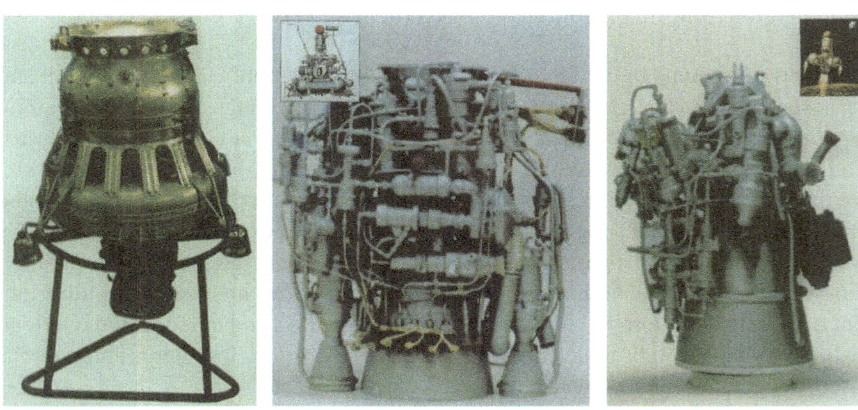

(a) (b) (c)

Fig. 8.1 Demonstration of lunar eclipse. **a** Luna 9 descent engine; **b** Luna 16 descent engine; **c** Luna 16 ascent engine

Table 8.1 Specifications for main engines of Luna series the lunar landers

<table>
<tr><th>Engine model</th><th>КТДУ-5А</th><th>КТДУ-417</th><th>КТДУ-417-В</th><th>КРД-61</th></tr>
<tr><td>Development time</td><td>1962–1963</td><td>1965–1969</td><td>1965–1969</td><td>1965–1969</td></tr>
<tr><td>Feed type</td><td>Pump-feed</td><td>Pump-feed</td><td>Pressurized-feed</td><td>Pump-feed</td></tr>
<tr><td>Combustion chamber</td><td>Monochamber+3 gas nozzles</td><td>Monochamber</td><td>Bi-chamber</td><td>Monochamber+2 gas nozzles</td></tr>
<tr><td>Nominal vacuum thrust/kN</td><td>$45.5 + 3 \times 0.245$</td><td>7.35~18.92</td><td>2.06~3.43</td><td>18.8</td></tr>
<tr><td>Specific impulse/s</td><td>287</td><td>308~314</td><td>249~254</td><td>313.3</td></tr>
<tr><td>Chamber pressure/MPa</td><td>6.28</td><td>8.3</td><td>0.89</td><td>9.22</td></tr>
<tr><td>Working time/s</td><td>43</td><td>650</td><td>30</td><td>53</td></tr>
<tr><td>Propellant</td><td>$N_2O_4/(CH_3CH_2)_2NH$</td><td colspan="3">N_2O_4/UDMH</td></tr>
<tr><td rowspan="2">Function</td><td rowspan="2">Luna 4~Luna 14 main descent engine with four turbine gas nozzles for attitude control</td><td colspan="2">Luna 15~Luna 24</td><td rowspan="2">Luna 16, Luna 20, Luna 24. Main ascent engine with 2 turbine gas nozzles for attitude control</td></tr>
<tr><td>Main engine on descent stage for decelerating</td><td>Small engines on descent stage for soft-landing</td></tr>
</table>

The propulsion subsystem of Apollo lunar module consisted of descent stage propulsion for deceleration in powered descent, ascent stage propulsion for acceleration in lift-off and ascending, and reaction attitude control propulsion for attitude control in descending/ascending.

The N_2O_4/mixed hydrazine bi-propellant propulsion subsystem pressurized with gaseous Helium of constant pressure was used in the descent stage. Total propellant loaded was 8919 kg. The block diagram of the propulsion subsystem is shown in Fig. 8.2. There were four identical 1900 L spherical titanium alloy fuel tanks with special equilibrium venting device, blade-type anti-vortexing device and propellant storage device to ensure equilibrium venting of parallel tanks and prevent gas entrapment under zero-gravity or −2 g condition. A 4655~44737 N throttling engine with throttleable ratio of 10~1 was used in the descent stage propulsion.

The N_2O_4/mixed hydrazine bi-propellant propulsion subsystem pressurized with gaseous Helium constant pressure was used in the ascent stage. Total propellant loaded was 2352 kg. The schematic of the ascent propulsion subsystem is shown in Fig. 8.3. Because the astronauts always stayed in ascending stage during descending/ascending process, high reliability and safety were required with multi-redundancy. In addition, propellant tanks could be connected with propellant tanks in reaction attitude control propulsion subsystem by pipelines to supply propellant for each other. A 15.58 kN constant thrust engine was used in the ascent stage.

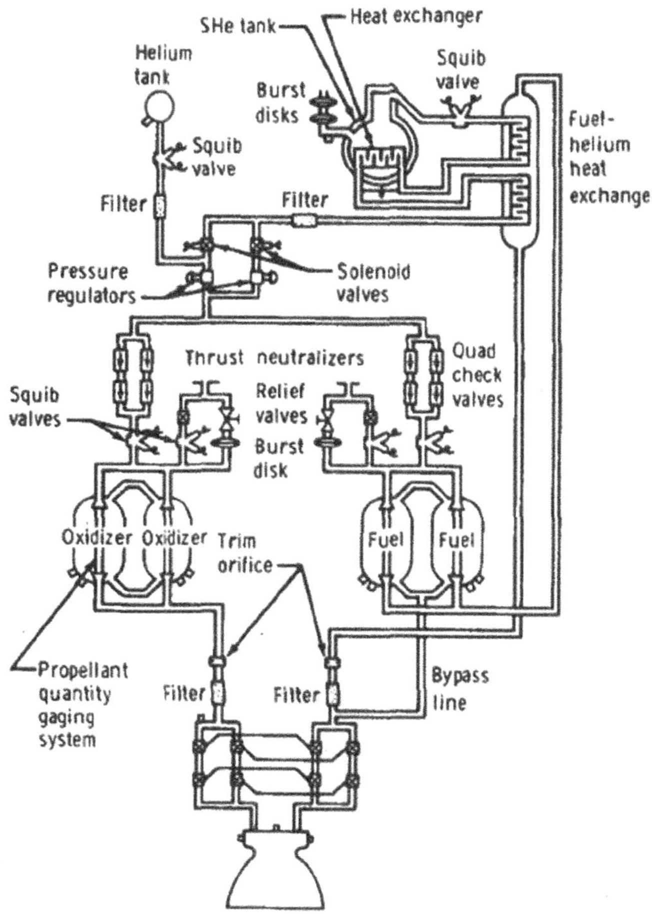

Fig. 8.2 Schematic of descent propulsion subsystem of Apollo lunar module

The N_2O_4/mixed hydrazine bi-propellant propulsion subsystem pressurized with gaseous Helium of constant pressure was used for reaction control. The propellant tanks of fluoro-plastic bladder with total propellant loaded 2352 kg were used. The schematic of reaction control propulsion subsystem is shown in Fig. 8.4. 2 sets of reaction attitude control propulsion subsystems (8 engines for each subsystem) were equipped independently, in which 16 engines distributed on 4 modules with 4 engines on each module (2-to-2 correspondent respectively to each independent subsystem) and their pipelines were connected to serve as redundancy for engines and propellant supply. Meanwhile, propellant tanks in reaction control propulsion subsystem were connected with that of ascent propulsion subsystem to supply propellant for each other. The thrust of all engines for reaction control was 445 N.

The specifications of main engines of Apollo lunar module are shown in Table 8.2.

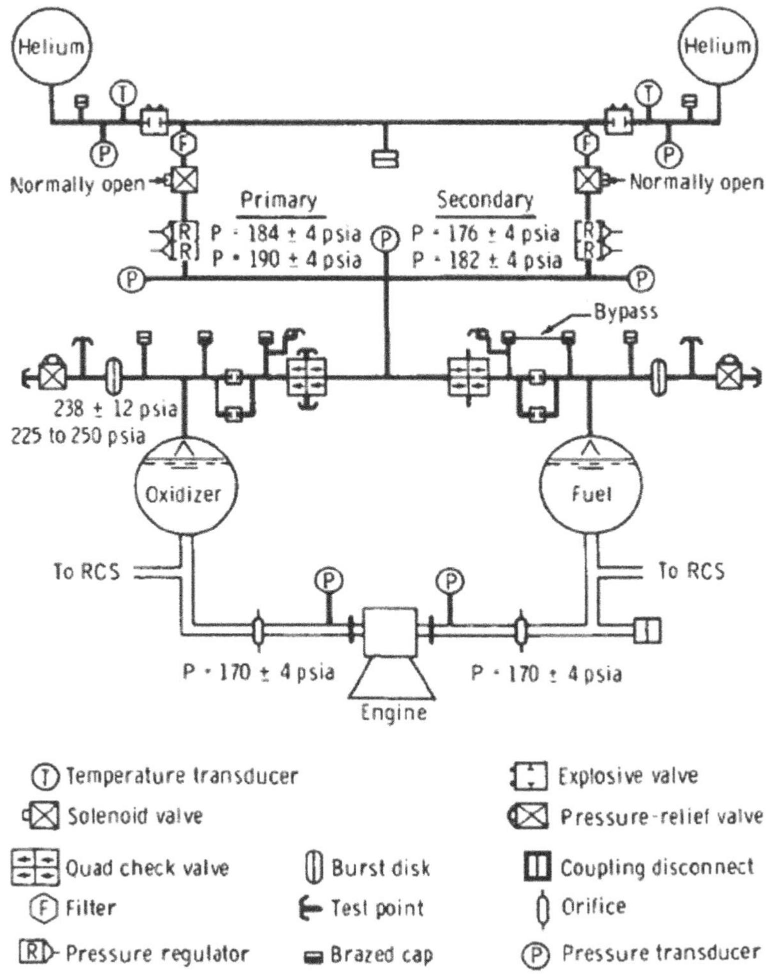

Fig. 8.3 Schematic of ascent propulsion subsystem of Apollo lunar module

The Altair the lunar lander was an important part of Constellation program initiated by the United States to return to Moon in the beginning of twenty-first century. The Altair the lunar lander consisted of ascent and descent stages. While the reliability for manned space mission was considered, the pressurized Liquid O_2/Liquid H_2 bi-propellant propulsion subsystems were applied in the ascent and descent stages. The main engine of ascent stage was a 24.5 kN AJ-10 modified engine. The descent propulsion subsystem consisted of 8 cryogenic tanks (4 Liquid O- tanks and 4 for Liquid H_2 tanks), while the main engines were 4 × 83 kN RL-10 modified Liquid O_2/Liquid H_2 bi-propellant engines with specific impulse of 444.4~451.0 s and mixed ratio of 5.1~5.5 for LOI and powered descent.

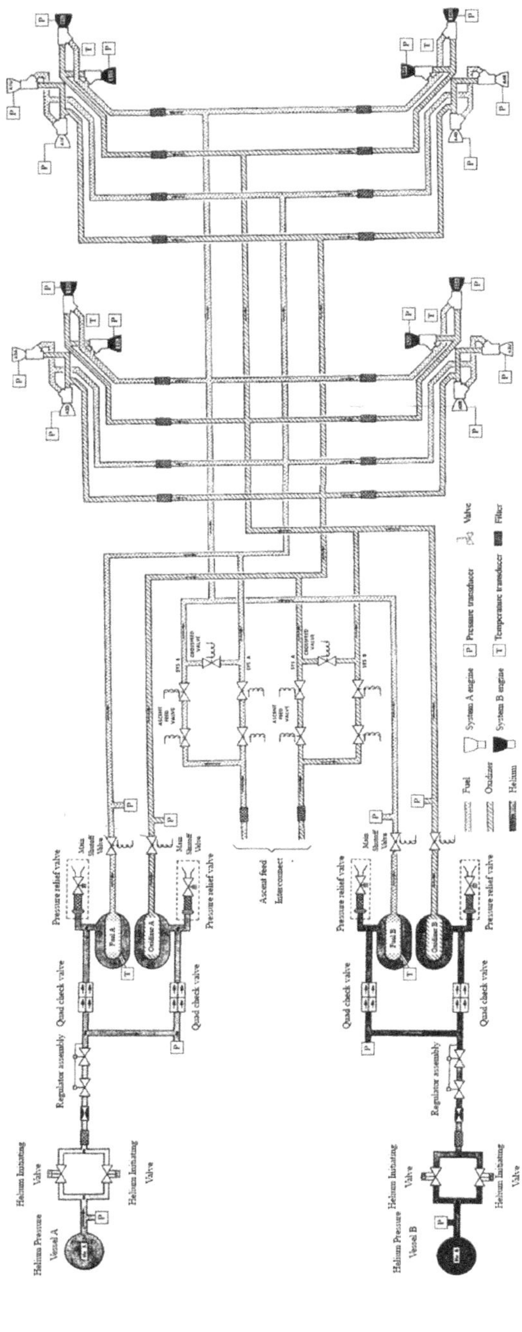

Fig. 8.4 Schematic of reaction control subsystem of Apollo lunar module

Table 8.2 Specifications of main engines of Apollo lunar module

Specifications	Descent engine	Ascent engine	Attitude control engines
Propellant	N_2O_4/AZ-50		
Vacuum thrust/N	4655~44737	15580	445
Specific impulse/s	298~305	310	120~290
Combustion chamber pressure/MPa	0.103~1.03	0.84	0.689
Mixed ratio	1.6	1.6	2.0
cooling	Ablation + film + radiation	Ablation + film	radiation + film
Gambled/(°)	±6(two direction)		

It can be concluded after typical the lunar lander propulsion subsystems were analyzed as follows.

(1) The major design was high performance, high reliable pressurized-feed bi-propellant propulsion subsystem with continuously throttling engine for descent stage, especially for manned the lunar lander to adapt overload requirements for astronauts in soft-landing.
(2) For a long term, the hypergolic, storable, toxic bi-propellant with oxidizer as nitrous oxide (N_2O_4, MON-1, MON-3) and hydrazine (N_2H_4, MMH, AZ-50, UDMH) was widely applied in propulsion subsystems of the lunar landers.
(3) Along with propulsion technology development, it becomes a trend to apply pressurized-feed propulsion subsystem based on high performance, environment friendly, cheap, resource exploited in situ propellant of LOX/LH_2, LOX/HC (Hydrocarbons) in order to adapt requirements of future larger scale manned the lunar lander exploration and manned Mars lander exploration missions.

8.3 Design Constraints and Analysis

8.3.1 Configurational Constraints

Generally, the mass and size of propulsion subsystem was the greatest part of whole the lunar lander. Therefore, it should be integrated with the propulsion subsystem when the configuration of the lunar lander was design. It could be seen that the truss and plate structure was usually applied after different the lunar landers were analyzed, for example, the structure of Luna 16 and Altair the lunar lander was truss type and the structure of Apollo lunar module was plate. Because of the constraints of the lunar lander mass, center of gravity, volume envelope, thermal protection, configuration, propellant load, structural types of tanks and etc., 4 symmetrically and horizontally distributed propellant tanks was usually applied for propulsion subsystems to lower the center of gravity of the lunar lander.

For sphere and cylinder type tank, it was difficult to be mounted by flanges (large flange size and heavy weight). Therefore it was usually to be mounted with the top and bottom, while the axial load was endured by the bottom and the transverse load was endured by the top (or the alternative option was the frames welded directly on the tank shells to transfer force). Such design would lead to large concentrated load in the tank bottom and instability of the shell. So the bearing capability of the tank should be considered when it was designed.

For the main descent engine, the engine usually was installed in the center of the lunar lander because of the landing safety. The thermal shield for the engine should be configured to isolate effectively the instruments of the lunar lander from high temperature thermal radiation of firing engine (the temperature of engine nozzle might reach 1300~1400 °C). But it would influence the normal heat dissipation of the engine and the operation reliability. So the temperature rising of the engine due to the thermal shield should be considered to ensure the working temperature of the engine within the safety margin when the engine was designed.

8.3.2 Functional Requirements

The lunar lander could be classified into non-recoverable and recoverable types. For recoverable the lunar lander, the ascent propulsion subsystem was necessary to implement ascending besides the descent propulsion subsystem for powered descent and soft-landing.

There were following functional requirements for propulsion subsystem of different robotic or manned the lunar landers.

(1) To provide force for orbit control such as LTO, LOI, circumlunar orbit, powered descent, ascending and de-orbit.
(2) To provide force for attitude control such as LTO, circumlunar orbit, powered descent, ascending and Earth Transfer Orbit.
(3) To provide force for translation, reaction control for near earth orbit, LTO, rendezvous and docking on lunar orbit.
(4) To provide force for emergency return of manned the lunar lander.

According to the difference between lunar exploration missions and design of the lunar landers, the configuration, demands and constraints of the propulsion subsystem were different. For example, the functional requirements of propulsion subsystem for non-recoverable the lunar lander were analyzed as follows.

(1) Usually the simple, reliable pressurized-feed orbit/attitude unified propulsion subsystem was used because of the lunar lander mass, mounting space, propellant sharing, orbit control engine reigniting, reliability and etc.

(2) The propulsion subsystem could not only be firing continuously with constant thrust for a long time but also be throttled to provide continuously variable thrust to satisfy the requirements of descending, hovering and obstacle avoiding during soft-landing.
(3) The allocation and grouping of attitude control engines should satisfy the requirements of GNC control, configuration, failure isolation/regrouping and redundancy for the orbit control engine.
(4) The propulsion subsystem should have self-management capability to some extent, especially in a long time without telecommunication to ensure the operation safety.

8.3.3 Technical Specification Requirements [4]

1. Structural Mass Requirement

The total mass of the lunar lander was limited by the carrying capacity of launch vehicle. Generally, more than 60% of the total the lunar lander mass was the propulsion subsystem. So there was strict constraint on propulsion subsystem mass. On the one hand, propellant mass could be reduced as result of improving engine specific impulse. On the other hand, the structural mass of the propulsion could be reduced by introduction of new technology, new material and system optimization. For example, the constraint of propulsion subsystem of the Chang'E-3 lunar lander was that the total mass of propulsion subsystem should be no more than 295 kg with propellant load of 2600 kg.

2. Configuration Constraint

Because of large-scale components with special shape (sphere, spherocylinder, bell shape and etc.) in propulsion subsystem like propellant tanks, gas tanks, orbit control engines and etc., most of available space in the lunar lander was used for installation of propulsion subsystem components, which had impact on utilization efficiency of the lunar lander configuration. For example, the diameter of the propellant tank of the Chang'E-3 lunar lander was 1050 mm, the length of the gas tank was 1400 mm, and the maximal radial envelope of propulsion subsystem was no more than 3650 mm.

3. Requirement for the Specific Impulse of Orbit Control Engine

The specific impulse of orbit control engine had direct impact on propellant load as well as total the lunar lander mass. Improving engine specific impulse could effectively reduce propellant load as well as volume of propellant tank and gas tank, so as to reduce propulsion subsystem structural mass. When the propellant option was determined, specific impulse of the engine could be improved by means of increasing combustion efficiency and enlarging nozzle expansion ratio. However, increasing combustion efficiency usually caused high working temperature of the engine and affected duration of anti-oxidization coating. Enlarging nozzle expansion ratio also led to increasing of engine volume, mass, and decreasing of basic structural frequency which was harmful to mechanical endurance. So it should be considered including propellant consumption, configuration, subsystem mass (wet and dry mass), coating lifetime, mechanical condition and thermal environment to properly define specific impulse of the engine. For example, specific impulse of the orbit control engine on the Chang'E-3 lunar lander should be higher than 308 s.

4. Throttling Thrust Ratio Requirements of Orbit Control Engine

Wide range of continuous throttling thrust for the orbit control engine was necessary for lunar soft-landing, which led to technical challenges for the design of propulsion subsystem and orbit control engine including follows: (1) Wide range of throttling thrust ratio led to large variation of flux conditions, which might cause variation of pressure and pipeline resistance, as well as dynamic instability in rapid throttling thrust. (2) A large deviation of throttling condition from nominal design les to decreasing of specific impulse, deviation of mixture ratio, which might reduce total impulse of the subsystem, especially when deviation of mixture ratio led to excess remaining of one type of propellant. For example, nominal thrust output of the orbit control engine of the Chang'E-3 lunar lander was 7500 N, and throttling thrust was 5000~1500 N. When the engine output was the nominal thrust, its specific impulse could be 308 s and mixture ratio was 1.64~1.65. However, when its output was lower than 5000 N, the minimal specific impulse was only 285 s and the minimal mixture ratio was 1.5. (3) In low thrust condition, because of low flux and small injector pressure drop, it is difficult to implement film cooling, spray, mixture and combustion. The larger the throttling ratio was, it was more difficult. (4) Constrained by propellant tank pressure of pressurized-feed subsystem, flow regulator (cavitation tube) for the orbit control engine with large throttling ratio was not capable of realizing complete cavitation in full range of thrust variation. When the operation mode of the engine changed from maximal thrust to throttling, flow regulator went through non cavitation, transitional cavitation and complete cavitation. Due to larger uncertainty and fluctuation of transitional cavitation, it was necessary for the engine to operate rapidly from non-cavitation to complete cavitation to avoid instability of subsystem.

From application of throttling engine in past the lunar landers, except the throttling ratio of the descent stage engine of Apollo lunar module was 10:1, the throttling ratio was generally no more than 5:1 (4:1 for the Surveyor the lunar landers, 2.5~1 for the Luna landers, 5:1 for the Chang'E-3 lunar lander). The larger throttling ratio was, it was more difficult to design.

5. Expulsion Efficiency Requirement for Propellant Tank

Tank propellant expulsion efficiency was closely related with amount of available propellant. Improving expulsion efficiency of propellant tank might increase propellant utilization efficiency. The tank expulsion efficiency was related with tank types. There were different types of propellant tank including non-metallic bladder, surface tension, metallic diaphragm (diaphragm or bellows) and piston. Except the surface tension tank was of gas liquid mixed type, other all were of gas liquid physical isolation type. Generally the expulsion efficiency for metallic diaphragm tank was 97%~98%, while for non-metallic bladder, surface tension tank, piston tank, the expulsion efficiency could be more than 99%. It was necessary to consider expulsion efficiency, flow rate, anti-vortexing, anti-sloshing, long term propellant compatibility, overload capacity, mechanical endurance, isolation complexity for two propellant components, structural mass and configuration, repeated venting times and etc. when the type of propellant tank was determined. The surface tension or metallic diaphragm tank was usually chosen for the lunar landers, for short term the lunar lander the non-metallic bladder type might be the choice. For example, the composite wrapped metallic diaphragm tank of expulsion efficiency more than 99% was chosen for propulsion subsystem of the Chang'E-3 lunar lander.

6. Parallel Tank Propellant Equilibrium Venting Constraints

Generally there were four propellant tanks symmetrically distributed in propulsion subsystem of the lunar lander (2 for oxidizer and 2 for fuel). Two symmetrical tanks for same propellant component were parallel with small deviation from equilibrium state in whole mission profile. Large deviation of equilibrium might cause large deviation of mass center, which would increase propellant consumption for attitude control and have impact on attitude control capability in severe case. In addition, for gas liquid mixed tank, equilibrium venting also had impact on propellant utilization efficiency (when all propellant in one tank was exhausted, remaining propellant in other tank was not available). Due to small margin of propellant load for the lunar lander, it was necessary to strictly control equilibrium propellant venting of parallel tanks. For example, the equilibrium propellant venting should not be more than 3% for the Chang'E-3 lunar lander. So it was necessary to diminish the impact of flow resistance deviation of parallel tanks on equilibrium venting, for example, to reduce the proportion of tank flow resistance in total subsystem flow resistance and adjust subsystem consistency according to tank flow resistance test results. Moreover, strict control of size, thickness, hardness of tank metallic diaphragm could improve flow rate consistency.

8.3.4 Flight Environment

The lunar lander propulsion subsystem would experience mechanical environment in launch phase, space environment in orbit, mechanical environment in descending phase, impact environment in landing and survival environment after landing and etc. So stringent requirements including mechanics, heat, radiation and EMC, especially mechanics and heat environment, should be adapted to by propulsion subsystem. Because most components of propulsion subsystem were an electromechanical product with the feature of load bearing, pressure bearing, movement, complicated pipelines and large loading mass, their function would be lost in case of failure. Therefore the most critical requirement was to ensure structural reliability of propulsion subsystem throughout whole mission profile. When the structure of propulsion subsystem was designed, the major concern was the mechanical environment in launch phase (overload, vibration, shock), while the orbital mechanical environment was relatively much lower and not necessary to be considered. For large-scale, heavy and high load components such as propellant tank, gas tank, orbit control engine and etc., the pressure bearing should be satisfied, as well as mechanical environment in launch phase, stiffness and strength of the lunar lander, otherwise there might be structural deformation, instability and even broken due to insufficient stiffness or strength.

For gas or liquid propellant in propulsion subsystem, there were requirements for temperature condition of individual components in order to prevent liquid propellant freezing in low temperature, vaporization in high temperature, pressure dramatic rise, propellant decomposition and even explosion, especially for components outside the lunar lander like engines, pipelines, valves, and even tanks, gas tanks which were directly exposed (outside environment: high temperature can be more than 150 °C in sun light, low temperature can be lower than −100 °C in the shadow). In order to ensure reliable work of subsystem components, it was necessary to control the temperature of individual components by thermal design. Generally, the temperature range of components for liquid such as pipelines, tanks, valves, engines should be 0~40 °C, while the temperature range of components for gas shall be −30~70 °C. For the engines/nozzles exposed outside the lunar lander experiencing the high/low temperature alternation which was much less than the temperature range of radiation-cooling nozzle, it had almost no impact on nozzle materials (nozzle is made of niobium-hafnium, niobium-tungsten, rhenium-iridium, carbon/carbon, carbon/carborundum with high temperature anti oxidizing coatings). It should be especially emphasized that external high/low temperature can transfer heat to engine head through nozzle, which might lead to temperature rise or decline in engine head and valves. In severe low temperature propellant in valves could be frozen with deflagration in cold starting. The high temperature might lead to propellant vaporization which had impact on normal opening of valves and engine firing. So it was necessary to conduct proper thermal design for engines to ensure reliable engine firing and in-orbit storage. Generally electric heating (auxiliary thermal multi-layer) was used to maintain the temperature of valves and injector with 5~15 °C for low temperature. The thermal isolation of engines was used to maintain the temperature of valves lower than 80 °C, such as titanium alloy materials with low heat conductivity, hollowed-out structure to increase heat resistance and non-metal between valves and injector.

8.3.5 Long-Term Storage Environment on Lunar Surface

After the lunar lander was landed on lunar surface, the propulsion subsystem finishes its job, but there was still remnant propellant and pressurized tanks including gas tanks and propellant tanks. During long term operation, the propulsion subsystem along with the lunar lander would experience alternation of lunar daytime and night where there might be some safety risks. So the remnant propellant and pressurized gas should be depleted. Generally, the propellant was depleted by engines firing, while pressurized gas was depleted by directly venting outside. In order to avoid large disturbing force to the lunar lander by engine firing and pressurized gas venting, the symmetrically distributed engines were simultaneously firing and shut down, as well as symmetrically gas venting vent was set to eliminate and diminish disturbing force.

When the remnant propellant was depleted by engines firing, due to the difference of two types of remnant propellant, depletion of all propellant at the same time might not be possible. It was necessary to control the time of shut down of engines by pressure variation during depletion, so as to avoid back flow of one propellant to the engine heads and explosion. In addition, due to dramatic pressure variation of propulsion subsystem when the propellant was depleted, it might lead to great uncertainty in engine combustion with wild variation of chamber pressure, mixed ratio and boundary film cooling, it was necessary to be verified by test in development to ensure the engine safety in depletion.

8.4 Propulsion Subsystem Design [5–7]

8.4.1 Fundamental Design Conditions

1. Environmental Condition

1) Ground environment

(1) Environment temperature: The environment temperature should be principally maintained at 20 ± 5 °C (if specially required, it could also be 12~20 °C), while more attention should be paid on temperature in propellant filling facility and launch vehicle fairing. When the balance was reached between the temperature of propulsion subsystem and environment, it was necessary to maintain enough safety ullage with no less than 3% of total volume of liquid cavity in propellant tanks and pressure in gas tank lower than its maximal allowed pressure.
(2) Atmospheric pressure: In case of propellant filling with tank evacuation, the tank shell should withstand ambient pressure (unnecessary for pressurized filling). During launch phase, it was also possible for the tank to bear external aerodynamic pressure. Therefore when the tank shell was designed, it was required that the shell must have enough margin resistant from external pressure with consideration of local atmospheric pressure level in launch site besides requirement for internal pressure, especially for composite wrapping tank. Generally, instability safety factor for external pressure should not be less than 5.0.

2) Flight environment during launch

The mechanical environments such as flight overload, vibration, impact and etc. during launch were crucial to structure of propulsion subsystem. While the structure and components bearing pressure of propulsion subsystem was designed, it was necessary to conduct the analysis of strength, stiffness and stability according to mechanical environmental conditions for propulsion subsystem and its components as well as comprehensive verification tests, especially for large-scale, heavy mass components such as propellant tank, gas tank, orbit control engine and etc. Due to its complication, the pipeline should not only be configured and fixed appropriately, but also be kept enough space for deformation of each part respectively to avoid pipeline broken caused by rigid constraints. The whole propulsion subsystem should be tested and verified under conditions simulated the flight environment as real as possible (if necessary, under more serious conditions) in order to satisfy flight mechanical environment requirements during launch.

3) Environment on orbit

There were environmental conditions including vacuum, zero gravity, temperature alternation and space radiation environment on orbit. The vacuum environment had no influence on propulsion system operation under internal pressure, but engine vacuum firing and cooling capability should be considered. At the same time, for moving parts exposed under vacuum environment, "vacuum welding" issue between metals should also be considered, which could be solved by remaining clearance, different metals, surface coating and so on. The zero gravity environment was an important factor when the propellant tank for space propulsion system was designed to ensure reliable management of liquid propellant inside tanks and avoid output of liquid with gas trapping. The methods for management of liquid propellant under zero gravity could be classified into two types: gas liquid mixed type and gas liquid physical isolation type. As an example of gas liquid mixed type, the surface tension tank and centrifugal tank (for spinning stabilization spacecraft) were designed to collect and expel liquid by surface tension effect. As an example of gas liquid physical isolation type, the metal diaphragm, non-metal bladder and piston tanks were designed to isolate pressurized gas from liquid propellant by metal diaphragm or non-metal bladder. For hypergolic bi-propellant propulsion subsystems, surface tension and metal diaphragm management tanks were generally used. When the propellant tank of the lunar lander was designed, it was necessary to consider large flux and wide range of flux variation of propellant (for orbit control engine with large flux and throttling ratio), lateral overload during soft-landing, liquid slosh, anti-vortex in EOL (end of life) large flux and etc. in order to determine the propellant management method. For example, the metal diaphragm tank was applied to isolate gas from liquid and eliminate uncertainty caused by propellant compatibility, slosh, adaptability, EOL disposal vortex and etc. for propulsion system of the Chang'E-3 lunar lander. Space radiation environment including ultraviolet rays, atomic oxygen, and charged ions, had significant impact on electronic devices of propulsion subsystem. Therefore it was necessary to strengthen anti-radiation design, especially anti-Single Event Effect. At the same time, it was important to avoid application of non-metals sensitive to ultraviolet irradiation, atomic oxygen erosion.

2. Propellant Load and Pressurized Gas Volume

Propellant load was determined by many factors such as specific impulse, orbit control, attitude control, tank expulsion efficiency, remnant propellant, mix ratio deviation and margin for propellant budget and etc. Once the propellant mass was determined, the volume of propellant tanks and the amount of pressurized gas could also be determined.

3. Nominal Operation Condition

Nominal operation condition for propulsion subsystem was up to the lunar lander system and GNC subsystem in accordance with flight control requirements. The specifications of propulsion subsystem were designed on the basis of design of nominal orbit control modes, as well as maximum output of delivery capacity, thrust drop amplitude for orbit correction/attitude control thrusters, thrust rise amplitude for attitude control thrusters under minimum mode of attitude control only. After the nominal operation condition was determined, it was necessary to give input design condition for each components and pipeline with synthetic parameter optimization of subsystem layout, pipeline direction and length, pipeline and valve flow diameter, supply pressure of propellant tanks, pressurized gas volume, structure mass and etc. For example, the nominal operation condition for propulsion subsystem of the Chang'E-3 lunar lander was determined as one orbit control engine with 7500 N thrust and six sets of attitude control thrusters with 150 N thrust.

4. Basic Configuration of Propulsion Subsystem

Basic configuration of the propulsion subsystem was up to system configuration design of the lunar lander. Due to large-scale and heavy mass of the propulsion subsystem, it should not only withstand its own load, but also transfer force for the lunar lander. Therefore the propellant tanks with welded frame were used as bearing structure of some the lunar landers. Layout and installation of large components like tanks and orbit control engines was up to system configuration of the lunar lander, as well as layout of pipeline, valves and especially pipeline direction. Different system configuration led to different load bearing design and had great influence on system structural mass. For example, the basic configuration of the Chang'E-3 lunar lander propulsion subsystem was central cross clapboards with 4 spherical propellant tanks symmetrically installed with their bottom fixed on the flange and the top supported by three horizontal rods, surrounded with peripheral side plates and a 7500 N orbit control engine located at the center of the lunar lander and 2 gas tanks hung out of side plates.

5. Structural Mass of Propulsion Subsystem

Structural mass of the lunar lander propulsion subsystem was regarded as one of key parameters for system design and was closely related with propellant load, subsystem configuration, components, system configuration and etc. Generally, it could be determined by repeating trade-off of subsystem design and system optimization of the lunar lander. Under strict constraints of structural mass, reduction of subsystem mass could be achieved by application of new technologies, new materials, integrated and lightweight design, and subsystem optimization. For example, the total structural mass of the Chang'E-3 lunar lander propulsion subsystem was required to be no more than 295 kg while the propellant load was 2600 kg under specific flight environmental conditions.

8.4.2 Determination of Major Specifications of Propulsion Subsystem

1. Propellant Option

The option of propellant was a comprehensive issue and generally determined when the concept design of the lunar lander was proposed. It should be considered including physical properties of propellant, mix ratio, vacuum specific impulse, density, cooling performance, vacuum ignition performance, storability, toxicity, safety, technical readiness, and etc.

1) Specific impulse of engine

The choice of propellant with high specific impulse and high density was an important factor to reduce propellant load, tank volume and subsystem structural mass. For the lunar lander propulsion subsystem, large total impulse and light structure weight were generally required. Usually bi-propellant based on N_2O_4 (MON-1 or MON-3 as an alternative)/MMH hypergolic propellant instead of mono-propellant would be more suitable with specific impulse performance between N_2O_4/N_2H_4 and N_2O_4/UDMH. Recently, along with increasing demand for high specific impulse and environmentally friendly propellant, propellant like LO_x/LH_2, LO_X/HC (LCH_4, LC_3H_8, RP-1) was developed very fast. The special advantage of such propellant with possibility of in situ production on Moon or Mars made it possible to support long term lunar base or Mars base and space transfer platform.

2) Long term storability in space

The long term storability of propellant was also an important factor for spacecraft. Due to restriction of the lunar lander mass, power consumption, lifetime, thermal control, engine and etc., there was still certain gap for application of liquid hydrogen/liquid oxygen, liquid oxygen/nontoxic hydrocarbon-based fuel, high specific impulse cryogenic propellant. After it was synthetically analyzed including physical properties such as frozen point, boiling point, density, saturated vapor pressure and technical readiness of engines, the conventional long term storable propellant was selected to simplify thermal control, reduce structural mass and development difficulties.

3) Re-ignition and pulse performance of engine

For attitude and orbit control of spacecraft propulsion subsystem, it was required to be capable of re-ignition in vacuum for orbit control engine and firing in pulse mode from tens of thousands to hundreds of thousands pulses for attitude control thrusters. For different propellant, the vacuum start-up performance, especially short pulse repeatable performance of engine might be different to some extent. Past research showed that hypergolic bipropellant based on N_2O_4(MON-1)/MMH had excellent vacuum ignition performance.

4) Engine cooling performance

For bi-propellant radiation-cooled engine, due to combustion temperature as higher as 3000 °C, the available metal could not withstand such high temperature. Therefore, small amount of propellant diverted from injector was sprayed directly onto the chamber internal wall to cool it, which was called liquid film cooling for boundary layer with 20–30% of total propellant flux. As liquid film coolant, fuel was usually used to dilute oxidant content near chamber wall and prevent oxidation of metals and coatings in high temperature. When fuel film coolant was selected, better physical characteristics such as broad liquid phase, large latent heat of vaporization, large heat capacity, small high temperature self-decomposition and high density should be considered preferentially. With comparison of three kinds of hydrazine fuel like N_2H_4, MMH and UDMH, the MMH had best cooling effect. Partial physical properties for four kinds of traditional propellants are given in Table 8.3 [8].

5) Uniform of propellant density ratio and mixture ratio

In radiation cooling engine design, the mixture ratio in core region with propellant theoretical optimal mixture ratio could lead to high specific impulse. At the same time, relative low mixture ratio in boundary layer can ensure effective cooling in this region. For radiation cooling engine using N_2O_4(MON-1)/MMH, the mixture ratio was basically chosen to be 1.65 which was the same as ratio of oxidizer to fuel density to achieve high specific impulse and better effective cooling. Such design was beneficial for realizing equal tank volume of oxidizer and fuel, simplifying the tank categories and layout, and reducing development cost. But for N_2O_4/UDMH, N_2O_4/N_2H_4, if the mixture ratio was the same as oxidizer and fuel density ratio, specific impulse will be penalized. Partial chemical properties for three kinds of traditional bipropellants are shown in Table 8.4.

2. Tank Volume

Tank volume was determined by many factors like propellant budget, tank expulsion efficiency, volume of propellant management device, propellant temperature, pipe

Table 8.3 Partial physical properties for four kinds of traditional propellants

Physical properties	N_2O_4	$(CH3)_2NNH_2$	N_2H_4	CH_3HNNH_2
Boiling point/°C	21.15	63.1	113.5	87.5
Freezing point/°C	−11.23	−57.2	1.4	−52.2
Density/($\times 10^3$ kg/m^3)	1.4460 (20 °C)	0.7911 (20 °C)	1.008 (20 °C)	0.8744 (20 °C)
Specific heat/($\times$ 4186.8 J(kg·K))	0.362 (20 °C)	0.653 (20 °C)	0.738 (20 °C)	0.70 (25 °C)
Heat conductivity/($\times$ 1.163 W(m·K))	0.132 (20 °C)	0.1376 (20 °C)	0.43 (25 °C)	0.2164 (29.7 °C)
Latent heat of vaporization/($\times$ 4186.8 J/kg)	99.0 (boiling point)	139.9 (20 °C)	299.8 (boiling point)	209.2 (20 °C)

Table 8.4 Partial chemical properties for three kinds of traditional bipropellants

Chemical properties	N_2O_4/UDMH	N_2O_4/N_2H_4	N_2O_4/MMH
Theoretical mixture ratio	2.57	1.30	2.17
Engine operation mixture ratio (remaining oxidizer coefficient 0.75)	1.928	0.975	1.628
Average density/($\times 10^3$ kg/m^3)	0.742 (20 °C)	0.585 (20 °C)	0.720 (20 °C)
Vacuum specific impulse/(N·s/kg)	3012	3064	3046
Density impulse/($\times 10^3$ N·s/m^3)	2234 (20 °C)	1792 (20 °C)	2193 (20 °C)
Density ratio of oxidizer to fuel	1.83	1.43	1.65
Deviation of density ratio from operation mixture ratio	−5.08%	46.67%	1.35%

Note Calculation in accordance with combustion chamber pressure 0.8 MPa, nozzle area ratio 100, specific impulse efficiency 91.2%

discharge volume, mixture ratio deviation and ullage volume ensuring safety for propellant temperature rise. For example, the volume of single tank of the Chang'E-3 lunar lander propulsion subsystem was 587 L (the sum of gas and liquid cavity) with total volume of 2348 L for four tanks.

3. Tank Operation Pressure

Tank operation pressure was determined by inlet pressure of orbit control engine/attitude control thruster, pressure drop of piping/tank (including valve, filter, orifice, and etc.) and static pressure rising of regulator. For example, the nominal tank operation pressure for the Chang'E-3 lunar lander propulsion subsystem was approximately 1.90 MPa. After the static pressure rising of the regulator was considered, it was selected to be 2.0 MPa.

4. Option of Pressurized Gas

When the pressurized gas was determined, it should be considered such as gas activity, relative molecular mass, compression, throttle-temperature and etc. For surface tension tank in which the gas was coexisted with the liquid, it should also be considered such as the compatibility of gas and propellant, as well as gas solubility in propellant. But for metal diaphragm or bellow tank with completed gas isolation from liquid, the issue could be neglected. The common pressurized gas was nitrogen or helium, the later one was more often used to minimize pressurized gas mass (under equal volume and pressure condition, nitrogen's mass is seven times heavier than helium).

5. Gas Tank Operation Pressure and Volume

Gas tank operation pressure and volume were comprehensively determined and optimized by propellant tank volume, operation pressure, flux, pressurized gas type, gas tank structural mass and size, difficulties and cost for development of gas tank. Along with rapid development of carbon fiber material, the application of high performance and light weight composite wrapping gas tank became a prevail tendency. Gas tank operation pressure reaches 60 MPa and the gas tank mass and size were effectively reduced. For example, two high pressure composite wrapping gas tanks with 100 L volume for a single one and pressure as high as 35 MPa with T1000 carbon fiber wrapped and seamless aluminum insider liner, were installed on the Chang'E-3 lunar lander propulsion subsystem.

8.4.3 Design of Propellant Feed System [4]

1. Design of Pressurized Gas Feed and Management

The deviation of nominal thrust of orbit control engine should be no more than 3% and the structural mass restriction was stringent for the lunar lander propulsion subsystem. Blow-down system couldn't satisfy the requirements. The gas pressurization system with regulated pressure could ensure thrust of orbit control engine with precision of 1%. For precisely control of mixture ratio, generally only one regulator was used (regulators can also be set in serials or parallel with consideration of regulator failure) to ensures equal pressure of two different propellant tanks by simultaneous pressurization. For surface tension tank, check valves were needed to prevent unexpected mixing of two propellants or its vapor in gas system. For metal diaphragm tank, check valves were not necessary and elimination of them can avoid gas pressure deviation caused by flow resistance and low frequency vibration due to gas flow in check valves. Because the tank pressure might be over pressurized if there was regulator lock up pressure rise under non-flow conditions, the latch valves were installed at upstream of regulators to realize overpressure regulation. When overpressure occurs, the latch valve would automatically shut down and cut off high pressure gas feed. After the pressure recovers, the latch valve would automatically open again and restore gas feed. To avoid tank rapid overpressure caused by regulator sealing failure and valve rod stuck, the pressure relief valves with automatic opening and venting under over pressurized condition were installed at downstream of regulators to release pressure and shut-off when the pressure recovered to avoid continuous vent.

For fill and vent of gas tanks, leak test and condition monitoring, fill and vent valves, test interface, pressure and temperature sensors should be configured.

After landing, it was specially required for propulsion subsystem to conduct depletion to vent completely all residual propellant and gas. Generally, a pair of attitude control thrusters fired simultaneously to deplete residual propellant and the pressurized gas was directly vented by solenoid valve or pyro valve at downstream of pressure regulator till the pressure in propellant and gas tank dropped to 0 MPa.

The pressurized gas line of the Chang'E-3 lunar lander propulsion subsystem were configured with 2 high pressure composite wrapped gas tanks for storing 100 L, 35 MPa pressurized Helium gas, feeding simultaneously through a large flow rate regulator to 4 metal diaphragm tanks (2 for oxidizer and 2 for fuel). And high pressure latch valve and relief valve were configured as double safety methods for tank overpressure management.

2. Propellant Storage and Management

Design of propellant storage and management consisted of propellant fill and vent method, storage and safety isolation from downstream pipeline, liquid management under zero-gravity environment, propellant feeding and condition monitoring and etc. Design of propellant storage and management was up to tank type. For surface tension tank with pressurized filling, tank outlet was safely isolated by latch valve or pyro valve. For diaphragm tank, due to its closed liquid cavity, propellant filling could be realized by evacuation filling and tank outlet can be safely isolated by diaphragm valve, latch valve or pyro valve. When it was isolated by diaphragm valve without electric control, the valve would be opened by tank pressurization. When the burst pressure of diaphragm valve was designed, it should be considered including saturated vapor pressure, liquid level, flight overload, burst pressure deviation, tank expulsion pressure and etc. to ensure not only reliable isolation of diaphragm prior to tank pressurization, but also effective burst after tank pressurization. Fill and vent valves, pressure and temperature sensors should be configured to monitor tank pressure, temperature and propellant filling. For example, there were 4 identical spherical metal diaphragm tanks (2 for oxidizer and 2 for fuel) in the Chang'E-3 lunar lander propulsion subsystem. The pressurized gas cavity was separated from propellant cavity by metal diaphragm with special profile, while the metal diaphragm was compressed tightly on liquid by pressurized helium gas metal diaphragm tightly and deformed to expulse the propellant in the propellant tank. Because the metal diaphragm deformed under certain pressure difference and the force applied on propellant was much larger than sloshing force, the propellant slosh could be eliminated and there was not constrain by overload direction, which made it more suitable for complicated overload environment of the lunar lander.

3. Propellant Feed and Management

Design of propellant feed and management consisted of orbit and attitude control engine management, and propellant feed system monitoring. Generally in order to minimize the interaction between orbit and attitude control engines, latch valves were used for orbit and attitude control engine to operate independently. For attitude control feed pipelines, solenoid latch valves were used directly, while solenoid or pneumatic latch valves can be used for orbit control feed pipeline depending on the thrust-level of engines. If the thrust was greater than 3000 N, the pneumatic latch valve was generally used because the solenoid valve was heavier. Because of constraints of structural mass and size for the lunar lander, basically there was only one orbit control engine and two branches for attitude control thrusters, where single branch could be isolated under failure conditions. For example, there were one 7500 N orbit control engine and 28 attitude control thrusters, while the orbit control engine was managed by 4 pneumatic latch valves and attitude control thrusters were managed by 8 electromagnetic latch valves divided into A/B two branches according to the same function and numbers. Since the orbit control engine was fired only few times and shut off during long term orbital flight, the propellant feed could be cut off by shutting down upstream latch valve to avoid permanent propellant leak because of leak failure of the orbit engine. When an attitude control thruster could not be fired, it could be switched to the corresponding attitude control thruster on the other feed line. When there was leak failure in one branch of attitude control thrusters, the whole branch could be switched off by shutting down upstream latch valve on a certain feed line. In order to monitor liquid pipelines and operation status of orbit/attitude control engines, there were pressure sensors on inlet pipelines of orbit control engine and attitude control thruster on A/B two branches, and temperature sensors on liquid pipelines, injector of engine and thrusters, valve and etc. The diagram of gas and liquid feed system of the Chang'E-3 lunar lander propulsion subsystem is shown in Fig. 8.5.

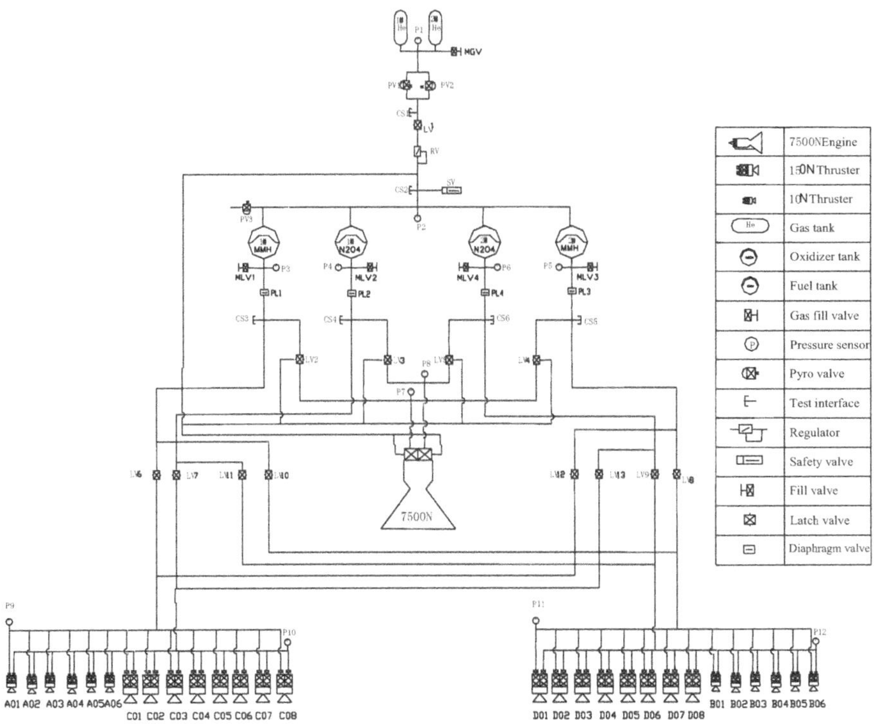

Fig. 8.5 Diagram of gas and liquid feed system of the Chang'E-3 lunar lander propulsion subsystem

8.4.4 Electronics Design of Propulsion Subsystem

Propulsion subsystem electronics which converted different types of control commands to drive propulsion valve and collected telemetry downlink data of different types of status parameters, was a "key node" between spacecraft control subsystems (GNC, power supply, telecommunication) and propulsion subsystem, and was generally integrated in the control unit. Because of the importance of control unit, dual or triple redundancy design could ensure reliable operation. The control unit of propulsion subsystem was designed for power control of different instruments, DC/DC secondary power conversion, valve (including solenoid and latch valve) drive control, stepping motor control for throttleable engine, sensor data collection and processing, autonomous on/off control of thermal heater, autonomous failure detection and switch control, internal status monitoring for control unit, telemetry communication and etc. Generally the relay was used to control on/off switch of instruments. CMOS was widely used to control on/off switch of valve and heater. H-bridge inverter based on intelligent chip was used to control stepping motor. RS-422 asynchronous serial interface or 1553 bus interface was used to implement command communication. For pressure, displacement and contact switch, voltage output is 0~5 V. For temperature sensor, branch voltage circuit was used. For example, the control unit of the Chang'E-3 lunar lander propulsion applied A/B branch with cold redundancy design based on 80C32 as the central unit, one of which could conduct all control functions of propulsion subsystem. The stepping motor was controlled by chopping constant current control mode based on H-bridge driver chip LMD18200T with RS-422 asynchronous serial command interface. Two subdivision in step angle could enhance operation resolution and stability of the stepping motor.

8.4.5 AIT Design of Propulsion Subsystem

1. Configuration Design

Configuration design of the lunar lander propulsion subsystem was closely related to configuration design of the lunar lander system. On the basis of the lunar lander configuration and synthesis of component position & pipeline layout, the principle of propulsion subsystem configuration was symmetry, easy to assembly, convenient for test, modular and integration. Analysis of past the lunar lander configuration showed that truss or plate structure was generally used with 4 tanks horizontally, an orbit control engine in the center, gas tanks outside of the lunar lander or between the tanks, attitude control thrusters outside the lunar lander, and other parts inside such as valves, pipelines, electronic unit, cable and etc. For example, it was cross-like plate structure with 4 propellant tanks symmetrically located between cross-like plates, an orbit control engine in the center, 2 gas tanks hung outside lander, 28 attitude control thrusters outside of the lunar lander respectively at upper, middle and bottom, tanks fixed by flanges and rods in propulsion subsystem of the Chang'E-3 lunar lander.

Fig. 8.6 Configuration of propulsion subsystem of the Chang'E-3 lunar lander

The configuration of propulsion subsystem of the Chang'E-3 lunar lander is shown in Fig. 8.6.

2. Seal Design of Pipelines in Propulsion Subsystem

There were two methods of pipeline seal design for propulsion subsystem including welding and screw thread connection. In order to enhance reliability, welding connection was generally used, especially for gas line with nitrogen or helium which was easy to leak. For liquid pipeline, it should be considered particularly such as engine/thruster orifice plate test, dissimilar materials welding difficulties, pressure sensor calibration, welding operability and etc. Therefore, more than two sealing should be used for screw thread connection. In order to satisfy the leakage rate requirements for propulsion subsystem, the leakage rate at each connection should be strictly controlled less than 1×10^{-7} Pa · m^3/s for welding connection and 5×10^{-7} Pa · m^3/s for screw thread connection. For the Chang'E-3 lunar lander propulsion subsystem, except screw thread connection for gas tank, latch valve, tank, pressure sensor, orbit/attitude control engine and test interface, other connections were all welding to ensure the leakage level of propulsion subsystem.

3. Alignment Accuracy for Orbit/Attitude Control Engine

Alignment accuracy for orbit/attitude control engine was one of key factors to ensure orbit and attitude control accuracy. Alignment accuracy for orbit/attitude control engine includes accuracy of position and nozzle geometric axis orientation. Installation reference of orbit control engine was head flange surface and that of attitude control engine was head flange surface or bracket installation surface. Accuracy of nozzle geometric axis orientation for orbit/attitude control engine was generally ensured through design or manufacture process. If necessary, accuracy test should be conducted. For orbit control engine, due to large thrust, actual thrust vector would deviate from the geometric axis. Because of the significant impact of thrust vector deviation to attitude control, the deviation was practically measured in engine hot-fire test, and then the engine was installed and adjusted according to test results with its force point located on the lunar lander longitudinal axis and coincidence between thrust vector and the lunar lander longitudinal axis.

For orbit/attitude control engine, the accuracy of nozzle axis orientation relative to head mounting surface was ensured by a combination manufacturing process, wherein the precision of nozzle axis was ensured by high precision mold and the precision of manufacturing process was ensured by processing equipment and device.

For orbit/attitude control engine, the accuracy of nozzle axis orientation was measured by optical theodolite or 3-D Coordinate Measuring Machine. Generally the optical method was used in the lunar lander test with cubic mirror installed on the reference surface of the measured object. The measuring data as a reference of the lunar lander was acquired from that relative to the cubic mirror through coordinate conversion. If the accuracy was not met with the lunar lander requirements, the installation method of single engine or engine complex could be adjusted to satisfy the accuracy requirements.

8.4.6 Major Components Design for Propulsion Subsystem

1. Orbit Control Engine [9]

Generally, continuously throttleable engine was used to adapt variable thrust requirements during soft-landing for orbit control in the lunar lander propulsion subsystem. During LTO, LOI, circumlunar orbit de-orbit and main deceleration phase of powered descent, fixed thrust operation mode was applied (according to the demand of orbit control, constant thrust mode could be applied in any thrust level). During powered descent process, real-time throttling thrust should be output by the demand of autonomous control of the lunar lander.

The methods of continuously throttling thrust included three types: regulator output pressure feed-back, injector throttling and "injector-cavitation tube" combined throttling. The advantage of the first method was structural simplicity and possibly to be designed by constant regime. The disadvantage was delay of thrust variation response inappropriate for rapid thrust throttling, especially because the ullage volume of large propellant tank was large and the change of pressure was slow. Meanwhile, it was difficult for the engine to be capable of outputting wide range thrust. The advantage of the second method was rapid thrust variation response. But for bi-propellant engine, the mixed ratio was difficult to be maintained precisely and the propellant utilization efficiency was affected when the engine works in throttling mode deviated from nominal design point. The third method could overcome the disadvantages of above two methods and was widely applied. For example, orbit control engine with thrust throttling ratio of 10:1 was applied in descent propulsion subsystem of Apollo lunar module and the engine with throttling ratio of 5:1 was applied in propulsion subsystem of the Chang'E-3 lunar lander.

Combined throttling for bi-propellant orbit control engine generally applied "stepping motor and pintle" mechanism, which drove the variable pintle of cavitating-venturi and injector to modulate flux and pressure drop synchronously.

Design of bi-propellant orbit control engine generally included 3 types of cooling: radiation cooling, regenerative cooling and ablative cooling. For large ratio throttling engine, because the propellant flux varied widely, it was difficult to maintain reliable cooling over all operation regime due to wide variation of coolant for regenerative cooling. For the orbit control engine which was used for LTO, circumlunar orbit and powered descent with long lifetime, the disadvantages of ablative cooling included short lifetime, unstable performance (uncertainty in ablation), heavy structural mass and etc., which made it suitable for short term scenario. Therefore radiation cooling was more suitable. The common materials used for engine chamber of radiation cooling engine, such as niobium, rhenium-iridium, platinum-rhodium alloy and carbon fiber ceramics materials, were capable of resisting high temperature, oxidation and erosion. For example, the niobium-tungsten alloy was used in thrust chamber of the 7500 N throttling engine on the Chang'E-3 lunar lander.

2. Propellant Tank

Propellant tank was used for storing propellant (oxidizer, fuel) required for orbit/attitude control engine operation, managing propellant in zero gravity condition and supplying propellant with capabilities of pressure-bearing, load-bearing and long term compatibility. There were generally three types of propellant tank for bi-propellant propulsion subsystem including surface tension tank based on titanium alloy (full management or local management), metal diaphragm with aluminum alloy and fluoro-plastics bladder, whose technology readiness was all good. For the lunar lander propulsion subsystem, aluminum alloy diaphragm tank had advantages such as easily realizing large flux, sloshing free, unconstrained by accelerate direction, long term propellant compatibility, simplicity of gas line (no need using check valve for isolation) and etc. The main disadvantage was low strength of aluminum alloy which might cause heavier structure, and limited overturn times of diaphragm (actually once, overturn capabilities no less than 3 times). Currently, the approaches to solved heavy weight of aluminum alloy diaphragm tank include: (1) application of high-strength aluminum-scandium alloy instead of traditional aluminum alloy to make tank shell, (2) wrapping with carbon fiber to strengthen tank shell structure, (3) using high-strength titanium alloy diaphragm tank. The key part of metal diaphragm tank was internal metal diaphragm. In order to ensure regular deformation and reliable overturn of diaphragm, it was necessary to apply variable thickness and curvature design. For example, the design of aluminum insider liner and carbon fiber wrapping was applied to reduce weight in propellant tank of propulsion subsystem of the Chang'E-3 lunar lander. Fully welded aluminum inside liner with good propellant compatibility was used to seal pressurized gas and propellant, and bear load (propellant mass, acceleration, vibration and etc.). Carbon fiber with multi-direction alternative wrapping was capable of withstanding tank inner pressure. Specially configured

Inner diaphragm separated gas cavity from liquid cavity to prevent propellant from gas. When the propellant tank worked, propellant was expulsed along with deformation of metal diaphragm. During flight, propellant sink operation was not necessary and there was no propellant slosh. In addition, there should be pressure differential to deform metal diaphragm, so it could endure acceleration along various directions.

3. Gas Tank

Gas tank was used to store pressurized gas with high pressure and good seal performance. The major types of high pressure gas tank in spacecraft propulsion subsystem were titanium gas tank and metal liner with composite wrapping. Under the same pressure and volume, the mass of composite wrapped gas tank was lighter. The main failure mode was leaking mode prior to burst, so it was safer than metal gas tank. Along with the development of carbon fiber materials, wrapping technology and non-destructive examination method, composite wrapped gas tank was developed rapidly. The lightweight, large volume of high pressure gas tank was widely used in propulsion subsystem. According to different structures of inside liner, composite wrapped gas tank was classified into welded titanium alloy inside liner and seamless aluminum alloy inside liner. The former one was easy to be welded with titanium pipelines and has good sealing in connection. Due to weld seam and heat affected zone vulnerable to failure in titanium inside liner, it was necessary to match the deformation between inside liner and wrapped layer and specify high requirements of welding process. The latter one was easy to be formed, manufactured with equal strength design (especially for hemisphere-cylinder structure) with excellent fatigue resistance. The disadvantage was that it cannot be welded with titanium pipelines, so double-sealing screw thread connection was generally applied. For example, a 35 MPa and 100 L composite wrapped gas tank with seamless aluminum inside liner for sealing pressurized gas and wrapped layer for bearing load of gas pressure was used in propulsion subsystem of the Chang'E-3 lunar lander.

4. Attitude Control Thruster

Attitude control thruster was used for attitude control, stabilization, adjustment, orbit correction, hover and translation during LTO, circumlunar and powered decent, which was also used to deplete remnant propellant after landing. Generally the specific impulse of attitude control thruster should be high and the thruster should be stable in continuous firing for thousands of seconds or more, hundreds of thousands of pulse firing, and especially small pulse of 10~20 ms minimal duration. In order to satisfy requirements of attitude control thruster, not only efficient internal combustion and boundary film cooling was necessary for long term reliable work, but also reducing filling volume of injector head and enhancing fast response of valve was necessary to conduct short pulse. At the same time, because the number of pulse was large, anti-jamming, anti-leak and anti-clogging should be emphasized in thruster design. For example, for the 150 N thruster in propulsion subsystem of the Chang'E-3 lunar lander, film and radiation cooling was applied. High temperature resistant niobium alloy sprayed with anti-oxidization coating is used for thruster chamber cylinder part. Oxidizer and fuel valves were installed on the head of thrust chamber to minimize the propellant filling volume behind valve and ensure minimal pulse impulse and rapid response. Redundant seat and seal structure were designed to ensure reliable seal of valve.

5. Large Flux Latch Valve

Because of large thrust and flux of orbit control engine in the lunar lander propulsion subsystem, large flux latch valve was generally used to control propellant feeding for orbit control engine and avoid propellant loss in case of engine leak failure. The large flux latch valve should be light weight, large flux, low flow resistance, reliable seal, reliable action, long term compatibility with propellant and etc. Traditional solenoid latch valve was not qualified for the task, so solenoid pneumatic latch valve was applied more frequently. For example, the solenoid pneumatic latch valve was used in propulsion subsystem of the Chang'E-3 lunar lander, which consisted of two parts: pilot valve and master valve. Pilot valve was a kind of solenoid latch valve to control gas driving master valve on/off switch. In order to ensure reliability of dynamic seal in oxidizer, metal bellow instead of rubber seal was used as dynamic seal. The swelling of rubber seal and compatibility with oxidizer were avoided.

6. Propulsion Control Unit

Propulsion control unit was a key node between GNC and OBDH subsystems, which was also key driver of components in propulsion subsystem. Through timely and accurate reception of control command from GNC and OBDH subsystem, it could drive components of propulsion subsystem, and sends telemetry, status data back to GNC and OBDH subsystem at the same time.

Based on reliability requirements, propulsion control unit was generally designed according to dual or triple-redundancy principle. It consisted of DC/DC power, CPU, flip latch, address decoder, PROM, RAM, I/O extension, RS422 or 1553B bus interface, power-driven circuit, watch-dog circuit and etc. For the two-phase stepping motor for thrust throttling, chopping constant current control based on H-bridge drive chip as the core was generally used to control stepping motor precisely, rapidly and stably. For example, A/B dual redundancy with cold stand-by design was applied in propulsion subsystem of the Chang'E-3 lunar lander. The LMD18200T based H-bridge drive chip as the core combined with 8-bit D/A converter AD558, LM139 comparator and time-based circuit LM555 were applied to establish closed-loop control of two-phase stepping motor for 7500 N throttling engine. The power for driving orbit/attitude control engine valve or solenoid valve could be designed in series, parallel or series-parallel combination to ensure reliable on/off switch.

8.4.7 Ground Test and Propellant Filling of Propulsion Subsystem

1. Ground Test of Propulsion Subsystem

The flight hardware of propulsion subsystem should be tested and verified to check function, performance, environmental adaptability and interface correctness. There should be tests in component, modular, subsystem and system level. The test items included electric performance, parameters adjusting, leakage rate, accuracy, polarity, engine/thruster simulated hot firing test for high altitude, hot firing of propulsion subsystem, tank expulsion, gas tank burst and environmental test and etc. The tests for propulsion subsystem of the Chang'E-3 lunar lander are given in Table 8.5.

2. Propellant Filling

Propellant filling was one of key operation for propulsion subsystem in launch site. Because of properties such as high toxicity, strong corrosion, flammable and explosive and etc., reliable and safe requirements for propellant filling were very high. At the same time, the operation was irreversible once propellant is filled, so it was highly emphasized. Propellant and gas filling are both implemented by ground equipment. According to different tank type, rational option for concrete filling method would be made. For example, propellant filling with tank evacuation was chosen for propulsion subsystem of the Chang'E-3 lunar lander because metal diaphragm tank was applied. Before filling, the internal cavity was evacuated. After required vacuum degree (generally vacuum pressure below 133 Pa) was satisfied, the propellant was pressurized and filled into tank while the propellant load was weighted. In particular, in the process of propellant filling, MON-1 oxidizer should be cooled for strictly maintaining the pressure in the liquid cavity of the tank to be less than restricted reverse pressure on metal diaphragm.

8.5 Testing and Verification of Propulsion Subsystem

The components of the propulsion subsystem were usually types of mechanic, electro-mechanic or electronic products with the feature of pressure bearing, load bearing, movement, fluid, combustion, thermal control and etc. System composition was complex. The products were closely related to manufacture process and materials used. Furthermore, a lot of working processes were difficult to be accurately analyzed by means of theoretical simulation. And past experience and experimentation were important basis of design. For example, engine combustion performance was basically verified by a great many tests. So a great many ground tests in component, subsystem and system level should be conducted for propulsion subsystem of the lunar lander. Tests were generally classified into acceptance, qualification, batch sampling and marginal test. Tests covered electric performance, parameter adjusting, leak test, accuracy measuring, environmental tests (centrifugal, vibration, shock, high and low temperature, moisture and hot, thermal vacuum, thermal circle and stress test), actuation characteristics, engine/thruster hot firing, pressure fatigue circle, burst strength, subsystem integration test, EMC and etc. Qualification test was applied for comprehensive validation of product function, performance, reliability, environmental adaptability, quality and etc. According to different types of products,

Table 8.5 Test items for propulsion subsystem of the Chang'E-3 lunar lander

Test item	Test content	Note			
		Component	Modular	Subsystem	System
Insulation test	Insulation resistance of detonator, solenoid valve, latch valve, stepping motor coil and pressure sensor	√	√	√	√
Electric current of solenoid, latch valve	Opening/shutdown response performance for solenoid valve, latch valve	√	√	√	√
Regulator adjusting	Pressure, flux performance	√	–	–	–
Engine/thruster cold flow test	Flow resistance of 7500 N engine and 150 N, 10 N thruster	√	–	–	–
Pipeline cold flow test	Flow resistance matching of orbit control engine	√	–	–	–
Helium leak test	Leakage rate of each component, modular, subsystem	√	√	√	√
Relief valve opening/shutdown pressure test	Opening/shutdown pressure characteristics	√	–	–	–
Electronic unit test	Function, interface, software and control of electronics	√	√	√	√
Installation accuracy test	Nozzle geometry accuracy of 7500 N engine and 150 N, 10 N thruster	√	√	√	√
Polarity test	Polarity for orbit/attitude control engine, latch valve, stepping motor, limit switch, pressure sensor, pipeline connection	√	√	√	√
Flight simulation test	The correctness of telemetry, command interface, software, flight procedure	–	–	–	√
Engine/thruster batch sampling hot firing	Validation of engine/thruster performance	√	–	–	–

(continued)

Table 8.5 (continued)

Test item	Test content	Note			
		Component	Modular	Subsystem	System
Tank batch sampling vent test	Diaphragm pressure drop, tank expulsion efficiency, burst pressure	√	–	–	–
Gas tank burst test	Fatigue and burst	√	–	–	–
Pyro valve test	Blasting and seal	√	–	–	–
7500 N engine hot firing	Specific impulse, thrust, mixed ratio, thrust throttle, thrust vector	√	–	–	–
Parallel tank cold equilibrium flow test	Equilibrium vent of parallel tank	–	–	√	–
propulsion subsystem hot firing	System performance and component matching	–	–	√	–

qualification test should be conducted on one or more prototypes. Acceptance test should be fully conducted. The flight hardware should pass acceptance test. Batch sampling test emphasized on batch quality, performance stability and uninspectable items of flight hardware. Marginal tests were specific verification tests for design and actual working conditions of propulsion subsystem. The marginal test was generally conducted for verification of adaptability and margin, which was conducted for only 1 or 2 times. For example, tests were conducted including subsystem hot firing, tank vibration and vent test, gas tank long life storage test, high altitude simulation hot firing for 7500 N orbit control engine/150 N,10 N attitude control thruster, 7500 N orbit control engine coating thermal shock test, stepping motor lock-rotor test, attitude control thruster composite bracket high temperature static test, pyro-valve pyro shock test, relief valve pressure bearing test, pressure sensor pressure bearing test, high-altitude simulation hot firing of orbit control engine with GNC, propellant filling test and etc. for propulsion subsystem of the Chang'E-3 lunar lander.

The typical tests of propulsion subsystem are listed as follows.

8.5.1 Subsystem Ground Hot Firing

Subsystem ground hot firing was a synthetic test to verify properties of propulsion subsystem and validate the correctness, consistency and reliability of design solutions comprehensively. The test could assess function, performance and status adaptability of hardware and software. It could acquire performance for different operation conditions (including failure mode) and provide data used for flight operation. Subsystem ground hot firing was an irreplaceable test using real propellant to fire instead of simulant to simulate (also known as cold test).

Basic concept for subsystem ground hot firing was that the configuration of propulsion subsystem (layout, pipeline orientation, interface, parameters, propellant load, gas tank pressure and etc.) was quite the same as that of flight hardware, and the orbit/attitude control engines were short nozzle version to adapt ground test without changing internal process of engines. The test procedures and conditions would simulate operation characteristics during different flight phases. The combustion chamber pressure of orbit/attitude control engines was measured to acquire static thrust or pulse impulse. The tank expulsion efficiency, system mixture ratio and equilibrium expulsion of parallel tanks could be evaluated by weighting residue propellant after test. Generally the parameters measurement and control procedures of test were simulated by ground test facility, otherwise the TT&C and control subsystem of the lunar lander could be applied. Test articles were installed and fixed on ground test facility. For subsystem reliability test, before test the test articles should also pass vibration test under propellant loading (actual loading simulant) to evaluate structural reliability during launch.

Because of complexity, challenge and new technology for propulsion subsystem of the lunar lander, there were usually 2 or 3 subsystem hot firing tests during different phases when the lunar lander was developed. For propulsion subsystem of manned the lunar lander, 1 or 2 additional test should be conducted to adequately validate function, performance, reliability, safety of the subsystem. For example, because the propulsion subsystem the Chang'E-3 lunar lander was totally new with great challenges in new technology, requirements and complexity, three hot firing test had been conducted in phases A(concept design), B/C(design and fabrication) and D(system AIT) respectively to examine and verify the subsystem operation.

8.5.2 Throttling Engine Test

The Throttling orbit control engine was the most critical component of the propulsion subsystem of the lunar lander. It served for orbit control in different flight phases, especially in powered descent. In case of ignition failure, flameout or shutdown failure, it would lead to mission failure. Because the throttling engine with large throttling ability, high performance and reliability was new to China space industry, it was necessary to be verified fully by ground test and high altitude simulation test. For example, a great many tests for performance, margin, reliability and lifetime were conducted for the 7500 N throttling engine of propulsion subsystem of the Chang'E-3 lunar lander as follows.

1. Test Deviated from Normal Conditions

Input pressure (1.6~1.9 MPa) and mixture ratio (1.6~1.7) were deviated from nominal values (1.7 MPa for input pressure and 1.65 for mixed ratio) in 7500 N engine test. Thrust was deviated up to 9000 N and down to 1200 N respectively and mixture ratio was deviated to 1.85 in 9000 N thrust output. The test fully evaluated the operation margin of the engine.

2. Hot Firing with Wet Stored Propellant

Hot firing with wet stored propellant could simulate the engine restart performance and reliability after long term contact with propellant during flight, especially smoothness of moving parts. Several sets of engines conducted multiple fire tests when the engine was developed, including short nozzle engines after 30 days, 64 days propellant wetted storage restated and large nozzle engines after 19 days propellant wetted storage restated. The engines all worked normally, reliably and stably. The thrust could be modulated flexibly and operated with good performance consistency.

3. Hot Firing after Engine Neutralization

Hot firing after engine neutralization was conducted to verify the performance and reliability of engine after its firing and subsequent neutralization of disassembling, cleaning, drying, adjusting and reassembling. There were 3 sets of short nozzle and 4 sets of large nozzle conducted hot firing after neutralization. The flight hardware of throttling engine after neutralization was reassembled and installed on the Chang'E-3 lunar lander, which was successful in flight mission. Ground and flight test results showed that the engine after neutralization worked reliably and stably.

4. Hot Firing under High and Low Temperature

Hot firing under high and low temperature was conducted to verify the engine adaptability to environmental temperature change. Multiple fire tests were conducted under different propellant temperature of −1~33 °C with normal, stable and consistent operation.

5. Hot Firing with Heat Shield

In order to verify the performance and reliability of the engine with heat shield, hot firing of 2 sets of large nozzle engines was conducted with accumulated firing time more than 4000 s for each set and continuous firing 2000 s once. The duration with nominal 7500 N output was 1000 s with maximal temperature of 1280 °C at throat. Test results showed that the engine with heat shield had suitable temperature and worked reliably.

8.5.3 Propellant Tank Test

The characteristics of propellant tank for propulsion subsystem of the lunar lander was large propellant load, high expulsion efficiency, small structural mass, special installation configuration, harsh vibration condition, wide flow variation, high consistency of pressure drop in parallel tank and etc. Furthermore, it should endure zero-gravity and lateral overload condition for propellant discharge in soft-landing. So it was very difficult for design of propellant management, shell pressure-bearing strength and stiffness. A great many ground tests should be conducted including shell load bearing, management device, discharge performance and reliability. For example, the lightweight composite wrapped aluminum alloy diaphragm tanks fixed at both ends were applied in propulsion subsystem of the Chang'E-3 lunar lander, for which centrifugal, vibration, discharge, long term inverse pressure of diaphragm tests were conducted.

1. Centrifugal Test

A real tank with actual installation interfaces and propellant load was used in centrifugal tests in 3 status and 7 operation conditions totally. The test levels could be classified into acceptance, qualification and marginal test level. Test conditions covered longitudinal/transversal, longitudinal, transversal and reversal condition. After tests, leak test and expulsion test were conducted. Parameters such as leakage rate, expulsive efficiency and diaphragm overturn times satisfied the requirements, and the anti-overload capabilities of tank shell and metal diaphragm were fully verified.

2. Vibration Test

Two real tanks with actual installation interfaces and propellant load were used to conduct vibration tests in acceptance and qualification level. One of the tanks was simulated for propellant filling with tank evacuation and then passed vibration test with simulant. After test, the leakage rate, expulsion, fatigue and burst performance were examined to be normal. After 8 tanks individually passed full load acceptance tests, they were installed on propulsion subsystem integrated on the lunar lander and passed one acceptance test and one qualification test respectively. After then, propulsion subsystem hot firing test was conducted. Through vibration test and expulsion test for several propellant tanks, the adaptability and reliability of the tank shell and diaphragm were verified.

3. Expulsion Efficiency Test

In order to simulate influence of flow variation on tank expulsion during orbit control engine operation in each flight phase, two tanks were used to conduct large flux expulsion within 0~80% of propellant consumption (simulating near LOI, main deceleration phase with constant thrust output of 7500 N throttling engine), tank expulsion with flux varied continuously within 80~90% of propellant consumption (simulating throttling thrust modulation in powered descent), and tank expulsion with constant flux within 90~100% of propellant consumption (simulating 2000 N constant thrust at the end of powered descent). The tank expulsion performance and reliability with flux as in flight were fully verified.

4. Inverse Pressure-Bearing Test of Metal Diaphragm

To simulate long term inverse pressure-bearing capability of metal diaphragm under saturated vapor pressure of MON-1 oxidizer after filled, two tanks were used to conduct inverse pressure-bearing test of metal diaphragm. One of the tanks was applied with different inverse pressure drop, and then the change of diaphragm perimeter was measured: Under 0.2 MPa, the perimeter was almost not changed; Under 0.2~0.25 MPa, the perimeter was slightly changed; Under 0.8 MPa, the perimeter was significantly enlarged, but no burst leaking happened. Another tank filled with oxidizer to simulate long term storage from propellant filling to launch (no more than 10 days) with endurance of saturated pressure (maximally no more than 0.2 MPa). After 10 days of storage with 0.2 MPa, if it was examined to be normal, the pressure was enhanced to 0.3 MPa for 10 days. Finally, the tank conducted expulsion test normally. The pressure drop, expulsion efficiency and overturn times of metal diaphragm satisfied the requirements.

8.5.4 Long-Term Pressure-Bearing Storage Test of Gas Tank

Gas tank of propulsion subsystem for the lunar lander was used to store pressurized gas for propellant expulsion. Because of mass constraints, gas tank of high pressure, light weight and composite wrapped structure was widely used. Besides leak test, vibration, shock, thermal vacuum, thermal circle, fatigue and burst pressure tests, in terms of ground storage time after gas filling and operation time in orbit, it was also necessary to validate long term pressure-bearing capability and reliability. For example, composite wrapped gas tanks (aluminum inner) with 1.5 of safety factor, 35 MPa of working pressure, 100 L of volume and T1000 wrapping materials was applied in propulsion subsystem of the Chang'E-3 lunar lander. In order to simulate storage process from filling to launch, one gas tank was used to store helium for 41 days with 35 MPa pressure. During the storage process, 6 helium leakage rate detection were conducted totally with satisfactory results. Long term pressure bearing capability and reliability were fully verified.

8.5.5 Propellant Filling Test

Propellant filling for the lunar lander was the most critical process in launch campaign. Because of toxicity of propellant, it couldn't be verified with real propellant when the flight hardware was developed, examined and tested. Therefore, during the process of development, special propellant filling test was usually conducted to assess the consistency and correctness of filling, parameters, safety, ground equipment and etc. For example, evacuation and synchronous equilibrium filling for the two tanks of same propellant was applied for propulsion subsystem of the Chang'E-3 lunar lander. Because there would be NO out of MON-1 oxidizer during vacuum filling, it was necessary to monitor and control filling temperature, pressure, speed and etc. Therefore, special propellant filling tests of propulsion subsystem with different conditions such as different propellant temperatures and filling pressures were conducted to formulate appropriate filling technical solution in launch site.

The propellant filling for the Chang'E-3 lunar lander in launch site showed that: The filling design was suitable and correct; The filling ground equipment worked reliably; The control of filling parameters satisfied design requirements.

8.6 Summary

The propulsion subsystem of the lunar lander had complicated operation modes and shall pass complex flight environment, which made it difficult to be developed. The propulsion subsystem should be designed on the basis of similar technology according to mission characteristics and requirements, current status and etc. Its function, performance and robustness should be fully validated through a great many ground test.

Special attention should be paid in design of propulsion subsystem.

(1) The subsystem and component should be designed to eliminate negative effect of different uncertain factors in flight, such as reliable expulsion under propellant sloshing and lateral overload, cooling and performance degradation in low thrust mode of orbit control engine with large throttling ratio, impact of propellant to sealing rings in case of dynamic sealing between moving parts and etc.

(2) The design parameters of propulsion subsystem should be optimized to improve subsystem performance. For example, propellant option should be optimized for boundary cooling flux, characteristic length of combustion chamber, nozzle expansion ratio. The lightweight and high performance composite materials could be applied to effectively reduce mass of propellant and gas tank.

(3) Under resource constraints, system redundancy, fault reconfiguration and management should be reasonably designed to increase system reliability. For example, there should be redundancy for attitude control engine branch, independent liquid pipeline management of orbit/attitude control engine, tank overload safety management, redundancy design for winding of stepping motor, propulsion control unit and etc.

(4) Test solution should be properly formulated to improve measurement precision of subsystem performance. For example, there should be flow resistance consistency test of tank and pipeline to increase expulsion precision for parallel tanks, calibration hot firing test for orbit control engine, multi-flux regulation for regulator and etc.

(5) Ground tests should be conducted fully and adequately. For example, there should be orbit control engine robustness test, tank vibration, expulsion and pressure-bearing test, tank vibration test with pressurized gas, long term pressure-bearing storage test, subsystem hot firing, propellant filling test and etc.

As more and more countries concern deep space exploration and implement deep space exploration missions, future large-scale space exploration activities like lunar or mars exploration will become hotspots and new landers shall be developed. Analysis of such mission requirements shows that the traditional propulsion technology is no longer suitable, it is an inevitable trend to develop new propulsion technology, such as environmental-friendly high performance LO_2/LH_2, LO_2/hydrocarbon fuel, reusable propulsion, propellant refueling in space, high power electric propulsion, nuclear propulsion and etc.

It is necessary to make breakthrough of key technology for propulsion subsystem as follows when new lander is developed:

(1) Non-toxic high performance LO_2/LH_2, LO_2/hydrocarbon propulsion subsystem for orbit/attitude control.
(2) Reusable propulsion subsystem.
(3) LO_2/LH_2, LO_2/hydrocarbon orbit/attitude control engine.
(4) Long term thermal management of cryogenic propellant tank.
(5) Small-scale, high-speed cryogenic electric pump,
(6) Throttleable engine with throttling ratio 15:1,
(7) Hall electric propulsion with hundreds of kilowatt,
(8) Cryogenic propellant refueling in space,
(9) Small-scale nuclear electric propulsion and nuclear thermal propulsion.

References

1. Chu G, Zhang H (2007) Technology of lunar probes. China Science and Technology Press, Beijing
2. Wu W (2007) To the Moon. China Astronautic Publishing House, Beijing
3. Ouyang Z (2004) International lunar exploration progress and chinese lunar exploration. Geol Sci Technol Inform 23(4):1–5
4. Jin G, Cao W, Wei Y et al (2014) Key technologies and flight performance analysis for Chang'E-3 lunar lander propulsion system. Sci Sin Tech 44(4):385–390
5. Huzel DK (2004) Modern engineering for design of liquid-rocket engine (in Chinese). China Astronautic Publishing House, Beijing
6. Liu G, Ren H, Zhu N et al (1993) Principle of liquid rocket motor. China Astronautic Publishing House, Beijing
7. Zhu N, Liu G, et al (1994) Design of liquid propellant rocket engine. China Astronautic Publishing House, Beijing
8. Gao S (1989) Liquid propellant. China Astronautic Publishing House, Beijing
9. Zhang Y (2001) Variable thrust liquid propellant rocket engine and its control techniques. National Defense Industry Press

Chapter 9
OBDH Technology of Lunar Lander

9.1 Introduction

The OBDH subsystem is an important subsystem of spacecraft. It is designed to monitor and control the spacecraft in a computer-based integrated system to provide synthetic services and management for the spacecraft platform and payload [1].

The information acquired by the spacecraft comes from its own, other spacecraft, and the ground station. The major information includes the downlink telemetry data and the uplink telecommand. Through the acquisition, processing, storage, transmission and distribution of these information, the monitoring and control of the spacecraft can be implemented. The functions of telemetry, telecommand, program-controlled, autonomous control and management of spacecraft can be integrated by data management technology in a microprocessor-based system and the computers are connected by the onboard data bus (or computer network). The telemetry and telecommand computers are connected and then all the onboard computers are connected to the network to implement the maximum centralized and unified management of the spacecraft information, sharing information and supporting the autonomous close-loop control of the spacecraft. Along with the development of OBDH system, the on-board software system has been developed to enhance scalability, maintainability and autonomy of the spacecraft.

New requirements including high-performance, high-reliability, intelligence, and integration of electronic systems have been come up for lunar landing mission [2]. For example, the OBDH subsystem of the Chang'E-3 lunar lander should accomplish TT&C, operation scheduling and comprehensive information processing. It also managed and controlled the operation of each task efficiently and reliably including monitoring of the status of the lunar lander, coordination of tasks, and management of thermal control, energy and mechanisms, as well as management and data processing for all onboard payloads, scientific instruments to implement unified onboard processing and sharing of information.

In order to meet the special requirements of mass and power, the integrated electronic technology was applied in OBDH subsystem of the lunar lander to implement

D.-Y. Yu et al., *Technology of Lunar Soft Lander*, Space Technology Library 38,
https://doi.org/10.1007/978-981-15-6580-9_9

high integration and miniaturization. The technical features, design methods and verification of OBDH are introduced in this chapter.

9.2 Development Status

Before 1970s, there was no concept of data management system on Chinese spacecraft. Telemetry subsystem, telecommand subsystem and tracking subsystem were relatively independent, which were called TT&C (Telemetry Tracking and Command) system. Since 1980s, along with the growing complexity of spacecraft, many autonomous task requirements had been put forward, such as program control, time-delay command, payload operation management, energy management, thermal control management, system level security management, system level reconfiguration, onboard time system, and data exchange and sharing among subsystems. At first, the traditional TT&C system of some remote sensing satellite was designed to be an on-board data management system (OBDH/C&DH). A hierarchical distributed computer system and modular structure was applied. Its function was extended from relatively simple TT&C to autonomous management of the whole spacecraft. The on-board data could be comprehensively utilized. At the same time, program control and autonomous control function of spacecraft could be implemented, which enabled autonomous capability of flight and the maintainability on orbit so as to greatly improve capabilities of spacecraft TT&C and survivability of spacecraft.

According to the development of space applications, the types and number of spacecraft were greatly increased, while the tasks and functions were becoming more and more complex. The success of space missions depended on whether the spacecraft could reliably implement task management, data processing and information exchange [3]. While more attention was paid to the information flow of spacecraft, the OBDH subsystem played more important role and extends its scope. The design of OBDH subsystem had evolved into a system design which was intersecting with many specialized fields including the design of top lever information flow, bus network and communication protocol, thermal control, autonomous management of power supply, telemetry format, and etc. The proposal and design of OBDH played an irreplaceable role in spacecraft system design and subsequent development.

Recently, following trends were shown in development of electronic systems of spacecraft.

(1) The function of autonomous management and intellectuality of spacecraft on orbit was improved. The functions of traditional telemetry and telecommand, data management, attitude and orbit control, thermal control, power management, and payload management were integrated, as well as autonomous management functions such as information management, resource management, operation management and crisis management (fault detection, reconfiguration, task downgrade and safe mode operation). Such on-board functions could reduce the burden of long-term on-orbit management, and effectively improve the autonomous diagnosis and autonomous survivability of spacecraft.

(2) The protocol, bus and interface standardization were focus on information technology. The information processing and communication protocol system, such as CCSDS, was widely used. It played an important role in the information transmission in spacecraft, spacecraft to ground and spacecraft to spacecraft, as well as spacecraft information fusion. In order to satisfy different data transmission requirements, there were a variety of bus types including the low speed buses such as 1553B, OBDH bus, CAN bus, and the high speed buses such as IEEE1394, SpaceWire and etc. In order to enable different bus and interface system to interact, CCSDS had developed spacecraft on-board interface service (SOIS) to define onboard interface reference model and the standard service of each layer, and standardize protocols and interfaces [4].

(3) The modularization, standardization, integration, and miniaturization were development trend of hardware. Modular design ideas were applied to achieve functional reduction and extension. The standardization and modularization was the trend in structure, i.e., standard board, standard chassis and unified standard internal bus were applied to improve the integration, extensibility and versatility of the system [5]; FPGA, ASIC, COB and SOC were widely used to improve the integration of the system and reduce the power consumption. The computer processing capability was constantly strengthened. The platform had good scalability and compatibility so as to adapt to the extension requirements of spacecraft.

(4) The layered software architecture and the component-based design concept were applied to realize the standardization of software interface and the high reusability of software components [6]. It was necessary to improve the function and performance of application software, as well as the intelligent level of system by autonomous task planning and intelligent self-management. The current representative researches included the flight software architecture defined by NASA GSFC and the space electronic open interface architecture SAVOIR established by ESA based on SOIS and the Package Utilization Standard (PUS).

The rapid development of electronic technology, information technology and software technology not only helped to improve the capability of spacecraft, but also laid a technical foundation for the development of spacecraft electronic system to integrated design. From the perspective of the development of onboard electronic system in advanced space countries, the open architecture was adopted, and high performance computers were used as the core. Each electronic device was integrated into standard chassis of an integrated electronic system in the form of a standard interface module, or embedded in the form of a chip. Interconnections with the standard bus, unified data management and control of spacecraft by an integrated electronic system had become a common practice. By such methods, a lot of chip-based technologies were used to greatly reduce the amount, weight, and volume of the devices, as well as improve the capabilities of electronic systems [7].

9.3 Analysis of Technical Characteristics

9.3.1 Mission Analysis

The development trends showed that the electronic system played an increasingly important role in spacecraft. More requirements for electronic system were come up in the lunar lander. Beside the conventional telemetry, tracking, telecommand and data management functions, the following requirements should be satisfied by OBDH subsystem of the lunar lander.

(1) Payload management and data processing including the operation mode management and data acquisition, storage, playback and image extraction processing of various scientific exploration equipment. For example, there were many operation modes and new technologies in the Chang'E-3 lunar lander such as image extraction during powered descent, recording and playback synchronously in access time of ground station, payload dynamic/static mode switch and etc., which brought up new requirements for payload data management.

(2) More complex control and management functions for the platform computer to implement unified control of the drive of all mechanisms, and more complex autonomous thermal management functions including special thermal control of the orbit control engine. For example, the OBDH of the Chang'E-3 lunar lander should implement power supply and distribution management, pyro device management, heating loop control, solar panel mechanism control, biaxial mechanism control of directional antenna, control of lunar rover transfer mechanism, control of camera pointing mechanism, control of –Y module cover mechanism, and two-phase fluid loop valve drive, and etc.

(3) Constraints of volume, power consumption and mass. The system integration method and advanced microelectronic technology should be applied to minimize the volume, power consumption and mass, so as to meet requirements of the lunar lander.

(4) Application of international standards. The CCSDS and other standards should be applied to design onboard electronic systems so that the downlink channels and international space resources could be effectively utilized to realize data transmission in space missions.From above all, the electronic system of the lunar lander should be designed to be optimized and reconfigure resources. Based on this, a space data system could be designed to satisfy requirements of the lunar lander and the relative key technologies should be developed. It was necessary to improve the capabilities of integrated electronic systems and reduce mass, power consumption and volume, which was the key to successful implementation of lunar soft-landing mission.

9.3.2 *Constraints*

Compared with past spacecraft, the most challenging requirement for OBDH subsystem of the lunar lander was the design of miniaturized and highly integrated electronic system. For example, the mass and power consumption constraints for OBDH subsystem of the Chang'E-3 lunar lander were similar to the Chang'E-1 probe. But there were many new tasks and devices integrated, such as payload management (including load bus controllers, multiplexing, and large-capacity storage), power supply and distribution controllers, pyro device controllers, heating loop controllers, solar panel mechanism controllers, directional antenna biaxial mechanism controllers, lunar rover transfer mechanism controllers, camera pointing mechanism controllers, –Y module cover mechanism controllers, and two-phase fluid loop valve controllers. At the same time, the mass and power consumption were strictly constrained, so it was very difficult to design a reliable OBDH sub system with enough reliability. There the new integrated electronic technology should be applied.

9.4 Design Methodology

9.4.1 *Principles of Design*

When the OBDH subsystem was designed, the modular design should be emphasized to support an open system. By adding and removing modules to form a system that adapted to different spacecraft flight missions, the reliability of the system was improved by using the hardware and software modules verified by flight tests as much as possible. The standardization of function and interface should be emphasized too. The standardization of function could help to implement the integration of electronic system. The standardization of the interface could greatly improve the standardization level of design, matching and reliability of the interface, which could not only shorten spacecraft development cycle and save cost, but also help to improve the versatility of OBDH subsystem and its equipment so as to meet the needs of different space missions and avoid duplication of development. There were following principles of design.

(1) The system was a modular structure that allows the same basic module to be used for a variety of spacecraft, which could achieve standardization of subsystem structure, operating procedures and testing.
(2) The open system architecture was applied in both hardware and software with standard interfaces, standard remote terminals and standard program to implement commands and data distribution function, and data acquisition function. It could be used repeatedly in development and conducive to applying more mature and even validated parts and software by flight validation.

(3) The data bus was used to realize data transmission between the terminals so as to reduce the internal harness of spacecraft, and make spacecraft integrated with separate modules and tested as separate modules.
(4) The real-time multi-task operating system was used to support the parallel structure of multi process software. It had the characteristics of layered structure and modularization. It was easy to cut, expand, reuse, and easy to maintain running on orbit, while following the open and rolling design guidelines.
(5) The use of data bus (sometimes multi-layer data bus or computer network) provided channels for information exchange between subsystems, which could not only facilitate the rapid exchange of large amounts of information between subsystems, but also provide the interaction support between computers and between ground station and spacecraft.
(6) The OBDH subsystem could support all level tests from component, software and subsystems to the spacecraft system as a comprehensive system.
(7) Targeting at light-weight design of the lunar lander, on the basis of reliability, comprehensive methods were taken from various aspects, such as function integration of equipment, selection of parts and components, selection of shell material, design of shell, internal and external connectors, so as to achieve miniaturization requirements.

9.4.2 System Architecture and Configuration

The architecture of OBDH subsystem was up to the size and complexity of the spacecraft, mission, orbit, lifetime, survivability, function and performance requirements of TT&C, support for other subsystems, other additional requirements such as number and distribution of TT&C channels, information exchanged and date rate between subsystems and parts, design constraints such as lifetime and corresponding radiation resistance and reliability, safety requirements, constraints of volume, mass and power consumption. The scale and complexity of OBDH subsystem was determined by all above factors.

The bus in the distributed architecture connected all the processor units with a common data bus (or network). Compared with other distributed architectures, it had many advantages such as modularity, easy extension, fault tolerance and simplicity. It was currently widely used in spacecraft electronic systems.

A typical diagram of integrated electronic system based on bus structure is shown in Fig. 9.1.

The system network was composed of a middle/low speed serial data bus and the terminals connected to the bus to realize allocation of uplink tele-commands, control of spacecraft, onboard time generation and management, the collection, packaging and formatting of telemetry data, and sending to ground station. At the same time, it supported handling of failure and emergency. In addition, the platform data was transmitted to the multiplexer through the low speed data bus. The communication network was composed of high speed data interface, payloads, data multiplexer,

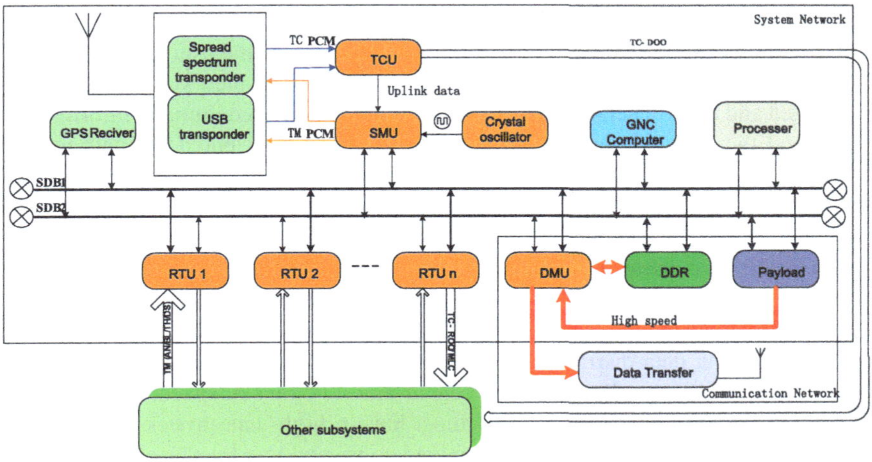

Fig. 9.1 Architecture diagram of integrated electronic system

solid memory and data transmission system to realize acquisition of payload data, data compression and formatting. The data multiplexer combined payload data and platform data into unified serial data stream. The data stream would be transmitted to ground stations or telecommunication relay spacecraft through data transmission channel. When the spacecraft could not be accessed by ground stations, the payload data and the platform data were recorded in a certain data format in the solid memory and played back when the spacecraft was accessed by ground stations.

The central unit was connected with all kinds of remote terminals through the data bus. The data collection and data processing capability of the spacecraft was allocated to several distributed modules, or different subsystems, payloads and scientific instruments. Compared with the centralized system, the interface, data transmission, software and testing were simplified. It was conducive not only to modular design, but also to facilitate the extension and reconfiguration of the functions. It was suitable for the case that the requirements for single spacecraft are few and the requirements for different spacecrafts were different. The distributed and autonomous operation could be realized, which made it easy to realize fault isolation and fault tolerance design, and help to improve the reliability of spacecraft and support the autonomous operation of spacecraft.

The OBDH subsystem was generally composed of Telecommand Unit (TCU), Central Terminal Unit (CTU), Remote Terminal Units (RTU), Data Multiplexing Unit (DMU), Digital Data Recorder (DDR), Data Bus (DB) which connected the terminals, and the software which controlled the operation of the electronic system.

TCU was a relatively independent component of the computer system. It received the uplink PSK (or FSK) signal from the Telemetry and Tracking radio frequency channel, and realized the subcarrier demodulation, and completed recovery of the code synchronization signal and the telecommand PCM signal, and completed the telecommand information and frame verification. The telecommand data would be sent to CTU, and direct on/off commands would be sent to the user after decoding.

CTU was the upper computer of the double-level distributed system. It was the distributed and control center of spacecraft telemetry and telecommand information. It received the telecommand data from the TCU, and distributed the data to user directly through the DB or indirect on/off commands and serial loading instructions. It also collected data from different remote terminals to process and store. The data generated by the DB and / or special high speed interface and the data generated by the CTU software could formulate the telemetry frame, format or source package. After the channel coding and subcarrier modulation, the serial data was sent to the downlink telemetry channel. By controlling the running state of DB, CTU could realize the uplink data distribution and spacecraft data acquisition in time division mode. CTU would complete integrated electronic functions of spacecraft according to mission requirements, so the CTU must have a high data processing capacity, enough running speed and storage space, and corresponding interface.

RTU was the interface between OBDH subsystem and user subsystems to complete the telemetry data (including analog, temperature, double level, and serial digital) acquisition, indirect on/off commands, serial loading commands and data distribution.

DMU received the telemetry source package and the remote sensor data bit stream. The unified organization and management of the data flow of the spacecraft could be realized by using the CCSDS standards. The platform data and the payload business data were combined for transmission through the digital physical channel.

DDR provided the storage medium for platform data and payload business data. The payload business data and the platform data were recorded in the digital data recorder in real time through the DMU, and were read from DMU when the spacecraft was accessed by ground station or relay spacecraft, and then transmitted through DMU to the telecommunication system.

All kinds of intelligent remote terminals were connected to the DB, and responded to CTU only when their addresses were in accordance with the addresses specified in the CTU command. DB was the only channel connecting CTU with all remote terminals, and the operation of DB was controlled by CTU.

9.4.3 Miniaturized and Integrated System Design

Lightweight and miniaturization had always been the focus of spacecraft electronic system design. The issue was particularly prominent in lunar soft-landing missions. One of the effective ways to solve this issue was to integrate electronic systems. Targeting at the mission requirements of lunar soft-landing mission, based on the intelligent modularization, the data information flow was optimized and the resource allocation was integrated. The integrated and optimized design of electronic devices was conducted with the guarantee of reliability, including the function integration of devices, the selection of components and the shell material, the design of housing, the selection of internal and external connectors and etc. Synthetic measures were taken to meet the requirements of lightweight and miniaturization of OBDH devices.

According to the characteristics and constraints of the lunar lander, through comprehensive utilization and analysis of spacecraft resources, the integrated design and management of function and information flow of the probe were conducted. The centralized management and classification of information were realized, and the integrated design of the function of electronic devices was realized to the maximum extent. According to the configuration of probe and the communication requirements of each part, it was also considered that the cable connection among different modules and devices could be maximally saved.

For example, the OBDH subsystem devices of the Chang'E-3 lunar lander were installed in the + Y module and –Y module to manage telemetry and telecommand. Based on analysis of function requirements, the integrated design of mechanism, electricity and thermal control, as well as design of hardware and software could achieve the optimization of resources allocation and the reduction of interface and hardware resources among subsystems. The traditional OBDH devices (CTU, TCU, RTU, DMU, DDR) and other subsystem devices (payload bus controller, power supply distributor, pyro controller, thermal control heater, ± Y solar panel mechanism controller, transfer mechanism controller, directional antenna bi-axis mechanism controller, the camera pointing mechanism controller, –Y cover mechanism controller and etc.) were integrated into two devices. In the selection of parts, components and materials, the lightweight materials of aluminum magnesium alloy were applied as the houses of key electronic devices for the first time, and a lot of integrated devices such as surface attachment chips, FPGA and high-density connector were used to reduce mass and power effectively. The electromagnetic compatibility and anti-interference issues of highly integrated devices were solved at the same time, so as to realize lightweight and miniaturization of the subsystem.

The system was composed of a combination of centralized and distributed architecture. A single device was integrated with several functional modules and connected by the internal bus, while the devices were connected by a serial data bus.

The OBDH subsystem was composed of a system management unit (including telecommand demodulation processing, control computer, command unit, telemetry unit, multiplexing, DDR, + Y power distribution, pyro controller, thermal controller, + Y solar panel mechanism controller), and a data interface unit (including computer, indirect command unit, thermal controller, transfer mechanism controller, –Y solar panel mechanism controller, camera pointing mechanism controller, –Y hatch cover mechanism controller, the dual axis mechanism controller, –Y power distribution unit and etc.), a double redundant serial data bus and the subsystem software. The architecture diagram of the Chang'E-3 lunar lander OBDH subsystem is shown in Fig. 9.2.

The System Management Unit (SMU) was the core of OBDH subsystem. Through the TT&C transponder, it received and performed all direct commands from ground stations, and implemented the collection of telemetry data and the distribution of telecommand indirect commands. SMU managed the power distribution of the + Y module, the center module and the –Z module, pyro devices, scientific payloads, data

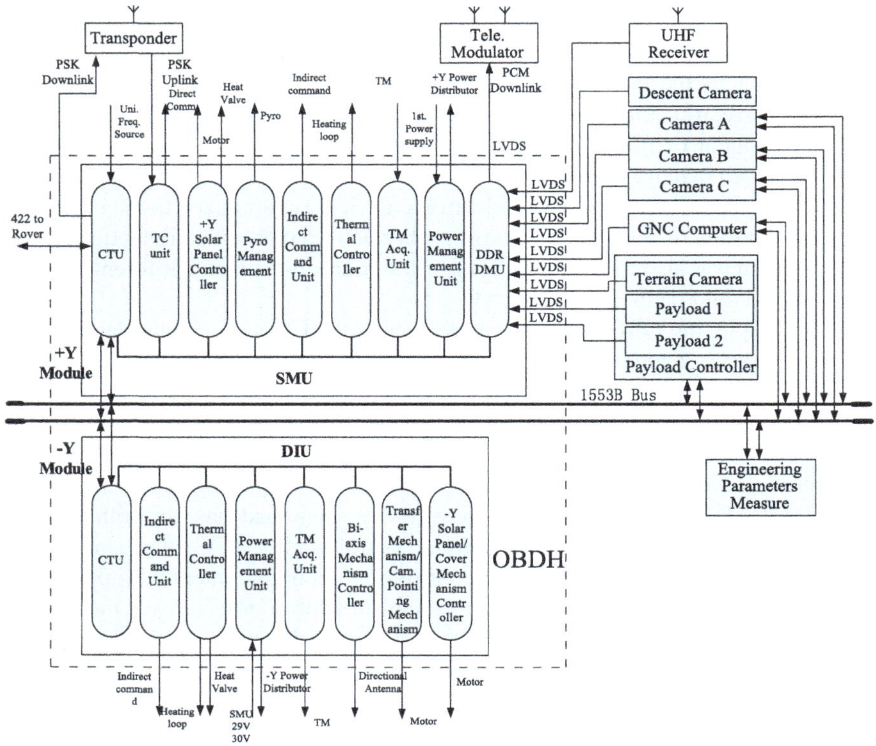

Fig. 9.2 Architecture diagram of Chang'E-3 OBDH subsystem

multiplexing and storage, driving of heating loops controller and mechanism control. SMU should receive data from the commands sent by ground stations, control the operation of serial data bus and lower computer, provide onboard time reference, collect all kinds of telemetry data and format downlink telemetry frame, and provide multiple data multiplexing, large capacity storage and etc.

Data Interface Unit (DIU) was an intelligent terminal based on a single chip microcomputer. It directly served the users in –Y module. DIU received commands from SMU and sent telemetry data to SMU through the serial data bus. DIU performed the power distribution of –Y module, the acquisition of telemetry data and the distribution of indirect commands, as well as the control and driving of heating loops and mechanisms.

The Serial Data Bus (SDB) was the channel for data transmission among SMU, DIU, GNC computer, payloads and other terminals.

9.4.4 Bus Network Design

There are many devices on the spacecraft to exchange data and control information. The bus is the carrier to connect all information nodes and transmit data. Each device and component involved in the exchange of data information is connected together via a bus, which sets up an onboard information network. Usually centered on the onboard computers, the data and commands communication between subsystems and related payload devices is realized on the onboard data bus. In the interior of OBDH subsystem, the bus is also necessary for data exchange because of the distributed architecture.

As the carrier and hub of data communication, the spacecraft bus has not only the function of data transmission, but also the ability to work in real time, high reliability in complex space environment and adapt to the requirements of data transmission service. Therefore, the spacecraft bus is not simply the same as the common data I/O cable.

The data bus can be classified into two categories: serial and parallel. Up to date, the serial bus is widely used in spacecraft. Some of the buses used in spacecraft were developed specifically for military and aerospace applications, such as the MIL-STD-1553B bus, the SpaceWire bus, and etc. Some are industrial or commercial buses that have been used in other fields, such as CAN bus, RS422/485 general line, Ethernet, IEEE1394, LVDS and etc. It can be selected according to the scale, speed, power consumption and distance of data transmission.

The MIL-STD-1553B bus (hereinafter referred to as 1553B) was first proposed by the United States military. It was a standard bus for military aircraft. It was a digital command-response time-division multiplexed serial bus. The bus controller was responsible for the bus scheduling and the management control of up to 31 terminals for data communication. The bandwidth was 1 Mb/s. Since 1970s, standard bus had been developed as well as standardized bus communication protocol, and the mechanism of bus communication was standardized, then redundant configuration of hardware and prevention of terminal short-circuit fault were designed. Because of its high reliability, it was widely used in the aerospace field. Now it was a mature spacecraft bus, which was applied in many spacecraft all over the word.

The 1553B bus was applied in the Chang'E-3 lunar lander as the main channel of information. The double redundant hot backup design was applied, and the interface was transformer-coupled. The terminal signal output terminals had isolation resistors, and the minimum non-response timeout time was set. The communication timeout process could avoid the failure of a node of the bus network from affecting the security of onboard data communication, so as to ensure reliable data transmission and non-proliferation of faults.

LVDS (Low Voltage Differential Signaling) was a general peer-to-peer physical interface technology that is suitable for high-speed data transmission with low power consumption. The technology was based on the ANSI/TIA/EIA-644 LVDS interface standard. It had the advantages of high speed, low power consumption and low noise. The LVDS had 330 mV low voltage differential signal and fast transition time. It could achieve a high speed transmission rate from 100 Mb/s to 1 Gb/s. In addition, the low voltage swing could reduce power consumption and have the advantage of differential transmission. LVDS was used in simple circuit drivers and receiver physical layer devices as well as complex interface communication chipsets. These chipsets could significantly save the cost of the system's cable and connectors, and reduce the physical space required for the connector's footprint. Because it could make the power supply voltage of the system low to 2 V, LVDS could enable mw/GB solutions for present and future high-bandwidth data transmission applications, so it was very suitable for the need of high speed and low consumption in the future deep space exploration. Up to date, LVDS interface had been widely used as a high-speed data transmission channel on spacecraft. LVDS interface had been widely used in high speed payload (such as camera, sensor), large capacity memory, multiplexer and data transmission devices.

The high speed image data flow on the lunar lander, such as the landing camera, the surveillance camera, the topographic and geomorphic camera, was only stored, and was not transmitted in real time. So it could be transmitted to the SMU after being multiplexed and stored through LVDS high-speed data interface.

9.4.5 Information Management Design

Through the uplink and downlink channels, the OBDH subsystem connected the platform data and the T&T information of the ground station, and made up a closed loop system to realize the operation management and maintenance of spacecraft on orbit, such as orbital test, the management and maintenance including monitoring and control of operational status, the diagnosis and elimination of faults, the switching control and reconfiguration of subsystems and components, and the modification and updating of flight software. The data flow of the integrated electronic system in spacecraft is shown in Fig. 9.3.

1. Platform Data Flow

The platform data flow included telemetry data and navigation/positioning data.

(1) Telemetry data: telemetry data collected by remote terminals or other subsystems are transmitted to CTU through SDB to form a telemetry transmission frame, and the telemetry video signals and telemetry PCM data were sent to the TT&C transponder, from which the data would be sent to ground station or relay spacecraft through the TT&C channel. In addition, the telemetry transmission frames were transmitted to Data Multiplexer Unit when the spacecraft could be accessed by ground station, and are delivered via the data transmission channel. When the spacecraft cannot be accessed by ground station, they are transferred to a large-capacity memory for storage.

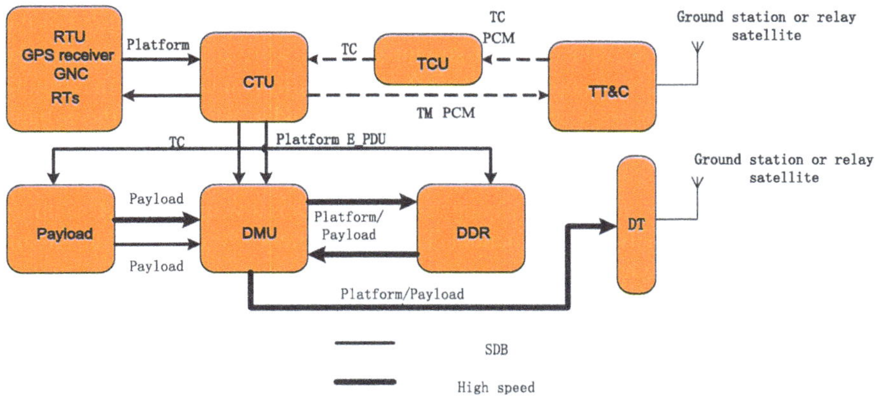

Fig. 9.3 Data flow of integrated electronic system in the Chang'E-3 lunar lander

(2) Navigation/positioning data: It included the data by the navigation/positioning system and the calculated derived data. The transmission of such data was the same as the telemetry data.

2. Telecommand Data Flow

The telecommand radio frequency signal transmitted by ground station was demodulated by the TT&C transponder. The telecommand video signal or PCM data were sent to TCU. After decoding and verification, the direct command was directly output by TCU, or the telecommand injection data was transmitted to CTU(SMU), then it was executed by CTU or transmitted to other subsystems via SDB.

3. Payload Business Data Flow

Scheduled by CTU, the real-time data generated by payloads was transmitted to the large capacity memory when the spacecraft could not be accessed by ground station or relay spacecraft. When the spacecraft could be accessed by ground station or relay spacecraft, it was transmitted to Data Multiplexer Unit via SDB or the high-speed data interface. The real-time data generated by CTU was transmitted to Data Multiplexer Unit via SDB in the form of a telemetry data source packet. In addition, the business data and the platform data recorded in the large capacity memory were transmitted Data Multiplexer Unit via a high-speed interface. After channel coding, it was serial output to data transmission system. Then these data would be modulated and sent to ground station or relay spacecraft.

The payload data rate of Chang'E-3 was high. The data rate of telemetry was relatively low to the payload data, as well as that of data size. The CCSDS application that was more suitable for the Chang'E-3 lunar lander was the downlink of combined telemetry and payload data in X-band, i.e., telemetry data at low data rate and payload data at high data rate were combined. The downlink was AOS. Because the telemetry data rate was low and the data volume was small, the telemetry data generally applied packaging service and multiplexing service. The payload data was streamed in the form of high-speed downlink, and bit-stream services were often used. Since the data channel was not always open and the antenna was directional, while the ground operation should often monitor the telemetry of the lander, sometimes the S-band low-speed telemetry and the PCM TT&C system should be reserved. The telemetry of X-band and S-band could be mutually backed up.

There were two channels for the Chang'E-3 lunar lander: low speed TT&C channel and high-speed data transmission channel. The management of the lander data stream by OBDH subsystem included:

(1) Uplink command and data processing on the T&T channel.
(2) Management and transmission of downlink data on the T&T channel.
(3) Management and transmission of downlink data on the data transmission channel.
(4) Storage management of platform and payload data.

The data flow of the Chang'E-3 lunar lander is shown in Fig. 9.4.

Because the TT&C ground station was independent from ground application station in China, the platform data was transmitted via TT&C channel and data transmission channel at the same time according to mission requirement. The data transmission downlink channel applied the CCSDS advanced orbit system (AOS). AOS used a unified high-speed channel to integrate all data sources, which had advantages of high reusability, adaptability to different data, easy standardization and interaction support so as to improve the success of data transmission.

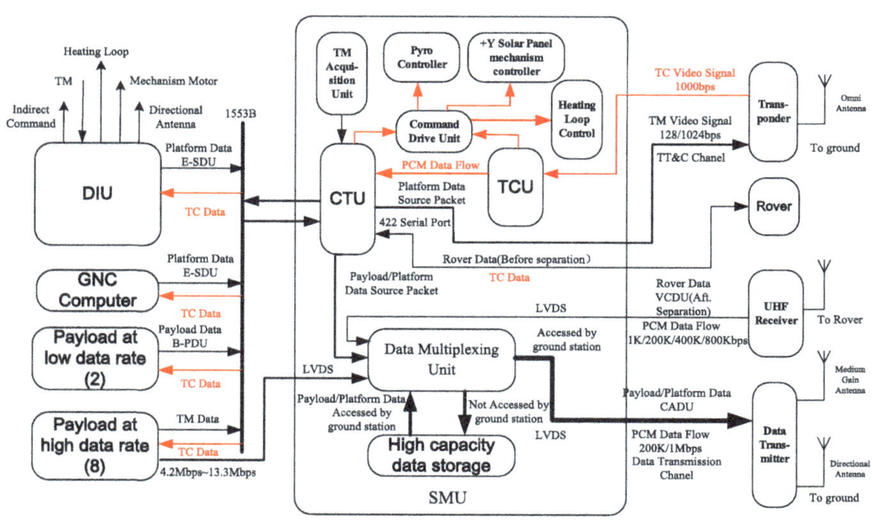

Fig. 9.4 Data flow of the Chang'E-3 lunar lander

9.4.6 Design of Spacecraft Autonomous Management

The autonomous capability was the one of most unique and effective functions of the OBDH subsystem. It was based on the fact that the OBDH subsystem serves as a hub for monitoring and control of spacecraft and had strong data processing capabilities. Spacecraft autonomous capabilities could not only reduce the cost of spacecraft operation and improve reliability, but also be extremely important for autonomously maintaining spacecraft normal operation and autonomously saving spacecraft under failure conditions, and greatly improve the survivability of spacecraft.

The autonomous capability meant that the spacecraft could monitor and judge the status of operation and perform data processing independently, automatically record abnormity and fill in the problem correction report, inform ground station to request support only when necessary. Many onboard controls were closed-loop processing, including routine processing and troubleshooting. It could not only reduce the time of communication with ground station and the labor of ground station to monitor spacecraft, but also support the lunar lander in a long term of some intermediate modes between the independent operation mode and the safe mode, in which some subsystems and devices could be shut down if necessary, so as to maintain survival of spacecraft and wait for support from ground station.

In the case of failure, autonomous control would firstly detect, diagnose and locate the fault parts of spacecraft, and then send a series of commands for redundancy and reconfiguration to switch spacecraft to a safe mode or downgrade operation. Obviously, the real-time requirements for fault processing were very prominent. Usually the time that the lunar lander was accessed by ground station is limited, and it was difficult to completely handle faults by ground station. It was simpler, quicker and more effective to implement closed-loop by onboard autonomous control than by ground station.

The OBDH subsystem of the Chang'E-3 lunar lander was the center of information and management, which was responsible for autonomous management of the lunar lander. The automatic thermal control of the lunar lander was a typical autonomous control. The temperature of the controlled components was detected by telemetry channel. When the set control threshold was reached, the control commands for switch-on or switch-off of the heater were sent by OBDH subsystem. The autonomous management of lithium battery pack discharge protection, high-pressure latch valve self-monitoring control, special temperature control of the orbit control engine, the electromagnet control of + Y/–Y solar panel mechanism before landing and the automatic judgment of the directional antenna back to zero, and etc., were different from traditional spacecraft. The OBDH software was specially designed for the lunar lander to extend and enhance the capability of autonomous management. In addition, the OBDH software could also supported to be on-orbit modified in autonomous management strategy. It could be injected and modify the autonomous management strategy without stopping other onboard software running. For sophisticated control, the detected parameters could be calculated according to certain algorithm, and the sending of commands could also be determined according to judgment of calculation results.

9.4.7 Reliable Design of Highly-Integrated and Lightweight Devices

Because of the requirements of autonomous survivability and lightweight, there were following difficulties and characteristics when the electronic devices of the lunar lander were developed. The increase of the function density made higher complexity of system and raises the level of hardware and software design. The increase in the degree of integration made the degree of coupling between the functions to increase too, leading to the possibility of fault diffusion in operation, which made the reliability of devices more important. The design of electronic devices should consider following factors: feasibility, thermal characteristic (heat dissipation and parts temperature), electromagnetic compatibility and constraints of volume and mass. Because of improvement of system integration and operation speed, heat dissipation and EMC performance became more and more prominent. It was necessary to integrate the mechanical, electric, thermal and EMC design.

The high-power modules such as power distribution, drive of heating loop control, and mechanism control drive had been integrated in the SMU and DIU of the Chang'E-3 lunar lander. The large electric current might interference other weak signals and high-speed signals in devices. The issue of concentrated heat dissipation in high-power units was particularly prominent. The design of devices involved cooperation and communication of many specific and manufacturers. The interface between boards in a device was more complicated. The key point of reliability design was to solve the issue of anti-interference and fault tolerance under the condition of unit combination.

1. Fault Tolerance and Redundant Backup Design under Severe Mass Constraint

The on-board computer was the core component of spacecraft OBDH subsystem, which performed important tasks such as spacecraft control, autonomous management, and information sharing. The success of spacecraft mission depended directly on normal operation of OBDH subsystem, so the reliability design had always been the focus. To date, fault-tolerant design was generally applied to improve the reliability of onboard computer. The fault tolerant technology was realized by two methods: external intervention and autonomy. The autonomous fault tolerance of onboard computer was to implement fault screening and processing without ground or manual intervention, so as to ensure continuous operation of onboard computer to complete the task. The basic technologies included fault detection, screening, switch processing, and fault recovery. The implementation means included hardware and software, which is closely related to architecture of onboard computer.

Up to date, fault-tolerant technology had been widely used in Chinese spacecraft on-board computers. There were memory EDAC error detection and redundant architecture on the single-board or functional circuit level. In the device level, methods such as dual-system hot backup and double-machine cold backup and three-machine hot backup were applied, while 1553B bus dual redundancy and etc. were applied in system-level. The nature was essentially utilization of hardware redundant resources for backup to achieve different level of fault tolerance according to design of cold and hot backup, as well as different costs paid.

The high integration of electronic circuits made it possible to change the design ideas of double redundancy or triple redundancy used in current OBDH system. The redundancy design could be conducted in device level, board level and chip level according to the importance of units. Each chip and function unit could be designed with double redundancy or triple redundancy. The cross reconfiguration of system could be realized to improve the success of system reconfiguration, so as to greatly improve reliability of system.

According to analysis of task requirements and system reliability, the SMU of OBDH subsystem of the Chang'E-3 lunar lander should perform all platform management, while there were many operation modes and the functional design was complex. There were internal cold backup and the hot backup circuit simultaneously, and the cross control requirement in internal interface, so the cold and hot cross backup based on fault isolation was applied in SMU redundancy design. The CTU and the Data Multiplexer Unit was cold backup. The TCU, the indirect command unit and the thermal control unit were hot backup. The telemetry acquisition unit was always online. The drive unit of mechanism control was online only when working, and circuit-level hot backup of main and backup windings was applied. The interface design between them for cross backup, hot/cold backup was difficult. The drive control circuit design for cross backup, hot/cold backup between units in the SMU effectively solved the issue of complex interface reliability between internal cold/hot backup function units in highly integrated devices, as well as the issue of fault isolation, anti-sneak path design, and other issues between cold/hot backup units.

The dual-system hot backup design was applied in DIU.

2. Thermal and Electromagnetic Compatibility Design in Highly Integrated Device

Since many functions were integrated in the OBDH devices of Chang'E-3, miniaturization and integration of devices induced issue of significant amounts of heat in operation. Stringent requirement for thermal design of the device was comp up. The focus of thermal design was not only on the power components mounted on structure, but also the cooling effect of critical control circuits to ensure accuracy of control. In the design, it was necessary not only to apply various thermal design methods for reducing thermal resistance, but also to effectively combine thermal simulation work with computer simulation to find overheated parts of each unit and adopt scientific cooling measures. Thermal analysis based on software simulation was of great significance to design rational layout of each circuit part and improve structural design level of devices. It was necessary to combine advanced manufacturing technology to solve heat dissipation of power units, such as the use of ceramic substrates and metal substrates (including aluminum substrates, copper substrates), and the use of surface mount devices instead of discrete components to improve heat dissipation conditions.

The EMC design of device was an important guarantee for long-term and reliable operation. The three elements of EMC design were shielding, filtering and grounding. Solving issue of anti-interference under condition of modular combination was the focus of electromagnetic compatibility design. The main source of interference within a device was the conduction of signals and electromagnetic interference between components of circuit. The anti-interference design of a device in OBDH subsystem could be classified into different levels including component, circuit, unit and device. The anti-interference design in all levels should be conducted.

(1) Component level design

Generally, the guideline of components selection was to select low-speed integrated circuits as far as possible, such as HC series and additional filter capacitor in power input.

(2) Circuit and unit level design

The digital circuit was the most common broadband interference source, and the transient ground current and the transient load current were the initial sources of interference. Through the rational design of power and ground, decoupling of power supply filter, layout and wiring, partition isolation, circuit layout, multi-layer circuit board, and enhancement of signal integrity, it could be ensured that circuit board-level functions were performed normally while the external disturbance of circuit board was minimized as far as possible.

(3) Device level design

The ground connection should be rationally dealt with. The ground should be classified and managed according to primary ground, secondary power ground, common secondary ground and command ground. The order of unit combination should be distributed according to electromagnetic interference, sensitivity and power.

The main approach of electromagnetic interference into devices was the I/O interface and the power line input. At the same time, the internal electromagnetic interference in devices could be coupled to the wire or cable connecting I/O interface and power line, which could produce interference current and then conduct or radiate outside the shield to cause conduction interference and radiation interference. A good chassis shell had a good inhibition effect on electric field or plain wave so as to increase interference loss. Therefore, the design of chassis should be integrated to reduce the type and number of parts and ensure integrity of connection. The connection surfaces of chassis should be in concave and convex groove structure used to block electromagnetic radiation channel and the thickness of metal material should be enough to ensure good electric field shielding in all frequency bands.

There was electromagnetic coupling between lines. The best approach to reduce interference and sensitive circuit was application of twisted pair and shielded line with shield layer grounded to reduce input resistance of sensitive lines. The sensitive circuit should apply the balanced line input to overcome interference from interference source to sensitive line by the inherent common mode suppression ability of the balanced line. The input and output of the DC/DC should be twisted as far as possible.

3. Anti-SEU Design of FPGA Chip

The development of microelectronics technology had created a good opportunity for solving problems. FPGA was favored by aerospace designers for their high functional density, small size, low power consumption, and semi-customized flexible configuration capabilities. Different types of FPGA chips had been widely used in Chinese aerospace electronics products. The main reason was that user's requirement for spacecraft technology was improved, and onboard devices should have advantages of high functionality, small size, light weight, and low power consumption, at the same time, some functions could not be realized by general IC or computer software. In addition, some new technologies and user requirements were difficult to be quickly solidified, and technical solutions could only be determined through repeated trials and technical coordination between users and contractors. So the project designer could only use a powerful, flexible and convenient FPGA to develop the prototype. Furthermore, once the project had been initiated, the FPGA application had become a final decision. ASIC was certainly one of the important approaches of miniaturization, but its full custom development model that could not be flexibly modified was difficult to adapt to the development requirement of new device.

There were currently two popular types of FPGA architecture: one was based on SRAM and another was anti-fuse FPGA.

The anti-fuse type FPGA had the ability to resist Single Event Upset (SEU). By permanent internal connection and program, the internal logic of chip could only be burned once. After being programmed, the logic could not be changed and there was no memory configured. Therefore, its anti-SEU performance was better than that based on SRAM, and it was unnecessary to be configured with PROM to avoid external PROM SEU problem. The anti-fuse FPGA had strong resistance to SEU, but the number of gates was relatively fewer and the speed was relatively slow for such type.

SRAM-based FPGA was configured by loading status information into SRAM to setup memory control, wiring, features, timing, I/O driver and etc. SRAM-based FPGA provided users with maximum flexibility. When the FPGA was working, the operation status of FPGA was setup by program stored in on-chip SRAM. During operation, the SRAM on chip should be programmed in different program modes according to different configuration modes. On power-up, the FPGA chip read the configuration data from PROM into SRAM on chip. After the configuration was completed, the FPGA was working. After the power was off, the FPGA recovered to a white chip and the internal logic relationship disappears. Therefore, FPGAs based on CMOS-SRAM technology could be repeatedly programmed and used, and provide different functions by different program data in the same FPGA. From analysis of design architecture, there might be SEU, SEFI, and SEL for SRAM type FPGA.

For SRAM-based FPGA chip SEU faults, there were three ways to mitigate or restore: reading configuration information for comparison, periodic or command-controlled reload, and triple-mode redundant design (TMR).

(1) Reading configuration data for comparison was to read data of configuration register inside FPGA through chip reading mode, and compare the data with internal configuration data of chip or original configuration data by data compare device outside the chip. The FPGA would be configured all or partially after the error is found.

The main feature of this method was that the logical fault in FPGA can be repaired purposefully and normal operation of FPGA would not be interrupted as well as interruption of the device operation, which could hardly influence on the system. However, additional hardware support was necessary such as peripheral control circuits, computers, configuration data memory and etc.

(2) Periodic or command controlled reload was to reconfigure internal logic of FPGA to correct possible SEU errors. This method did not need a lot of additional hardware support. The FPGA configuration data was not required to be read and judged before the logical refresh, so the design was simple. Reload time could be determined according to requirement or reload could be implemented by command control.

However, the disadvantage of this method was that normal operation of FPGA would be interrupted during the configuration process, which would have a negative impact on system operation. It could be loaded when there was no task or interruption was acceptable according to requirements of operation mode. For example, the DMU should be reloaded before the payload was powered on to ensure that the task could be performed normally.

(3) The triple-mode redundancy was a SEU mitigation method. In the FPGA design, all the functional units and circuits of logic circuit were duplicated three times, and then the output of three circuits was passed through the decision circuit as the final output, and all logic was implemented by one FPGA chip. This design method would bring more than 3 times consumption of logical resources. If one of three circuits had a SEU error, the other two could still output correct logical value, and the entire circuit could still maintain the correct output through the majority of three circuits. From above analysis, it could be seen that TMR technology could make FPGA obtain a single SEU immunity. The basic architecture of TMR is shown in Fig. 9.5.

A lot of FPGA integrated circuits design for miniaturization was applied in OBDH devices of the Chang'E-3 lunar lander. The relatively simple and key circuits were mostly anti-fuse type FPGA, for example, telecommand decoding circuits. For circuits with more complicated logic, large scale FPGA parts based on SRAM technology were selected and combined with corresponding anti-SEU measures such as multiplexing and memory circuits according to practical applications and hardware support conditions.

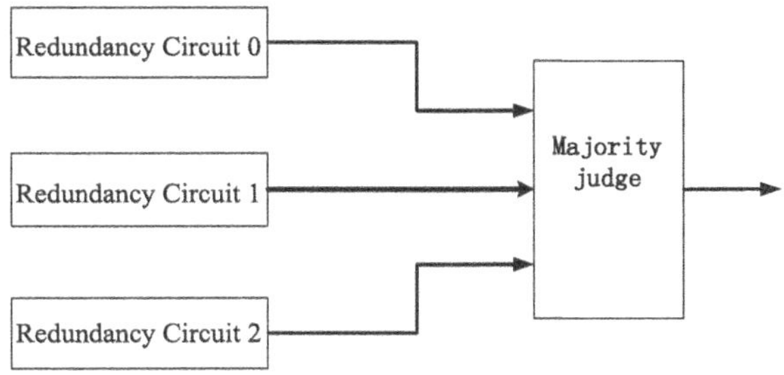

Fig. 9.5 Basic architecture of TMR

9.4.8 *Software Design*

Onboard OBDH system had become a key part of spacecraft TT&C and application. The OBDH system was composed of software and hardware. In the system, software and hardware were equally important, and neither was indispensable. Hardware was the material foundation of computer system, while the software was the soul. Recently the processing ability requirement of onboard computer was becoming more and more demanding, intelligent, automatic and accurate, which made the complexity and scale of software increase dramatically. The software development was no longer the work that the one or two people can accomplish.

Up to date, the OBDH software was usually composed of real-time operating system and application software. Along with the continuous development of spacecraft technology and the increasing demand for practical applications, the tasks to be handled by the computer were becoming more and more complex and the computation was becoming larger and larger. It was necessary that the computer software should be highly portable and inheritable, and the software architecture should be modularized and hierarchical. Software was designed according to idea of hierarchy. In hierarchical architecture, software was organized by layers, and each layer of software was built on lower layer software. Many software systems on same layer have same universality, and the software in the lower layer was more versatile than the upper layer software.

The first layer of the four layers architecture was oriented to hardware, which depended on the hardware architecture and configuration, and provided hardware management and interface drivers. The second layer was a real-time operating system that implements software scheduling and system calls, which could apply commercial products. The third layer was basic service software, including monitoring, data transmission, file transmission and command processing. The fourth layer was specific software related to specific flight missions, from simple command processing to complex data processing. The architecture of onboard OBDH software is shown in Fig. 9.6.

1. Real Time Multitasking Operating System

Embedded operating system is the operating system used in embedded environment. In the embedded environment, the processor generally does not have much available

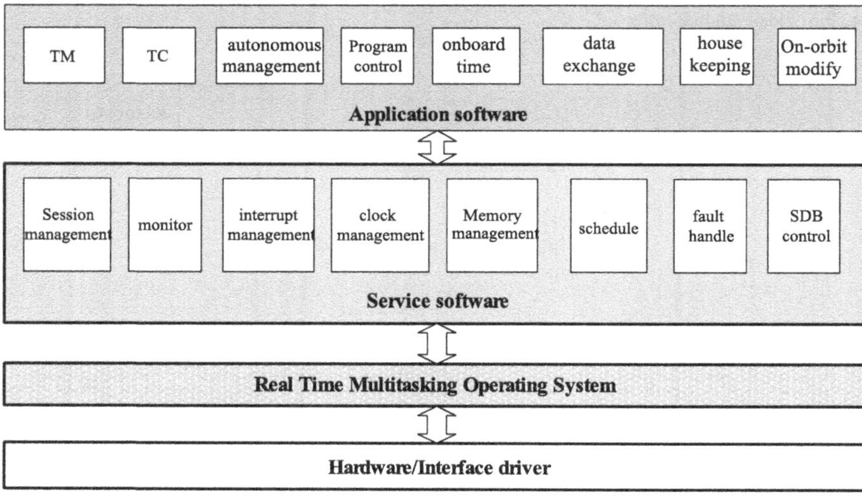

Fig. 9.6 Architecture of onboard OBDH software

memory or more available external memory, and the operating system is installed in a limited memory (usually ROM or FLASH) which is called an embedded operating system. Generally embedded operating system is composed of a very small kernel and some system modules that can be customized according to requirements. The characteristics of embedded operating system are small size, highly flexible, real-time and multitasking. The embedded real-time operating system is a program embedded in the object code and executed first after system reset, which is equivalent to user's resident program. Other user applications are built on the Real Time Multitask Operation System (RTOS). RTOS is a standard kernel that packs the resources of CPU time, interrupt, I/O, timer and etc., which provides user with a standard API interface and assigns CPU time reasonably among different tasks according to the priority of each task. RTOS is an efficient, real-time multitasking kernel optimized for different processors.

A real-time multitasking operating system is an operating system that can perform specific functions within a specified time range or respond to external asynchronous events. The results of session execution on the real-time operating system depend on not only the correctness of logic judgments and logic calculations, but also on the performing time of process. So the real-time operating system must be able to ensure that its time requirement of process shall be satisfied, i.e., to ensure that the specified task is completed within the required time.

The real time multitasking operation system provides operation interface for application software to complete real-time multitask concurrent scheduling of application software. The operating system applies the architecture of micro kernel which has the function of dynamic linking and downloading. It can support preemptive scheduling and support priority rollover protocol. After the function, performance, reliability, maintainability and inheritance of the operating system was fully considered, the OBDH SMU software of the Chang'E-3 lunar lander applied ASOS as the central computer operating system. The major software functions are as follows.

(1) To support multitasking.
(2) To support task scheduling mechanism based on priority preemption.
(3) To support the operation management and maintenance of ERC32 processor.
(4) To provide timers.
(5) To support synchronous and asynchronous interrupt management for ERC32 processors.
(6) To support multi-task communication.
(7) To support static memory allocation and management mechanism.
(8) To support management and maintenance of common serial port operations.

The system software was composed of three main components: BSP support package, kernel and interface manager. The idea of hierarchical design was applied. The architecture diagram of the system software is shown in Fig. 9.7.

Three levels of system software design were described in above figure. The entire supporting software was composed of three modules, which provided a bridge for low level hardware and upper application software. It packed low level hardware operation and provided system call services. It also provided multi-task operating environment for application task, and support interrupt management, communication mechanism between tasks, and timer.

2. Application Software Design

The application software included the flight mission data processing software and the system management software. The flight mission data processing software should collect, receive and process a lot of data, including the data acquired during flight and the data transmitted from ground station. The requirements of data processing were different in different flight missions. System management software included fault

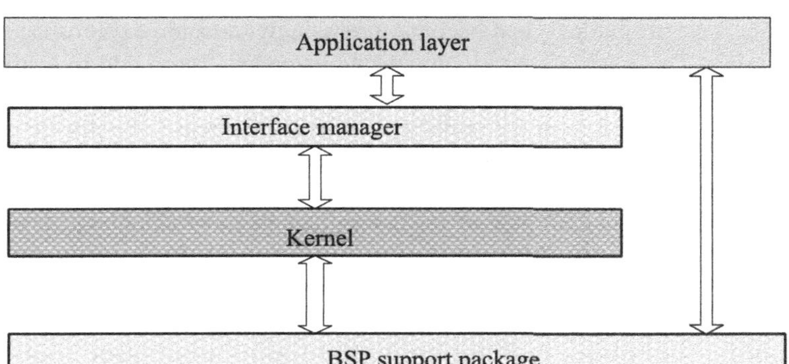

Fig. 9.7 Architecture diagram of the system software

detection and correction, scheduling of long term continuous events and management of spacecraft payloads. In general, it could be classified into several software sessions according to software architecture and functional requirement, such as the telecommand data processing and distribution session, the telemetry data acquisition and processing session, delay telemetry data compression session, downlink telemetry data generation session, program control session, the autonomous control sessions including thermal control, power supply monitoring and mechanism control, special data processing session, auxiliary data generation and transmission session, routine session, program loading session and etc. Configuring different sessions was not only conducive to development and maintenance of software, especially during orbital maintenance, but also easy to satisfy requirements of different spacecraft missions.

Along with extension of spacecraft application, the function of application software was becoming more and more complex. Besides traditional telecommand data processing, telemetry data acquisition and processing, onboard data exchange, system time management, autonomous flight management (program control), the data record, system routine management and etc., more functions were added such as spacecraft fault diagnosis and redundant switch, automatic control and autonomous management, multimedia data processing, and CCSDS data service processing. In addition, the time of spacecraft development was becoming shorter and shorter and number of spacecraft is increasing, which made life cycle of software development was becoming shorter and shorter. Therefore, the design level of application software should be improved from the following aspects.

(1) Software design technology: On the basis of consolidating current software engineering achievements, new software engineering methodology and new technology were continuously researched, including development and application of software configuration management, project management, top-down structured software design technology, and object-oriented software design technology, software fault tolerance technology and etc.

(2) Software reuse: As software development costs became higher and higher, software reuse technology had become increasingly valuable to aerospace engineering. The form of reuse components extended from code to everything that could be used to construct software systems. The reuse of aerospace software had evolved from the initial code reuse to the reuse of software configuration items such as design, planning, test data, and documents. Reuse of code could only bring minimal productivity and reliability gains. The benefits were achieved by multiplexing the design and documentation associated with "reusable source code." Now that software reuse technology was implemented through component development. Software engineers could adapt to new changes by selecting, deleting, modifying, or repeating any subcomponents.

(3) Software testing technology: Software testing was a very important and essential process in software development, which played a key role in the discovery of software failure to improve the quality of software. There were many methods

of software testing, including code checking, dynamic testing on hard platform and soft platform, black-box test, white-box test and etc. Generally, it was better to apply combination of multiple methods. It was one of the most used and most effective methods to use special test device or simulation device to interconnect with the object to be tested. By application of the latest technology of current ground computers to establish a test system, the level and effectiveness of software test automation were improved.

The development process of OBDH SMU software of the Chang'E-3 lunar lander applied a waterfall model. The data flow oriented method was applied when the software was developed. According to requirements of software users, each functional requirement was decomposed one by one, and the task modules were classified according to flow direction.

The application software used the interface and service provided by the system software and was finally compiled together with the system software, which was stored in PROM of SMU. After the hardware was powered on or restarted, the system software was loaded and executed. When the CPU register and software module was initialized by the system software, it was running in the multitasking mode. The operation of application software was initialized by switching the execution flow to the session of application software. The application software implemented creation of multiple tasks through system calls and was running by the support of the system software scheduling. The connections with the system software was established in the form of system calls, as well as the creation of message queues, interrupts service routines, and timer management, so as to coordinate all tasks to implement all the functions of application software.

The OBDH SMU software of the Chang'E-3 lunar lander was a multi-task and large-scale spacecraft software. Besides the function of CTU software in traditional OBDH subsystem, the SMU software should also provide the functions of RTU software and data multiplexer software. Beside functions of the traditional OBDH software such as telemetry, telecommand, program control, data processing, self-management and routine management, the OBDH SMU software should also be responsible for special control and management functions of the lunar lander, such as the unfolding/folding of the solar panels, deployment of the lunar rover, self-monitoring and control of the high-pressure latch valve, special thermal control of the orbit control engine, directional antenna pointing control and etc. The SMU application software was upgraded to abstract level from system level design. The task flow diagram was established. The task modules and interrupt service program were classified based on system level optimization with limited hardware resources. The methods including cooperative design of hardware and software, accurate timing division and concurrent operation process setting were applied to implement rational scheduling and efficient operation. A lot of complex functions were integrated in a software configuration which satisfied the complex functional and performance requirements of the lunar landing mission and provided a strong support for integration and miniaturization of the system.

3. On-Orbit Maintenance Design

The OBDH software with high autonomy and flexibility had become a key part between the ground station and spacecraft operation. In order to ensure that the flight software could provide safe and continuous service during spacecraft operation up to several years and even more than ten years, it was indispensable to implement on-orbit maintenance of spacecraft software. It meant that the ground station could make some modification of onboard program which were running on spacecraft, including some modifications, additions and deletions of the software, or even redefinition, which could be used to modify the integrated electronic performance based on main status of the electronic system and overcome the problems encountered in operation, as well as to eliminate the errors that had not been found and excluded in ground tests. The requirement for on-orbit maintenance of space software might be proposed at any time, from the launch of the spacecraft to the end of the mission, in order to modify the operation to satisfy the updated status after launch, or to satisfy any new requirement added at any time during flight mission, as well as remedial measures that might lead to functional failure due to degradation of hardware status after running for a period of time.

Compared with ground software, on orbit maintenance of onboard spacecraft software had more constraints, and generally it was not easy to maintain in full sense. In order to implement on orbit maintenance of flight software, at least the following conditions should be met:

(1) The architecture of flight software must be easy to maintain.
(2) The operational status of the flight procedure, especially the failure, shall be known as much as possible for the ground.
(3) The onboard computer shall have sufficient design allowance.
(4) The support of the uplink and downlink channels.
(5) The ground simulation environment.

First of all, the flight software should be design for on orbit maintenance. For example, the real time multi-task operating system used in OBDH software supports the parallel software architecture of multi sessions operation, which made the software system have good disassembly performance. It could be controlled flexibly and activated any session by ground station when it was confirmed. When a session was wrong and needs to be suspended, a new program could be injected to replace the original program in RAM as long as it was temporarily suspended by ground station control. The function could also be checked by memory unload to verify the old and reloaded programs and data in memory. It was necessary to copy the program from ROM or EEPROM to RAM, otherwise it was impossible to update the program. It was also possible to design the standby process in advance so that the new process could be loaded and activated to realize the new requirements when it was necessary.

On orbit maintenance referred to that the functional defects of spacecraft system were corrected by telecommand injection through uplink channel, or the functions of spacecraft system were changed/added to meet user requirements. The onboard maintenance of OBDH subsystem was generally implemented through software design. The software should be designed to be able to inject the program uploading blocks and be commanded by ground to correct or replace a software module that had been identified as faulty, change the flight procedure or operation parameters, and hang, delete, activate, create process and etc.

Change or modification of operation parameters of flight software could be considered as the simplest on orbit maintenance of software. Some modifications, additions, deletions, updates and even redefinitions of flight software are the means of on orbit maintenance. Loading a new software module from ground station was implemented in the original flight program by "screw up". It could modify the original sessions, add and activate new session in order to modify the functions and operation manner of OBDH software based on hardware status to overcome the difficulties encountered during flight.

9.5 Testing and Verification

9.5.1 Simulation and Test Environment for OBDH Subsystem

The subsystem simulation test environment was designed to meet requirements of the functional and performance testing of OBDH subsystem. The system was designed to be used in test of OBDH subsystem, which could also support software debugging, software confirmation and test, simulation verification of entire information flow, fault diagnosis, onboard-ground interface test and joint test with other subsystems.

The test system was composed of the hardware platform and the software running in devices to control and perform simulation test, which could improve automation of test process. The composition of subsystem simulation and test environment is shown in Fig. 9.8.

9.5.2 Content of Verification

The content of test and verification of the OBDH subsystem in all development phases of the Chang'E-3 lunar lander included electrical performance test of devices, environmental test, software test, subsystem test, joint interface test with other subsystems, as well as electrical performance test, environment test and onboard-ground interface test of the lunar lander.

Because of strict constraint of mass, OBDH subsystem devices of the Chang'E-3 lunar lander were designed with high integration, powerful function, many external interfaces and complex interface between internal functional units. The reliability

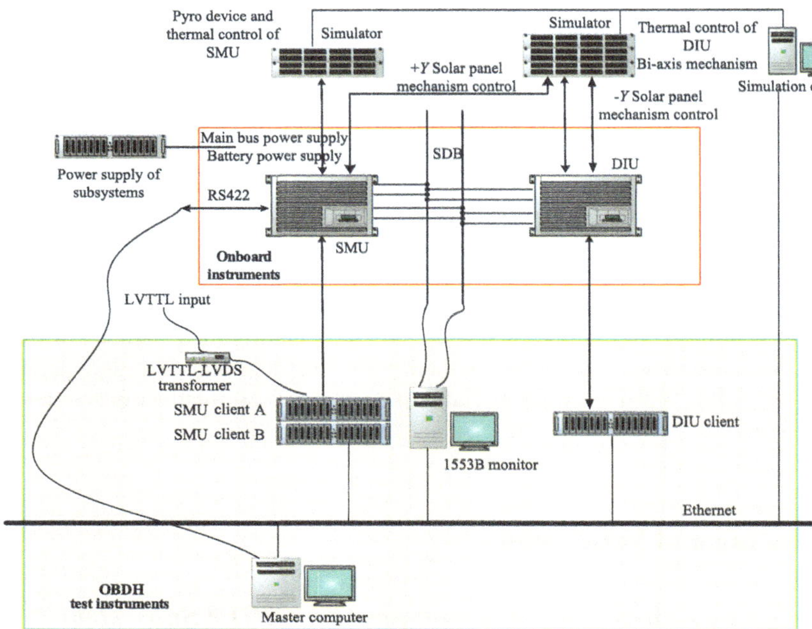

Fig. 9.8 OBDH subsystem simulation and test environment

design of anti-interference under modular combination was the focus of test and verification.

During final design phase, electrical models and qualification models were manufactured. Through electrical performance test and software/hardware matching test of devices, subsystem test, and joint test with other subsystems (especially the mechanism subsystem, antenna subsystem and payload subsystem), as well as electric test, EMC test, onboard-ground interface test, EMC test with launch vehicle, and all qualification environmental tests of the lunar lander, the design of reliability and new function of the OBDH subsystem under high integration were fully verified.

During phase D, the tests of the OBDH subsystem devices were the acceptance level test, and the environmental test was no longer the focus of device test, while there was no special test. In electrical performance test and software/hardware matching test of the devices, subsystem test, and joint test with other subsystems, all tests and interface tests with ground station of the lunar lander, especially the joint test with the mechanism subsystem and the antenna subsystem, the tests under real flight modes and motors with load were conducted to fully verify the design.

Because of the criticality and complexity of the OBDH software, besides normal software test, the special ground rigorous test was designed and conducted before launch. The key of the rigorous test was to test the performance of the software running under a higher intensity or a special bias status than normal operation conditions, as well as the impact of external faults or input errors on the software function, to further fully verify the reliability of the OBDH software.

9.6 Summary

The OBDH subsystem of the Chang'E-3 lunar lander was designed to realize on orbit telemetry, tracking and command, operation and scheduling and comprehensive information processing, which was also an important part of the autonomous management function of the lunar lander. According to the characteristics of lunar soft-landing mission, the application of onboard OBDH technology in deep space exploration is described in this chapter.

The development of spacecraft electronic system has gone through the process from simple to complex, from device, subsystem to spacecraft. Compared with the past, the spacecraft electronic system shall be enhanced to improve the ability of data processing, and be gradually converted into information processing. Through the fusion of information, autonomous task planning and management of spacecraft can be implemented autonomously to reduce the burden of ground users and improve the intelligent level of spacecraft. At the same time, through the integration and standardization of electronic products, the volume, mass and power consumption of the product have been greatly reduced, and the generality and scalability are greatly strengthened. With the continuous reuse of the products, the reliability can be continuously improved.

In order to meet new requirements of deep space exploration missions, the system analysis capability of data management technology and R&D platform shall be enhanced. The design and analysis platform of spacecraft function and information flow shall be further developed, and the advanced electronic system design technology shall be induced continuously, so that the system design of spacecraft electronic system can be induced in mainline of spacecraft development process. From the early stage of spacecraft system design, the design of electronic system shall be integrated with the mechanical and thermal design, so as to realize the rational design, classification and management of the functions and information flow of the spacecraft. Through the simulation and iterative verification of spacecraft function and information flow, the function design, simulation, verification and perfection of the electronic system can be achieved in the process of spacecraft development, so that the design of the electronic system can satisfy the system requirements of the spacecraft and the shortage can be found as early as possible to provide reliable basis for spacecraft system design. By this way, the integrated design, simulation and verification of spacecraft electric, mechanical and thermal can be realized, while the software and hardware coordinate simulation and design of the electronic system can be realized, so as to improve integrated and efficient management of the information flow of the spacecraft.

References

1. Weichi TAN, Yingqi GU (2004) Space data system[M]. China Science and Technology Press, Beijing
2. CHU Guibai, ZHANG He (2007) Technology of lunar probes[M]. China Science and Technology Press, Beijing
3. Weichi TAN, Jingang HU (2009) Spacecraft system engineering[M]. China Science and Technology Press, Beijing
4. Joseph FS, Chris P, Patrick P (2003) Standardized spacecraft onboard interfaces: AIAA space 2003 Conference and Exposition[C]. AIAA, USA
5. Walls B, McClelland M, Persyn S et al (2002) Leveraging flight heritage to new compact PCI space systems: a fusion of architectures[J]. IEEE Aerosp Electron Syst Mag 17(7):9–13
6. Randy B, Mitch F (2005) Next generation space avionics: layered system implementation[J]. IEEE Aerosp Electron Syst Mag 20(12):9–14
7. Randel CB, Gary SB (2003) X2000 Advanced avionics project: development of a new generation of avionics for space applications: IEEE aerospace conference proceedings[C]. IEEE, USA

Chapter 10
TT&C and Telecommunication Technology of Lunar Lander

10.1 Introduction

Radio technology is the main approach to establish telecommunication between spacecraft and ground station. According to its functions, space radio technology can be classified into four types: tracking, telemetry, telecommand and telecommunication, while Telemetry, Tracking and teleCommand are called TT&C in China [1]. The tracking is the real time positioning of spacecraft and measurement of velocity and orientation by automatic tracking of ground station antenna aiming at spacecraft. Telemetry is to transform all kinds of measured physical parameters on spacecraft into electrical signals, which are transmitted to ground station by radio wave to provide all working status and data of on orbit spacecraft after reception and demodulation. The telecommand is to send the control information on the ground (for spacecraft control) in the form of electrical signal which is modulated and transmitted on radio carrier to spacecraft. After received and demodulated, the radio signal is processed by relative subsystems of spacecraft to control spacecraft by ground requirements. The telecommunication is to transmit the information acquired from measuring instruments, observation instruments and instruments that collects and forwards all kinds of information to the ground [1].

The TT&C and Telecommunication subsystem of the lunar lander was used to perform the information transmission between lander and ground or between landers during mission, which also played an important role in orbit determination. The functions of onboard TT&C and Telecommunication systems included: receiving and demodulating telecommand information transmitted by ground station to provide OBDH subsystem for further processing; modulating and transmitting engineering telemetry signals from OBDH subsystem; receiving, demodulating and forwarding ranging baseband signals in downlink carrier phase modulation; cooperation with ground station for precise orbit determination in downlink carrier modulation beacon; transmitting scientific payload data to ground station; telecommunicating with other landers, and etc.

D.-Y. Yu et al., *Technology of Lunar Soft Lander*, Space Technology Library 38,
https://doi.org/10.1007/978-981-15-6580-9_10

In the typical lunar soft-landing mission, the implementation of telemetry/telecommand and telecommunication function should overcome technical challenges including remote telecommunication during powered descent and on lunar surface. To ensure that the lunar lander could be monitored in all mission phases, the TT&C and Telecommunication subsystem should be capable of telemetry/telecommand and telecommunication in 4×10^5 km. Telecommunication during powered descent and on lunar surface was peculiar in lunar soft-landing mission. The process of powered descent was one of the key phases to mission success. In the process of powered descent, the attitude of the lunar lander varied in a large range. It was required that the information of operation status of the lunar lander and the images of descent camera in powered descent process should be transmitted to ground in real-time. Telecommunication on lunar surface was different from the telecommunication link among the traditional spacecrafts or between ground and spacecraft, where the telecommunication link performance was affected by the lunar surface. Telecommunication on lunar surface was a technology challenge that has never been faced before the Chang'E-3 mission in China.

There are three parts in his chapter based on telemetry/telecommand and telecommunication instruments onboard the Chang'E-3 lunar lander. The first one is analysis of the characteristics of telemetry/telecommand and telecommunication technology for the lunar lander. Then the design and key technology principles of telemetry/telecommand and telecommunication subsystem are described based on the analysis. The second one is the latest development of new technologies in telemetry/telecommand and telecommunications for deep space exploration in China. The third one is design and verification of the Chang'E-3 lunar lander.

10.2 Development Status

In early space missions, the carriers of tracking, telemetry and telecommand were independent. From 1950s to mid of 1960s, it was found that more carrier frequencies were easy to interfere with each other and there were difficulties in spacecraft configuration, while the ground facilities were large and complex, difficult to operate. In the mid of 1960s, the unified carrier telemetry/telecommand system was come up. Only one uplink carrier and one downlink carrier were required to realize the telemetry/telecommand of spacecraft [3]. In 1980s, along with application of spacecraft, different carriers were used in spacecraft telemetry/telecommand and telecommunication. A unified carrier telemetry/telecommand system was applied in the telemetry/telecommand. Generally, suppressed carrier modulation was applied for telecommunication and the baseband signal was directly modulated to the carrier [1].

Since Luna 1 was launched by the Soviet Union in January 1959, the United States, Russia, Japan, ESA, China and India had successfully launched lunar probes up to date. The telemetry/telecommand and telecommunication subsystem used by lunar probes was developed from multi-band and multi-carrier analog baseband signal to digital baseband signal, digital modulation and unified carrier system.

The S-band unified carrier modulation system was applied in Apollo spacecraft for the first time. For Apollo spacecraft, in circumlunar orbit, double frequencies were applied, where the S-band unified carrier system was applied for telemetry/telecommand at one frequency and the suppressed carrier modulation system was applied for high-speed telecommunication at another frequency. In the unified carrier modulation, the uplink and downlink carriers are coherent. The uplink carrier modulation had range pseudo code, telecommand and voice, while the telecommand and voice were firstly modulated on two subcarriers to realize the frequency separation and then to modulate the carrier with the ranging pseudo code. The downlink carrier was modulated with ranging pseudo code, telemetry and voice, while the telemetry and voice were also modulated on the subcarrier at first [4].

In twenty-first century, NASA launched a series of Mars surface soft-landing probes such as Spirit, Opportunity, Phoenix and Mars Science Laboratory, and implemented soft-landing on Martian surface successfully. The X-band unified carrier telemetry/telecommand system and the suppression carrier modulation system were applied for all above missions. The telecommunication band was improved from S-band in Apollo missions to X-band. The X-band unified carrier telemetry/telecommand system is the main system of global deep space exploration. Compared with S-band, the use of X-band can improve tracking accuracy and realize the miniaturization of onboard instruments [6]. At the same time, the UHF band telecommunication instruments onboard above Mars probes were used to relay telecommunication with Mars orbiters such as Mar Odyssey, Mars Express and Mars Reconnaissance Orbiter to forward the uplink data from ground station to Mars probes and the downlink data from Mars probes to ground station, which could improve the volume of engineering and scientific exploration data sent back by Mars probes.

On October 24, 2007, Chinese first lunar probe, the Chang'E-1 was launched successfully and the S-band unified carrier telemetry/telecommand system was applied in the Chang'E-1. The ground telemetry/telecommand facilities were upgraded on the basis of the original Earth orbit spacecraft telemetry/telecommand and telecommunication facilities. In order to improve the capability of long distance orbit measurement, the Very Long Baseline Interferometry technology in radio astronomical observation was introduced into the ground telemetry/telecommand system [5]. On October 1, 2010, the Chang'E-2 was successfully launched. An X-band onboard telemetry/telecommand instrument was installed on the Chang'E-2, which successfully implemented the first on-orbit demonstration of X-band unified carrier telemetry/telecommand technology in China. The X-band telemetry/telecommand experiment verified the high sensitivity carrier capture technology and other technologies that were suitable for deep space exploration [6]. On February 18, 2014, the distance of the Chang'E-2 to Earth exceeded 7×10^7 km, which was a new record of deep space telemetry/telecommand and telecommunication in China.

On December 2, 2013, the Chang'E-3 was successfully launched. the Chang'E-3 lunar lander was the first X-band as the main frequency band that provided telemetry/telecommand and telecommunication support in China up to date. The X-band unified carrier telemetry/telecommand architecture was researched for the lunar lander. The X-band telemetry/telecommand and telecommunication subsystem that adapted to lunar soft-landing mission was designed. The X-band telemetry/telecommand and telecommunication instruments were developed and verified in flight. In addition, the analysis and subsystem design of UHF band telecommunication model on lunar surface was conducted. Then after the Chang'E-3 was landed on lunar surface, telecommunication between the lunar lander and the lunar rover was successfully implemented.

The deep space network on ground is the dedicated telemetry/telecommand and telecommunication network that is especially designed for telemetry/telecommand and data transmission of deep space probes. The large aperture parabolic antenna with high-power transmitter, high sensitivity receiving system, signal processing center, high precision and high stability time frequency system was established, which can be used to support telemetry/telecommand and telecommunication of deep space missions [5].

To date, the United States NASA, ESA, Russian Federation Space Administration (RFSA), Japanese space development group (JAXA), and India Space Research Organization (ISRO) had also established a deep space telemetry/telecommand station or telemetry/telecommand network. The deep space network consisted of deep space ground station with large aperture antenna, deep space network management center and ground communication system. In the implementation of deep space telemetry/telecommand and telecommunication, the mission control center carried out the resource scheduling through the deep space network management center. The deep space ground station connected with deep space probes through the radio link, and interacted with the mission control center through the ground communication network.

For requirements of future deep space missions, deep space telemetry/telecommand and telecommunication network was developed and built up, including the Jiamusi S/X dual-bands deep space station and the Kashi S/X/Ka triple-bands deep space station. At the same time, S/X/Ka triple-bands deep space station would be built up in South America, which would form the global network of deep space telemetry/telecommand and telecommunication of China together with the other two deep space stations.

10.3 Analysis of Technical Characteristics

10.3.1 Requirements

The mission of the lunar lander was to realize soft-landing on the surface. Generally there were 5 phases in flight profile.

(1) Launch: from lift-off of the launch vehicle to separation of the lunar lander and the launch vehicle. The lunar lander was injected into LTO by the launch vehicle.
(2) Lunar Transfer orbit: from separation of the lunar lander and the launch vehicle to LOI.
(3) Circumlunar orbit: from LOI to the beginning of powered descent.
(4) Powered descent: from predefined altitude of circumlunar orbit to landing on lunar surface.
(5) Operation on lunar surface: exploration on lunar surface.

During launch, a telemetry and telecommand channel should be established between the lunar lander and ground to complete telemetry monitoring, command transmission and orbit determination. During such flight phase, usually the lunar lander could not be accessed by deep space network, and the telemetry and telecommand should be implemented by other ground stations and telecommunication ships. During LTO phase from separation of the lunar lander with launch vehicle to LOI, there were crucial actions such as trajectory maneuver correction when the lunar lander could be accessed by ground stations. Before LOI, precise orbit determination by means of telemetry and telecommand channel was necessary. After LOI, the lunar lander was inserted into circumlunar orbit. When the lunar lander was above predefined landing site and accessed by deep space network, the powered descent was initiated and telecommunication channels should be established. After landed, the lunar lander would conduct exploration on lunar surface.

The realization of telemetry/telecommand and telecommunication for lunar soft-landing mission required the mutual coordination of the lunar lander and ground. The ground telemetry/telecommand and telecommunication terminals generally were composed of ground stations of telemetry/telecommand system, telecommunication ships and ground stations of ground application system. The telemetry and telecommand system were mainly based on deep space network with large aperture antenna, high-power transmitter and high sensitivity receiving system to realize the telemetry and telecommand of the lunar lander. The ground stations of ground application system were designed to receive scientific payload data of the lunar lander.

In order to ensure normal implementation of telemetry/telecommand and telecommunication, it was necessary to make sure the coincident match of lander-ground telemetry/telecommand and telecommunication interface, including telemetry/telecommand and telecommunication architecture, frequency band and antenna polarization.

The telemetry and telecommand architecture of the Chang'E-1 and the Chang'E-2 lunar probes were unified S-band (USB). Compared with the independent carrier radio telemetry and telecommand equipment (such as radar, telemetry, telecommand and telecommunication instruments of continuous wave), USB could implement multi telemetry and telecommand tasks simultaneously with only one carrier. Therefore it could alleviate the radio frequency congestion and was of great importance to increasingly strained frequency band resources, while the onboard radio frequency instruments could be simplified so as to reduce volume, mass and power consumption of onboard instruments. It was more suitable for requirements of function integration and miniaturization for deep space missions.

The USB architecture was the main telemetry and telecommand architecture of the Chang'E-1 and the Chang'E-2 lunar probes. On the Chang'E-2 lunar probe, a unified X-band (UXB) telemetry and telecommand instrument was onboard to demonstrate UXB architecture on orbit. The transition from USB to UXB of deep space mission in China was achieved by the Chang'E-3 lunar lander, on which UXB was the main telemetry and telecommand architecture. As one of the development trends of deep space missions, UXB architecture had some advantages: (1) The frequency band resources of X-band were more abundant than those of the S-band, which could ensure the signal isolation among space mission in term of frequency division and improve the capacity of the channel to make it easier to design; (2) When the frequency was higher, the accuracy of velocity measurement for the lunar lander could be higher; and (3) The radio instruments of X-band was generally smaller than those of S-band, which was more conducive to lightweight and miniaturization design of passive radio frequency instruments.

The onboard telecommunication system was used to complete downlink of the massive exploration data which is generally scientific exploration data of the lunar lander. It was not necessary to realize multifunction reuse relative to the telemetry and telecommand architecture (such as telemetry, ranging and angular measurement signals modulated in the telemetry and telecommand downlink channel simultaneously). Furthermore, the telecommunication data of lunar missions was relatively lower than that of remote sensing spacecraft on near Earth orbit, and the frequency band resource was not the main contradiction, so the suppressed carrier modulation architecture was commonly used around world. The BPSK modulation architecture of the Chang'E-3 was the same as the Chang'E-1 and the Chang'E-2, except that the radio frequency was changed from S-band to X-band just like the telemetry and telecommand architecture.

In order to minimize of interference of Earth atmosphere on radio signals (it could be approximately regarded as a vacuum environment on lunar surface) and other factors such as the possible polarization mismatch, the telemetry/telecommand and telecommunication antenna was generally circular polarized and different polarization directions (left or right) for receiving and transmitting antennas should be designed to increase isolation of reception and transmission. At the same time, in order to ensure the real-time telemetry and telecommand of the lunar lander and improve the reliability of the mission, the telemetry and telecommand antenna should guarantee the coverage of the full space, so that the telemetry and telecommand of the lunar lander in any gesture could be ensured. Because of shielding of the lunar lander body to the antenna, two pairs of omnidirectional antennas to cover each half space should be configured. Two antennas were installed symmetrically to cover the full space.

10.3.2 Mission Characteristics

There were characteristics including long distance, complex channel model and operation mode for lunar soft-landing mission compared with Earth orbit spacecraft mission.

1. Long Distance

Compared with Earth orbit spacecraft, the telecommunication path loss in the deep space missions is higher because the telecommunication distance becomes further. One of technical challenges in deep space missions is to make up for such loss to achieve telemetry/telecommand and telecommunication [9]. The farthest telecommunication distance and telecommunication delay of the lunar probe are shown in Table 10.1.

Compared with Earth orbit spacecraft, if same ground station with same configuration was applied for lunar landing missions, it was necessary to enhance the receiving and transmitting capability of the lunar lander by about 20 dB to implement same telemetry/telecommand and telecommunication. In order to reduce the requirements for the lunar lander, more powerful deep space network should be built up. There was large aperture antenna, high-power transmitter and high sensitivity receiving system for deep space network compared with common ground station.

2. Complex Channel Model

Besides the telemetry/telecommand and telecommunication channel between the lunar lander and the ground station, there might be telecommunication channel among lunar probes on lunar surface in lunar missions. The telecommunication channel model of lunar probes was more complex than probe-ground link.

The propagation media of telecommunication on lunar surface was different from that of the probe-ground telecommunication. In the probe-ground telecommunication, radio waves could be seen as straight line propagation, and the propagation media are vacuum environment and earth atmosphere. In vacuum environment, the quality of telecommunication system is affected by free space loss and additive Gauss white noise. When the radio wave enters Earth atmosphere, it is necessary to consider the attenuation of precipitation, vapor in atmosphere and oxygen. In telecommunication on lunar surface, the propagation media is extremely tenuous

Table 10.1 Farthest telecommunication distance and telecommunication delay of lunar probe [9]

Farthest distance/km	Increased path loss (to geostationary orbit spacecraft)/dB	Delay/s
0.4055×10^6	21.03	1.35

lunar atmosphere (almost vacuum) and lunar reflector. Due to existence of lunar reflector, the field distribution is complex and variable. The channel model cannot be simply equivalent to the probe-ground link model.

3. Complex Operation Modes

Compared with Earth orbit spacecraft, there were flight phases including LTO, circumlunar orbit, powered descent and operation on lunar surface for the lunar lander. The antenna of the lunar lander should be able to ensure full space coverage, so as to achieve telemetry and telecommand in all mission phases for reliable mission success. At the same time, because of relatively complex surface status of the lunar lander, there might be negative affection to antenna pattern which imposes stringent constraint for layout of antenna.

Powered descent was one of the key steps in soft-landing mission. It was generally required to ensure that the information such as image during landing process can be transmitted to ground station in real time. In order to achieve the high data rate transmission during powered descent, corresponding telecommunication channel should be configured. Generally, the telecommunication channel shall be provided by high gain antenna with mechanism to implement telecommunication to ground. However, the landing impact was severe during soft-landing and it was not suitable for mechanism. For example, the telecommunication antenna was fixed on outer surface of the Chang'E-3 lunar lander. At the same time, because of attitude control error of the lunar lander, a wide beam antenna was applied to implement downlink transmission of image data in powered descent.

10.3.3 Design Constraints

1. Long Distance

The telecommunication distance of the lunar lander was further than that of Earth spacecraft to ground station and there might be more link loss about 20 dB. In order to overcome the large path loss in long distance, deep space network could be built up on ground, and high sensitivity receiver and high power amplifier should be onboard the lunar lander. In the second phase of China Lunar Exploration Program, a global deep space network was initiated to be built up, including 35 m deep space station in Kashi, 66 m deep space station in Jiamusi and 35 m deep space station in the west Argentina, which could realize the 90% coverage of telemetry/telecommand and telecommunication for the lunar lander [5].

2. Frequency

According to the recommendation of CCSDS, the S/X/Ka frequency band could be applied for the lunar lander-ground telemetry/telecommand and telecommunication. The Ka band telecommunication was limited by the engineering difficulty of large aperture antenna on ground, which was usually used only as a test link, not the main telecommunication band. The specific frequency band of S/X-band was divided into uplink band 2025~2110 MHz (S-band) and 7190~7235 MHz (X-band); downlink band 2200~2290 MHz (S-band) and 8450~8500 MHz (X-band). The S-band telecommunication technology was developed earlier and was widely applied, while the frequency resources were relatively limited. The application of X-band was conducive to lightweight and miniaturization design of antennas and other instruments. In the Chang'E-3 mission, the X frequency band was applied for the telemetry/telecommand and telecommunication. The S-band was applied for telemetry/telecommand near Earth and the backup channel of X-band telemetry and telecommand channel in other phases.

3. Full Space Telemetry and Telecommand

In order to ensure stable and reliable TC/TM channel during all flight phases of the lunar lander, the X-band telemetry and telecommand of the lunar lander should meet full space telemetry and telecommand requirement. Generally, two half space (or quasi half space) covered telemetry and telecommand antennas were applied in Earth orbit spacecraft, which were installed symmetrically on spacecraft surface to achieve quasi full space coverage. The uplink signal was received by two telemetry and telecommand antennas, and then was combined by antenna network and other passive devices, which would cause interference between two antenna patterns. In order to overcome the same frequency interference phenomenon, antenna array with two different frequency half space covering antennas was applied in the X-band telemetry and telecommand of the Chang'E-3 lunar lander to achieve full space coverage. Frequency point 1 channel was for half space and frequency point 2 channel was for another half space, so as to avoid the blind area caused by interference effectively and achieve full space telemetry/telecommand coverage in all flight phases. After landing, the telemetry/telecommand channel to lunar surface was shut off and only the ground channel was reserved.

10.4 Design of Telemetry/Telecommand and Telecommunication Technology

10.4.1 Overview

During the mission, the telemetry/telecommand and telecommunication link between the lunar lander and the ground station was established. As shown in Fig. 10.1, the Chang'E-3 lunar lander and the lunar rover communicated directly with the ground station, including receiving uplink telecommand, sending engineering telemetry and payload data. the Chang'E-3 lunar lander could also cooperate with the ground station to measure the orbit by the tracking signal (ranging and interferometric angle measurement signal) in the uplink/downlink signal. Besides conventional

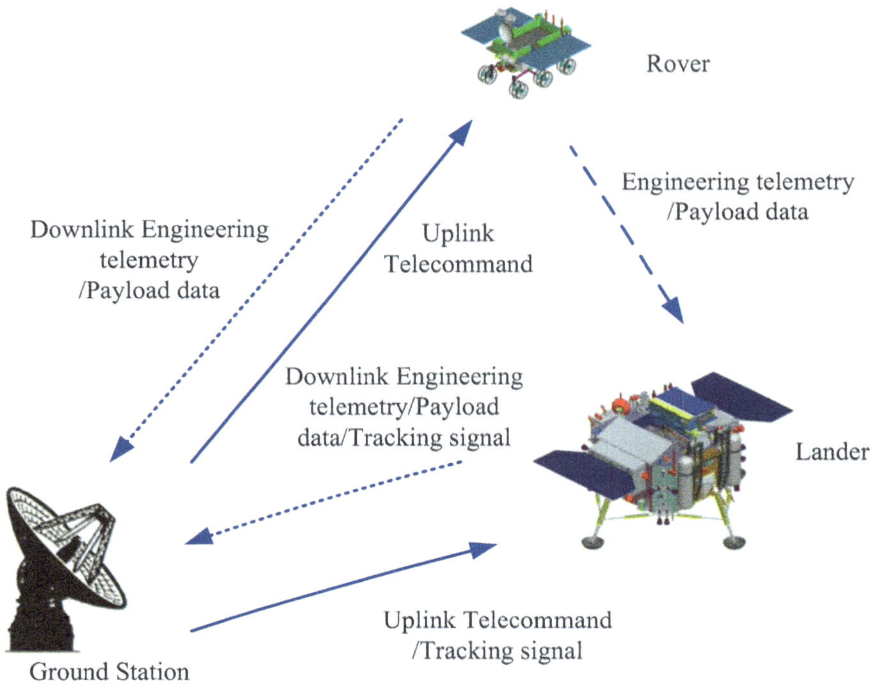

Fig. 10.1 Telemetry/Telecommand and telecommunication link of the Chang'E-3 lunar probe

telemetry/telecommand and telecommunication links, there was another communication link between the lunar lander and the lunar rover, which was applied to receive the engineering telemetry and payload data of the lunar rover so as to serve as a backup of the direct downlink to ground.

During launch phase, the lunar lander-ground telemetry and telecommand channel was established. During Lunar Transfer phase after separation with launch vehicle to arriving at the perilune, the telemetry and telecommand was implemented when the lunar lander was accessed by the deep space network and the lunar lander could not be accessed by the deep space network. When the lunar lander could not be accessed by the deep space network, it was implemented by ground stations of non-deep space network.

After the lunar lander arrived at the perilune, it was still necessary to maintain the telemetry and telecommand channel during LOI until the lunar lander entered circumlunar orbit. When the lunar lander was above the landing area and could be accessed by the deep space network, the powered descent was initiated. The telecommunication channels were established for the ground stations to monitor and receive images acquired by the lunar lander during powered descent.

After the lunar lander was landed, the rover was deployed. There were ground telemetry/telecommand and telecommunication channels both for the lunar lander and the lunar rover. According to the requirements, there was a UHF telecommunication channel on lunar surface between the lunar lander and the lunar rover. The lunar lander could receive the exploration and telemetry data from the lunar rover and send it back to the ground station through its own telecommunication channel.

The orbit determination of the lunar lander consisted of velocity measurement, ranging measurement and angle measurement. Because of the long distance between the lander lunar and the ground, the interferometric measurement technology was widely used in global deep space missions to ensure the accuracy of orbit determination [8].

The Very Long Baseline Interferometry (VLBI) developed in 1960s has a great advantage in deep space exploration at long distance. The VLBI system consisted of two or more radio telescopes at several thousand to tens of thousands kilometers away to realize the precise measurement of the extragalactic radio celestial bodies, the deep space and the artificial celestial bodies at high altitude. Because of the high measurement accuracy, it has been widely applied in deep space exploration [12].

VLBI was the most accurate deep space angle measurement technology. In order to eliminate or weaken the clock deviation between stations, the station site deviation and the deviation caused by signal propagation path in the measurement process, the ΔDOR technology was generally applied. An extragalactic radio source with a small angular distance from the lunar lander (the separation angle is generally less than 10°) as the second observation was introduced in the ΔDOR technology. By the difference of two kinds of observations, the errors such as the station site deviation, atmospheric deviation, delay deviation of receiving equipment and clock deviation could be basically eliminated and the accuracy of time delay measurement could be improved. The principle of ΔDOR technology is that the DOR beacon from the spacecraft is tracked by two remote ground stations at the same time and the angle relation between the spacecraft and the ground station was calculated according to the difference value of time used for beacons to reach two ground stations (delay time).

When two spacecraft (such as the lunar lander and the lunar rover) were very close in the angle, they could be observed in the same beam of the observation station antenna. Observing two spacecraft by two stations at the same time could generate differential interferometry observation, which was called the same beam interferometry measurement. From observation on Earth, the angle between two spacecrafts was much smaller than that between the spacecraft and the radio star (10°). Because many measurement deviations were proportional to the angle, the measurement between the spacecraft could be more accurate than the ΔDOR interferometry. If two spacecrafts were located in the same antenna beam, the carrier phase of two spacecraft could be tracked simultaneously. By using carrier phase instead of group delay (or delay rate), the accuracy of measurement could be further improved. The Same Beam Interferometry (SBI) had some operational advantages over traditional spacecraft interferometry: it was not necessary to calibrate the system with radio source, and there was no related step in data processing, so that real-time phase measurement of spacecraft could be realized. The processing of phase measurement data was more like traditional Doppler data. The same beam interference data combined with bidirectional Doppler or mono/bi-directional Doppler might be a powerful orbit determination data type and could track multiple spacecraft at the same time [5].

10.4.2 Design Method

1. Architecture Design of Telemetry, Telecommand and Telecommunication

The maximum communication distance of the lunar soft-landing exploration mission was about 4×10^5 km. The S-band, X-band and Ka band could be selected for the telemetry, telecommand and telecommunication of the lunar probe to the ground. In the design of telecommunication on lunar surface, it was necessary to consider the diffraction characteristics of radio waves, the constraints of antenna size and other aspects. Generally the UHF band was selected.

The common telemetry and telecommand modulation architecture for space missions included unified carrier modulation and spread spectrum modulation. The telemetry/telecommand and telecommunication of lunar soft-landing mission had the characteristics of long distance and weak signal. The unified carrier telemetry and telecommand architecture could realize high sensitivity reception and integration of interferometric measurement signals, which was more suitable for the lunar mission than the spread spectrum modulation architecture. It was necessary to maintain certain orbit determination accuracy in a long distance for lunar soft-landing missions. Generally coherent ranging, coherent velocity measurement and interferometric angle measurement were applied to achieve high accuracy of orbit determination for the lunar lander.

Generally, the suppressed carrier modulation architecture was applied in space mission telecommunication. Compared with the unified carrier modulation architecture, there was no residual carrier after the suppressed carrier modulation. In case of certain transmission power, it was beneficial to improve the transmission energy of unit code element and increase the transmission code rate.

2. Specification Analysis

There were coverage requirements for antenna pattern of both telemetry/telecommand and telecommunication. Generally, full space coverage was necessary for telemetry/telecommand antenna pattern. The final coverage of telecommunication antenna pattern should be determined after the factors are considered comprehensively such as manufacture, installation, pointing control accuracy, and etc. According to the antenna pattern coverage requirements, antenna efficiency and other factors, the antenna gain could be determined.

The channel budget should be regarded to determine the specifications of the telemetry/telecommand and telecommunication subsystem. It could be checked whether the ratio of signal power to noise power (signal to noise ratio) or the ratio of signal power to noise spectrum density (signal to noise spectral density ratio) at the receiving end was above the threshold by the channel budget in form of table, as well as the proper margin. At the same time, the channel budget table could make designers understand the subsystem and provide theoretical data for the adjustment of specifications and the selection of some hardware.

10.4.3 Telemetry/Telecommand and Telecommunication of Lunar Lander to Ground

The telemetry/telecommand and telecommunication architecture for deep space probe generally included unified carrier and suppressed carrier. The unified carrier architecture was applied in telemetry/telecommand and the suppressed carrier modulation was generally applied in telecommunication. The telemetry/telecommand and telecommunication subsystem of spacecraft was composed of telemetry/telecommand instruments and telecommunication instruments. According to requirements of different missions such as telecommunication on lunar surface of the Chang'E-3 lunar lander, the relay telecommunication instruments for telecommunication between the lunar lander and the lunar rover were onboard.

1. Telemetry/Telecommand Instrument

The telemetry/telecommand instruments generally included receiving antenna, telemetry/telecommand transponder, power amplifier and transmitting antenna.

1) Antenna

The antenna was an important component of spacecraft telemetry/telecommand and telecommunication subsystem, and its task was to form a stable and reliable radio frequency transmission channel with the ground antenna to meet predetermined requirements. The function of spacecraft antenna was to convert the guided wave into the space wave and transmit the downlink radio frequency signal to ground station, or to receive the uplink radio frequency signal transmitted from ground station to the spacecraft and convert the space wave into the guided wave as required [2]. Generally, the telemetry/telecommand antennas should be full space coverage antennas for lunar soft-landing missions. The configuration of telemetry/telecommand and telecommunication antennas of the Chang'E-3 lunar lander was shown in Fig. 10.2.

The telemetry/telecommand antennas of the Chang'E-3 lunar lander included S-band antenna A/B, X-band transmitting antenna A/B and X-band receiving antenna A/B, where the S-band antenna A/B were transmission and reception common antennas.

Generally, two half space (or quasi half space) covered telemetry/telecommand antennas were applied for Earth orbit spacecraft. Two antennas were installed symmetrically on the surface of spacecraft to achieve quasi full space coverage. The pattern of the antenna array is shown in Fig. 10.3.

According to the analysis results of spacecraft attitude and the relative position to Earth in the mission process, two X-band telemetry/telecommand antennas were

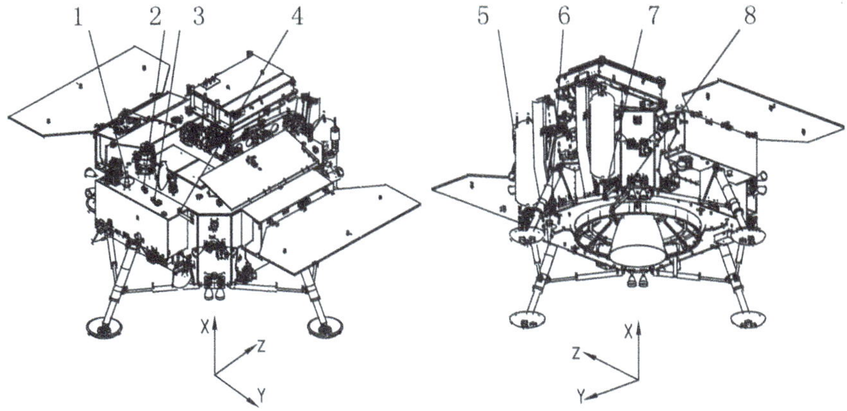

Fig. 10.2 Configuration of TM/TC and telecommunication antennas of the Chang'E-3 lunar lander 1-S-band antenna A; 2-medium gain antenna; 3-X-band receiving antenna A; 4-X-band transmitting antenna A; 5-UHF band antenna; 6-S-band antenna B; 7-X-band transmitting antenna B; 8-X-band receiving antenna B

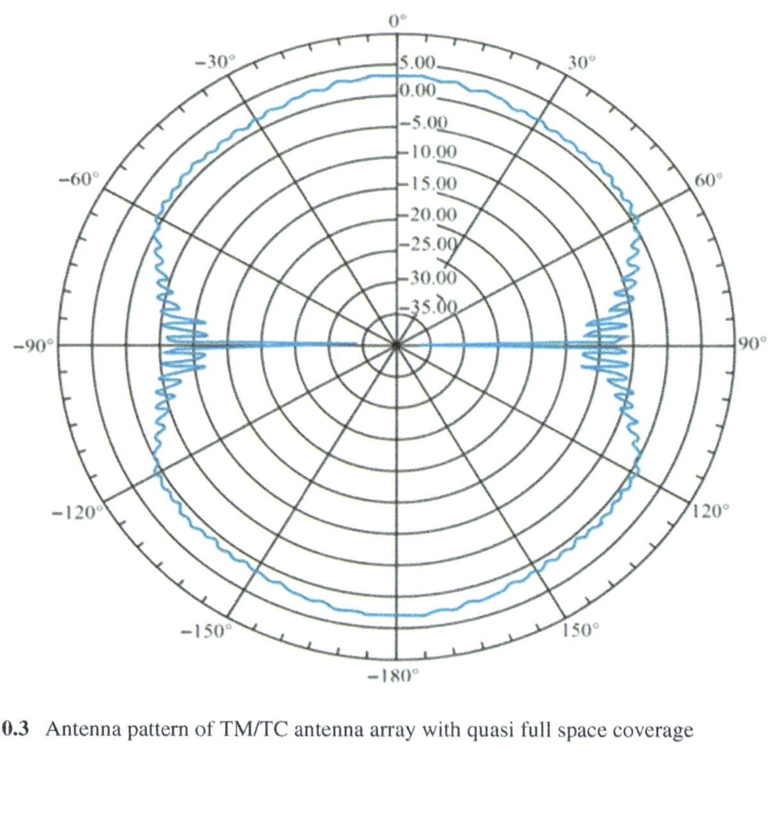

Fig. 10.3 Antenna pattern of TM/TC antenna array with quasi full space coverage

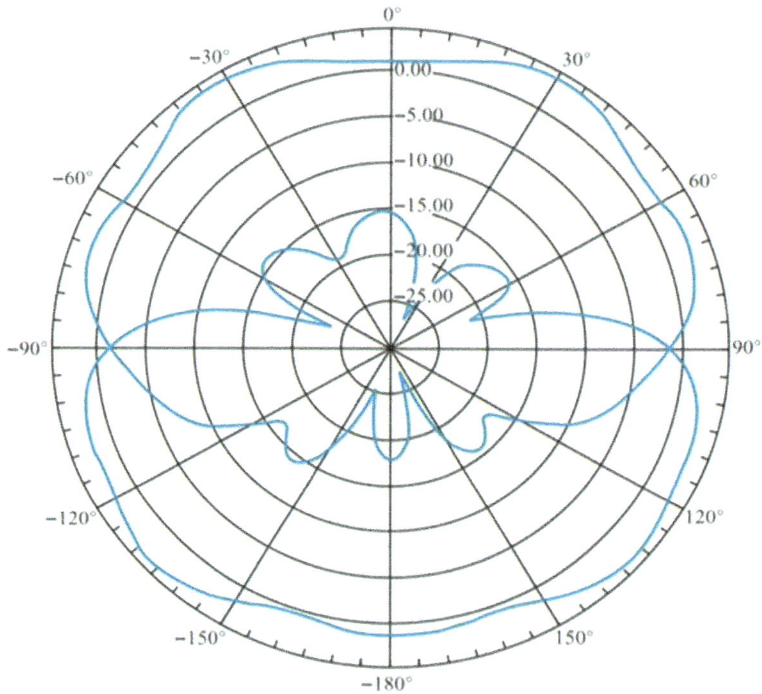

Fig. 10.4 X-band TM/TC antenna pattern with quasi full space coverage of the Chang'E-3 lander

necessary to cover each half space so as to ensure full space coverage during key steps like LOI and powered descent for the Chang'E-3 lunar lander. The antenna pattern is shown in Fig. 10.4.

The telemetry/telecommand antennas of the Chang'E-3 lunar lander were all semi space covered antennas. The S-band antenna A and S-band antenna B were installed vertically on the +X and −X planes of the Chang'E-3 lunar lander, pointing to the +X direction and −X direction respectively. The diffraction ability of X-band radio wave is relatively worse than that of S-band radio wave. In order to reduce the influence of surface instruments on the antenna radiation pattern, the X-band transmitting/receiving antenna A pointed to the angle of +X direction as α, as well as X-band transmitting/receiving antenna B with the −X direction, and the X-band transmitting/receiving antenna A was parallel to the X-band transmitting/receiving antenna B (shown in Fig. 10.2). Then the full space coverage of the X-band telemetry/telecommand antenna was achieved.

2) Telemetry/Telecommand Transponder

As a main component of the radio frequency system of the lunar lander, the function of the transponder was to receive and demodulate the uplink signal, modulate and emit the downlink signal. It was generally classified into three types: coherent transponder, incoherent transponder and hybrid transponder [3].

The phase of receiving and emitting signals in the coherent transponder was coherent, i.e., the frequency of receiving and emitting signals was designed according to the specified coherent forwarding ratio. The carrier tracking phase locked loop and coherent demodulation technology was applied in the receiver which has high receiving sensitivity. The coherent forwarding method could ensure that the frequency stability of the transponder transmitting signal was consistent with the ground frequency source and achieved higher measurement accuracy. The phase of receiving and emitting signals in the incoherent transponder was incoherent. The frequency conversion receiver was applied in receiving channel of the incoherent transponder and the emitting channel was based on the onboard frequency source, while the receiving frequency and emitting frequency were incoherent. Such method would lead to large Doppler shift in the intermediate frequency channel and introduce large ranging error. The hybrid transponder was a combination of coherent transponder and incoherent transponder. In deep space mission, the coherent transponder was generally applied.

The telemetry/telecommand transponder of the Chang'E-3 lunar lander consisted of phase-locked receiver, phase modulation transmitter and power supply module. Its principle block diagram is shown in Fig. 10.5.

The transponder receiver locked the uplink carrier by the phase-locked loop, as well as received and demodulated the uplink telecommand instructions (or injected data) and ranging signals transmitted by the telemetry/telecommand ground station. The demodulated telecommand instructions (or injected data) were BPSK modulated subcarrier signals, and would be processed further by the OBDH subsystem of the lunar lander.

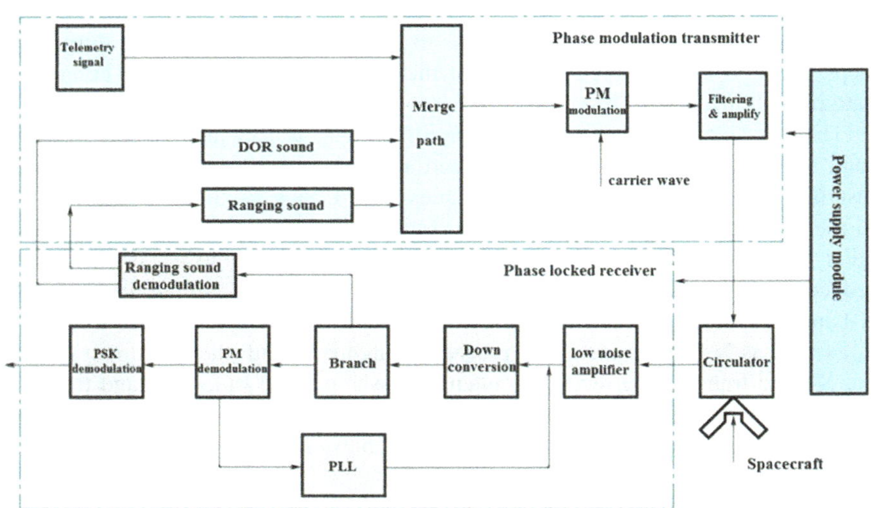

Fig. 10.5 Principle block diagram of the transponder of the Chang'E-3 lunar lander

Another important function of the phase-locked receiver was to provide the phase modulation transmitter with a reference frequency signal of the downlink transmitting carrier frequency. There were two operation modes for the phase modulation transmitter: coherence and incoherence. Under the coherent mode, the reference frequency signal of the downlink transmitting frequency was output by the voltage controlled crystal oscillator in the phase-locked loop of the receiver, and the frequency stability was up to the uplink signal of the telemetry/telecommand ground station. Under the incoherent mode, the reference frequency signal of the downlink transmitting frequency was output by the temperature compensated crystal oscillator in the transmitter. The downlink signal of the phase modulation transmitter could be used as the target acquisition beacon of the ground station, and provide telemetry downlink, ranging signal forwarding or regeneration.

The ranging sound signal demodulated by the phase-locked receiver and telemetry sub carrier generated by the OBDH subsystem of the lunar lander were sent to the ground station after modulated by the phase modulation transmitter. The downlink telemetry signal was from OBDH subsystem of the lunar lander, which was BPSK modulated subcarrier signal. The telemetry signal and forward ranging sound signal were linearly added in the pre-modulator of the transmitter and then modulated to the downlink carrier. In order to lower the ranging error, it was necessary that the phase noise should be small enough when the transmitter was forwarding the range sound. Therefore, the forward signal was not deeply processed, resulting in the uplink ranging sound signal and remote control command signal transmitted to the downlink carrier simultaneously. In the process of downlink signal modulation, the general method was applied that the forward telecommand subcarrier signal was isolated from the telemetry subcarrier signal and the range signal in the frequency domain, so as to ensure that the processing of the telemetry subcarrier signal and the ranging signal by the ground station was not affected.

The function of the power supply module was to convert the primary electric power supplied by the lunar lander into DC power in all voltages required by the telemetry/telecommand transponder, and to provide the electric current limit protection according to the requirements of actual operation current and derating design. At the same time, the input filter must be inserted at the input of the DC converter to ensure that the output voltage interference pulse was small enough and the isolation of the input power was improved. The primary power ground of the transponder was isolated from the second power ground, and there was no direct connection.

3) Power Amplifier

Generally, the Solid State Power Amplifier (SSPA) or Traveling Wave Tube Amplifier (TWTA) was used to amplify and transmit downlink RF signals in power amplifier of the lunar lander. The basic functions of the power amplifier were almost the same as those of the amplifier in other radio systems. However, high reliability, high efficiency and small size were necessary due to the special mission requirements of the lunar lander and specific operation environment [5].

In order to be in accordance with the development of new technologies and new standards in modern telecommunications, electronic countermeasures, radar and spacecraft instruments of telemetry/telecommand and telecommunication, the power amplifier should be higher in output power, efficiency, reliability and linearity in wider frequency band. In recent years, the development and progress of solid state power amplifier (SSPA) were prompted. Remarkable progress had been made in microwave power devices, power amplifier design, microwave circuit design and process.

SSPA had many advantages, such as small size, low operation voltage, good stability, high reliability, long life and good repeatability. It was widely used in the field of spacecraft telemetry/telecommand and telecommunication. The configuration of solid state power amplifier in Mars Exploration Rover (MER) is shown in Fig. 10.6.

The TWTA was composed of a space traveling wave tube (TWT) and an Electronic Power Controller (EPC) as shown in Fig. 10.7. The TWT implemented amplification of microwave power. The EPC provided various voltages for normal operation of TWT and interfaces including all telemetry/telecommand signals of the traveling wave tube.

For example, the principle block diagram of the X-band telemetry/telecommand of the Chang'E-3 lunar lander is shown in Fig. 10.8.

The operation process of onboard telemetry/telecommand instrument was as follows: The transponder received and demodulated the telecommand transmitted by the telemetry/telecommand ground station, then provided it to OBDH subsystem for further processing; the engineering telemetry signals from OBDH subsystem was phase-modulated and transmitted; the ranging baseband signals was received, demodulated, phase-modulate in the downlink carrier and forwarded; the one-way differential ranging sound was modulated in the downlink carrier to cooperate with the ground station for accurate orbit determination. The downlink signal generated by the transponder was amplified by the power amplifier and radiated to the ground by the transmitting antenna.

The telemetry/telecommand channel A (including receiving antenna A, transponder A, power amplifier A and transmitting antenna A) and channel B (including receiving antenna B, transponder B, power amplifier B and transmitting antenna B) respectively covered each half space to form a full space coverage, which could realize full space telemetry/telecommand coverage in all flight phases.

Fig. 10.6 Configuration of SSPA in MER

Fig. 10.7 Structure of TWTA

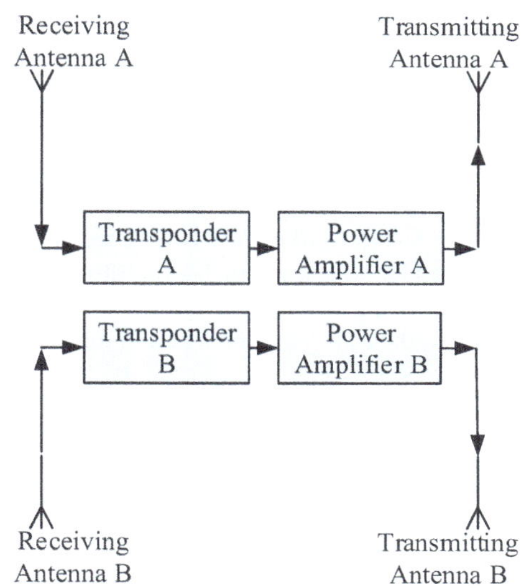

Fig. 10.8 Principle block diagram of X-band TM/TC of the Chang'E-3 lunar lander

2. Telecommunication Instrument

Compared with the telemetry/telecommand channel, the function requirement of telecommunication instrument was relatively simple, which was to send high speed data back to the ground. Therefore, the suppressed carrier modulation architecture was generally applied to realize higher utilization rate of downlink signal. Generally the telecommunication instruments included modulator, power amplifier, telecommunication antenna and etc.

1) Modulator

The modulator was one of the major active devices of telecommunication, which was used to implement suppressed carrier modulation of the data on the downlink carrier. Compared with the unified carrier modulation architecture, the suppressed carrier modulation architecture had no residual carrier. In case of certain transmitting power, it was more effective to improve the transmission energy of unit code and increase the transmission code rate.

2) Telecommunications Antenna

Compared with the telemetry/telecommand antenna, the telecommunication antenna had a higher antenna gain and a narrower antenna beam, which could achieve higher effective total radiation power and improve the channel capacity. Generally there were two types of telecommunication antennas in the lunar lander: one was fixed antenna for telecommunication during powered descent; another was an antenna with a driving mechanism for telecommunication during operation on lunar surface. For example, there were one mid-gain antenna and one high gain antenna in the Chang'E-3 lunar lander.

The mid-gain antenna was used when the Chang'E-3 lunar lander was at about 1 km altitude from lunar surface during powered descent until landing on the Moon. At the phase, the +X axis of the lunar lander was roughly perpendicular to the Y'Z' plane of the northeast coordinate system (O' −X'Y'Z') at landing site, and the −Z axis was pointed to local south (to lunar equator).

The X-band link of the mid-gain antenna was established when the Chang'E-3 lunar lander was at about 1 km altitude from lunar surface during powered descent. During the phase, the attitude of the lunar lander was always kept, and the angle between the mid-gain antenna and the direction of the lunar lander to Earth was kept unchanged (as shown in Fig. 10.9). And the beam direction of the X-band mid-gain antenna was constant.

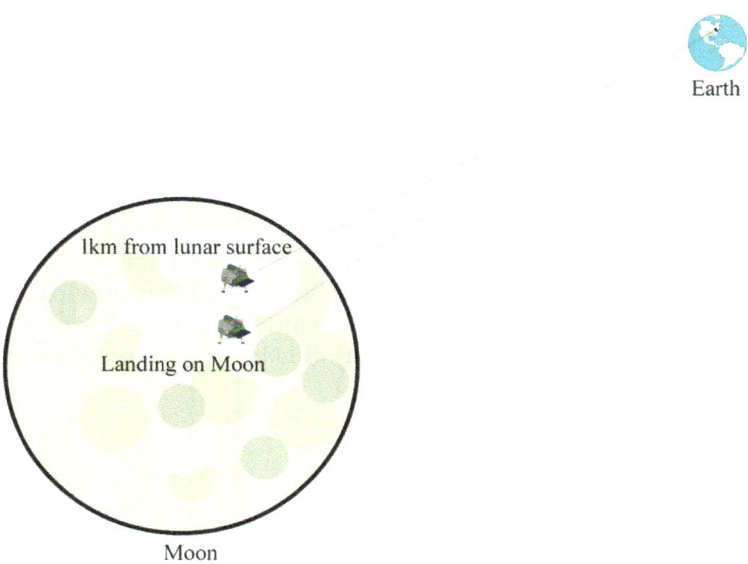

Fig. 10.9 Principle block diagram of X-band TM/TC of the Chang'E-3 lunar lander

On different launch windows, the pointing angle of the X-band mid-gain antenna was different during powered descent. Therefore the configuration and beam coverage of the X-band mid-gain antenna should be carefully designed to adapt different launch windows. Analysis was conducted for predefined landing area, Sinus Iridum and different launch windows. The half-beam angle of the X-band mid-gain antenna was designed to meet the coverage requirement of Earth during powered descent.

The high-gain antenna of the Chang'E-3 lunar lander was a fixed reflector antenna driven by 2- DOF. After landing, the relationship between the ground station and the Moon varied, so the high-gain antenna should be pointed to Earth by the driving mechanism. The direction to Earth of the high-gain antenna was determined by the pointing accuracy of the driving mechanism. The beam angle of the high-gain antenna should not be less than the antenna pointing accuracy.

The pointing accuracy of the high-gain antenna of the lunar lander was determined by various factors including the following aspects.

(1) Error of Earth station location: Earth was regarded as a point in orbit measurements because of the distance between Earth and the Moon. However in practice, there was a field angle of Earth when observed on lunar surface, which could induce a reference error.
(2) Measurement error: orbital measurement error, ephemeris error, and attitude measurement error.
(3) Open-loop control error of antenna orientation: there was the control error of the high-gain antenna when it was pointed to a certain direction by bi-axis driving mechanism.

The self-factors that affected pointing accuracy of the high gain antenna included the installation accuracy of the high-gain antenna, the electromechanical axis error of the RF component, the pointing error of the bi-axis driving mechanism, and the thermal deformation error.

The principle block diagram of onboard telecommunication instrument is shown in Fig. 10.10. The high-stability clock source could be shared with other subsystems of the spacecraft. While providing a high-stability downlink frequency, it could also provide a reference clock for the lunar lander.

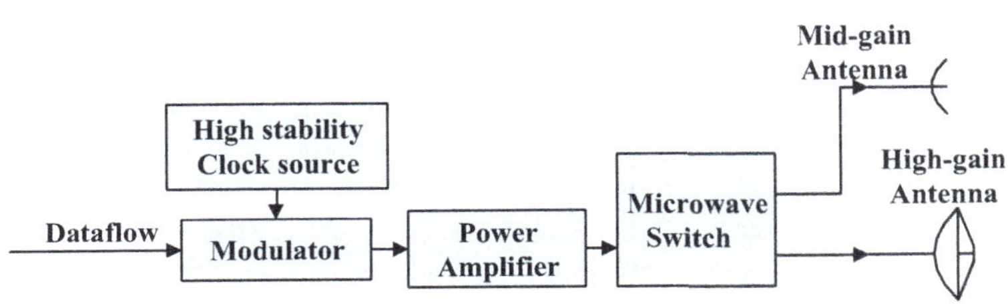

Fig. 10.10 Principle block diagram of X-band TM/TC of the Chang'E-3 lunar lander

The modulator digitally modulated the carrier with the telecommunication data stream, amplified power with the power amplifier, and radiated to the ground with a telecommunication antenna (the mid-gain antenna or the high-gain antenna). There were two channels for telecommunication. During landing, the mid-gain antenna could be selected by the microwave switch to transmit the telecommunication modulation signal. After landing, the high-gain antenna could be selected by the microwave switch to transmit the telecommunication modulation signal rapidly to the ground.

10.4.4 Telecommunication Between Lunar Lander and Lunar Rover

1. Telecommunication between the Lunar Lander and the Lunar Rover on Lunar Surface

For the telecommunication requirement between the lunar lander and the lunar rover on lunar surface, telecommunication instruments were onboard on the Chang'E-3 probes. The telecommunication instruments on lunar surface consisted of a receiving antenna, a receiver, a transmitter and a transmitting antenna. The transmitter and the receiver respectively implemented the modulation and demodulation of telecommunication signals between the lunar lander and the lunar rover. The receiving antenna and the transmitting antenna established a radio channel. The principle block diagram is shown in Fig. 10.11, where the receiving antenna and the receiver were installed on the lunar lander, and the transmitter and the transmitting antenna were installed on the lunar rover.

The engineering telemetry or payload data sent by the lunar rover was received by the receiving antenna of the lunar lander, and then was sent to the receiver for demodulation by the receiver. The demodulated data was sent to the OBDH subsystem of the lunar lander. After processed by the OBDH system, it was sent to the telecommunication channel of the lunar lander to the ground for forwarding of engineering telemetry or payload data of the lunar rover.

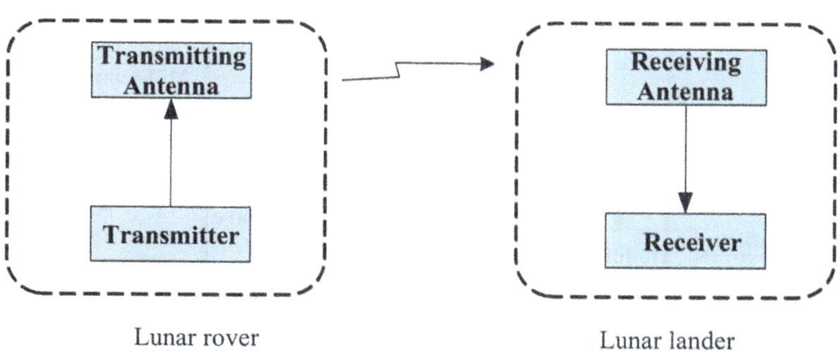

Fig. 10.11 Principle block diagram of telecommunication instruments on lunar surface

Table 10.2 Classification of radio frequency band [7, 10]

Band	Frequency	Band	Wavelength
VLF	3~30 kHz	Very long wave	100~10 km
LF	30~300 kHz	Long wave	10~1 km
MF	300 kHz~3 MHz	Medium wave	1 km~100 m
HF	3~30 MHz	shortwave	100~10 m
VHF	30~300 MHz	ultrashort wave	10~1 m
UHF	300 MHz~3 GHz	Decimeter wave	1~0.1 m
SHF	3~30 GHz	Centimeter wave	10~1 cm
EHF	30~300 GHz	Millimeter wave	10~1 mm

2. Selection of Frequency

Because there is no atmosphere and ionosphere on lunar surface, the radio wave in telecommunication between the lunar lander and the lunar rover on lunar surface was transmitted in the form of spatial wave and ground wave. Because of the irregularity of lunar surface, it was necessary for the radio wave to be capable of certain diffraction in case of over-the-horizon. When the radio wave propagates in the over-the-horizon range, the diffraction loss of the radio wave was higher along with increasing of radio frequency. Therefore, an appropriate telecommunication frequency band should be selected for telecommunication between the lunar lander and the lunar rover on lunar surface. The classification of radio wave band is listed in Table 10.2.

The antenna size was generally proportional to the wavelength of the radio wave. Since the instruments on the lunar rover were relatively small, it was not possible to set up an antenna below the short wave frequency band, so the frequency band below the short wave was not considered.

The SHF band could provide higher data transmission date rate, but it could only be used for line-of-sight telecommunication and occlusion (such as rock, and etc.) between two probes may cause trouble in telecommunication. Similarly, higher frequencies did not apply to mutual telecommunication between two probes.

The antenna size of the VHF band and the UHF band was generally short, and it could fully meet the requirements of high emission efficiency, certain diffraction and installation on the lunar rover. In addition, the ultrashort wave could be designed with a relatively large bandwidth and provide higher telecommunication data rate, which would meet the requirement for data transmission between the lunar rover and the lunar lander.

After the size and mass of the antenna was considered, the UHF frequency band was selected for telecommunication between the Chang'E-3 lunar probes on lunar surface.

3. Selection of Antenna

When the radio wave propagates on lunar surface, the loss of vertical line-polarization signal was minimal. The vertically polarized antenna was selected for telecommunication between the Chang'E-3 lunar probes on lunar surface. At the same time, the antenna gain pattern should cover the variation in relative position of the lunar lander and the lunar rover under nominal condition, i.e., the horizontal plane was non-directional (360° coverage) and the vertical direction was insensitive to the axial deviation of the antenna.

The receiver antenna in form of a monopole oscillator was selected for telecommunication between the Chang'E-3 lunar probes on lunar surface. Simulation software was used to optimize the design and analysis of the receiving antenna unit. The simulation model of the antenna unit in high-frequency simulation software is shown in Fig. 10.12. The unit antenna pitch pattern is shown in Fig. 10.13.

After the lunar lander was landed, the lunar rover was deployed and implemented exploration. During the phase, the lunar lander could receive telemetry or exploration data from the lunar rover with the UHF link. Due to uncertainty of moving direction of the lunar rover, the orientation of the UHF antenna of the lunar lander varied from 0° to 360°. Following aspects were considered when the pitch angle coverage of the UHF receiving antenna of the lunar lander was determined.

(1) According to the equivalent height of UHF antennas of the lunar lander and the lunar rover, the variation of the pitch angle of the lunar lander to the lunar rover was analyzed in telecommunication range.
(2) After the lunar lander was landed, the angle between the + X axis of the lunar lander and the opposite direction of local gravity was analyzed.

Fig. 10.12 Simulation model of receiving antenna

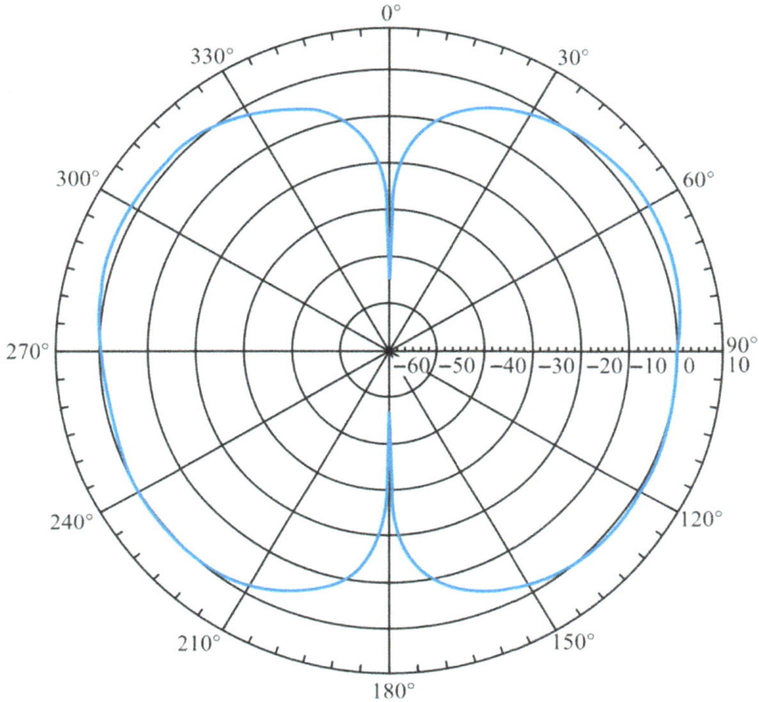

Fig. 10.13 Pattern of UHF receiving antenna

After the pitch angles of (1) and (2) were considered, the pitch angle covered by the UHF antenna pattern was determined. The monopole antenna was installed on the lunar lander for reception of the telemetry exploration data of the lunar rover. The monopole antenna pattern was fully covered in the horizontal direction, and the elevation pattern was in "∞" shape, which could satisfy the requirements of telecommunication between the lunar probes on lunar surface.

4. Telecommunication Model between Lunar Probes

The radio wave in telecommunication between the lunar probes on lunar surface was transmitted in form of space wave and ground wave. The received signal strength at the receiver was affected by direct wave, reflected wave, and the scattering effects of the reflected wave on rough surface.

1) Reflection and scattering models on lunar surface

When the radio wave encountered a smooth reflective surface in the transmission path, it would be reflected, and scattering occurs when it encountered a rough reflective surface. According to different roughness of the reflective surface, the reflective surface could be classified into smooth reflective surface and rough reflective surface. Electromagnetic wave was reflected on smooth reflective surfaces and scattered on rough reflective surfaces.

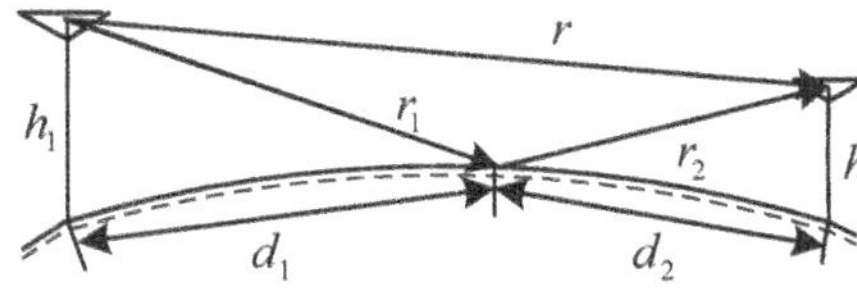

Fig. 10.14 Reflection of radio wave on lunar surface

To be strict, the mirror reflection principle was only applicable to the case when the reflecting surface was a plane. The lunar surface is actually spherical. The obvious difference between the plane reflection and the spherical reflection is that the diffusion of the two beams is different. The spherical emission signal can be equivalent to a plane reflection by the spherical emission diffusion factor.

The loss formula caused by lunar surface reflection and scattering is:

$$L_{mp} = \left[1 - D|R|\left(\frac{r}{r_1 + r_2}\right)\right]^2 + 4D|R|\left(\frac{r}{r_1 + r_2}\right)\cos^2\left(\frac{1}{2}\varphi'\right) \tag{10.1}$$

where: D is the spherical reflection diffusion factor; |R| is the lunar surface reflection coefficient, or the equivalent reflection coefficient in case of scattering; r is the distance of the effective radiation source of the receiving and transmitting antenna; $r_1 + r_2$ is the distance between the transmitting and receiving antennas (shown in Fig. 10.14); φ' is the phase difference between the reflected wave and the direct wave.

2. Propagation model in free space

When there is no obstacle between the transmitting end and the receiving end, i.e., when there is a direct path, the free space model can be used to predict the strength of the received signal.

When radio wave propagates in free space, the energy is attenuated by diffusion. This attenuation is called the transmission loss of free space [11]. In order to calculate in practice, the telecommunication distance d is taken as km, and the wave-length is replaced with the corresponding frequency f in MHz. Then the expression of the transmission attenuation in free space is:

$$L_0 = 32.45 + 20\lg d + 20\lg f \tag{10.2}$$

The total link attenuation can be calculated by plus of Eqs. (10.1) and (10.2).

10.5 Testing and Verification

A series of test and verification should be conducted for the onboard TT&C and telecommunication subsystem of the lunar lander before launch to ensure that it could meet mission requirements. Besides general tests such as environmental adaptability tests, special tests such as compact field test and interface test between spacecraft and ground system should also be conducted.

10.5.1 Compact Field Test

In recent years, various antenna measurement technologies had been widely used. Generally, the electrical test was a general term that studies the radiation performance of the antenna by means of experiments with source and probe when the antenna was regarded as an unknown space radiation [13].

The radiation measurement of the antenna could be classified into near field measurement, far field measurement and compact field measurement according to the position of spacecraft. The probe was located near the antenna to be measured so as to sample field data in the near field measurement, and the far field parameters were acquired by transformation. According to the sampling method, it was classified into plane near-field measurement, cylindrical near-field measurement and spherical near-field measurement, which were suitable for antennas with different radiation characteristics. The far field measurement was a direct measurement of the parameters such as the antenna pattern, gain, electric axis, phase center, polarization and etc., when the test distance satisfied the far-field free space condition. The compact field measurement was a method for the intermediate distance between the near field and far field measurements. Generally, the compaction field was classified into single reflective surface compaction field, double reflective surface compaction field, and triple reflective surface compaction field. The double reflective compaction field was classified into cylinder-shaped compaction fields and Cassegrain-compensated compaction fields. The triple reflective compaction field was classified into double Cassegrain, Cassegrain-Gregorian, Gregory Cassegrain and double Gregorian.

When the antenna was installed, its electrical performance such as the gain pattern would be degraded to some extent compared with the unit antenna because of the influence of surface status of spacecraft. In particular, along with the increase of spacecraft function, the surface status of spacecraft became more sophisticate. Therefore, the performance of the antenna installed on spacecraft was an important specification to check compliance of the antenna.

The basic principle of the compaction field was to convert the spherical wave emitted by the test feed into a plane wave by large aperture hyperboloid and parabolic reflector to provide an ideal plane wave test environment within a short distance. The antenna to be tested or the spacecraft was located in the quiet zone of the plane test wave to acquire the ideal and accurate test results. In an indoor environment, real-time measurements of the entire spacecraft's radiation characteristics could be conducted fast with high accuracy.

In the compact field test, the transmitting or receiving antenna and the spacecraft were mounted on the turntable and located in the test quiet zone in the center of compaction field. In the operation status of continuous wave, the continuous wave signal was transmitted from the source of compact field and the antenna to be tested was in receiving. The spacecraft antenna to be tested was connected with the receiver system of compaction field with the test system. The turntable was controlled by computer software to rotate the spacecraft. The radiation pattern of the antenna installed on the spacecraft was acquired through related instrument and equipment for data acquisition and processing.

During the antenna pattern test, the antenna gain of the lunar lander could be calibrated by means of the comparison method. After processed by software, the gain coverage map of the antenna coverage area could be obtained. The principle of the comparison method for calibration of the antenna gain was as follows: In the same test, the antenna to be tested and the gain standard antenna were tested to acquire the received power P_{T} of the tested antenna and the received power P_{S} of the gain standard antenna respectively. Then the power gain of the tested antenna is:

$$G_{\mathrm{T}} = G_{\mathrm{S}} + 10\lg\frac{P_{\mathrm{T}}}{P_{S}} \tag{10.3}$$

where: G_{S} is the power gain (dB) of the gain standard antenna.

During the test of antenna radiation pattern, for each specified azimuth and pitch angle, two main polarizations of horizontal and vertical feeds was switched by a switch. Therefore, the amplitude and phase of two main polarities of horizontal and vertical could be acquired simultaneously in one test. The amplitude and phase information were referred to calculate the polarization ellipse and axis ratio test results for the specified azimuth and pitch angle. For the axis ratio test of the circularly polarized antenna, since accurate amplitude and phase test results could be acquired in the compact field test, the polarization ellipse could be acquired by data processing of computer software to get axis ratio test information directly, which could ensure the test accuracy of antenna axis ratio effectively.

10.5.2 Testing of Interface to Ground System

Generally in China, the engineering system of space mission included spacecraft, launch vehicle, launch site, ground telemetry/telecommand and ground application. The ground telemetry/telecommand system was to measure and control launch vehicle and spacecraft. The ground application system cooperated with spacecraft on orbit to perform specific functions [14]. Before launch, the spacecraft should be conducted tests of interface to the ground telemetry/telecommand system and the ground application system. The objective was to verify the interface compliance between the spacecraft and the ground telemetry/telecommand, as well as the spacecraft and the ground application system.

Generally, tests of interface to ground system included wired interface test and wireless interface test. The wired interface test focused on the check of uplink and downlink radio signal characteristics as well as orbit determination (speed measurement, ranging and angle measurement). The wireless interface test focused on the interface check such as antenna polarization compliance, zero distance in telemetry/telecommand channel.

Generally, the uplink and downlink radio signal characteristics check included the modulation characteristics such as the modulation mode and modulation degree of the uplink and downlink signals, and the format check of the uplink telecommand, injected data, downlink telemetry and telecommunication data, and the check of the demodulation performance, etc. Tracking was one of the keys for success of spacecraft mission. Therefore, the other major work in the process of wired interface test was check of the joint operation of spacecraft onboard telemetry/telecommand instruments and ground telemetry/telecommand system for the measurement of speed, distance and angle.

In order to avoid the influence of Earth atmosphere on the radio propagation channel, the antenna of spacecraft telemetry/telecommand and telecommunication subsystem was generally selected as circular polarization antenna. The modes of circular polarization included left handed and right handed circular polarizations. The antenna polarization method would be settled by spacecraft, ground telemetry/telecommand and ground application to avoid influence on the normal reception of uplink and downlink radio frequency signals between spacecraft and ground station. Therefore, the verification of antenna polarization was one of the important jobs in wireless interface test. And the time delay characteristics of the distance measurement signal sent by ground telemetry/telecommunication station and the received ranging signal was used by ground telemetry/telecommunication system to calculate the distance of spacecraft to ground. The distance measurement signal would introduce the delay variation of the fixed value during transferred with the spacecraft onboard telemetry/telecommand instruments, which was called the zero distance value of telemetry/telecommand channel. During test of interface, the zero value must be calibrated, which was one of the bases for distance measurement of spacecraft on orbit.

10.5.3 UHF Field Test

In order to verify the correctness of UHF band telecommunication system on lunar surface, according to environment characteristics on lunar surface, a test field on Earth desert was selected for telecommunication verification on lunar surface when the Chang'E-3 lunar probes were developed. The transmission characteristics of UHF band telecommunication link under desert environment were tested, and UHF band telecommunication subsystem was verified. The block diagram of ground test system for telecommunication on lunar surface is shown in Fig. 10.15.

According to actual terrain of the test field, three paths were planned as the UHF band transmitter's moving direction with UHF band receiver of the lunar lander as the center and 3 km as the radius. The top view of the terrain of the test field for UHF band telecommunication on lunar surface and planned paths is shown in Fig. 10.16.

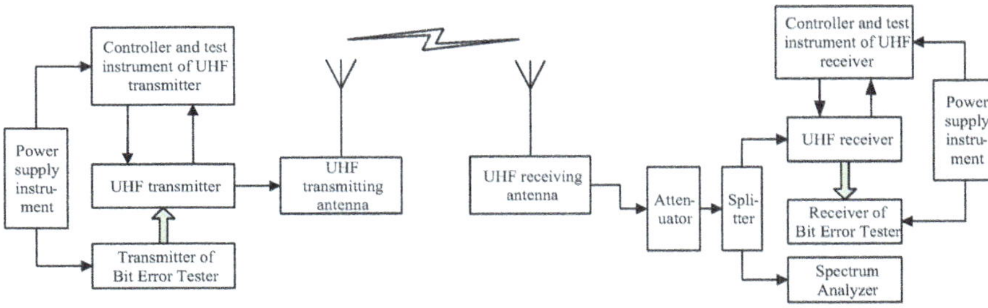

Fig. 10.15 Block diagram of ground test system for telecommunication on lunar surface

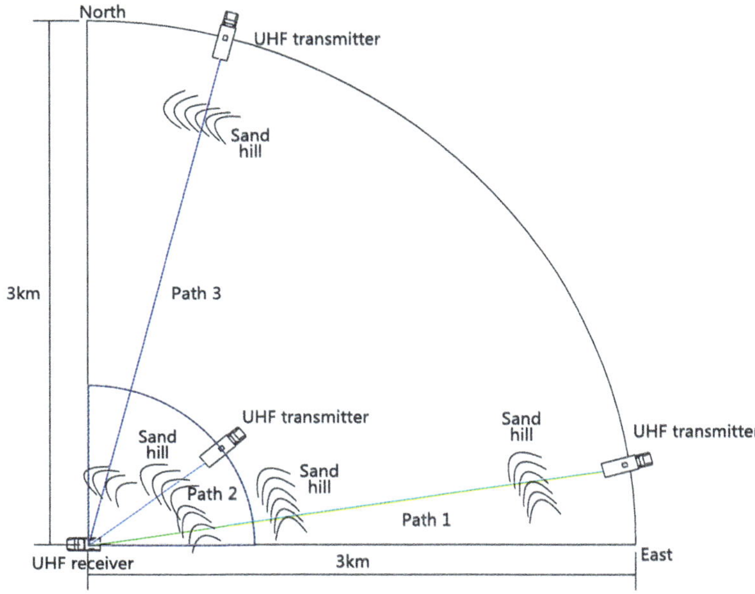

Fig. 10.16 Block diagram of ground test system for telecommunication on lunar surface

On the movement path of the UHF band transmitter, the signal received by UHF receivers was monitored at regular intervals with spectrum analyzers and Bit Error Testers. The power of radio signal received by the UHF band receiver was tested by the spectrum analyzer to calculate the actual link loss varied with distance. The actual value was compared with the value calculated by the attenuation model used in simulation analysis to verify the correctness of the attenuation model. The Bit Error Tester compared the data demodulated by the UHF band receiver with the data transmitted by the UHF band transmitter to verify the operation of telecommunication subsystem.

The test results showed that the transmission characteristics of telecommunication link were affected by the distance between two lunar probes and the terrain. The actual loss of UHF telecommunication link was consistent with the simulation analysis of telecommunication link. The model applied in the simulation analysis was consistent with actual results.

10.6 Summary

The telemetry/telecommand and telecommunication subsystem was designed to implement the information transmission between the lunar lander and the ground, as well as the lunar lander and the lunar rover in the Chang'E-3 lunar mission, which was also an important part of orbit determination. The characteristics, design and verifying of the onboard telemetry/telecommand and telecommunication technologies was described according to mission features of lunar landing exploration with development status of telemetry/telecommand and telecommunication technology in deep space mission.

Along with continuous improvement of space science and technology, deep space exploration such as lunar exploration is constantly extending. Higher requirements for deep space telemetry/telecommand and telecommunications are come up. The development trends of telemetry/telecommand and telecommunication technology in future deep space mission include the following aspects.

(1) Software Radio. At the end of twentieth century, the concept of software radio was come up for the first time. The central idea was to construct a general hardware platform with openness, standardization and modularization to implement various telecommunication functions such as modulation/demodulation, data format, encrypt mode and telecommunication protocol by software. At the same time, the broadband A/D and D/A was made to be near the radio frequency and extended from the baseband to intermediate frequency or radio frequency, so as to develop a new generation of wireless telecommunication system with high flexibility and openness. Once the concept of software radio was put forward, it gained wide attention in radio field around the world, and became the future development trend of telecommunication and even radio technology [15].

(2) Large aperture antenna. The main challenge in deep space telecommunications is that the RF signal is usually very weak due to long distances in deep space mission when it reaches the receiver. In order to increase the power utilization as much as possible when the signal power is limited, the method can be selected to increase the antenna aperture. The antenna gain can be improved when the antenna aperture increases. The antenna gain is an important specification that shows the capability of the antenna to concentrate and radiate energy [16].

(3) Ka band telecommunication. The telecommunication in deep space exploration has the characteristics of long distance and large path loss,

which limits the channel transmission capability. Since the 1970s, X-band telemetry/telecommand and telecommunication technology was widely applied in deep space exploration. Compared with X-band, Ka band antenna can acquire higher antenna gain at same aperture, so as to improve the channel data capacity [17]. Up to date, the deep space telemetry/telecommand and telecommunication network in China is capable of Ka band signal reception.

(4) High stability oscillator. The telecommunication in deep space exploration has the characteristics of long distance and signal delay, which is not conducive to real-time and effective navigation of the lunar lander through radio measurement technology. The one-way radio measurement technology has the characteristics of relatively short signal delay compared with two-way radio measurements. The accuracy of one-way radio measurement was lower because of the lower stability of onboard frequency reference on the lunar lander. The onboard high-stability oscillator can effectively improve the stability of the frequency reference to meet the requirements of radio measurement technology for future deep space exploration [5].

References

1. Jiang C, Fan X (2003) Introduction to aerospace communications tracking technology[M]. Beijing University of Technology Press, Beijing
2. Chen Y (2007) Spacecraft radio measurement and control technology[M]. China Aerospace Press, Beijing
3. Han K (1984) Introduction to the design of microwave unified measurement and control system[M]. National Defense Industry Press, Beijing
4. Chu G, Zhang H (2007) Technology of lunar probes[M]. China Science and Technology Press, Beijing
5. Wu W, Dong G, Li H et al (2013) Deep space observation and control communication system engineering and technology[M]. Science Press, Beijing
6. Huang X, Xu B, Ji X et al (2013) Design and verification of the X-band satellite-borne on-board measurement and control system for the Chang'E-2 spacecraft[J]. Chinese Science 43(5):499–505
7. Chen F (1992) Spacecraft measurement and control manual[M]. Science Press, Beijing
8. Hu Q, Yang F (2010) Introduction to aerospace[M]. China Science and Technology Press, Beijing
9. Zhang N, Li H, Zhang Q (2007) Development trend and thinking of deep space exploration communication technology[J]. Journal of Astronautics 28(4):786–793
10. Xiong H (2000) Radio wave propagation[M]. Publishing House of Electronics Industry, Beijing
11. Tan W, Hu J (2009) Spacecraft system engineering[M]. China Science and Technology Press, Beijing
12. Tang G (2012) Measurement and control radio measurement technology in deep space[M]. National Defense Industry Press, Beijing
13. Ye Y (2007) Spacecraft antenna—engineering and new technology[M]. China Science and Technology Press, Beijing
14. Peng C (2011) Spacecraft system design[M]. China Science and Technology Press, Beijing
15. Jon H, Marvin KS (2009) An autonomous software radio receiver for deep space[M]. Yan C, translated. Tsinghua University Press, Beijing

16. William AI (2013) Spaceborne antennas for planetary exploration[M]. Li H, Yu Y, translated. Tsinghua University Press, Beijing
17. Zhou X (2009) Deep space communication[M]. National Defense Industry Press, Beijing

Chapter 11
Landing Gear Technology of Lunar Lander

11.1 Introduction

Lander cushion is one of key technologies in soft-landing on celestial bodies other than Earth. It can attenuate the landing shock to acceptable level to ensure stable landing of the lander and safety of onboard crews and devices by means of energy transformation or dissipation.

During a typical soft-landing process, the lunar lander fired its retro-rocket engine to decelerate at first. When the velocity relative to the landing surface was dozens of meters per second or even more preferable several meters per second, the landing impact energy would be absorbed by the landing gear to ensure safe landing. Up to date, there were two types of landing cushion system in past lunar landing missions: landing airbag [1] and landing gear. As for the airbag, impact energy was absorbed by gas compression in the bag, gas venting from one chamber to another chamber, and elastic-plastic deformation of airbag material [2]. For the landing with airbag, there might be bouncing, rolling and turnover of the lunar lander during landing impact. It was difficult to achieve high landing accuracy and predefined landing attitude. Therefore the, airbag was preferred to the lander with higher landing velocity, but light landing mass and small size, which would not return from the Moon. The landing gear system absorbed the impact energy by the deformation of one or more struts in the landing gear mechanism. With the landing gear system, it was easy to control the attitude after landing and avoid bouncing during landing. The landing gear system was more suitable to the lander with lower landing velocity, heavier landing mass and large size, especially a strict attitude control requirement after landing. Meanwhile with the support of landing gear system, the lander could serve as a launch platform for lift-off on lunar surface. Therefore, the landing gear system was more suitable to the landing and return mission. A typical landing gear system is shown in Fig. 11.1.

The crucial function of the landing gear was to ensure stable and safe landing, i.e., not only the lander would not turn over when landing but also the landing impact on the structure was within the specific range. In practice, generally there were several sets of landing gear, symmetrically installed on the bottom of the structure. The

D.-Y. Yu et al., *Technology of Lunar Soft Lander*, Space Technology Library 38,
https://doi.org/10.1007/978-981-15-6580-9_11

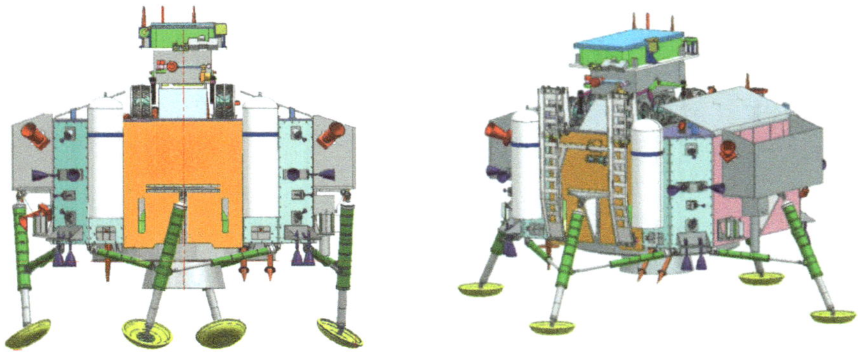

(a) (b)

Fig. 11.1 Typical landing gear system. **a** Stowed; **b** Deployed

landing gear was stowed (shown in Fig. 11.1a) during launch and deployed after orbit injection (shown in Fig. 11.1b). After the lunar lander was inserted into circumlunar orbit and the powered descent was initiated, there would be several phases including main deceleration, quick attitude adjustment, approaching, hovering, hazard avoidance and free falling. At last, the landing impacting was absorbed by the landing gear to implement stable and safe landing, and support the lunar lander on lunar surface. Therefore, the reliability of the landing gear system function and its performance was fatal to the lunar soft-landing mission.

In this chapter, the development status of landing gear technology is introduced at first. Then the technical characteristics of landing gear are analyzed. It is also introduced including description of some common landing cushion methods, design methodology of landing gear, typical design, simulation analysis of landing impact, methods of ground test and verification.

11.2 Development Status

Since the 1950s, the Soviet Union and the United States launched a series of deep space probes for lunar and Mars exploration. In 1966, Luna 9 of the Soviet Union was landed on the Moon with the airbag system, while Surveyor 1 of the United States was landed on the Moon with the landing gear system. In the 1970s, Luna 16 of the Soviet Union and Apollo lunar module [3] of the United States implemented successful landings on the Moon with the landing gear system [4]. Meanwhile, several successful missions to the surface of Venus and Mars were implemented by the Soviet Union and the United States. In 2004, Mars rovers of the United States, Spirit and Opportunity were landed on Martian surface with the airbag, and achieved outstanding exploration. In August 2012, Mars rover Curiosity of the United States was landed on Mars surface with the sky crane, which was a brand-new method of soft-landing. Recently along with the booming development of space exploration, deep space exploration became the focus of all space countries and organizations. Landing missions to the Moon, Mars and asteroids are implemented or planed by the United States, Russia, European Space Agency (ESA), Japan and India.

Development of landing cushion technology was initiated relatively lately in China. As CLEP was implemented, research and development of landing cushion technology was conducted by many academies and universities in China. Since the 1990s, the seat cushion mechanism for astronaut was studied and applied in ShenZhou series manned spacecraft. On the basis, the landing gear system for the Chang'E-3 lunar lander was developed successfully. During past years, technologies were developed including several cushion methods, design and analysis of landing cushion mechanism, ground tests and evaluation technology. Above research and development provides solid basis for engineering implementation of landing cushion technology for spacecraft in China.

11.3 Configuration and Technical Features of Landing Gear System

Generally, there are more than one set of landing gears in the lunar lander to absorb landing impact. The landing gears were symmetrically installed on the bottom of the lander structure. There was one primary strut, two secondary struts, footpad, Hold Down and Release Assembly in one landing gear of the Chang'E-3 lunar lander as shown in Fig. 11.2. The upper end of primary strut was connected to the main structure

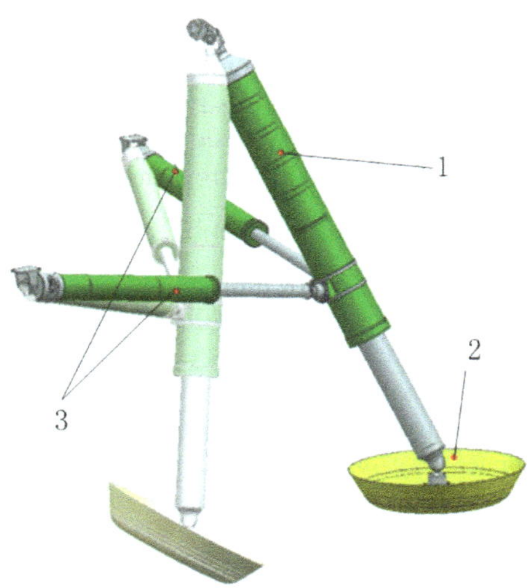

Fig. 11.2 Configuration of landing gear (stowed configuration in light color, deployed configuration in dark color). **1**-Primary strut; **2**-Footpad; **3**-Secondary struts

of the lunar lander by universal joint, and the lower end was connected to the footpad by ball joint, while the middle was connected with two secondary struts by spherical hinge. Moreover, the other end of each secondary strut was also connected to the main structure of the lunar lander by universal joint. The cushion cells inside the primary and secondary struts could absorb the landing impact energy. Most of vertical impact was absorbed by the primary strut, while most of horizontal impact was absorbed by the secondary struts. The primary strut could only be compressed and two secondary struts could be compressed and tensioned. The footpad was applied to adapt to diverse landing surface and avoid excessive penetration into lunar regolith, which was also benefit to horizontal slide for the lunar lander at large horizontal velocity to improve landing stability. The landing gears were held by the Hold Down and Release Assembly during launch. When the lunar lander was injected into predefined orbit, the landing gears were deployed and locked up. The landing gear was a kind of typical spatial multiple DOF mechanism during the landing process and it could also be taken as a special parallel mechanism.

11.4 Common Cushion Methods and Their Features

Up to date, there were a lot of methods for cushion. But only few of them are suitable to landing cushion mechanism for space application, such as cellular material deformation, plastic deformation of a thin-walled metal tube, hydraulic damping and tensile deformation of the special material, etc.

11.4.1 Cellular Material Deformation

Aluminum honeycomb and aluminum foam were two kinds of typical cellular material. Because of high porosity, high specific strength, high specific modulus, excellent heat resistance and corrosion resistance, they were usually used as shock cushion material of irreversible deformation.

1. Aluminum Honeycomb

The mass density of aluminum honeycomb is lower, while its compression ratio (ratio of compressible length to initial length) is higher with reasonable compression strength and specific energy absorption up to 30 kJ/kg. Compared to fluid or gas filled damper, the advantages of aluminum honeycomb are steady damping force, low mass density, insensitive to space environment. Furthermore, according to different impact load input during landing process, several chained aluminum honeycomb blocks with different compression strength can be applied to ensure the stability of impact response.

The correlation between load and deformation of aluminum honeycomb could be acquired through static compression test and dynamic impact test. In application or

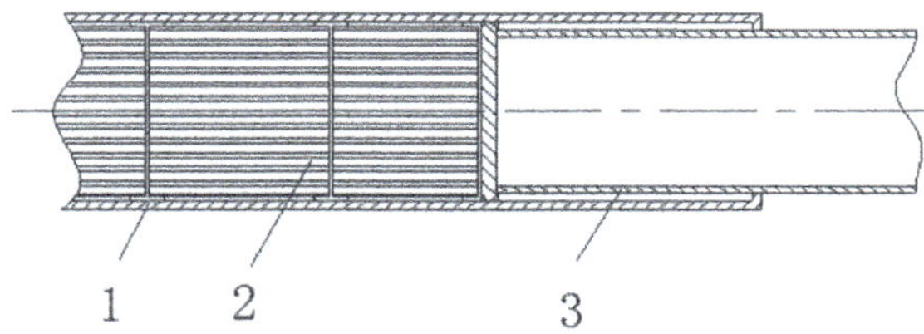

Fig. 11.3 Aluminum honeycomb absorber. **1**-Outer tube; **2**-Aluminum honeycomb; **3**-Inner tube

test of aluminum honeycomb, the direction of the impact force should be in line with the longitudinal axis of aluminum honeycomb as shown in Fig. 11.3.

In dynamic impact test, ballast weight was installed on the test device and dropped down to the aluminum honeycomb article. The response acceleration on center of gravity of the ballast weight was acquired. Motion equation of the ballast weight is as follows:

$$m\,\ddot{y} = F - \mathrm{mg} \tag{11.1}$$

where, m is the mass; $\ddot{y}$ is coordinate of the center of gravity; F is damping force of the aluminum honeycomb; g is acceleration of Earth gravity.

According to Eq. (11.1), the damping force F can be given by the formula:

$$F = m\,(\ddot{y} + g) \tag{11.2}$$

A typical curve of load to deformation of aluminum honeycomb is shown in Fig. 11.4. It could be seen that the load almost remains as a constant during plastic deformation procedure. Meanwhile, at the beginning of compression, there was a peak load, which could be effectively avoided with a pre-compression before use.

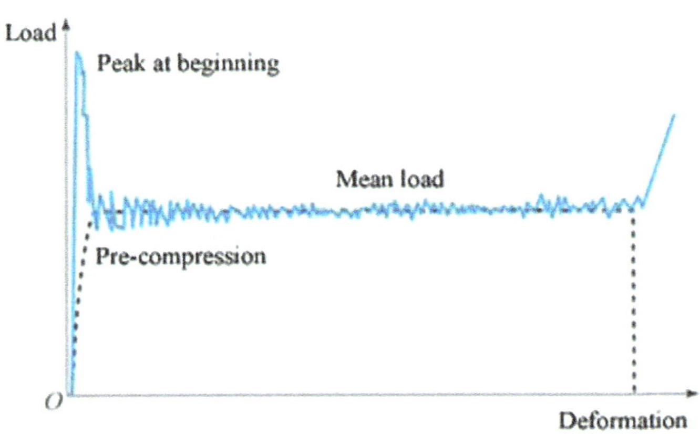

Fig. 11.4 Typical experiment result of aluminum honeycomb material

There were features of aluminum honeycomb as follows.

(1) Aluminum honeycomb material with same size and foil gauge showed remarkable consistency.
(2) The damping force was almost same when the aluminum honeycomb material deformed.
(3) Occasional instability of some part of honeycomb material would lead to a fierce oscillation of the damping force, which should be avoided in application.

The application of aluminum honeycomb absorption technology was very mature, such as the landing gear system of Apollo lunar module and Viking Mars lander.

2. Aluminum Foam

Aluminum foam is one kind of light material with specific mechanical properties. The difference of damping feature for different aluminum foam material is significant. Compared with aluminum honeycomb, the density and crushing strength of aluminum foam is higher. Unlike aluminum honeycomb, the load of aluminum foams varies with stroke when compressed. The highlighted advantage of aluminum foam is its isotropic damping capability, which is fit to application on the occasion that the direction of impact force is not certain.

11.4.2 Deformation of Thin-Walled Metal Tube

The thin-walled metal tube can deform in different patterns. Various energy absorption buffers can be designed by different patterns of deformation. An absorber of metal plastic deformation is shown in Fig. 11.5. It absorbs the impact energy by both friction and radial plastic deformation of metal tube. The buffer components consist of a hollow conical cylinder and a series of expanding tubes with diverse thickness. When the input load exceeds specific threshold, the buffer is activated with every expanding tube successively going through the conical cylinder. Because the diameter of outer conical cylinder with larger end is greater than the diameter of inner expanding tubes, the diameter of conical cylinder is expanded when the expanding tubes are getting through the conical cylinder, as shown in Fig. 11.6. In

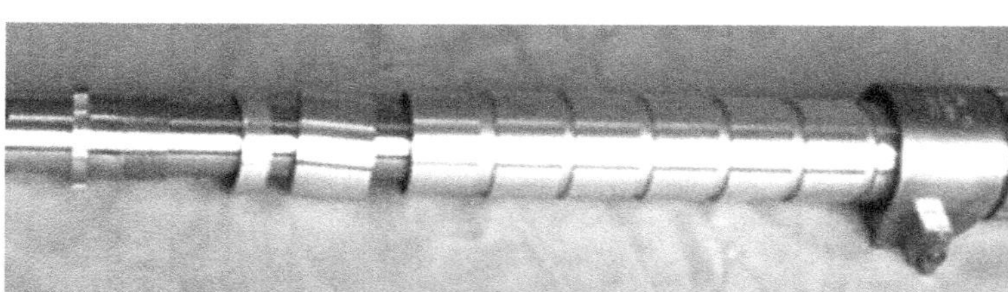

Fig. 11.5 Aluminum honeycomb absorber

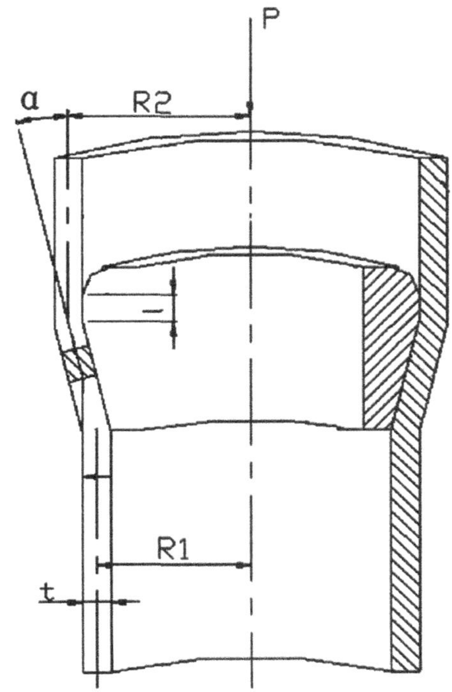

Fig. 11.6 Diameter of expanding tube is expanded

such process, the impact energy is transformed to elastic-plastic deformation energy of expanding tubes and the friction heat between conical cylinder and expanding tubes to be exhausted.

The process that the expanding tubes go through the conical cylinder can be taken as a plastic molding issue, and the damping force is as follows [5]:

$$P = 2.3\,\pi\,\sigma_s\left[1 - \left(\frac{R_1}{R_2}\right)\frac{\tan\alpha + f}{(1 - f\tan\alpha)\cdot\tan\alpha}\cdot e^{\frac{fl}{R_2}}\right]R_1\,t \tag{11.3}$$

where: R_1 is initial central diameter of the expanding tube; R_2 is central diameter of the expanding tube after deflection; t is thickness of the expanding tube; α is one half of the taper angle; l is length of the conical cylinder with maximal diameter; σ_s is yield limit of the material of the expanding tube; f is coefficient of the friction between conical cylinder and expanding tubes.

There are features of deformation of thin-walled metal tube as follows.

1. The damping force of the expanding tube with same material, specification and deformation is well consistent.
2. Cushioning features are insensitive to environments such as temperature alternation and humidity, etc.

3. Lubrication\between expanding tubes and conical cylinder is the key to consistency of cushioning features.

Such method was applied used in seat cushion mechanism for astronaut of ShenZhou spacecraft.

11.4.3 Hydraulic Damping

The hydraulic buffer absorbs impact energy by hydraulic damping force. When the external force acts on the buffer, hydraulic fluid flows through the orifice resulting in pressure difference between two sides of the valve, so as to generate damping force to absorb impact energy. Generally, the hydraulic buffer takes both fluid and gas as the medium. Because the mass density, pressure, coefficient of viscosity of the fluid and gas is closely related to temperature, it is necessary to decrease influence of temperature over these parameters as much as possible to achieve stable cushioning performance. Furthermore, sealing of fluid and gas medium has to be solved under temperature alternation. It is a great challenge to design the hydraulic buffer with proper parameters in deep space environment. The hydraulic buffer [6] of Surveyor lunar lander of the United States is shown in Fig. 11.7.

The damping force is as follows:

$$F = (p_f - p_0) \cdot \frac{\pi d^2}{4} \tag{11.4}$$

where: p_1 is pressure of internal fluid; p_2 is pressure of ambient pressure; d is the cross-section diameter of piston rod.

There are features of hydraulic buffer as follows.

(1) It is suitable for relatively lower load. Compared with other damping methods, the mass and size are larger when the compact load is higher.

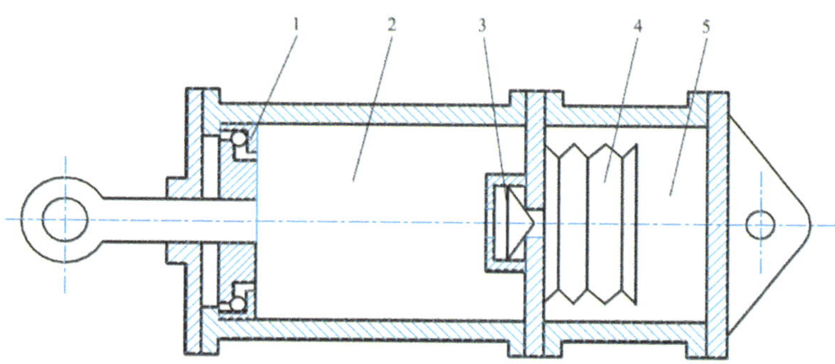

Fig. 11.7 Configuration of hydraulic buffer. **1**-Damping orifice; **2**-Fluid; **3**-Valve; **4**-Bellow; **5**-Gas

(2) The attitude of the lunar lander after landing can be adjusted. The damping system will not be damaged by cushion process and can gradually recover. It is suitable to missions with strict attitude requirement after landing.
(3) The cost of research and development is higher, and it is very difficult and low reliable. The parameters such as density and coefficient of viscosity of hydraulic medium are closely related to environment temperature. Therefore, technical challenges such as sealing, temperature control and control of viscosity coefficient, shall be overcome before application. At the same time, the requirement for manufacture accuracy is high, which is definitely costly.

The Surveyor lunar lander of the United States with hydraulic buffer implemented successful soft-landing on lunar surface. The hydraulic buffer was also applied in Apollo spacecraft. The buffer came to leakage once in Surveyor missions, but a pyro mechanism was fired to realize a lockup for the landing gear system and ensured the attitude of the lunar lander. Because of potential risk of hydraulic buffer, it was not applied in the landing gear system of Apollo lunar module.

11.4.4 Tensile Deformation of Special Material

The impact energy is absorbed through tensile deformation of a slender rod made of special material. In the process of damping, the metal rod is stretched. When the yield strength of the material is exceeded, the metal rod has a plastic deformation. The impact energy is transferred into plastic deformation energy. The advantages of such method include reliable cushion process, high specific energy absorption and small size. The material of tensile rod is generally super plastic in normal temperature, whose specific elongation is over 50%. The conventional metal is hardly to meet the requirement.

Such method was applied in secondary struts of the Chang'E-3 lunar lander to absorb impact energy by tensile deformation.

11.5 Design of Landing Gear Mechanism

11.5.1 Requirements Analysis

1. Functional Requirements and Configuration Analysis

Generally, there were following function requirements for the landing gear mechanism.

(1) It could be stowed, held down, deployment and locked up. Due to limited volume of launch fairing, the landing gear mechanism was usually stowed and held down during launch. It would be deployed and locked up after the lunar lander was injected into predefined orbit.

(2) It could effectively absorb the impact load during landing. In the landing process, the maximum impact load on the lunar lander should be limited in specific range to ensure the safety of onboard instruments.
(3) It could ensure landing stability. Under complex landing conditions, the landing gear mechanism could prevent the lunar lander from turning over to ensure landing stability.
(4) It could provide effective long-term support for the lunar lander. After landing, the landing gear mechanism should provide effective long-term support for the lunar lander with required attitude. Sometimes it was also employed as a stable launch platform for lift-off of sample return mission.
(5) Other functions. For example, the landing gear mechanism should give an indicating signal when the footpad touches lunar surface to judge whether the lunar lander was touched down.

In order to meet above function requirements, generally there were following components in the landing gear mechanism.

(1) Primary strut. It was to absorb the vertical impact load and support the lunar gravity of the lunar lander on lunar surface.
(2) Secondary struts. It was to absorb the horizontal impact load and help to support the lunar gravity.
(3) Footpad. It was to prevent excessive penetration due to exceeding impact load. Meanwhile, it was benefit to horizontal slide of the lunar lander if lunar terrain was right. If the footpad impacted on an obstacle and horizontal impact load was high enough while sliding on lunar surface, peripheral part of the footpad could plastically deform to absorb part of energy.
(4) Hold-down and Release assembly. It was to hold down the landing gear mechanism during launch and release it after orbit injection.
(5) Deployment and lock-up assembly. It was to deploy and lock up of the landing gear mechanism after it was released.
(6) Lunar surface sensing probe. It was installed in the footpad or other part of the landing gear mechanism and could output a triggering signal when the footpad touched lunar surface.

2. Analysis of Initial Design Conditions

The initial design conditions of landing gear mechanism generally included the following aspects.

(1) Mass properties of the lunar lander. It was the input for analysis and evaluation of cushion performance and touchdown stability including mass, position of the center of mass and moment of inertia in three directions.
(2) Touchdown velocity. The velocity was generally presented in local coordination system of the Moon, including the vertical and horizontal components. Together with total touchdown mass, it determined impact energy to be absorbed by the landing gear mechanism.

(3) Attitude angles and angular velocities in three directions. These terms were also generally presented in the said local coordination system and related to touchdown stability and attenuation performance.
(4) Landing surface. It referred to the slope, depth of pit or height of rock, and mechanical properties of lunar surface. It was closely related to touchdown stability, and cushion performance to some extent.
(5) Maximal envelope of launch faring. It referred to radial and axial envelope. The axial envelope was often expressed relative to a reference plane of the lunar lander. It determined whether the landing gear mechanism should be stowed during launch.

3. Requirements Analysis for Performance and Interface

The performance specifications of the landing gear mechanism included following aspects [7].

(1) Mass, strength and stiffness. The mass referred to maximal acceptable mass. Requirements of mass, strength and stiffness are one of core issues in design process of the landing gear mechanism. The performance of stable landing should be improved as far as possible on condition that the requirements of mass, strength and stiffness were satisfied.
(2) Deployment and lock-up. If the landing gear mechanism should be deployed on orbit, such requirement would be come up. It included the static torque (force) margin, deploy time and peak of impact load in deployment process. The method to realize hold-down, release, deploy and lock-up was determined by such requirements.
(3) Cushion characteristics. It was core requirement of performance which included peak response and time span of impact load, landing stability and attitude change after landing, sometimes even cushion method and length of the stroke.
(4) Lifetime and reliability. The requirement of lifetime generally included storage time on ground and operation time in space. The requirement of reliability generally referred to the degree of reliability in the mission profile under a specific confidence.
(5) Interfaces. Interface requirement included mechanical interface, thermal interface (thermal environment) and electrical interface (including energy consumption). It was determined by these interfaces including the method and timeline of assembly for the landing gear mechanism in AIT phase, thermal control methods, material, type of electrical component and part, and etc.

4. Environment Analysis

Besides general ground environment, following environments should be endured by the landing gear mechanism.

(1) Vibration. The vibration environment during launch might cause deformation or broken of mechanism because of stress. For the landing gear mechanism, the hold-down device should be avoided to be loosened. Moreover, the stress concentration due to load cycling of vibration should be avoided, which might cause deformation or rupture of parts of the landing gear mechanism.

(2) Radiation. The charged-particle radiation and ultraviolet radiation might significantly affect nonmetal materials such as rubber seal rings and adhesives, but little on metals. For the landing gear mechanism, rubber seal rings and adhesives should not be exposed directly. Material of outer multi-layer insulation should be good at anti-radiation and processed by surface coating.
(3) Vacuum. The polymeric material was prone to sublimation and decomposition under vacuum, which might result in degradation of material performance. In order to avoid sublimation and decomposition of material under vacuum, metallic material was better choice for parts of landing gear mechanism. For the parts like footpad and adhesive of aluminum honeycomb, materials that had been successfully applied in space missions are preferred. Moreover, under ultrahigh vacuum condition, the joints of mechanism might suffer from cold welding at a certain pressure and fail. Effective measures were lubrication for joint surfaces and limiting the friction resistant force into an acceptable range.
(4) Gravity. Acceleration of lunar gravity is about one-sixth of Earth gravity. The lower gravity may lead to turning over during landings. This gravity difference should be sufficiently considered in design and ground test of the landing gear mechanism.
(5) Lunar terrain of landing zone. It included slope of lunar surface, dimension and distribution of pit and rock, and etc. The large slope might cause declination or even turnover of the lunar lander. Pit and rock can significantly affect the landing stability.
(6) Mechanical properties of landing surface and lunar dust. Mechanical properties of landing surface referred to bearing strength and coefficient of friction under variable dynamic impact velocity. Different properties led to different bearing strength for the footpad on landing surface and different cushion performance in landing process. Before landing, the retro rocket engine was usually applied to decrease touchdown velocity. Therefore, the dust might be raised by the engine plume when the lunar lander approached the dusty surface. The dust adhering to joint surfaces had adverse effect on movement of mechanism. Measures such as coated protection for the movable parts should be designed, which could be implemented together with thermal control.
(7) Thermal environment. There were thermal effects on the lunar lander during LTO and powered descent. The mechanical parts of relative movements should be in same kind of material or materials with approximate expansion coefficient to make sure that the fit grade of joint surfaces would not change, so as to adapt such environment. Meanwhile it was suggested to select the cushion method insensitive to temperature. The proper thermal control should be applied if necessary. Moreover, the landing gear mechanism was generally installed on the bottom of the lunar lander, which was close to the main engine. Therefore, special attention should be paid to effect of engine plume on the landing gear mechanism to ensure sufficient cushion performance under such thermal environment.

11.5.2 Design and Simulation

1. Configuration and Cushion Parameters
1) Analysis of layout and configuration

(1) Layout. Generally, there were several uniformly distributed sets of landing gear mechanisms installed on the bottom and lateral plates. The attachment points of struts and structure should have enough stiffness and strength due to huge concentrated load of landing impact. Therefore, integrated layout design should be conducted for landing gear mechanisms and main structure of the lunar lander.

(2) Quantity. The polygon of landing stability was a polygon with bottom center of footpads as vertex, shown in Fig. 11.8. In landing process, the landing was stable if the center of mass of the lunar lander projects into the stability polygon along gravity vector of the landing surface, i.e., the lunar lander will not turn over. Under same condition, more sets of landing gear were configured, the larger the area of stability polygon was, and the better landing stability was. But meanwhile it would lead to complex mechanical interfaces and increasing of mass. Therefore, other aspects like system mass and mechanical interfaces should be considered as well when the layout and quantity of the landing mechanism was determined.

(3) Configuration. There were two types: cantilever-type landing gear (as shown in Fig. 11.1) and inverted-tripod landing gear (as shown in Fig. 11.9). The landing stability and reliability of cantilever-type landing gear was better than inverted-tripod landing gear. The composition and stress status of cantilever-type landing gear was complex and there is some sort of difficulty in manufacture and assembly. Generally, a small lander preferred inverted-tripod landing gear while a large-scale lander preferred cantilever-type landing gear.

2) Analysis of key geometric parameters

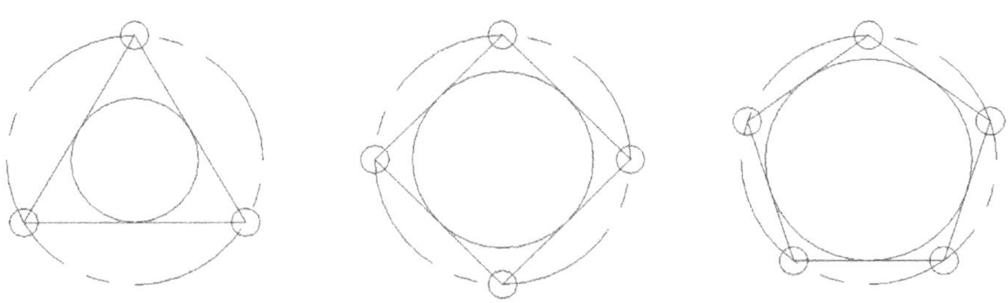

(a) (b) (c)

Fig. 11.8 Stability polygons of different quantity of landing gear mechanism. **a** Three; **b** Four **c** Five

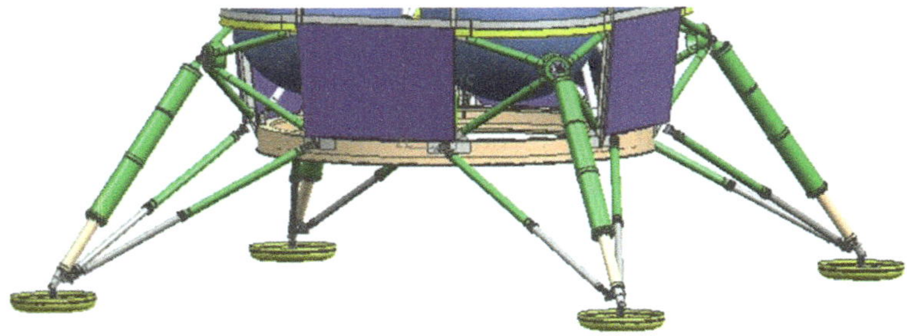

Fig. 11.9 Inverted-tripod landing gear

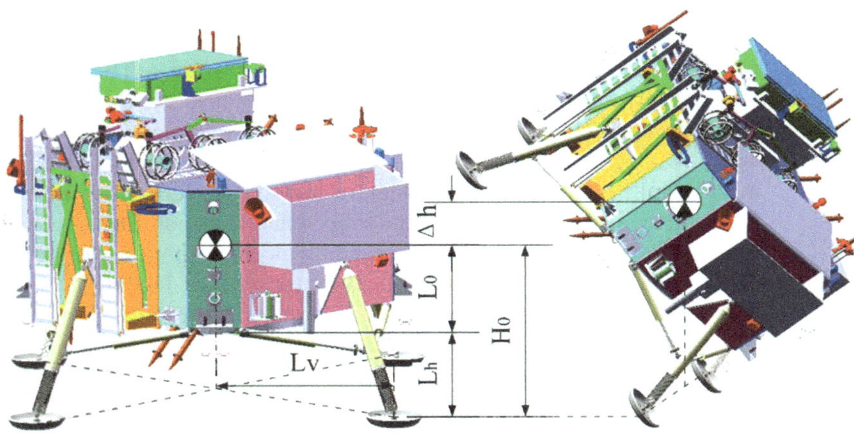

Fig. 11.10 Potential energy variation of the lunar lander in turn over process

(1) Key geometric parameters. There were several key geometric parameters including initial height from the lowest point of lander structure to bottom face of footpad L_{h}, and the shortest distance between the line connecting the bottom center of two adjacent footpads and the central axis of the lunar lander structure L_{v} (as shown in Fig. 11.10), which had significant effect on landing stability and system mass.

(2) Determination of parameters. In order to minimize system mass, value of L_{h} and L_{v} should be minimized as far as possible on condition the requirements of mass, strength and stiffness are satisfied. The minimal value of L_{h} is:

$$\mathrm{L_{h\,min}} \approx \mathrm{S} + \mathrm{S_0} + \mathrm{T} + \Delta\mathrm{H} \tag{11.5}$$

where: S is vertical displacement of primary strut; S_0 is reserved safety distance from the lowest point of the lunar lander structure to landing surface; T is height of the footpad; ΔH is depression depth of the lunar lander after landing relative to landing surface.

Generally, the distance from center of mass of the lunar lander to the lowest point of main structure L_0 is known. So, after L_{hmin} is determined, height of center of mass of the lunar lander H_0 could be derived from:

$$\mathrm{H_0 = L_{hmin} + L_0} \tag{11.6}$$

The value of L_v is determined through simulation of landing stability. At preliminary design, a preliminary value of L_v is estimated with principle of energy transformation. In order to assure that the lunar lander would not turn over in landing process, the following formulation should be met.

$$W_D \leq W_H = m_{\max} g' \Delta \mathrm{h} \tag{11.7}$$

where: W_D is kinetic energy of the lunar lander when the lunar lander starts to turn over; W_H is increasing of potential energy of the lunar lander relative to that at touchdown when center of mass of the lunar lander moves into vertical plane including the connecting line of spherical joints of two adjacent footpads; m_{max} is possible maximum landing mass at touchdown; g' is gravity acceleration on landing surface; Δh is increasing height of center of mass of the lunar lander in turn over process.

Deduced from Eq. (11.7), the lunar lander would not turn over if kinetic energy at touchdown was less than increasing of potential energy in landing process.

Assuming the primary strut is not compressed and the footpad does not subsidence, the value of Δh is estimated as follows:

$$\Delta h = \sqrt{\mathrm{H_0^2 + L_V^2}} - \mathrm{H_0}$$

In actual landing process, the struts would absorb some part of energy and height of center of mass will decrease due to compression of the primary strut. Therefore, above estimation was conservative.

3) Analysis and determination of cushion capability

Total energy to be absorbed by the landing gear mechanism consisted of two parts: one was kinetic energy of the lunar lander at touchdown and the other was potential energy reduction of the lunar lander due to strut cushion and footpad subsidence.

(1) Hypothesis of cushion energy. At preliminary design, it was assumed that potential energy reduction due to footpad subsidence was equal to deformation energy absorbed by landing surface. Such hypothesis was physically similar with situation of landing on a rigid surface.

The energy absorbed by the primary strut in landing process is:

$$W = \frac{1}{2} m_{\max} v_{\mathrm{V}}^{2} + m_{\max} g' \Delta \mathrm{H} \tag{11.8}$$

where: v_v is maximum vertical velocity of the lunar lander relative to landing surface.

(2) Critical parameter condition. If the lunar lander with a horizontal velocity was landed on a slope with pit and rock, energy absorbed by one set of the landing gear was remarkably larger than that of other sets. At preliminary design, it was assumed that the maximum cushion of every primary strut was:

$$A_{\max} = a_0 W \tag{11.9}$$

where: a_0 is a constant correction factor less than 1. The more landing conditions were uncertain, the larger the value of a_0.

The secondary strut is generally applied to absorb horizontal impact load. It was suggested that every single secondary strut could absorb all horizontal kinetic energy of the lunar lander, i.e., the maximum energy absorption of one secondary strut should be:

$$B_{\max} = \frac{1}{2} m_{\max} v_{\mathrm{h}}^{2} \tag{11.10}$$

where: v_h is maximal horizontal velocity of the lunar lander.

The energy absorbed by tension and compression of secondary strut was generally different for different configuration of the landing gear mechanism. The tension was usually larger than the compression, and even larger than the value calculated from Eq. (11.10). The larger the angle between the axis of primary strut and the axis of the lunar lander at touchdown, the more easily secondary struts stretched, and the more energy was absorbed by secondary struts.

After key geometric parameters including L_h, L_v cushion force were analyzed, the requirement of maximum impact load should be met.

$$a' \geq \frac{F}{m_{\min}} \tag{11.11}$$

$$F \geq m_{\min} a' \tag{11.12}$$

where: a' is maximum acceptable acceleration response; F is maximum acceptable cushion force; $m_{\min}$ is minimum possible landing mass.

Based on Eqs. (11.9) and (11.10), the minimum stroke of primary strut could be estimated.

$$S_{\min} = \frac{A_{\min}}{\frac{F}{n}} \tag{11.13}$$

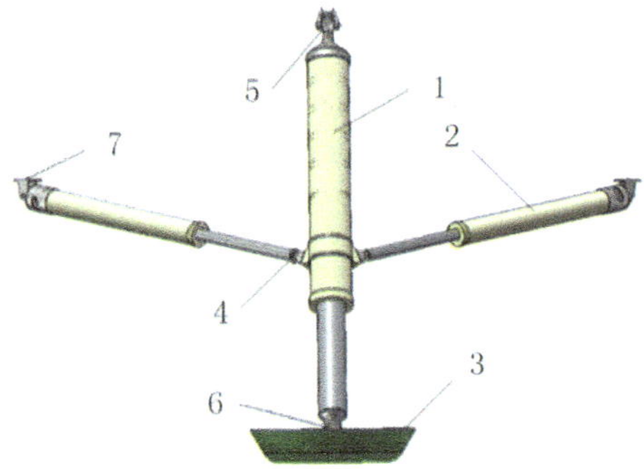

Fig. 11.11 Connection of parts. **1**-Primary strut; **2**-Secondary strut; **3**-Footpad; **4**-Ball joint II; **5**-Universal joint I; **6**-Ball joint I; **7**-Universall joint II

where: n is quantity of the landing gear mechanisms.

2. Design of Landing Gear Mechanism
1) Subsystem design of landing gear

(1) Joint design. As sown in Fig. 11.11, the primary strut was attached at the upper end to main structure of the lunar lander by a universal joint I, while at the lower end to a footpad by a ball joint I. One end of each secondary strut was connected to primary strut through a ball joint II, while the other end was connected to main structure through a universal joint II.
(2) Strut design. Every strut absorbed impact energy by cushion material inside. The primary strut generally absorbed vertical impact load while secondary strut absorbed horizontal impact load. The primary strut could only be compressed while secondary strut could be both compressed and stretched.
(3) Design of hold down, release, deploy and lock up mechanism. The primary strut could be held down by additional joint between the primary strut and the structure, or compression of the secondary strut. The deployment spring and lock up mechanism at the joint connecting strut with the structure or inside the secondary strut were designed for deployment and lock up. The minimum static torque (force) margin of the mechanism should be not less than 1. In order to reduce locking impact load, a damping cell was usually included.
(4) Support and anti-subsidence design. The footpad with a larger base area could effectively support the lunar lander and prevent it from subsidence on soft-landing surface. In conditions with larger horizontal velocity, the lunar lander with footpad could smoothly slide on landing surface so as to improve landing stability.
(5) Design of lunar surface sensing probe. The lunar surface sensing probe was designed to output signal at touchdown, which might be a probe beneath the footpad or triggering switch on the bottom surface of the footpad.

2) Design of primary strut.

(1) Keys. It included cushion design and strength design. The cushion design included cushion material selection, determination of cushion force and stroke. The strength design was to meet requirement of strength margin by structural design.
(2) Cushion material selection. Generally, the material like aluminum honeycomb was preferred because of the stably compression process [8]. In some special cases, stretchable wire made of super plastic material like high manganese steel is alternative.
(3) Cushion force and stroke design. A combination of two different levels of cushion force was generally preferred as shown in Fig. 11.12. The aluminum honeycomb with lower cushion performance served as first level of cushion force F1 to absorb most of impact energy under ideal conditions that all sets of landing gear simultaneously contact landing surface. At the same time, the height of center of mass of the lunar lander can rapidly decrease to improve landing stability and reduce impact response under ideal condition. The aluminum honeycomb with higher cushion performance served as second level of cushion force F2 to absorb more energy with higher impact response under critical conditions. The area under the curve of cushion force versus stroke should meet Eq. (11.9) and the maximum cushion force shall meet Eq. (11.12).
(4) Structural design. The primary strut was wan a piston structure. It consisted of outer tube, inner tube, cushion material, guide rings, universal joint and ball joint, and etc. as shown in Fig. 11.13. The outer tube and the inner tube were both made of same type of aluminum alloy, while the universal joint and ball joint were made of titanium alloy.
(5) Limit design. Pins in the ball joint and between the outer tube and the inner tube could prevent the footpad and tubes from rotation, so as to keep the footpad at expected position and avoid to contact or impact with launch fairing. Due to vibration load during launch, the aluminum honeycomb

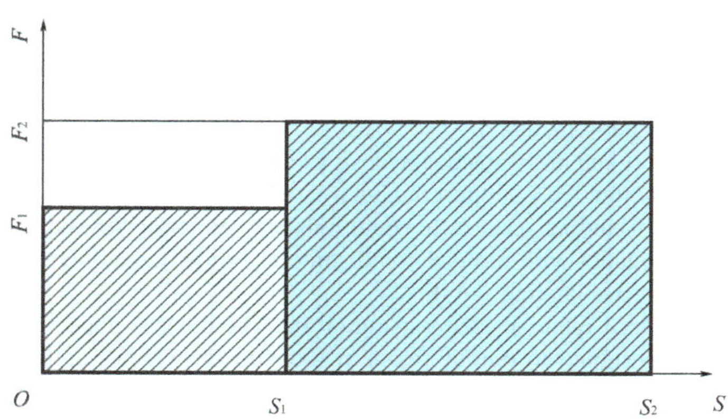

Fig. 11.12 Cushion force and stroke of primary strut

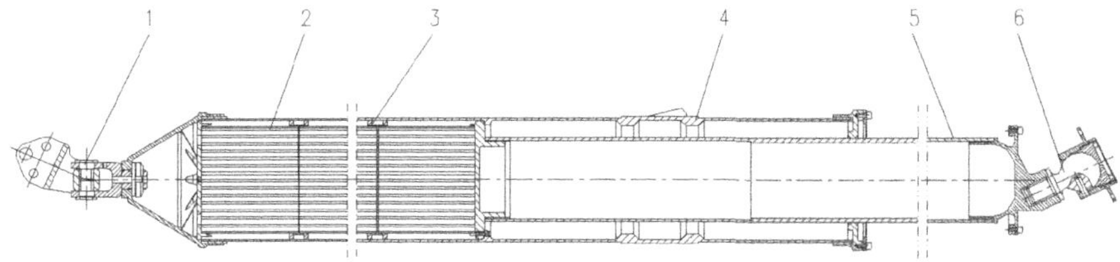

Fig. 11.13 Structure of primary strut. **1**-Universal joint; **2**-Aluminum honeycomb; **3**-Guide ring; **4**-Outer tube; **5**-Inner tube; **6**-ball joint

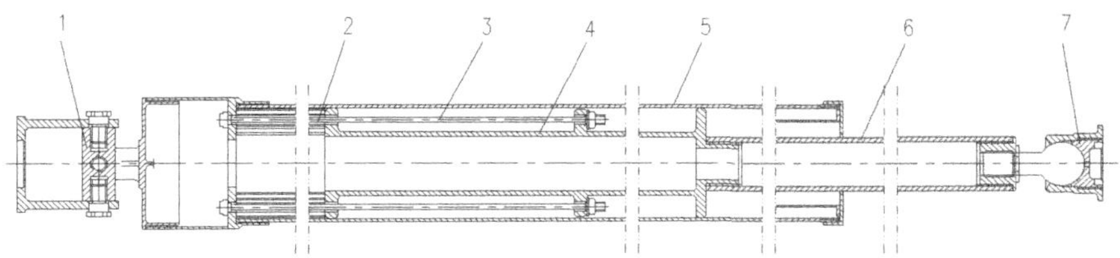

Fig. 11.14 Structure of secondary strut. **1**-Universal joint; **2**-Aluminum honeycomb; **3**-Strethable rod; **4**-Middle tube; **5**-Outer tube; **6**-Inner tube; **7**-ball joint

in the primary strut might be compressed with plastic deformation in a few millimeters, which might cause gap between inner and outer tubes and increase vibration magnitude of the footpad. Therefore, elastic cell should be attached between inner and outer tubes to compensate axial deformation.

3) Design of secondary strut

(1) Keys. It included cushion design and strength design. The cushion design included cushion material selection, determination of cushion force and stroke. The strength design was to meet requirement of strength margin by structural design.
(2) Cushion material selection. Generally, the material like aluminum honeycomb was preferred because of the stably compression process. In some special cases, stretchable wire made of super plastic material like high manganese steel was alternative.
(3) Design of cushion force and stroke. The energy-absorption capability of compression should meet Eq. (11.10). The energy-absorption capability of tension should be 1.2–1.5 times of that of compression. The larger the angle between axis of the primary strut and axis of the lunar lander was, the larger this ratio was.
(4) Structural design. The secondary strut was also in a piston structure. It consisted of outer tube, middle tube, inner tube, aluminum honeycomb, stretchable rods, universal joint and ball joint, etc., as shown in Fig. 11.14. The outer tube, middle tube and inner tube were all made of same type

of aluminum alloy, while the universal joint and ball joint are made of titanium alloy.

4) Footpad design.

(1) Keys. It included anti-subsidence design, strength design, cushion design and design of lunar surface sensing probe. Anti-subsidence design refers to selection of a reasonable supporting area. Strength design was structural design to meet requirement of strength margin. Cushion design was to ensure that the footpad periphery has reasonable capacity of deformation.
(2) Anti-subsidence design. The base area of the footpad is determined by anticipated impact load, relation of landing surface strength versus depth of subsidence and acceptable depth of subsidence.
(3) Cushion design. Aluminum honeycomb with certain strength was selected to fill in the footpad periphery so as to provide auxiliary cushion for horizontal impact load.
(4) Design of lunar surface sensing probe. The lunar surface sensing probe was a probe or switch installed on the bottom surface of the footpad.
(5) Structural design. Generally, the footpad consisted of basin, flange and trigger as shown in Fig. 11.15. The flange and basin were connected to each other through cementing and screw joint. The basin was made of honeycomb sandwich with aluminum face-sheets. Both the outer face-sheet and inner face-sheet were made of thin plate of aluminum alloy by spinning. An embedded hat beam fits into where the basin and the flange screw joint each other. Aluminum honeycomb with high strength was filled in the central zone of hat beam, which afforded an effective support. Aluminum honeycomb with low strength was filled in the peripheral zone to realize horizontal cushion. The lateral of basin inclined outside with an advisable angle and certain height, which could help to push away lunar regolith during sliding and reduce the sliding resistant friction, so as to ensure stable landing. The flange with reinforcing ribs served for reliable connection between the ball joint of the primary strut and the footpad to ensure effective support for the lunar lander.

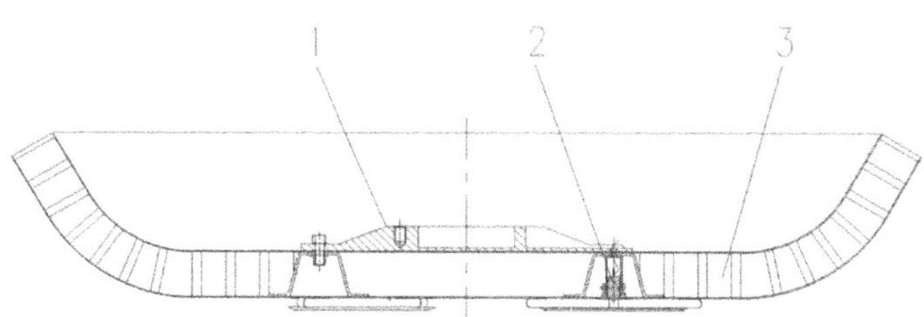

Fig. 11.15 Structure of footpad. **1**-Flange; **2**-Trigger; **3**-Basin

In order to output landing signal at touchdown, there was triggering switch in the bottom surface of the footpad. There could be several triggering switches because the touchdown attitude of the footpad was uncertain on complex landing surface. The landing signal could be output when any of these switches was on. In order to avoid wrong touchdown signals, logical judgment of landing signal was necessary.

5) Lubrication design. The joint surfaces of the landing gear were exposed to a high level of vacuum for the lunar lander on orbit. The vacuum might induce phenomenon of cold welding on joint surfaces, which might cause friction increment or unstable mechanical movement, even failure of the mechanism movement. To reduce probability of above malfunction, proper lubrication on the joint surfaces of landing gear mechanism was necessary.

After all factors were considered, such as environment experienced by the landing gear from launch to touchdown on lunar surface, load, requirement of lifetime, characteristics of mechanism movement, and compatibility of materials and the lubricant, generally solid lubricant MoS_2 was sprayed or sputtered to joint surfaces of landing gear mechanism.

6) Thermal design. The temperature of all parts of the landing gear mechanism was controlled in specific range to ensure normal operation by thermal design. The focus of thermal design included install position of pyrotechnical devices, important joints and load-bearing parts [9].

For anticipated extreme temperature under high and low temperature environment, different units of low temperature and high temperature multi-layer insulation were wrapped over the parts, which were fastened by nylon buckles, temperature resistant wires and silicon rubber. Additional insulation on the side adjacent to the engine was necessary to avoid thermal effects of the engine plume during flight.

There should be ground connection on parts with the multi-layer insulation to release static electricity.

The temperature sensors were necessary to monitor real time temperature during flight and evaluate effectiveness of thermal control.

3. Simulation of Landing Dynamics

Because of special lunar surface environment, it was very difficult to establish precise model of landing impact dynamics, which involved mechanical properties of lunar soil, landing gear mechanism and landing impact process. Analysis and verification if landing impact dynamics were a complex and systematic task, which played an important role in research and development of landing gear mechanism [10].

The common procedure of analysis and verification of landing dynamics is shown in Fig. 11.16.

(1) The landing dynamic simulation model for landing gear mechanism should be established, which was a parametric model for simulation of landing dynamics under random conditions and different lunar surface.

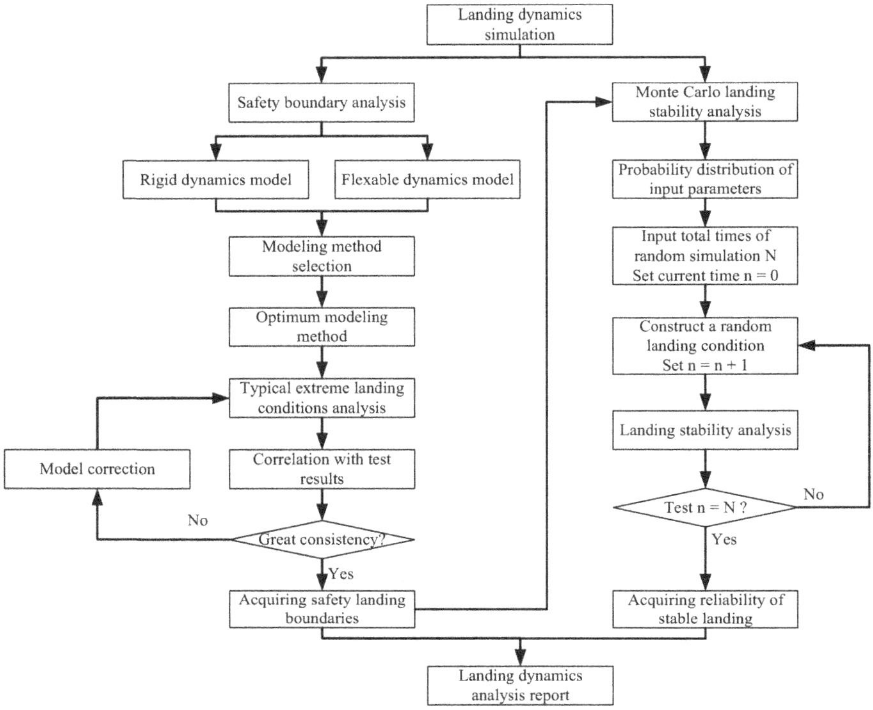

Fig. 11.16 Procedure of analysis of landing impact dynamics

(2) The landing impact simulation analysis of single landing gear should be conducted to verify and correct simulation model through comparison and analysis of simulation and test results. At the same time, an optimum modeling method could be determined based on comparison of rigid and flexible model.

(3) Typical extreme landing conditions should be determined, such as extreme turn over condition, extreme compression condition for primary struts, extreme tension and compression condition for secondary struts, and etc. The safety landing boundaries could be acquired through analysis of above extreme conditions.

(4) Typical 3-D model of lunar surface (including slope, pit/rock and its distribution, etc.) as shown in Fig. 11.17 should be created. The initial landing conditions should be created randomly to conduct landing stability analysis with Monte Carlo method.

Generally, landing impact dynamic analysis of the lunar lander included 3-D model (plane motion), six DOF (spatial motion) and single landing gear impact model. In order to improve calculation efficiency, it was assumed both the lunar lander and lunar surface were rigid. But the flexibility of the lunar lander and landing gear mechanism and the plastic deformation of lunar surface had a great effect on

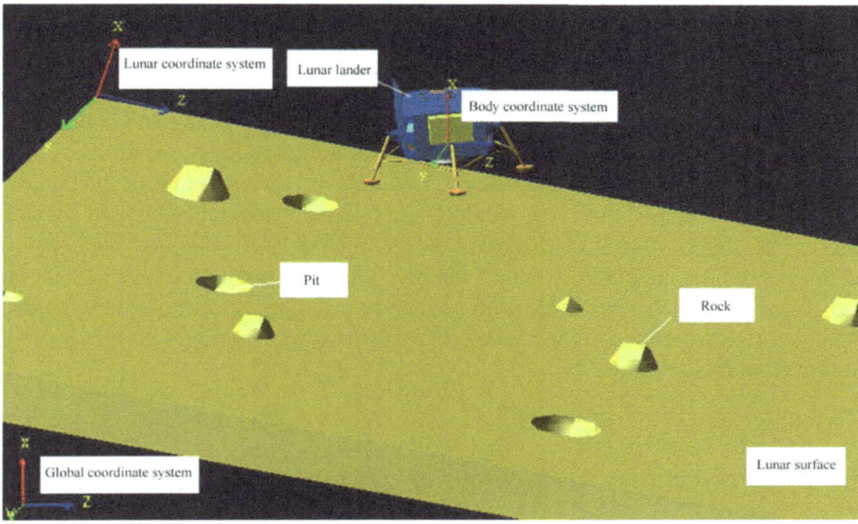

Fig. 11.17 Typical model of landing site and the lunar lander

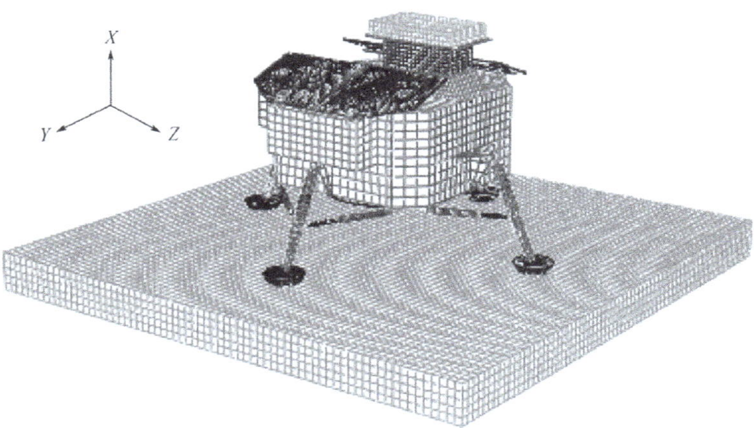

Fig. 11.18 Finite element model for analysis of landing impact dynamics

landing impact characteristics. Therefore, based on theory of transient dynamics, the finite element model of landing gear mechanism (as shown in Fig. 11.18) was built with mature commercial software like MSC and Patran Dytran to verify design of the landing gear mechanism. Different landing process was simulated according to landing condition and lunar terrain of landing site, so as to improve design and guide ground landing impact test.

11.6 Testing and Verification

The landing gear mechanism was a multi-function mechanism including hold-down, deploy and lock-up, landing impact cushion, lunar surface sensing and support for the lunar lander, which was totally different with common mechanism, especially in ground testing and verification. During research and development of the landing gear mechanism, tests should be conducted, such as static and impact test of cushion material, assembly static test, friction test, deployment test, landing cushion test and environmental adaptability test of landing gear mechanism, as well as system-level cushion performance verification and landing stability test of the lunar lander, which were very complex in test status. Only some typical test and verification are introduced in this section.

11.6.1 Performance Verification of Cushion Material

Performance verification of cushion material included static test and impact test. In preliminary design, static test and impact test were conducted to acquire cushion characteristics and bearing strength and to determine dimension or size of cushion material to meet specific requirements. In manufacture process of flight hardware, such tests should be conducted for cushion material to check the stability and consistency of cushion performance.

1. Static Test of Cushion Material

Static test of cushion material was generally conducted on universal test machine, as shown in Fig. 11.19.

For example, the aluminum honeycomb was machined to demanded profile at first; then put them into a dedicated loading tool; at last the load was applied on the aluminum honeycomb by universal test machine and the curves of load versus deformation is recorded.

A typical load versus deformation curve of aluminum honeycomb is shown in Fig. 11.4. It was shown that the load climbed rapidly to a peak at initial phase and significantly dropped down, then maintained a level stage as deformation increases, and in the end, the curve drastically climbed up which imply that the aluminum honeycomb was compressed into a solid block.

2. Impact Test of Cushion Material

Impact test of cushion material was generally conducted on dedicated impact test bed. Under specific load condition, the impact test bed provided sufficient impact velocity to ensure that the cushion material was fully stretched or compressed. The cushion impact test on the landing impact test bed was shown in Fig. 11.20. For example, at first, the aluminum honeycomb was pushed into piston cylinder; then the piston cylinder was connected with the gantry and ballast; the gantry was lift up

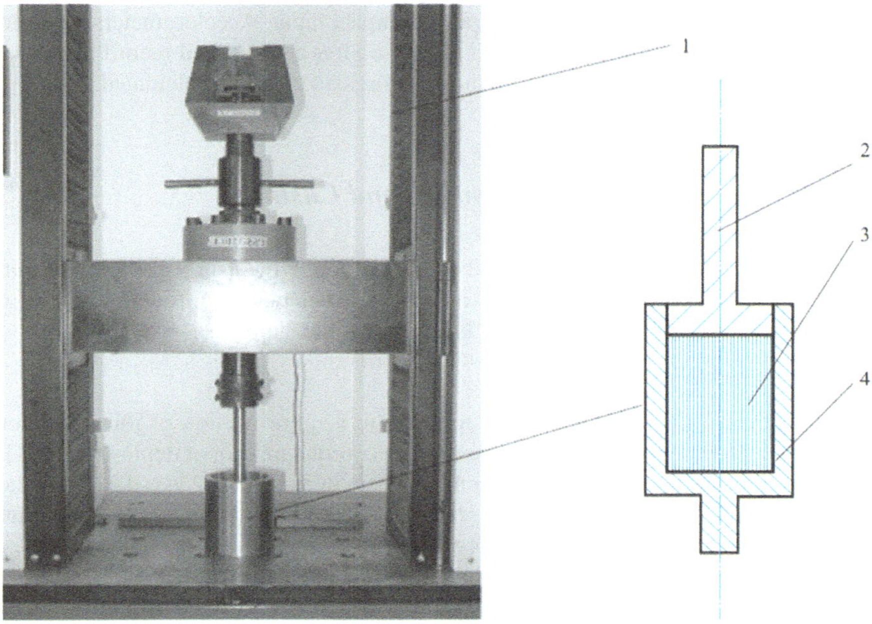

Fig. 11.19 Static test of aluminum honeycomb. **1**-Universal test machine; **2**-Piston; **3**-Aluminum honeycomb; **4**-Piston cylinder

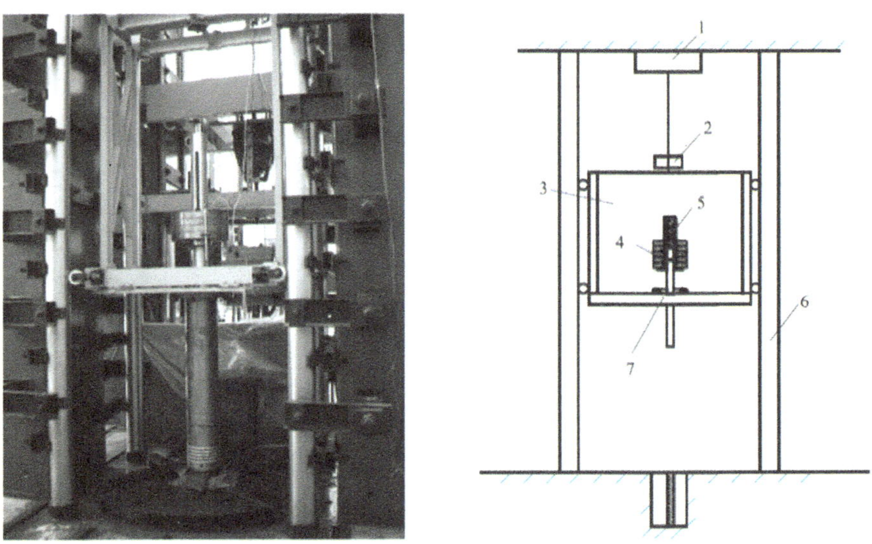

Fig. 11.20 Impact test of cushion material. **1**-Lift device; **2**-Release device; **3**-Gantry; **4**-Ballast; **5**-Cushion material; **6**-Guide rail; **7**-Bottom board

to specific height and release to get expected impact input. Accelerometers attached to the ballast could acquire the impact response after cushion and record the curve of acceleration versus time, and then the cushion load could be calculated.

11.6.2 Friction Test of Deployment and Cushion

Friction between joint surfaces would affect on deployment process and cushion performances. Therefore, friction test of joint surfaces played an important role in design and performance evaluation of landing gear mechanism.

1. Deployment Friction Test

The objective of deployment friction test was to acquire friction of joint surfaces related to deployment, so as to evaluate the static torque margin of deployment [11].

For example, when the landing gear mechanism in Sect. 11.5.2 was deployed, the deployment spring in one of the two secondary struts, i.e. the multi-function strut, pushed out the inner tube to bring forward primary strut and the other secondary strut to specific deployed position. Friction on this process included friction during stretching out of the multi-function secondary strut, and friction of all joints. It was very difficult to acquire precise friction data at assembly level because the deployment of landing gear mechanism was spatial movement. Therefore, the test should be conducted in several steps.

The principle of friction test for the multi-function secondary strut was shown in Fig. 11.21. The multi-function secondary strut was placed vertically on the universal

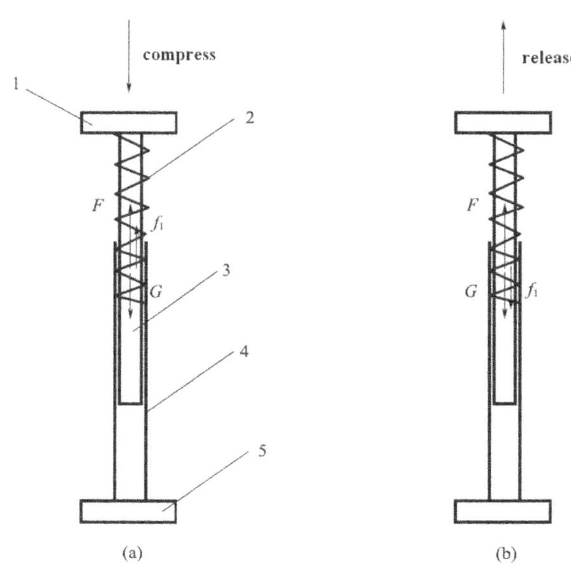

Fig. 11.21 Principle of friction test for multi-function secondary strut. **1**-Upper plate of test bed; **2**-Deployment spring; **3**-Inner tube; **4**-Middle tube; **5**-Lower plate of test bed

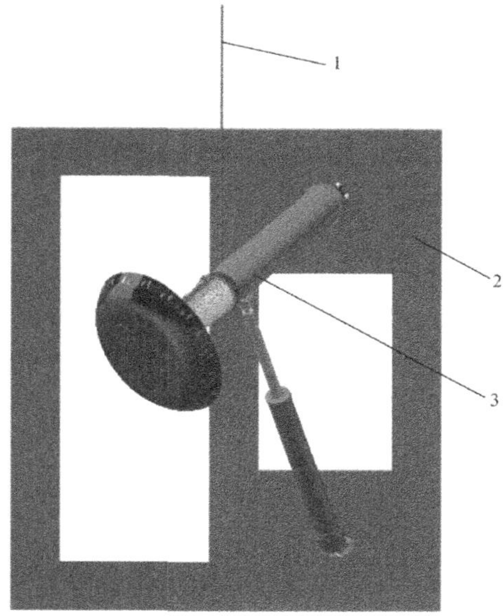

Fig. 11.22 Principle of friction test for joints. **1**-Hanging device; **2**-Fixed tool; **3**-Test article

test bed. Then the multi-function secondary strut was ascended and descended slowly and uniformly to implement compress and release of the secondary strut. Assuming that the force of deployment spring is F, friction force of deployment is f_1, the load is F_1 when the test bed was descending, and the load was F_2, when the test bed was ascending, the deployment friction of multi-function secondary strut is given by following equation.

$$f_1 = \frac{F_1 - F_2}{2} \tag{11.14}$$

The principle of friction test for joints is shown in Fig. 11.22. The primary strut with footpad was connected with the single function secondary strut, which was connected with the test tool through the universal joints of the primary strut and the single function secondary strut. The hanging device was applied to balance gravity in order to eliminate gravity influence on friction in deployment process. The dynamometer connected with above complex was applied to record friction force f_2 while the complex is pulled slowly and uniformly.

Because the joints in both ends of two secondary struts were identical, when the static torque (force) margin of deployment was analyzed, the joints friction force of both ends of the multi-function secondary strut can be conservatively taken as f_2. Generally, the static torque (force) margin of deployment should be not less than 1. The static torque (force) margin could be calculated by following equation.

$$M_s = \frac{F}{\alpha(f_1 + 2f_2)} - 1 \tag{11.15}$$

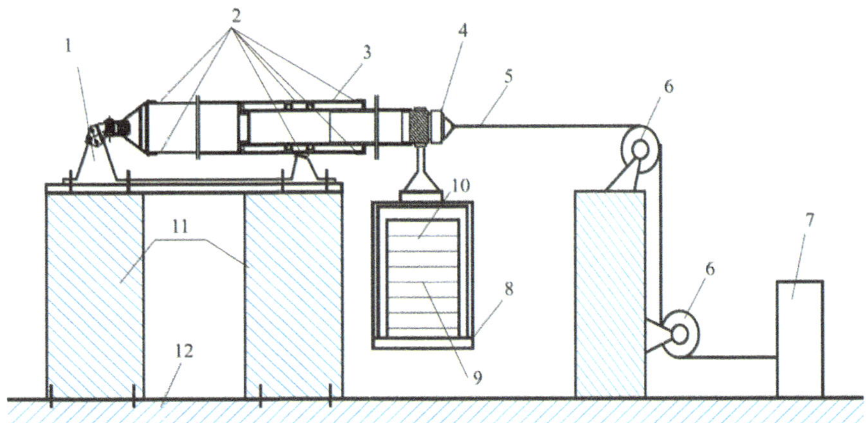

Fig. 11.23 Principle of cushion friction test. **1**-Tool; **2**-Strain gauge; **3**-Primary strut; **4**-Force sensor; **5**-Wirerope; **6**-Pulley; **7**-Loading tool; **8**-Basket; **9**-Accelerometer; **10**-Ballast; **11**-Test bed; **12**-Ground track

where: α is coefficient of correction.

2. Test of Cushion Friction

In landing cushion process, the landing gear mechanism was acted by bending load, which would result in remarkable friction force between joint surfaces involved. Therefore, test of cushion friction should be conducted in diverse bending load conditions to acquire data and its trends, so as to provide guide to design and simulation of landing gear mechanism.

For the landing gear mechanism with multiple struts, only the friction of primary strut was tested, which was acted by bending load. As shown in Fig. 11.23, the primary strut was horizontally fixed on test bed, and the tool was adjusted to ensure that both axes of inner tube and outer tube were horizontal. The upper end of inner tube was connected with the loading tool. The horizontal load was changed by altering weight of ballast on the loading tool. An accelerometer was installed on the ballast to measure acceleration along axis of primary strut. Moreover, the load tool was connected to the force sensor through the rope, and the other end of the force sensor was connected with inner tube.

The linear displacement sensor connected on one end of the loading tool was applied to measure the displacement of inner tube of primary strut. The strain gauges were on the upper, middle and lower surfaces of outer tube of primary strut is applied to measure deformation.

In the test, the force provided by the loading tool was acted on inner tube of primary strut through the rope. When the force was increasing to specific value, there would be relative motion between inner tube and outer tube. The data of all these sensors, such as force sensor, displacement sensor, accelerometers and strain gauges, would be recorded by data acquisition system for subsequent processing and analysis.

11.6.3 Performance Testing of Lunar Surface Sensing Probe

Multiple lunar sensing switches were installed on the bottom surface of the footpad to trigger touchdown signal when the lunar lander was landed on lunar surface. As any of these parallel switches was triggered, the touchdown signal is output. To ensure a reliable trigger, adequate ground test is necessary. Performance test of lunar surface sensing probe included manual triggering test, static triggering test and landing impact triggering test.

The multimeter was used in manual triggering test. The multimeter was switched to continuity test of ohm block, and two sticks were connected with output ends of switch. The lunar surface sensing switches were pressed with fingers. A buzz while pressing indicates that the signal was normal. For example, for 3 switches on one footpad, an adequate test was generally conducted as the sequence shown in Table 11.1.

The static triggering test could be conducted together with static test of the footpad. The specific monitoring device for lunar surface sensing probe was connected with output ends of the footpad to monitor and record triggering signal during the static test process, as shown in Fig. 11.24.

The landing impact triggering test was generally conducted with cushion performance test of single set of landing gear mechanism and combined cushion performance test for the lunar lander. In the process of cushion performance test, the specific monitoring device was applied for data recording.

Table 11.1 Performance test sequence of lunar surface sensing probe

Test sequence	Status of switch 1	Status of switch 2	Status of switch 3
1	disconnected	disconnected	disconnected
2	manual triggered	disconnected	disconnected
3	disconnected	manual triggered	disconnected
4	disconnected	disconnected	manual triggered
5	manual triggered	manual triggered	disconnected
6	disconnected	manual triggered	manual triggered
7	manual triggered	disconnected	manual triggered
8	manual triggered	manual triggered	manual triggered

Fig. 11.24 Static triggering test

11.6.4 Cushion Performance Test and Verification of Single Set of Landing Gear Mechanism

The cushion performance test of single set of landing gear mechanism was conducted to verify whether the cushion and support capacity meets design requirements, as well as emerge possible defects in design and manufacture process.

The test was generally conducted on specific cushion test bed. The test bed consists of bearing pillars, track, lift and release device, force measurement platform (or sandbox filled with analog lunar soil), as shown in Fig. 11.25. The impact cushion test bed could only provide vertical movement. Therefore, the landing gear was obliquely installed to simulate landing conditions with specific inclination.

The landing gear was vertically or obliquely installed on the gantry. The gantry was raised up to a specific height by the lift device. Then the gantry was released by the release device. The landing gear mechanism with the gantry was landed on landing surface (force measurement platform or the analog lunar soil) with specific velocity to simulate landing impact process. Test data was acquired by strain gauges, displacement sensors and accelerometers on the landing gear mechanism and the gantry. The data would be analyzed to evaluate cushion performance of single set of landing gear mechanism. The image of cushion performance test field for single set of landing gear mechanism is shown in Fig. 11.26.

Fig. 11.25 Principle of cushion performance test for single set of landing gear mechanism. **1**-Lift device; **2**-Track; **3**-Release device; **4**-Gantry and ballast; **5**-Test article; **6**-Force measurement platform

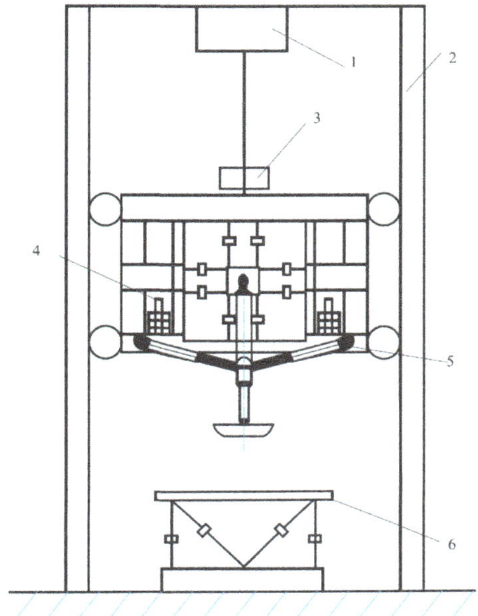

Fig. 11.26 Cushion performance test field for single set of landing gear mechanism

11.6.5 Simulation Verification of Combined Cushion Performance for Landing Gear Subsystem

The simulation test was conducted to verify combined cushion performance of landing gear subsystem by four set of landing gear mechanism on a simulated structure of the lunar lander. Because the test system was complex and the test is long term and highly costly, only a few tests for typical conditions were conducted.

Generally, the simulation test of combined cushion performance of landing gear subsystem was conducted in Earth gravity with equivalent method. The principle of test is shown in Fig. 11.27. The test system consisted of test article, lift and release device (tower crane), test equipment and landing surface of analog lunar soil with specific mechanical properties. At firstly, the assembly of simulated structure of the lunar lander and landing gear subsystem was raised up to specific height, and the assembly with the swing rods is dragged to a specific offset angle with the pulling rope. After the assembly with the swing rods is released, the assembly will be separated with the swing rods when the assembly swings to the vertical position (where the offset angle is zero). Then the assembly would be in horizontal projectile motion to simulate landing with horizontal and vertical velocity. In the landing process, high-speed video system would measure landing velocity and attitude, while sensors on simulated structure and landing gears would measure force load, impact load input on the bottom of landing gears, and impact load at mass center of simulated structure. The image of the cushion performance test field of combined landing impact is shown in Fig. 11.27.

(a) (b)

Fig. 11.27 Cushion performance test field for combined landing impact. **a** Before released; **b** After released

11.7 Summary

In this chapter, a typical landing gear mechanism is given as an example to introduce common cushion methods, design, analysis and verification of landing gear mechanism. The landing gear system was applied China's the Chang'E-3 lunar lander, and the function and performance had been comprehensively verified through flight, which meant that the key technology of landing gear mechanism for lunar landing was successfully developed and demonstrated.

Since the twenty-first century, space technology has been boomed all over the world and deep space exploration was the focus of all space countries. United State, Russia, European Space Agency, Japan and India, etc. are implementing or planning landing exploration project for the Moon, Mars and asteroids. Different technology of landing impact cushion is developed.

Landing impact cushion for deep space landing exploration is an important factor to all space countries. Along with progression of China's deep space exploration, the landing gear technology will be advanced in China. Due to diversity of exploration targets and mission objectives, the technology is diverse. For example, the landing gear system for asteroids shall be light miniaturization, while that for manned lunar or Mars landing shall be in large size, capable of high load and high stability for landing. Meanwhile, along with continuous progression of robotic technology, intelligent landing gear mechanism integrated with cushion, walking and exploration will be the new interest.

References

1. Sandy CR (1997) Development of the Mars Pathfinder inflatable airbag subsystem: collection of technical papers. AIAA, Reston VA
2. Waye DE, Kenneth J, Rivelini TP (1995) Mars pathfinder airbag impact attenuation system: aerodynamics decelerator systems technology conference (13th). AIAA, Washington, DC
3. Rogers WF (1972) Apollo experience report: lunar landing gear subsystem. NASA, Wahsington, DC
4. Kraemer R (2000) Beyond the moon. Smithsonian Institution Press, Washington
5. Wenbin XU, Peiqi GE et al (1996) Calculation for expanding deformation force of a thin-walled tube. Forging Stamping Technol 21(3):19–21
6. Sperling FB (1970) The surveyor shock absorber. Aerosp Mech Ser 1:211–217
7. Zeng F, Yang J, Jianfeng MAN et al (2011) Study on design method of landing gear for lunar lander. Spacecraft Eng 20(2):46–51
8. Yang J, Man J, Zeng F et al (2012) An aluminum honeycomb deformation cushion for lander. ZL201120551968.5
9. Yang J, Zeng F, Man J et al (2014) Design and verification of the landing impact attenuation system for the Chang'E-3 lunar lander. Sci Sin Tech 44(5):440–449
10. Wang C, Guo Y (2009) Dynamic analysis of lunar lander. J Beijing Univ Aeronaut Astronaut 25(2):183–187
11. Yu D, Yang J et al (2011) The spacecraft mechanism technology. China Science and Technology Press, Beijing

Chapter 12
Integration and Assembly Technology of Lunar Lander

12.1 Introduction

Spacecraft system integration and assembly is the whole process from the beginning of integration and assembly of subsystem to the delivery of qualified spacecraft for launch, which is generally called the final assembly of spacecraft. Spacecraft system integration and assembly shall meet the requirements of system design, final assembly design, functions and performance of subsystem design. It is an important phase of spacecraft development, which is also the final step to ensure system performance of spacecraft [1]. Spacecraft system integration and assembly involves multiple disciplines and systems. The major jobs include instrument assembly, cable assembly, pipeline assembly, and assembly of thermal control components. Generally, the installation and transportation of spacecraft are also included in spacecraft system integration and assembly [2].

The Chang'E-3 lunar lander was the first soft lander on celestial body by China. The Chang'E-3 lunar lander was a lander for lunar soft-landing mission with the features of compact product layout and cross-support module structure with external equipment modules. The characteristics of system integration and assembly of the Chang'E-3 included complex assembly of propulsion subsystem, high accuracy of multi-sensor assembly, high assembly requirements for movable parts, high risk of assembly of radioisotope device, and high requirements for assembly reliability. All above difficulties should be fully considered, analyzed and solved during integration and assembly design.

In this chapter, development status of system integration and assembly technology is introduced. As an example, the technical features of integration and assembly of the Chang'E-3 lunar lander is described, including the design of system integrated assembly of the lunar lander, typical assembly technology and application, such as the assembly of propulsion subsystem, instrument, cable and thermal-control components, and the transportation.

D.-Y. Yu et al., *Technology of Lunar Soft Lander*, Space Technology Library 38,
https://doi.org/10.1007/978-981-15-6580-9_12

12.2 Development Status

Along with continuous development of modern advanced manufacture technology, there was rapid development of spacecraft system assembly. The development level of spacecraft system integration and assembly technology was reflected in advanced design mode of concurrent design, advanced automated assembly and digital assembly technology, which was developed all over the world.

1. Concurrent Design Technology

Concurrent design technology was an important application of concurrent engineering in spacecraft system assembly. Concurrent design was a mode of parallel and integrated design of product and relative processes (including manufacture and integration). In traditional serial development process, problems or deficiencies in the design were found in processing, integrated and assembly, then the design and processing was modified to reintegrate and assemble. In contrast, the design, process and manufacture are combined together in concurrent design, and the computer and Intranet/Internet was applied to operate in parallel so as to greatly reduce the manufacture cycle. In the design phase of spacecraft system integration and assembly, the assembly process review and design were conducted in parallel, and the problems in design and process were fed back to design department in time, which could effectively avoid the flow of design defects or assembly process problems into implementation of spacecraft system integration and assembly. The concurrent engineering design flow was compared with the traditional serial design flow, as shown in Fig. 12.1, where the significant change was the transformation of the original iterative design-coordination mode into the mode of coordination in design, which could simplify the intermediate step so as to improve the efficiency of system integration and assembly [3].

Both research and development of Boeing 777 aircraft and TEAM-X/TEAM-I in NASA JPL were successful examples of concurrent engineering implementation in aerospace [4]. According to reports of NASA JPL and ESA ESTEC, because of application of concurrent engineering technology, the efficiency of spacecraft design had been improved by 4–10 times. JPL's working efficiency had increased from the original one to four spacecraft design to 40–50 spacecraft design per year [5]. ESTEC completed its tasks scheduled for 6 months in only 6 weeks [6]. In Jan. 1992, the United States Defense Department's Missile Defense Agency and NASA began to discuss plans of Clementine lunar mission. On Jan. 25 1994, the Clementine lunar probe was launched. It only took 24 months. The application of parallel technology greatly improved the efficiency of system integration and assembly.

2. Automated Assembly Technology

Along with continuous development of new spacecraft, especially spacecraft with the large-scale and high-accuracy instrument or mechanism such as large deployable antennas and space robotic arm, the automated assembly technology was an important solution to improve the assembly accuracy and ensure the assembly safety of instrument.

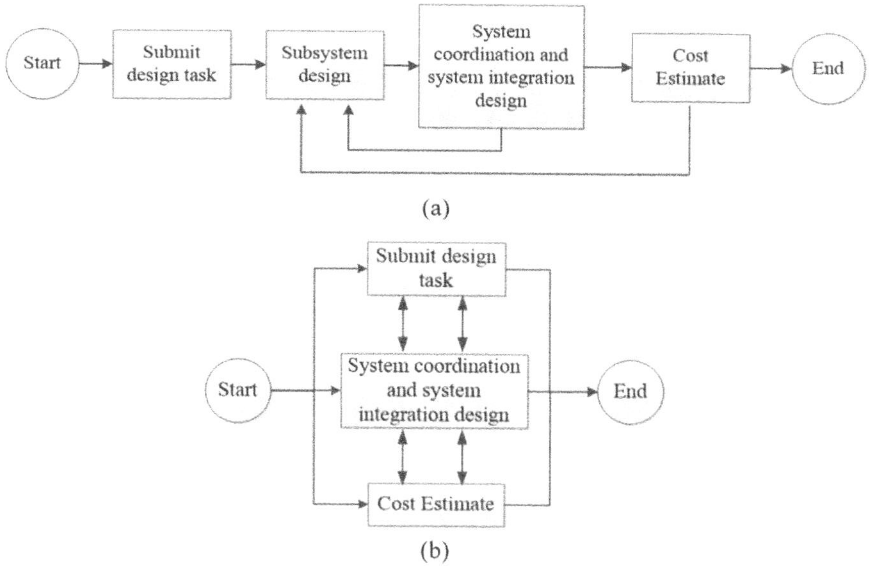

Fig. 12.1 Comparison of serial design flow and concurrent design flow (**a**) Serial design flow; (**b**) Concurrent design flow

Fig. 12.2 Supply module

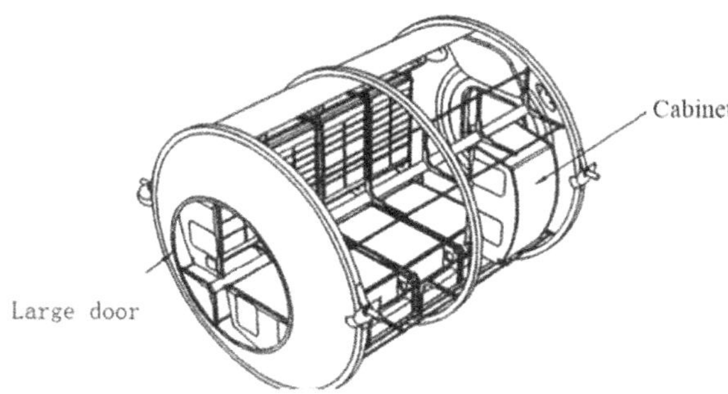

In the assembly process of spacecraft, generally the robotic arm was applied in the assembly of large parts. For example, an 816.5 kg cabinet should be assembled in space station assembly. The cabinet should be precisely aligned with the connection position of the module after it passed through the aisle, as shown in Fig. 12.2. The robotic arm could pick up the cabinet from the transfer truck and insert it in less than one hour, so the assembly time of the cabinet could be greatly saved. Nevertheless, saving time was not the main purpose of the robotic arm. It was a priority to reduce the operational risk of flight hardware and personnel in the assembly process. In the early time, the cabinet should be push onto a floor equipped with rollers by technicians. The safety of operation depended on the strength and coordination of

Fig. 12.3 Robot assembly system for assembly in compact space

technicians. However, the introduction of automated robotic arm greatly reduced the human-induced operational risks. [7]

In China, the spacecraft automated assembly technology had been developed in recent years. The automated assembly process equipment such as centroid automation adjustment hanger, Stewart posture adjustment platform and robot assembly system, could basically meet the requirements of spacecraft research and development. The application of robot assembly system for assembly in compact space of spacecraft is shown in Fig. 12.3. In system integrated assembly of the Chinese lunar lander, flexible and automated technological equipment such as electric push-rod dual-axis turntable was applied to accommodate high-accuracy assembly.

3. Digital Assembly Technology

A great many research had been conducted in digital assembly by major aerospace companies, including virtual assembly modeling, virtual assembly process planning, virtual assembly system planning, assembly ergonomic analysis, assembly process simulation and other virtual assembly technology. A series of prototype system were established [8]. The three-dimensional stereoscopic display simulation assembly process of Automatic Aircraft Fuselage Assembly Procedure (AAAP) developed by Northrop Corporation of the United States was applied to an improved version of United States Air Force F-18 fighter jet. In the design of Boeing 777 aircraft, application of digital prototype technology based on virtual assembly reduced the cost by 94% and the design change by 93%, and the manufacture cycle was shortened by 50%. Chryssolouris, a manufacture system laboratory at University of Patras, Greece, developed a virtual assembly work unit and took the assembly of a propeller of a speedboat as an example to analyze and evaluate the factors on assembly time

Fig. 12.4 Digital aided assembly of solar panel in NASA

(such as strength of assembler, layout of work units, etc.), and a semi-empirical time model was established [9]. Antoishek et al. also developed a new semi-immersive virtual assembly system based on the Virtual Workbench and data glove technology of CTC Corporation with the WTK development tool library on SGI workstations, which enabled two operators to virtually assemble a satellite [10].

Up to date, the digital virtual assembly technology in aerospace field was developed rapidly and applied widely. [11] The digital technology was applied in solar panel assembly in NASA to assist in assembly and improved assembly accuracy, as shown in Fig. 12.4. In China, digital system integration assembly technology had been in great progress, especially in virtual assembly simulation and verification technology including AIT full flow simulation in facility-level and workshop-level, and simulation of system integration and assembly with digital product prototype based on processing approach planning, human-computer interactive simulation based on virtual reality, virtual wiring simulation of flexible cable and ergonomics analysis, as shown in Fig. 12.5. In the system integration and assembly of the Chang'E-3 lunar lander, simulation of assembly process with full digital prototype was conducted and demonstrated.

12.3 Characteristics of System Integration and Assembly

The Chang'E-3 the lunar lander was a spacecraft with complex configuration, moving parts and new developed instruments [12]. At the same time, the schedule was tight for parallel development of the lunar lander and the lunar rover, which meant new that there were challenges and difficulties in system integration and assembly of the lunar lander. In this chapter, the requirements and characteristics of system integration and

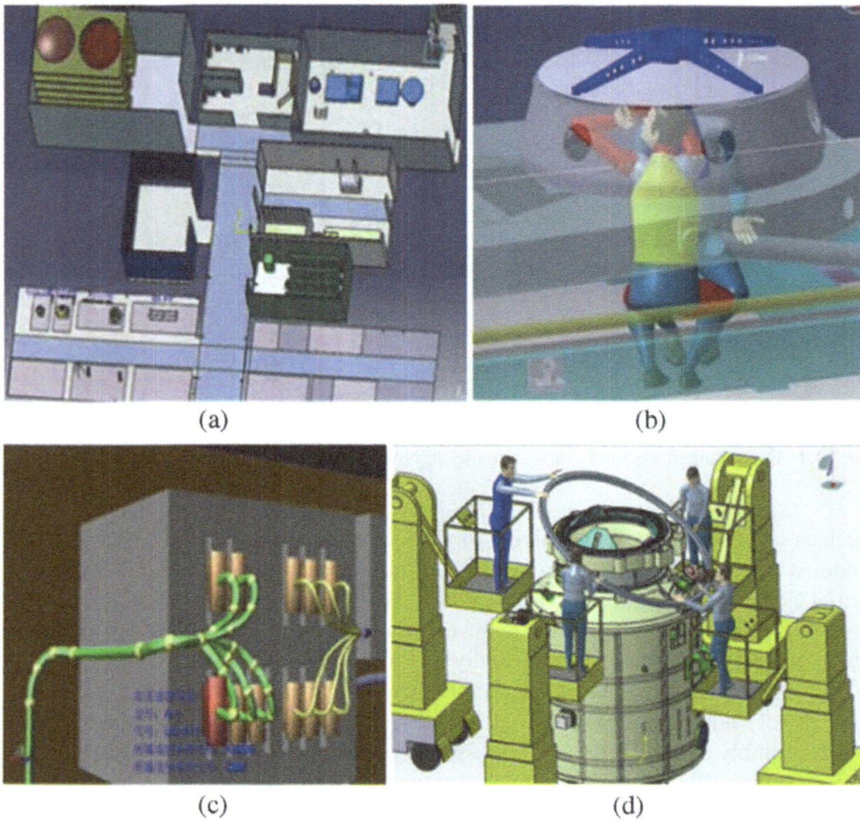

Fig. 12.5 Development trend of digital Integration and assembly technology (**a**) Digital design of facility layout; (**b**) Simulation analysis of operability (**c**) Digital wiring of Flexible cable; (**d**) Assembly simulation of full process

assembly is analyzed for the example of research and development of the Chang'E-3 lunar lander.

12.3.1 Task Analysis

There were not only common assembly tasks as typical spacecraft, but also special assembly requirements in integration and assembly of the Chang'E-3 lunar lander, such as follows:

(1) Assembly of common instruments, cables, and pipeline. As common spacecraft, common instruments, cables, pipeline and thermal control components should

be assembled for the lunar lander. During assembly process, the operating environment should be controlled and the safety precautions such as anti-static and collision prevention should be developed.

(2) Assembly of instruments with professional test requirements. There were over 20 sensors on the lunar lander, whose accuracy should be ensured in system assembly. At the same time, there were complex pipelines with thousands of solder joints and nearly 100 screw contacts, whose leakage rate requirements should be satisfied. In addition, the requirement for mass center of the lunar lander should be satisfied. Therefore, in the assembly of hardware with special requirements, it was necessary to ensure the polarity of hardware assembly, control foreign object, and conduct crucial confirmation in the key assembly steps.

(3) Assembly and test of moving parts. There were more than 10 sets of mechanisms on the lunar lander. The assembly quality of mechanism was directly involved the deployment performance of the mechanism. In assembly process of mechanisms, it was necessary to strictly control the interference inspection of the moving parts, including cables, multi-layers, ground test tools, and conduct ground verification tests.

(4) Assembly of radioisotope instruments. There were 4 sets (sets) of instrument with radioactive isotopes on the lunar lander. The assembly process flow must be carefully designed in order to ensure the operator's safety in the process.

(5) Assembly requirements for ground support equipment. Various special tests should be conducted for the Chang'E-3 lunar lander. In order to implement the tests, various types of technological equipment, such as flip trolley, flip hanger and impact hanger, should be developed. In the design of each ground support equipment for test, the assembly process should be determined, as well as protection for onboard instruments during assembly.

12.3.2 Features of Assembly Technology

The features of system integration and assembly technology for the lunar lander were as follows.

(1) There were a lot of engines/thrusters, and the piping system was complex, which made the assembly difficult. Because the flight profile of the Chang'E-3 lunar lander was complex, there were more engines or thrusters and complex pipeline on the lunar lander. Especially during the powered descent phase, the large engine should fire for a long time, which led new requirements of large engine assembly and thermal protection. There were 29 engine/thrusters including one orbit control engine and 28 attitude control thrusters on the Chang'E-3 lunar lander. Because of these engine and thrusters, there were a lot of propulsion lines and the configuration was complex, as well as large volume of propellant tanks, which caused that there were many key assembly steps. At the same time, the 7500 N throttling engine was applied for the first time in China, and new issues were faced with in assembly, measurement and precision adjustment of the engine.

(2) There were a lot of sensors with high assembly accuracy. There were launch, LTO, circumlunar orbit and powered descent in the flight profile of the Chang'E-3 lunar lander. The navigation and control of the lunar lander was crucial. In order to ensure smooth and safe soft-landing, high-accurate sensors such as star sensor, solar sensor, lunar terrain identification sensor, speedometer and ranging sensor, should be configured. The accuracy of such sensors had important implications on orbit control and obstacle avoidance for soft-landing of the lunar lander. There were 20 sensors on the Chang'E-3 lunar lander, for which there were special requirements, such as stringent assembly accuracy requirements (for example, the assembly accuracy requirement of the speedometer antenna should be less than 12 arcseconds in system assembly), as well as contamination protection for optical surfaces of sensors, collision prevention for hardware, and transportation damping.

(3) There were many mechanisms with high assembly requirements and complex verification. Besides conventional solar wing deployment mechanism as common satellite, generally there were landing legs mechanisms. At the same time, for roving on lunar surface and other scientific exploration requirements, there should be the lunar rover release and transfer mechanism, pointing mechanism for cameras and etc. The assembly quality of mechanism was directly related to the deployment and movement performance of the mechanism, which was crucial for success of the lunar lander. Meanwhile, for the newly developed mechanisms such as the landing gear mechanism and the lunar rover transfer mechanism, it was necessary to conduct verification and test during AIT process, which led new challenges for system integration and assembly.

(4) The high assembly risk of radioactive isotope equipment should be simulated and verified. In order to ensure that the lunar lander could survive in −180°C of lunar night, generally there should be multiple radioisotope equipment on the lunar lander, including radioisotope heat unit for key instruments during lunar night and gamma altimeter. There were four instruments with radioactive isotopes, which were installed on the top plate, bottom plate, and lateral plates of the Chang'E-3 lunar lander. These instruments with radioactive isotopes should be installed or assembled at the launch tower. It was an inevitable issue to determine the assembly time, assembly process, assembly reliability of these instruments and personnel operation safety during assembly.

(5) The reliability requirement for assembly process was higher because the operation environment on lunar surface was harsh. Compared with general satellite, the lunar lander should experience not only dynamic environment during launch and space environment during flight, but also landing impact environment, high temperature during operation in lunar day and extremely low temperature during lunar night. Therefore, during assembly process, the adaptability and reliability of the assembly process under special environment for the lunar lander should be considered, such as the elongation and shrinkage of the cable in high and extremely low temperature.

12.4 Design of System Integration and Assembly

Generally, in the design of spacecraft system integration and assembly, there are the design of systematic assembly process, systematic planning of technological equipment, process test planning and technical process flow. Because of system complexity of the lunar lander and specialty of system integration and assembly, in the design of system integration and assembly of the lunar lander, there should also be the design of digital assembly and automated assembly. In this chapter, it is described with the practice of system integration and assembly of the Chang'E-3 lunar lander.

12.4.1 Design of Systematic Assembly Process

The systematic assembly process is the top-level document for spacecraft system integration and assembly, whose objective is to ensure smooth manufacture and development. It is based on the design requirements, combined with actual manufacture capacity and constraints. Then the process inheritance, assembly difficulties and risks of the system integration and assembly process are analyzed, and the manufacture model is formulated, so as to establish support conditions in advance. Generally, the major contents of systematic assembly process include the follows.

(1) Analysis of product structure, performance and process characteristics.
(2) Process inheritance analysis of system integration and assembly.
(3) Analysis of difficulties in process, process research and support condition.
(4) Analysis of major examination, professional test and its implementation plans.
(5) Analysis of application of new technology, new equipment, and its implementation approaches.
(6) Identification of technical risk and analysis of reliability and safety.
(7) Process quality assurance measures of product, including process examination, technical status control, process review, tabular documentation, tools management, process document management and documentation requirements.
(8) Analysis of support condition, including Mechanical Ground Support Equipment (MGSE), resource planning and supply readiness of system assembly.
(9) Coordination and cooperation with partners.

The Chang'E-3 lunar lander was a complex spacecraft and it was necessary to design the system integration and assembly in the early stage of phase B (preliminary design and technology completion). In the design of system integration and assembly, it should be analyzed as follows.

(1) Analysis of the impact of the lunar lander configuration and scientific payloads on process of assembly and manufacture. The structure of the Chang'E-3 lunar lander was a cross-support structure, and more than 80% of onboard instruments were newly developed. It should be analyzed including how to assembly and conduct electrical

test, when to conduct special tests and what was the constraints of special tests, whether available AIT facilities and resources could satisfy requirements of system integration and assembly of the lunar lander.

(2) Based on technical requirements of the Chang'E-3 project, the compatibility of available technological approaches with process methods was analyzed. In the design of system integration and assembly, the process technology research should be analyzed, as well as necessary special process parameters and process methods, for the department of process technology to verify and qualify the process technology in advance.

(3) Based on technical requirements of the Chang'E-3 project, the planning of manufacture mode of system integration and assembly was analyzed. For the requirements for development of the lunar lander, several prototypes of the lunar lander were planned to be developed in parallel in phase B. The requirements of facility resource, process equipment resource and human resource were analyzed to help the management departments to organize manufacture and support conditions in advance;

(4) The development of technological equipment for system integration and assembly should be analyzed. According to manufacture management mode, a preliminary analysis of technological equipment development matrix for system integration and assembly was conducted, including newly developed equipment, old equipment to be used, borrowed, or modified. So, the manufacture management department could understand in advance the scale of manufacture of the technological equipment and the cost, so as to coordinate and organize the corresponding resources.

(5) Quality assurance measures and requirements for technological equipment development of system integration and assembly should be analyzed. According to the technical characteristics and manufacture management of the lunar lander, the quality assurance, process review plan and quality control forms should be established in advance, so that the quality assurance department could plan and prepare quality assurance and control in advance.

(6) Identification and requirements of risk in manufacture process should be analyzed. The safety control requirements for special transport, lifting, transportation and radioisotope equipment of the lunar lander were proposed. It could help the safety management department to conduct manufacture safety assurance and preparation in advance.

12.4.2 Planning of Technological Equipment

The planning of technological equipment was to analyze and evaluate the preliminary support conditions of technological equipment in design of system integration and assembly. Based on the system integration technology and technical procedure of the lunar lander developed in serial and parallel, requirements for technological equipment should be analyzed and evaluated, so as to ensure that the development process of the lunar lander would not be affected by lack of technological equipment. The technological equipment to be planned included hanger, parking trolley, transfer container, product turnover equipment, special test equipment, and ground special test equipment.

In the phase B (preliminary design and technology completion) of the Chang'E-3 lunar lander development, there were electrical model, structural model, thermal model and model for landing validation of the lunar lander developed, while electrical model, structural model and thermal model were developed in parallel. Therefore, when the technological equipment was planned, it should be planned according to the technical process flow. The types and quantities of technological equipment should be determined as well as the usage status.

12.4.3 Planning of Process Test

The planning of process tests was for the specific planning and analysis of the process verification, process test and process research involved in the design of systematic assembly process. The objective was to analyze the completeness of process tests, the approach of process tests, the support of process tests and the relevance with process flow.

Generally, there were simulated assembly test, digital simulation test and process parameters verification tests in assembly process tests. The simulated assembly test was conducted when the product constraints during assembly process could not be clearly defined, the assembly process flow was not determined, and it was difficult to conduct man-machine performance evaluation in virtual environment. In the simulated assembly test, the wood product model would be manufactured according to actual product conditions and constraints, so as to build a real assembly environment. Then the corresponding assembly operation was carried out to find the defects of assembly process methods and assembly tools to establish feasible assembly plan and process flow. The advantages of simulated assembly test included full process validation, good for practice and process training in advance. Nevertheless, the disadvantages included long term of test preparation and difficulties in constraint change. In the assembly digital simulation test, the three-dimensional model of product could be applied to virtually evaluate operation accessibility, product interface compatibility and human-machine ergonomics in virtual reality environment. The characteristics of assembly digital simulation test were convenient to verify and change constraint conditions. The process parameters verification test was to verify the specific assembly process parameters. For example, the tests for the process of heater stick on large curve surface and the process of thickness control of thermal grease coating on instrument were conducted for the Chang'E-3 lunar lander.

12.4.4 Design of Technical Process Flow

Technical process flow is the baseline of system integration and assembly. The efficiency and quality of manufacture management is directly decided by the design of technical process flow [13]. Because in the development of the Chang'E-3 lunar lander, there were great technical challenge, parallel development of multiple models, a large number of technological equipment and a lot of special validation tests, it was very important to design and optimize the technological process flow of system integration and assembly.

The technological process flow is the program for all work assigned in time during the process of system integration and assembly. It is generally described in a flowchart diagram and detail description in a table, including classification of spacecraft development phases, the tasks and work in each phase, as well as the sequence and relation between them [14]. The technical process flow of the system integration and assembly of the Chang'E-3 lunar lander is shown in Fig. 12.6. In the technical process flow, there should be detail description for each block, including support conditions, details of work, cautions, involved partners and completion time.

In the design of technical process flow of the lunar lander, attention should be paid to follows according to characteristics of spacecraft development.

(1) Stabilization and dynamic adjustment of the baseline of technical process flow. The technical process flow was the control baseline for system integration and assembly of the lunar lander. The stability of the technical process flow was helpful to smooth management of the development of the lunar lander. However, because of the complexity of the lunar lander, it was inevitable to adjust dynamically the technical process flow. Therefore, it was necessary to pay attention in time to the impact of dynamic adjustment of the technical process flow on technological equipment resources, facility resources and hardware status, so as to avoid resource conflicts through process optimization and separation between mainstream and auxiliary, as well as control the hardware implementation status effectively through control tables, key inspection points and key process controls.

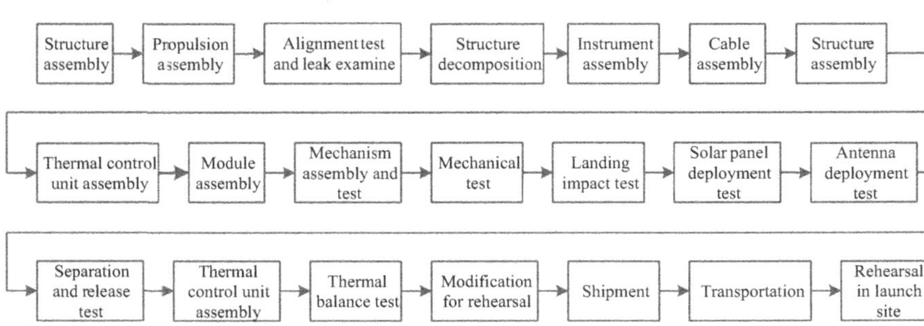

Fig. 12.6 Technical process flow of system integration and assembly of the Chang'E-3 lunar lander in phase B

(2) Coordination of special tests and routine assembly. Because there were new status and great challenges in the special test of the lunar lander, in the design of technical process flow, the special tests should be treated as the key steps in the process design, and the corresponding technical constraints, support conditions and risk plans should be refined and quantified.

(3) Compatibility of technological guidance documents and technical process flow. The technological guidance document was the criterion for assembly and manufacture. Therefore, according to the characteristics of technical process flow for the lunar lander, it was necessary to classify and manage the technological guidance documents.

12.4.5 Digital Assembly Simulation

AS an important part of the design of spacecraft assembly process under by digital simulation, assembly simulation is the key of digital design in parallel. With digital assembly simulation technology, the potential assembly defects in design can be eliminated, and assemblability and maintainability of spacecraft configuration can be evaluated, so as to identify the problems in the assembly process in advance and significantly improve the design level of digital assembly process. Along with application of digital technology, assembly simulation technology has become an important method to optimize process design, shorten development cycle and reduce assembly manufacture costs. The digital assembly simulation technology can be classified into three categories according to different functions and purposes: assembly simulation centered on technological evaluation of product design, assembly simulation centered on process planning and assembly simulation centered on ergonomics analysis.

1. Assembly Simulation Centered on Technological Evaluation of Product Design

The assembly simulation centered on technological evaluation of product design is to intentional computer-aided design technology that analyzes assembly relationships of computer data models in a virtual environment in order to make better assembly-related design decisions in design process. Combined with theory and method of design-assembly (Design For Assembly, DFA), its basic task is to start from the design principle and find the optimal solution for assembly under all kinds of constraints to formulate the assembly process approach. Its purpose is to improve the assemblability. Through trial assembly of models and quantitative analysis, the structural features that are not easy to assemble or with poor assemblability in the design of parts and components can be identified and modified to ensure feasible assembly of designed products [15].

2. Assembly Simulation Centered on Process Planning

For the assembly process design of product, based on product information model and assembly resource model, computer simulation and virtual reality technology are applied to design the assembly process and plan the assembly process, so as to establish a feasible and optimized assembly process to guide practical assembly manufacture. The assembly simulation centered on process planning includes the planning of assembly sequence, the planning of assembly path and the formulation of process apprcach. In the assembly simulation centered on process planning, the focus is the fidelity of operation simulation, which is reflected in the highly accurate product model in assembly simulation, exactly the same operation process as practical manufacture, and the highly consistent tool with tool model in practical operation. Through assembly simulation, the feasibility of process design and process approach can be accurately evaluated.

During the process design of the Chang'E-3 lunar lander, in order to evaluate the assembly sequence of the tanks and the engines, the lunar rover and solar panels of the lunar lander, the digital assembly simulation technology was applied to analyze and evaluate the interference and operation accessibility of assembly process in different assembly sequence to formulate a safe and reliable assembly approach.

3. Assembly Simulation Centered on Ergonomics Analysis

For complex operations and steps in assembly, through the introduction of three-dimensional human body model, dynamic simulation of assembly, operation and maintenance is conducted in a virtual environment, and the attitude and burden of the operator under such environment are analyzed, so as to verify the visibility, accessibility and operability of the assembly operation. At the same time, in term of ergonomics, the human attitude is evaluated and improved, or propose improvements to process assembly solutions.

For the difficult implementation of key components such as large engine, tanks and high-temperature heat shield of the Chang'E-3 lunar lander, simulation analysis of the safety, force-measuring space, and operation accessibility of such component assembly operations was conducted, so as to verify the feasibility of assembly process solution [16]. It is shown in Fig. 12.7 for the assembly simulation of 7500 N engine of the Chang'E-3 lunar lander.

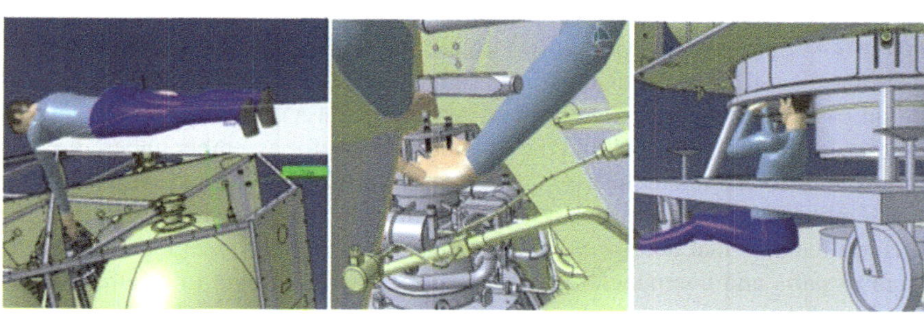

Fig. 12.7 Assembly simulation of 7500 N engine of the Chang'E-3 lunar lander

12.5 Typical Application of System Integration and Assembly

The system integration and assembly runs through the AIT process of spacecraft. The implementation sequence is based on the technical process of system integration and assembly. Generally, after the delivery of the main structure, the propulsion subsystem assembly, instrument assembly, cable assembly, thermal control unit assembly, structural assembly and mechanism assembly will be implemented in sequence. Then there will be operations related with various types of tests, such as turning and attitude adjustment, hanging and transshipment and etc. At the end, the spacecraft will be transported to the launch site for final assembly. As an example, the application of typical system integration and assembly of the Chang'E-3 lunar lander is described.

12.5.1 Propulsion Subsystem Assembly

Generally, the propulsion subsystem assembly was the first step of system integration and assembly, including assembly of propulsion pipeline, gas tanks, propellant tanks, engines and thrusters.

1. Manufacture and Assembly of Complex Pipeline

The pipeline of spacecraft is generally made of aluminum alloys, titanium alloys and stainless-steel pipe. There are two manufacture methods. One is the traditional manual method, characterized by a large burden, long cycle, low precision, and in serial installation with spacecraft assembly. It is a mature method with less cost and suitable for the development of single spacecraft pipeline manufacture. Another method is to apply CAD/CAM technology to automatically bend the pipe. The pipeline is designed in software and the data files that can be recognized by the banding machine through interface software of numerical control bending machine to implement CAD/CAM integration, which is characterized by high efficiency and high accuracy. It can be implemented before assembly and in parallel with spacecraft assembly. It is suitable for the manufacture of batches of pipeline of common spacecraft platforms. Before pipeline assembly, a series of examination and acceptance shall be conducted. The pipeline shall be fixed by pipeline support. After completion of propulsion subsystem assembly, the propulsion pipeline shall be examined by system leak test and single point leak test. The lunar lander was a single spacecraft. Therefore, the traditional manual method was applied to manufacture pipelines with less cost and mature technology. The good result was achieved in the design, manufacture and assembly of the pipeline in the Chang'E-3 lunar lander.

2. Assembly of Parallel Tanks on Cross Support Structure

The assembly method of tanks in the Chang'E-3 lunar lander was different from other Chinese spacecraft. The traditional tank is assembled on the spacecraft bearing cylinder through the lug. The structure of the Chang'E-3 lunar lander was a cross-support structure. The bottom of the tanks was assembled on the interface ring and the upper part was connected with the structure through the rod. The virtual assembly technology was applied in the tank assembly to simulate and analyze the process of tank assembly. The assembly sequence of the tank and the 7500 N engine was confirmed, and the risk in tank assembly process was identified. By planning and optimizing the lifting path of the tank, the implementation risk was effectively eliminated and the four tanks were assembled symmetrically in success.

3. Assembly of Gas Tank Fixed in Clamp Band

Generally, the gas tank of common spacecraft is fixed by rods on base, while the gas tanks of the Chang'E-3 lunar lander were fixed by the clamp band. During the process of filling and deflation in gas tanks, the outer diameter of the tanks would be changed with the inner pressure change. Therefore, the tanks should not be completely fixed in assembly. In the tank assembly, the test of filling and deflation test was conducted to verify the torque for the fixed fastener. At the same time, the film of special material was put between the tank and the clamp band to meet the requirements of the expansion and contraction of the tank.

4. Assembly of 7500 N Engine

Because the large engine was in the mid-bottom of the Chang'E-3 lunar lander, the upper part was the main structure. The assembly method was not the same as the tank assembly, which could not be assembled by ordinary lifting methods. At the same time, there was alignment requirement for the 7500 N engine assembly. At early stage of development, a new type of assembly process method was designed for the engine assembly of the lunar lander. A reversed docking process method was proposed, in which the large engine was aligned with the lunar lander by multiple lifting processes. The 5-DOF park attitude adjustment device was designed to connect and assemble the large engine with the lunar lander. At the same time, the alignment accuracy of the engine was examined during assembly process to avoid deformation caused by uneven force during the assembly process of the top and bottom fasteners on the engine. At last, the high-precision assembly and alignment of the large engine was achieved.

12.5.2 *Instrument Assembly*

The instrument assembly is the typical one in spacecraft system integration and assembly. The assembly of the instrument was to connect the instrument with the structure, measure the torque and prevent loosening. At the same time, according to the performance and characteristics of the instrument, necessary process design should be conducted to meet the different assembly requirements of different instruments.

1. Assembly of Instrument with Alignment Requirement

The instruments with alignment requirements included attitude control subsystem instruments, scientific payload instruments, propulsion subsystem instrument and separation instruments. The assembly process of instrument with alignment requirement was a process of installation and adjustment. During assembly, adjustment should be made while the alignment is measured. The adjustment gaskets were adjusted the alignment accuracy requirement was satisfied. In order to ensure the alignment accuracy of instrument assembly, it was also possible to apply a positioning pin which was replaced with a screw before launch. In the Chang'E-3 lunar lander, there were instruments with alignment requirement including gyro assembly, star sensor of attitude control subsystem, engine and thrusters of propulsion subsystem and EUV camera of scientific payload subsystems, whose alignment requirements were higher. Two methods including adjustment gasket and positioning pin were applied to meet the fine-tuning requirements and high-precision assembly was achieved.

2. Assembly of Instrument with Sealing Requirement

In spacecraft, there is sealed design in many positions, which is generally applied detachable sealed connection with sealing material including rubber sealing gaskets and sometimes other sealing materials, such as fluorine plastic, metal and material in special environment (such as high temperature and chemical corrosion environmental material). For seals with elastomeric seals, a groove shall be machined at the mounting flange or fitting to install the seal. In the Chang'E-3 lunar lander, the major seals were the sealing of propulsion subsystem pipeline. The sealing methods of welding and screw ball joint connection were applied with good performance.

3. Alternative Assembly

In system assembly process, sometimes the alternative of the instrument is required. For example, the structural alternative is necessary in mass property measurement or mechanical environmental tests. The thermal alternative to simulate thermal property is necessary in thermal environment tests. When the modules or projectile are connected in assembly facility, the pyrotechnic alternative is used. The key of alternative assembly is the control of technical status of alternative items. In the development in parallel of the Chang'E-3 lunar lander, the alternatives were often applied in tests. During the assembly process, the name and code number of products should be strictly controlled and the uniqueness should be checked and confirmed. Before launch, all alternatives in spacecraft should be strictly removed and confirmed.

4. Assembly of Ground Protection

The protection for the protrusions and key components on spacecraft surface is an important part of system integration and assembly process, so as to prevent damage or contamination of key components such as camera lens, plug, socket, pipe connectors, igniter of pyro, nozzle of engine and thruster with ground protection during ground assembly operations, transportation, test and storage. Ground protection shall be designed in red color and hollow for warning and visual purposes. The protrusions and key components on the surface of the Chang'E-3 lunar lander were protected in special. The camera lens was protected with red in AIT. The protection and red alert cloth for the protrusions should be planned and designed in detail. Before launch, all ground protection should be confirmed and removed by video record and table management.

12.5.3 Cable Assembly

The major tasks of spacecraft cable assemblies are cable assembly and connection, which is the typical one in spacecraft system integration and assembly. There are generally including secondary insulation before cable installation, cable support and nylon base installation, cable assembly, transition plug bracket and transition connector installation, cable sorting/binding/fixing, as well as free-end connection, electrical connector insertion, cable binding/fixing and electrical connector anti-loose treatment after the cable network test.

The key of spacecraft cable assembly is cable installation. Cable installation is generally implemented after instrument assembly. At first, the cabin is installed in single module, and then the cable between modules is connected. There are two modes of cable installation: The first one is to install the cable in position according to drawings and documents such as cable installation drawings. The second one is to manufacture and install the cable with templates. For the second mode, a spacecraft model is design and manufactured according to actual dimensions, as well as models of instrument, cable support, and electrical connector brackets, which is installed at theoretical position of the spacecraft model. According to the technical requirements of spacecraft design, manufacture and acceptance, the preparation of wire ends, the manufacture of cables and the welding of wires, cables and electrical connectors are conducted. After the manufacture is completed, the transplant is conducted. Before transplant, all electrical connectors are covered with protection caps. When transplanted, all the cables shall be protected from damage. After transplanted, the original configuration of all the cables shall be maintained. By such method, the cable length and branches are more accurate, so as to reduce the excessive length and cable mass. In spacecraft, the cable mass is about 5 ~ 15% of the total mass. The manufacture and installation of cables with templates will increase the complexity of the development, but is of great significance in terms of reducing spacecraft mass. In the development of the Chang'E-3 lunar lander, the cable was designed in form of template cable, which saved some mass for the lunar lander.

The cable implementation should be implemented according to the technological requirements.

(1) The low frequency cable was generally installed along the structural surface. When the cable was installed, the distance between the cable and the instrument, the fixed spacing and the inner bending radius should be bound according to requirements; When the cable passed through the cavity, corner of the structure or the instrument, near the high temperature source and the place where the cable was fixed

and bound, the cable should be wrapped with the protective layer to prevent abrasion during vibration.

(2) High-frequency cables should be installed separately and not be mixed with low-frequency cables. When it was installed, shaping and trial installation in cable installation were acceptable, but damage to metal surface was not allowed; Similarly, the distance between the cable and the instrument, the fixed spacing and the inner bending radius should be controlled during installation.

(3) When the electrical connector was inserted, the consistence of the code of the plug and the socket should be carefully examined, as well as cleanness and intact condition of the pin and the socket. When the electrical connector was inserted, it should be installed in place and anti-loosening treatment should be conducted after insertion.

(4) When the cable of moving parts was fixed, it was necessary to strictly control the bending radius of the cable to ensure its activity margin. Not only the cable length requirement during movement should be satisfied, but also interference with others during movement should be avoided.

For the Chang'E-3 lunar lander, because there were lots of moving parts and the environment on orbit was harsh, the focus of cable assembly was the cable length reserved for moving parts, as well as the margin for expansion and contraction under extreme temperature change. In order to satisfy the deployment requirements of the landing legs and solar panels, the appropriate movement margin for binding and fixing should be reserved. At the same time, because of extreme hot and cold temperature in lunar day and night, not only the process should be designed for firm fix, but also the binding material for the cable should be adaptive to the extreme temperature change.

12.5.4 Structural Assembly

In the AIT process, part spacecraft structures shall be disassembled to meet the requirements of system integration and assembly. At the end, the modules are reassembled in position. In the process of spacecraft system integration and assembly, generally only module or deck is reassembled to maintain the status of main structure.

1. Module Assembly

The layout of cable among the modules and main structure of the Chang'E-3 lunar lander was different from other spacecraft. For the complex cable layout in the lunar lander structure, a special technological method with a 6-DOF module attitude-adjusting and tilting carriage was proposed for the solution to dock and disassembly of the deck with the lunar lander. The new attitude-adjusting and tilting carriage was a side-opening tilting carriage, which could implement a smooth lateral opening and closing of the module and might provide a new idea for module assembly of other spacecraft.

2. Assembly of Deck (including Transfer Mechanism)

There was a special deck with the lunar rover transfer mechanism on the Chang'E-3 lunar lander. Since the transfer mechanism was designed to ensure the deployment of the lunar rover, it was strictly forbidden to touch during the assembly process. At the same time, there were protrusions such as gas tanks around the structural deck. Therefore, it was impossible for the operators or tilting carriage to lift. Through the parallel design of the equipment and technological process, the lifting method was applied to assemble the deck with the transfer mechanism. At the same time, the hanging point and hanging mode of the module were verified in the phase of structure assembly. At the end, in the process of system integration and assembly, the assembly difficulty of the large size and large mass deck with the transfer mechanism in compact space was solved.

3. Assembly of Lunar Rover

On the top deck of the Chang'E-3 lunar lander, there was the lunar rover. After the lunar lander was landed on lunar surface, the lunar rover would be deployed by the release and hold on mechanism and the transfer mechanism. The lunar rover was connected with the lunar lander with a transition bracket. During the process of separation, the moving path of the lunar rover was provided by the guide device, lifting mechanism and the track in the release and hold on mechanism. Therefore, the assembly of the transition bracket and the release and hold on mechanism were the key of the assembly of the lunar rover. In the design of assembly process for the lunar lander, the assembly simulation analysis of ergonomics of the transition bracket based on virtual simulation technology was conducted. At last, a reasonable technological process was designed, and a multi-degree of freedom attitude adjustment platform was developed to implement the docking operation of the lunar rover with the lunar lander at certain height. In addition, special measuring tools were developed to implement fine adjustment of the guide device of release and hold on mechanism, so as to ensure high efficiency and high reliability of the lunar rover assembly.

12.5.5 Mechanism Assembly

Mechanism and structure are all mechanical in spacecraft, with many of the same or similar features. But there are significant differences between them. The function of mechanism is to produce movement. In order to ensure the function of mechanism movement, it is very important to ensure the accuracy of mechanism assembly and verify the deployment and folding of mechanism. Generally, there are solar panel mechanism and deployable antenna in conventional spacecraft. The typical mechanism in the Chang'E-3 lunar lander was the solar panel mechanism and landing gear mechanism.

1. Solar Panel Assembly

There are two types of solar arrays in the assembly of spacecraft. The body-mounted solar array is installed on the outer surface of spacecraft, while the surface may be in shape of sphere, polyhedron and cylindrical. Such assembly method is the simplest and there is no moving part, therefore it is more reliable. The deployable solar array is connected with spacecraft, but is independent of the spacecraft body. During launch, the deployable solar array is folded in the spacecraft or on lateral plate. After orbit injection, it can be deployed. The assembly of deployable solar array needs help of air-floating and suspension for simulation of zero-gravity. In case of one-dimensional deployment, generally suspension is applied, while air-floating is applied to simulate zero gravity for dual or multiple dimension deployment.

There were both body-mounted and deployable solar panels on the Chang'E-3 lunar lander. When the body-mounted solar panel was assembled, the focus was insulation. The deployable solar panels were driven by motor to implement one-dimension deployment with high assembly accuracy requirements. Special attention should be paid to assembly accuracy and movable cables in the assembly process.

(1) Accuracy of solar panel assembly. When the solar panel was assembled, the output shaft of the reducer should be aligned with the rotation axis of the hinge after the drive mechanism (motor, reducer and transmission arm) was assembled, and the error of the shaft should be less than 0.15 mm. The deployment suspension method was applied to achieve low gravity environment in assembly of solar panel. Then the dual-axis turntable was applied to adjust the attitude of the lunar lander. Real time monitoring was implemented by measurement to ensure the assembly relationship and accuracy requirements between the lunar lander and the solar panel. The assembly of solar panel for the Chang'E-3 lunar lander is shown in Fig. 12.8.

(2) Assembly of movable cables. In assembly process of the solar panel cable, it was necessary to consider the interference between the cable and the solar panel during deployment and folding, in order to avoid the compression and scraping of the

Fig. 12.8 Assembly of solar panel for the Chang'E-3 lunar lander

cable, as well as movement margin of the cable to ensure smooth deployment of the solar panel. General, verification tests on deployment and folding of the solar panel on ground is necessary, as well as the cable interference and movement margin.

1. Assembly of Landing Gear Mechanism

The landing gear mechanism was one of the most crucial equipment of the lunar lander. There were four landing legs and landing buffer on the landing leg to ensure safe landing of the lunar lander.

The assembly of the lander gear mechanism included assembly of mechanical interface and electrical interface. Four landing legs were assembly respectively. There were primary strut, multi-function secondary strut and single function secondary strut on each landing leg. The mechanical assembly of the lander gear mechanism was relatively simple, where the main strut was connected to the main structure by special bolts, and the secondary struts were connected by bolts with the bottom plate of the lunar lander. The challenge in assembly of the landing gear mechanism was the implementation of the assembly of the cable. At first, the margin of the cable should be controlled to ensure that the landing gears were not interfered during deployment and landing. Secondly, the exposed cable should be thermally protected to avoid damage of high temperature cause by engine firing. After process design, strict control of the cable margin was implemented and special stainless-steel wire was applied to ensure the fixed strength of the cable, as well as adaptability to high temperature environment caused by engine firing.

12.5.6 Assembly of Thermal Control Unit

The lunar lander would experience extremely cold lunar night at −180°C and extremely hot lunar day at 120°C. Therefore, the requirement for thermal control was higher and thermal control design was more complex for the lunar lander, which resulted in strict requirement for assembly of thermal control units. Generally, there were assembly of multi-layer heat-insulation material coating, heat pipe, electric heater and temperature sensor in assembly of thermal control units.

1. Manufacture and Assembly of Low Temperature Multilayer Insulation Components

The material of low temperature multilayer insulation is generally formed by alternately stacking of high-emissivity reflective layer and low thermal conductivity layer, which can form a very high thermal resistance to radiative heat flow. Because of its good thermal insulation performance, it is a general passive thermal control method.

In order to adapt to the harsh environment with high and low temperature on lunar surface and reduce spacecraft mass, the manufacture and assembly of low temperature multilayer insulation components should be planned, elaborately assembled and managed. There were more than ten types of low temperature multilayer insulation components on the Chang'E-3 lunar lander, including insulation on surfaces such as external plate, tank, pipe, engine and cable. In order to ensure the detailed manufacture and assembly of low temperature insulation components, the digital method was applied for the Chang'E-3 lunar lander. The cutting of the drawing template, subsidiary and sewing of multilayer insulation components were implemented by automated cutting machine, which had greatly improved the shape accuracy and efficiency of manufacture of multilayer insulation components.

2. Assembly and Arrangement of Multilayer Around Optical Sensor

Inappropriate assembly of multilayer insulation components may obstruct the field of view of optical sensor such as camera. Therefore, assembly, arrangement and fix of multilayer around optical sensor shall be strict controlled. The multilayer insulation assembly on the fill/drain valve of the Chang'E-3 lunar lander appeared in the field of view of the descent camera. Therefore, the arrangement and fix status of multilayer insulation assembly on the fill/drain valve had important influence on camera imaging. The outer contour of the multilayer insulation assembly should be as small as possible and not peel off. After repeated confirmation and process verification, the 'lining' type assembly process was applied for the multilayer insulation assembly on the fill/drain valve, which not only achieved the minimization of the multilayer insulation assembly, but also ensured the shape and performance of multi-layer insulation assembly. The assembly status of the multilayer insulation on the fill/drain valve is shown in Fig. 12.9, which optimized the dimensions of the multi-layer insulation as much as possible. A photograph taken by the descent camera during landing is shown in Fig. 12.10.

3. Manufacture and Assembly of High Temperature Multilayer

The high temperature multilayer was applied for high temperature protection of the hardware surrounding the engine or thruster during firing. The high temperature

Fig. 12.9 Assembly status of multilayer insulation on fill/drain valve

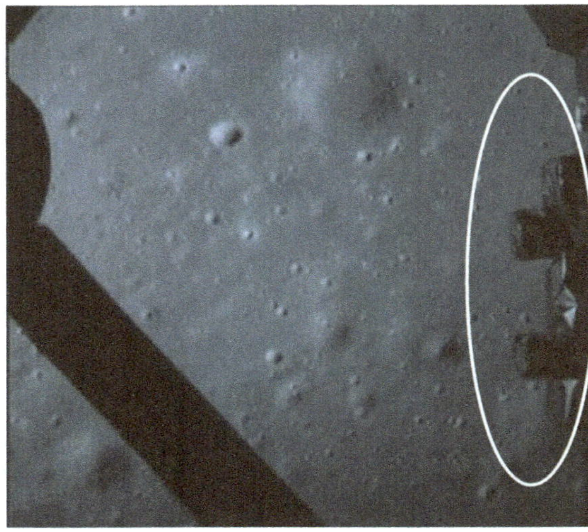

Fig. 12.10 Photograph taken by descent camera during landing

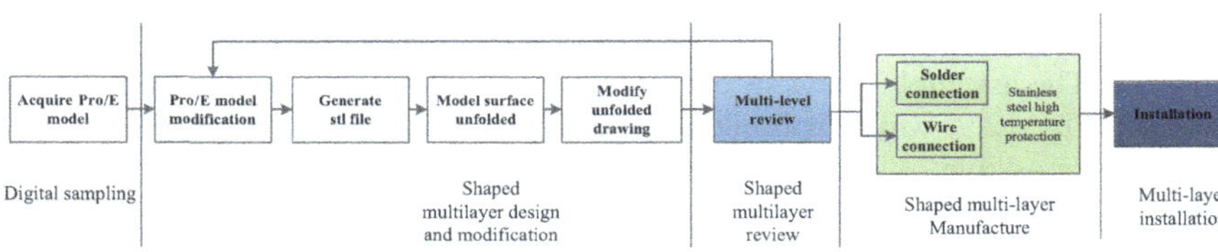

Fig. 12.11 Digital manufacture process method based on reverse engineering

multilayer should be assembled on the engine or thruster, surrounding plates and antenna, where were complex pipeline and hollow support bracket. The information of three-dimensional models of the pipeline and the bracket were incomplete and inaccurate, and the hollow bracket could not be exactly modeled. It was difficult to transfer the complicated curved surface of the assembly position to conventional two-dimensional drawing. If on-site sampling was applied, the manufacture and assembly of high temperature multilayer would not only spend long term with low efficiency, inaccurate manual sampling result, and poor consistency, but also have risk of collision during sampling and assembly near the engine or thruster. Through technological research and verification, the digital manufacture process method based on reverse engineering was applied in assembly of high temperature multilayer on the Chang'E-3 lunar lander, and good result was achieved. The technical flow of digital manufacture of high temperature multilayer is shown in Fig. 12.11. The high temperature multilayer insulation assembly is shown in Fig. 12.12.

4. Heat Pipe Assembly

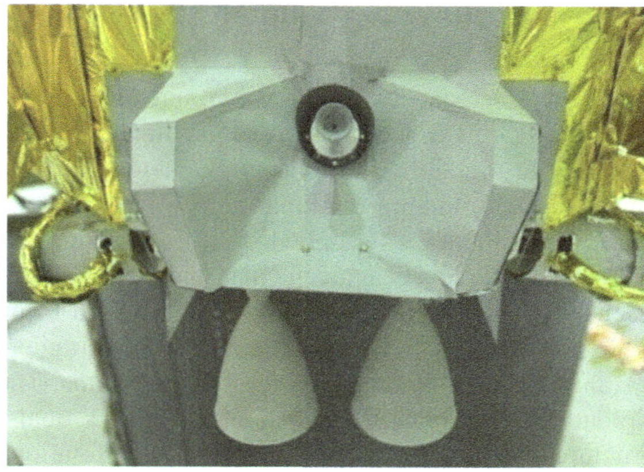

Fig. 12.12 High temperature multilayer insulation assembly on the Chang'E-3 lunar lander

The heat pipe can transfer the heat of high-heat-generating component to the non-heat-generating component, so as to make the spacecraft be isothermal. The pre-embedded and externally attached heat pipes were applied to meet requirement for temperature control. There were more than 20 heat pipes pre-embedded in the plates, which was implemented during structure assembly. There were nearly 10 outer heat pipes installed during system assembly of the Chang'E-3 lunar lander. The key of heat pipe assembly was good contact between the heat pipe and the assembly surface with heat conductive material filled to ensure good heat conductive performance. For the Chang'E-3 lunar lander, the heat pipe was long and across two plates. After assembled, the heat pipe was pressurized by special pressurizing tool to achieve good contact between the heat pipe and the assembly surface.

5. Temperature Sensor Assembly

Generally, the temperature sensor is thermistor. The types of thermistors are up to the temperature range. There were more than one hundred thermistors on the Chang'E-3 lunar lander for temperature measurement, which were classified into temperature measurement thermistors and telemetry thermistors. The temperature measurement thermistor was applied to monitor the temperature level of the lunar lander and acquire the temperature data during flight. The telemetry thermistor could not acquire temperature data of the lunar lander, but also provide temperature reference for the heater at the same time. The thermistor was fixed by stick. There were three types of requirements when it was assembled. The first one was to ensure good adhesion of thermistor and the component to be measured. The second one was to ensure the insulation performance between thermistor and the component to be measured. The last one was to measure the resistance between the leads after stick and the insulation resistance between the lead and the component to be measured, which should satisfy the requirements.

12.5.7 Turnover and Attitude Adjustment of Lunar Lander

When the solar panel or deployable antenna was assembled or deployed for the lunar lander, it was necessary to adjust the attitude of the lunar lander to match the assembly or test of the mechanism. Generally, the dual-axis turntable was applied to adjust the attitude of the lunar lander. Up to date, the screw rail structure in the dual-axis turntables was applied to implement turnover of spacecraft in China.

When the landing gear mechanism and the solar panel were tested for the Chang'E-3 lunar lander, the lunar lander should be tilted from a vertical status to a certain angle and then rotated around the axis. The attitude adjustment of past spacecraft was implemented by dual-axis turntable in China. The turntable can achieve 0–90° flip and 0–360° rotation for spacecraft. For the dual-axis turntable with screw rail structure, the turnover height is high and the rotation envelope is large, which was difficult to meet requirement for installation and adjustment of mechanism in the Chang'E-3 lunar lander.

In order to satisfy the special requirement for turnover and attitude adjustment of the Chang'E-3 lunar lander, a new type of dual-axis turnover platform based on electric push rod was applied. While the probe was turned over, the probe could be turned outside of the turntable. Then the turntable could get out the way of the landing gear mechanism, so as to lower the height of the turned probe on the turntable. In the turntable based on electric push rod, there were support platform, adjustable support, retractable electric push rod, sunk flip frame, anti-overturn device, adapter, rotation system and electronic control system. The application of the dual-axis turntable based on electric push rod in solar panel illumination test of the Chang'E-3 lunar lander is shown in Fig. 12.13.

12.5.8 Lifting of Lunar Lander

During assembly process of spacecraft, it will be assembled, tested, examined, packed and transported at different location, when the spreader is necessary equipment for spacecraft transfer. Generally, in order to facilitate the convenience and safety of the disassembly and assembly of the spreader, the lift point is at the top or the outer side of the spacecraft structure. The lifting device is generally connected to the spacecraft through pin, screw or simple adapters and fixed. The structure of the Chang'E-3 lunar lander was cross support. The main body was a decahedron. Because the lift point was on the bottom of main structure, it was difficult to lift and turn over the lunar lander in the air.

1. Lifting of the Lunar Lander

The Chang'E-3 lunar lander was vertically lifted by 4 lift points which were on the bottom of main structure and inside the structure. The interfaces of lift points were 4 through-holes uniformly distributed on a circle. The axis of through-hole was

Fig. 12.13 Application of dual-axis turntable based on electric push rod in solar panel illumination test

tangential to the circle. The position and interface of the lift point was quite different from conventional spacecraft. Therefore, conventional spreader could not satisfy the requirement. A new special lifting device should be developed, which could extent the lift point to the outside of the structure and be adjusted at multiple degrees of freedom, so as to implement the assembly of lifting device and the lift point interface and lift the lunar lander in special shape.

In the lifting of the Chang'E-3 lunar lander, a special-shaped spreader based on a lever structure was applied. The spreader consisted of four support levers, four support seats, rigid ring, pin shaft, lugs, and lift components as shown in Fig. 12.14.

2. Turnover of Lifted the Lunar Lander in The Air

In order to test the antenna of the Chang'E-3 lunar lander, the lunar lander should be turned over and adjusted the attitude in the air. There were three types of antennas to be tested, which were installed on different assembly surfaces of the lunar lander. Because the assembly angles of the antennas were different and the directions were different, different test supports were required to meet the test attitude requirements. During test process of different antennas, the spreader should be used to turn over and horizontally lift the lunar lander to adjust the attitude, so as to be assembled with different test supports. In the test, two sets of turnover spreaders, a set of horizontal spreaders, three test supports and a set of transitional parking support were used to implement attitude adjustment of the lunar lander. Because of complex test status of

Fig. 12.14 Chang'E-3 lunar lander in lifting

the lunar lander and a lot of tools involved, it was necessary to propose a set of special technological method for attitude adjustment and assembly, including test supports with different angles, attitude adjustment of lifted the lunar lander, and assembly of the lunar lander and test supports in horizontal or vertical status, so as to implement attitude adjustment of antennas to be tested [17].

Two sets of single beam spreader were applied for lift and turnover of the Chang'E-3 lunar lander in antenna test. Two sets of single beam spreader with tow lift points were connected in sequence with two sets of cranes and lift points of the lunar lander. The lifting distance of the corresponding spreaders was independently controlled by two sets of cranes to implement 90° turnover of the lunar lander by control of the lift distance (length of rope).

12.5.9 Transshipment and Transportation of Lunar Lander

1. Transshipment

The spacecraft and hardware need to be transported within the facility. Generally, the assembly parking bracket, air-cushion vehicle, omnidirectional wheel transport, AGV (Automated Guided Vehicle) intelligent transport technology and Mecanum wheel are applied.

(1) Transshipment with assembly parking bracket was the simplest and most reliable way. At the same time, it could also be applied in system assembly, test, experiment, examination and propellant loading. It was also the most common way of transshipment of spacecraft and hardware. The only shortcoming was that aid of operators was necessary in small turn.

(2) The Air-cushion vehicle could transship super heavy spacecraft, which was suitable to the development of large spacecraft. The air-cushion truck could be used in different facility and implement "zero" turning radius. The height of air-cushion vehicle was only several tens of millimeters, which would have almost no impact when it moves and stops. The disadvantage was that the requirement for flatness of passage in facility is high and there should be large gas source.

(3) AGV was a type of transshipment equipment with guide and high positioning accuracy. It could be manually controlled in all directions and move automatically along a predefined path. It should have adequate safety protection and could automatically identify obstacles in the path.

(4) Mecanum wheel could also implement transshipment in all directions. Its advantage was movement in all directions within compact space environment, flexible and easy to operate. However, its structure was too complex and the effective contact area of its wheels was small, which made that its effective load capacity was small, the efficiency was low, it was easy to be damaged, and the movement accuracy was relatively low.

The total mass of the Chang'E-3 lunar lander after propellant was loaded was about 3.8 tons, while less than 1.2 tons without propellant. At the same time, the transshipment distance in the facility was relatively short. Therefore, assembly parking bracket was applied for the Chang'E-3 lunar lander, which was simple and reliable.

2. Transportation

Generally, the transportation of spacecraft and hardware is transshipment between different assembly facilities, test sites and launch site. There are road transportation, railway transportation, air transportation and sea transportation. The characteristics of each transportation method are different.

(1) Road transportation. It is the major transportation method of spacecraft and hardware in close range. Its advantages are convenient and no time restriction. The disadvantage is that road transportation of large spacecraft may be affected by the height of roads and bridges and the width of highway toll stations. Generally, special transportation vehicle is applied in road transportation, which is very long, wide and lower chassis.

(2) Railway transportation. It is the general transportation method of spacecraft and hardware in long distance, and most ground support equipment is also transported by railway. The advantage is that it is not limited by road; the disadvantages are slow and poor transport environment. There are several types of railway carriages: instrument carriage, flatbed truck, gondola truck and tank truck. The crucial hardware shall be placed in the instrument carriage with temperature control.

(3) Air transportation. It is a general transportation method of spacecraft to launch site. The greatest advantage of air transportation is fast. Its disadvantage is that it is

Fig. 12.15 Unloading of the Chang'E-3 lunar lander at Xichang Qingshan Airport

limited by the airport location and the aircraft. Large transportation aircraft such as AN-124, AIRBUS-SUPER TRANSPORTER can transport large spacecraft.

(4) Sea transportation. It is only suitable to launch site near the sea. Its advantage is that large dimension and heavy object can be transported. Its disadvantages are slow and poor environment. The sea transportation is closely relative to the geographical location of launch site.

In the development process of the Chang'E-3 lunar lander, road transportation, rail transportation and air transportation were applied. The flight model of the Chang'E-3 lunar lander was transported to the Capital Airport by road transportation, then was transported to Xichang Qingshan Airport by AN-124 aircraft, and finally was transported to the launch site by road transportation. The ground support equipment was transported to the launch site by railway transportation. The unloading of the Chang'E-3 lunar lander at Xichang Qingshan Airport is shown in Fig. 12.15.

12.5.10 Assembly of Radioisotope

In order to survive under extremely low temperatures of −180°C on lunar surface, radioisotope devices were equipped on the Chang'E-3 lunar lander. In order to ensure the safety and process feasibility, the assembly process was designed with simulation, rehearsal and process optimization. The assembly of radioisotope device is referred as 'nuclear operation' in the chapter.

1. Nuclear Operation Process Design and Process Optimization

The nuclear operation was generally implemented at launch site. The final assembly of radioisotope device was crucial to the quality of the devices and the safety of operators. The process was multi-coupled, and the technological process was particularly important. In accordance with people-oriented principle, an assembly process design that combined time, distance and shielding was proposed. In term of implementation timing, the nuclear operation could be classified into two phases: the assembly phase in technical area and on launch tower. During the assembly phase on launch tower, the technological process was refined and optimized through rehearsal in launch site. The different nuclear operation were managed to be implemented at different timings, such as after opening of the transport container cover, before closing of the launch fairing, after closing of the launch fairing and before launch, so as to minimize the implementation risk of operators. In term of nuclear operation process, the process documents were refined. Methods were proposed, such as 'local operation in storage location' to improve the operation efficiency of removing the cover of the radioisotope devices, application of Velcro fastener for multilayer insulation assembly to rapidly complete restoration of multilayer insulation after implementation of electrical connectors on surface of the lunar lander.

2. Simulation and Rehearsal of Nuclear Operation

In order to further verify the feasibility of nuclear operation process and improve the familiarity of nuclear operation operators with the process at the same time, a simulation environment of nuclear operation that was basically consistent with the actual operation conditions was established, and a simulated tower and a simulated the lunar lander were developed. The nuclear operation operators repeatedly performed verification and rehearsal in a 1:1 simulation environment, so as to verify the assembly processability and improve the proficiency of operation, which ensured high-quality implementation of nuclear operation at last.

12.5.11 Reassembly at Launch Site

The re-assembly at launch site is the final step of system integration and assembly, which is also the most important step of space mission. According to the technological process, generally there are assembly of instruments, cables and multilayer insulation components. The hardware status of all subsystems shall be confirmed for launch. In term of location, the re-assembly at launch site can be classified into technical area assembly, propellant load area assembly and launch area assembly. In the technical area, the solar panel and pyrotechnics shall be assembled, measured and tested. In the propellant load area, the propellant shall be loaded and the gas shall be filled for spacecraft. In the launch area, there will be operation related to test of spacecraft and launch vehicle, including lift, protection cover disassembly, plug and unplug of electrical connectors on surface of spacecraft.

The Chang'E-3 lunar lander was a new developed spacecraft in China. There were a lot of tasks that should be verified and tested in the re-assembly at launch site, including instrument assembly, mechanism deployment, accuracy test, leak test, mass property test, fuel load and electrical performance testing. The technological process of re-assembly at launch site was refined and implemented successfully, which accumulated valuable experience for future re-assembly of Chinese deep space probes at launch site.

12.6 Summary

The development of spacecraft is a complex system engineering involving different locations, facilities and subjects. The system integration and assembly of spacecraft is the final step of spacecraft development. Its implementation quality directly determines the performance of spacecraft on orbit and success of mission. Therefore, the system integration and assembly is the key step to ensure smooth development of spacecraft, which is also guarantee for system performance and function of spacecraft.

In this chapter, the development status of spacecraft system integration and assembly technology is introduced as well as the characteristics of system integration and assembly of the Chang'E-3 lunar lander. The detail design of system integration and assembly of the Chang'E-3 lunar lander is described, including systematic process design, technological equipment planning, technological test planning, technological process design, digital assembly simulation, and etc. As an example, the development practice and typical application of system integration and assembly of the Chang'E-3 lunar lander is introduced, such as instrument assembly, cable assembly, propulsion subsystem assembly, lift and transshipment.

References

1. Tan W, Hu J (2009) Spacecraft system engineering[M]. China Science and Technology Press, Beijing
2. Xu F, Hou S et al (2003) Introduction to satellite engineering [M]. China Aerospace Press, Beijing
3. Liang S (2004) From CAX to Concurrent Engineering—Reported at the Chinese Academy of Sciences Conference in 1994. Liang Sili Collection[M]. China Aerospace Press
4. Boe,Boeing777 Program computing innovations: computing & design/build processes help develop the 777 [OL]. http://boeing.com/commercial/777family/compute/index.html
5. Smith JL (1998) Concurrent engineering in the jet propulsion laboratory: project design center[R]. SAE transactions, USA
6. Cullimore B, Genbery V, Kahan M et al (2002) Integrated analysis thermal/structural/optical system[R]. SAE transactions, USA
7. Dai S, Xiao W, Wang C (2013) Status and development trend of Japan's H-2 transfer vehicle[J]. Int Space 3:19–31
8. Chen Y (2009) Spacecraft digital design technology[M]. China Science and Technology Press, Beijing

9. George Chryssolouris,Dimitris Mavrikios,Dimitris Fragos et al. (2000) A virtual reality-based experimentation environment for the verification of human-related factors in assembly processes. Robot and Comput Integr Manufact,16:267~276
10. Antonishek B, Egts DD, Obeysekare U R. Virtual assembly using two-handed interaction techniques on the virtual workbench[C]. Proceedings of the 1998 ASME Design Technical Conference and Computers in Engineering Conference, Atlanta, 1998: CIE-5540
11. He S (2013) Large aircraft digital equipment technology and equipment[M]. Aerospace Industry Press, Beijing
12. Chu G, Zhang H (2007) Technology of lunar probes[M]. China Science and Technology Press, Beijing
13. Yang B, Tang B, Dai W et al (2006) Research and practice of process reengineering for returning satellite programs [J]. Journal of Astronautics 6:23–27
14. Sun G, Yi W, Dai W et al (2008) Analysis and consideration of optimization of spacecraft assembly process flow[J]. Spacecraft Environ Eng 25(4):381–383
15. Yi W, Ma Q, Li M et al (2010) Research on CAPP, MES and AVIDM integration technology based on product data structure[J]. Spacecraft Environ Eng 27(5):640–642
16. Feng W, Zhang Y, Yi W, et al. (2014) Spacecraft assembly process design for lunar exploration phase II based on virtual simulation technology[J]. Spacecraft Environ Eng 31(3):326–331
17. Feng W, Li X, Yi W et al (2014) Design and implementation of lunar probe tightening field test pose adjustment scheme[J]. Aerospace Manufact Technol 5:43–48

Chapter 13
Metrology Technology of Lunar Lander

13.1 Introduction

During system integration of the lunar lander, metrology technology was applied in alignment accuracy measurement, seal performance measurement and mass property measurement. The alignment accuracy measurement was to acquire the geometry information of structure size, position and angle. The seal performance measurement was to examine the leakage rate of seal components of propulsion subsystem. The mass properties measurement was to acquire the information of mass, center of mass and inertia after the lunar lander was integrated and assembled [1].

Metrology technology played an important role in development of system integration of the lunar lander. In order to ensure high accuracy of attitude determination, attitude control and orbit control of the lunar lander on orbit, during the system integration, it was necessary to apply geometry metrology technology to ensure the installation accuracy of highly accurate instruments, such as gyroscope, accelerometer, star sensor, radar ranging and speedometer and laser ranging sensor, as well as actuators including 10 N thrusters, 150 N thrusters and 7500 N throttling engine. At the same time, accurate pointing data after assembly should also be acquired for requirements of onboard computer to formulate strategy of attitude control and orbit control. The mass property measurement method should be applied to acquire mass property data including mass, mass center and inertia, as well as help to adjust during assembly, and then provide important parameters for attitude control of the lunar lander. In order to ensure effective utilization of liquid and gas and no leak in propulsion subsystem of the lunar lander on orbit, during assembly process, different methods should be applied to examine the seal performance of propulsion subsystem, so as to find defect in propulsion subsystem and take effective remedial measures.

In order to satisfy requirements for measurement and control in landing process of the lunar lander, higher technical requirements for assembly geometry accuracy, seal performance and mass property were come up compared with conventional spacecraft. In this chapter, the characteristics of metrology of the lunar lander are analyzed. Development status of metrology technology is introduced. Special

D.-Y. Yu et al., *Technology of Lunar Soft Lander*, Space Technology Library 38,
https://doi.org/10.1007/978-981-15-6580-9_13

metrology methodology and practical application in system integration of the lunar lander are described.

13.2 Development Status of Metrology Technology

Along with development of aerospace technology, new metrology-based technologies and instruments for advanced manufacture and test had been emerged. In global aerospace manufacture industry, there were geometry metrology technologies in assembly including theodolite metrology, laser tracker metrology and laser radar metrology, and etc. The theodolite metrology had been widely applied in spacecraft assembly process all over the world. In order to determine the relative alignment between spacecraft components, a measurement network of theodolites was usually built up. Because the operator was necessary to observe the target in measurement process, the accuracy was affected by environment and personal skill. Generally, the positioning accuracy could be as less than 0.2 mm and the angle accuracy could be a dozen of arc-seconds. The accuracy of laser tracker metrology could be within 10 ppm, which had been applied in Boeing and NASA. But it could only be applied to measure point. It cannot be applied to measure a normal direction of small mirror surface. Laser radar metrology was a three-dimensional metrology to scan object surface. Now many companies such as Metric Vision, Cyra in USA and Mensi in France had introduced different types of laser scanner systems, in which the largest measuring range was 60 m and the minimums distance was 2 m and the highest accuracy could be as higher as 0.05 mm, which could be applied to measure the surface of structural component [2].

Up to date, there were two measures to test seal performance of spacecraft, non-vacuum accumulation leak test technology and vacuum leak test technology. The features of non-vacuum accumulation leak test technology were low cost, no requirement for large vacuum vessel and vacuum system, but low sensitivity. The features of vacuum leak test technology were relatively high sensitivity and expensive. The non-vacuum accumulation leak test method by helium Mass Spectrometer had been widely applied in powered descent of the United States, ESA, France and Italy. For the British SB4000 satellite platform, accumulation leak test method was defined for system leakage rate in its leak test requirement. The test accuracy was about 50%. Such method was also applied for some satellite in France about 50% test accuracy.

In aerospace industry, three-point weighing method with centroid bed was applied for measurement of mass and center of gravity. The torsion pendulum method with torsion pendulum bed was applied for measurement of inertia. Along with continuous development of metrology technology, the traditional methods and equipment had gradually become inadequate in terms of efficiency and accuracy. The integrated test system was developed for measurement of mass, center of gravity and inertia, which had been applied in NASA, ESA and JAXA [3].

13.3 Technical Characteristics Analysis of Metrology Technology

The Chang'E-3 was the first lunar lander of China, which had new requirements for measurement of alignment accuracy, system seal performance and mass property.

In order to implement autonomous navigation and control during powered descent of the lunar lander, it was necessary for GNC instruments to measure the velocity, position and attitude of the lunar lander relative to lunar surface in real time. Therefore, when the lunar lander was assembled, on the one hand, the measurement accuracy of alignment of GNC instruments to main structure of the lunar lander should be improved; on the other hand, there were new requirements such as measurement of new GNC instruments like ranging and speedometer antennas and the structural stability of the lunar lander. Because there were a lot of new measurement systems introduced, it was difficult to unify the measurement data in different coordinate systems. For example, the laser tracker was applied in measurement of the new ranging and speedometer antennas, and the laser radar was applied in measurement of structural stability of the lunar lander, while the reference coordinate system of the lunar lander was established by traditional theodolite network. Therefore, it was difficult to transfer the measurement data of different systems and unify several measurement coordinate systems for measurement of alignment accuracy in system assembly [4].

The measurement of seal performance for the lunar lander was to test seal of the propulsion subsystem. The propulsion subsystem was an important subsystem which was crucial for successful soft-landing and key for the mission success. It would provide not only Delta-V for trajectory maneuver corrections in LTO and LOI, but also enough force during powered descent, as well as torque for attitude control. In order to implement above tasks, the propulsion subsystem was complex with large volume of propellant and gas tanks, high pressure of gas tanks, high pressure during leak test, high precision for control of gas fill and vent, lots of engines/thrusters and valves. There were a lot of seal in propulsion subsystem. So, the requirements for seal performance of propulsion subsystem were more stringent.

The characteristics of mass properties of the lunar lander were variable during the mission profile. After the lunar lander was separated with the launch vehicle, the mass property would change after the antenna, solar panel, landing legs and other mechanisms were deployed. With consuming of propellant in LTO, LOI and powered descent, the mass property was changing. In order to ensure the control during all flight phases, the changes of mass property caused by status changes should be simulated and tested on the ground, so as to acquire the change pattern of the mass property parameters and formulate detailed flight control strategy to ensure the mission success. The loading of propellant was the reverse process of propellant consumption. By measuring the dynamic change of the propellant during loading, the change pattern of the mass property during propellant consumption could also be acquired. In order to ensure the measurement accuracy of the mass property parameters, the measurement accuracy in dynamic loading process should achieve the same accuracy as the static measurement, which was very difficult.

In summary, because of mission characteristics of the lunar Lander, many new measurement methods and technologies were applied in metrology. In the alignment measurement, there were applications including the normal direction measurement of the ranging and speedometer antenna end surface based on common point transformation and three-dimensional shape scanning measurement of the interface ring based on Unified Space Metrology Network (USMN), which solved the issue of measurement data fusion and high accuracy requirements under complex conditions. The automatic leak test technology and quick leak positioning technology were applied in system leak test of propulsion subsystem, which effectively improved the accuracy, reliability and safety. In the measurement of mass property, technologies were applied including mass property test technology based on multi freedom transformation, test and correction technology during propellant loading, and center of gravity correction technology with ballast, so as to acquire inertia of the lunar lander with high accuracy and efficiency.

13.4 Application of Special Measurement Technology

13.4.1 Normal Direction Measurement of Ranging and Speedometer Antenna End-Surface Based on Common Point Transformation

1. Requirements

According to autonomous control of the Chang'E-3 lunar lander during soft-landing, the alignment of the Range and Velocity Sensor (RVS) antennas should be measured in system assembly phase of the Chang'E-3 lunar lander and the measurement accuracy should be better than 2′. There were several ranging and speedometer antennas with different pointing direction on the lunar lander. At same time, because it was impossible to install reference mirror on the antenna and its support to the structure design of the antenna, the optical cube reference cannot be installed on the antenna and its supporter, the conventional theodolite measurement could not be implemented. Therefore, an integrated measurement method with laser track and theodolite were applied to measure the points' coordinates of the circular antenna ring and determine the direction of electrical axis of the antenna, as shown in Fig. 13.1.

2. Technical Design and Principle

When the lunar lander was assembled, the position and attitude of instruments onboard was established to the reference coordinate system O_1-$X_1Y_1Z_1$ based on 3 pin holes center *I*, *II* and *III* on the interface between the lunar lander and the launch vehicle, as shown in Fig. 13.2. However, in assembly process, the 3 pin holes between

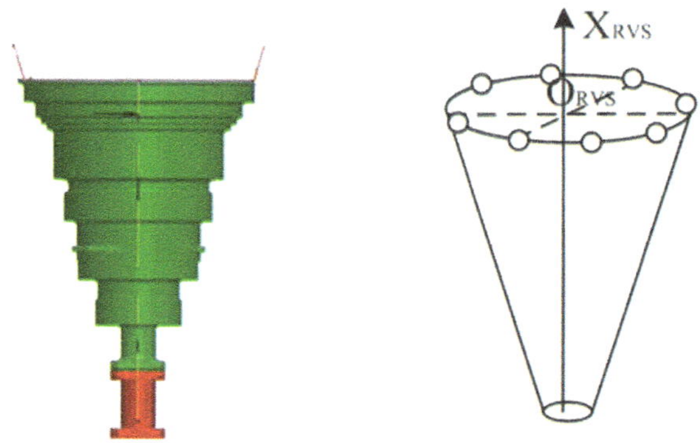

Fig. 13.1 Shape of antenna and definition of coordinate system

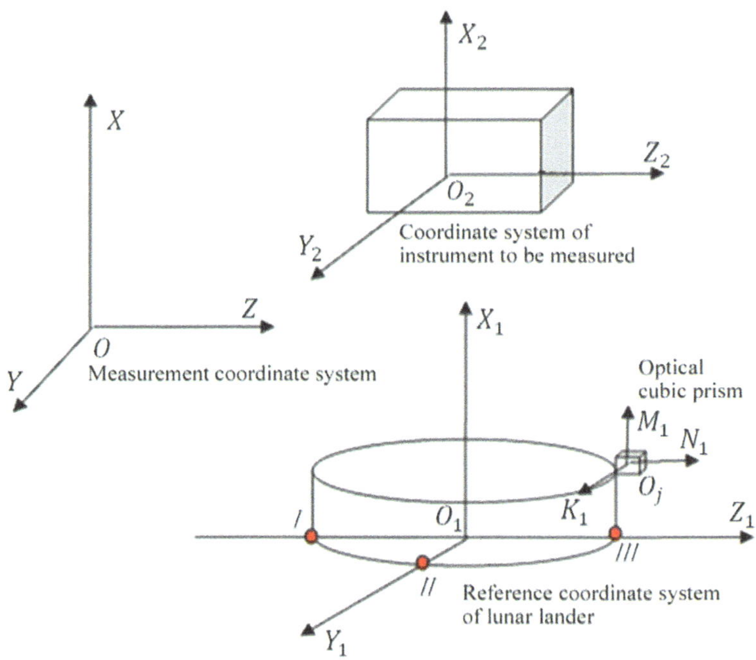

Fig. 13.2 Measurement principle of the lunar lander assembly accuracy

the lunar lander and the support bracket were not visible. Therefore, before assembly the coordinate system of the lunar lander by the 3 pin holes should be transferred to a visible coordinate system of optical cubic prism. I.e., the lunar lander coordinate system was replaced by a reference cubic prism coordinate system O_j-$M_1K_1N_1$. The relationship between the coordinate system O_2-$X_2Y_2Z_2$ of the instrument to be measured and the lunar lander coordinate system O_1-$X_1Y_1Z_1$ was equivalent to the relationship between O_2-$X_2Y_2Z_2$ and O_j-$M_1K_1N_1$.

Because the RVS was small and the side plate was thin, it was impossible to install the reference optical cubic prism. Therefore, the coordinates of several points on the RVS antenna end surface should be measured to fit its normal direction. The practical measurement accuracy of theodolite was only 0.2 mm, so the fitting error for normal direction of small aperture end surface was not adequate. Therefore, the laser tracker system was necessary. The laser tracker system consisted of laser tracker and spherically-mounted retro-reflectors (SMR), as shown in Fig. 13.3. During measurement, the laser beam was emitted from the laser tracker and the SMR was tracked in real time. A corner mirror was embedded in the SMR, and the laser beam was reflected back along incident direction. After the reflected beam was received, the center position of the SMR could be calculated. When the instrument was measured, the SMR would be close to the point or hole to be measured. After compensation of the diameter of SMR, the center coordinates of the point or hole could be determined. The measurement of three-dimensional coordinates by laser tracker system was implemented in spherical coordinates. As shown in Fig. 13.4, the coordinate

Fig. 13.3 Laser tracker system

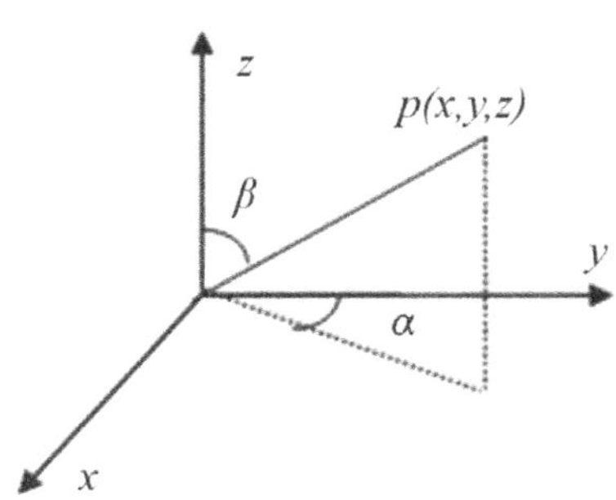

Fig. 13.4 Coordinate system of laser tracker system

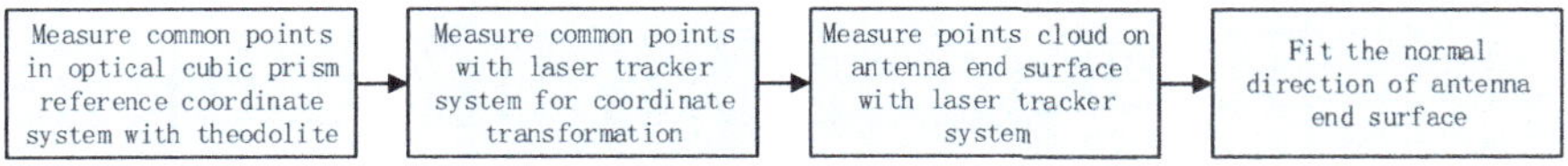

Fig. 13.5 Technical process of antenna measurement

system was set up by the vertical and horizontal axis of rotation and its intersection point. When the laser tracker was applied, a SMR was placed at the measured point p. The laser beam was emitted by the laser head and reflected by the SMR. By measurement of two angle parameters (α, β) and the distance S from the measured point p to the center of the laser head, the coordinates (x, y, z) of the point p could be given by the equation as follows.

$$\begin{cases} x = S\sin\beta\sin\alpha \\ y = S\sin\beta\cos\alpha \\ z = S\cos\beta \end{cases} \tag{13.1}$$

In the antenna of the lunar lander was measured, at first, the reference coordinate of the lunar lander structure was established. Then the coordinates of the point cloud on the end surface of each antenna were measured to the reference coordinate system. At last, the normal direction of each antenna end surface was fitted to determine the spatial angle between antennas' normal direction and the reference coordinate system. Because the size of optical cubic prism on the lunar lander was 20 mm × 20 mm × 20 mm, the axis could not directly be measured by the laser tracker system. So it was necessary to use the collimation theodolite to transfer the coordinate system to other reference points, i.e., the common points were measured in the coordinate system of optical cubic prism by theodolite, and then the common points were measured in the same order by the laser tracker system, so as to make coordinate transformation. The technical process of antenna measurement is shown in Fig. 13.5.

When the RVS antenna was measured, at first, a set of fixed common points A, B, C, D, E and F near the lunar lander was set up, as shown in Fig. 13.6.

The coordinates of fixed common points were measured with theodolite in the reference coordinate system of optical cubic prism, and the coordinates were recorded as (x_{A1}, y_{A1}, z_{A1}), (x_{B1}, y_{B1}, z_{B1}), (x_{D1}, y_{D1}, z_{D1}), (x_{E1}, y_{E1}, z_{E1}) and (x_{F1}, y_{F1}, z_{F1}). Then, the same 5 common points were measured with laser tracker system in the coordinate system O_3-$X_3Y_3Z_3$. The coordinates are (x_{A2}, y_{A2}, z_{A2}), (x_{B2}, y_{B2}, z_{B2}), (x_{D2}, y_{D2}, z_{D2}), (x_{E2}, y_{E2}, z_{E2}) and (x_{F2}, y_{F2}, z_{F2}), and the coordinates were recorded as (x_{A2}, y_{A2}, z_{A2}), (x_{B2}, y_{B2}, z_{B2}), (x_{D2}, y_{D2}, z_{D2}), (x_{E2}, y_{E2}, z_{E2}) and (x_{F2}, y_{F2}, z_{F2}). According to the least square algorithm [5], the relationship between the laser tracker coordinate system O_3-$X_3Y_3Z_3$ and the lunar lander coordinate system O_1-$X_1Y_1Z_1$ could be calculated, including the rotation matrix and translation vector of the coordinate system O_3-$X_3Y_3Z_3$ to the coordinate system O_1-$X_1Y_1Z_1$. After the coordinate systems were set up, the point coordinates of 8 points on the end surface of RVS antenna were measured by the laser tracker system. The 8-point coordinates were

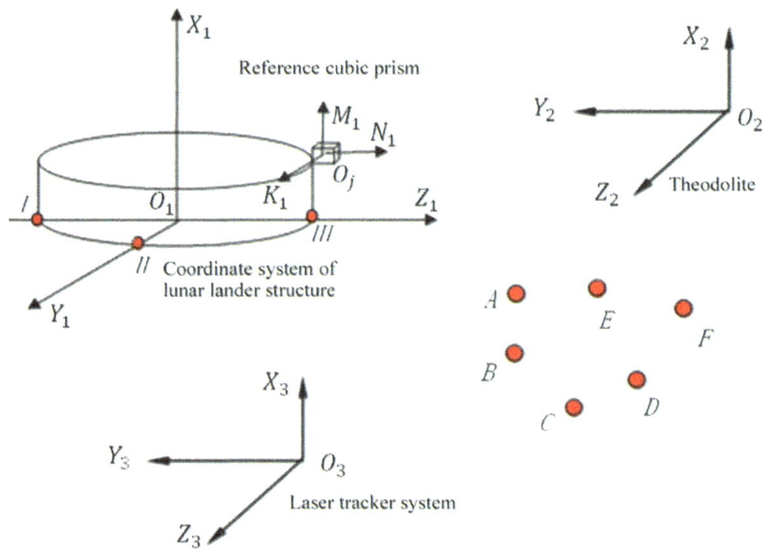

Fig. 13.6 Setup of common points

used for the plane fitting to calculate the normal direction and center position of the end surface.

3. Analysis of accuracy

In three-dimensional coordinate transformation of common point, the accuracy distribution model of coordinate transformation at any point is a quadric surface. According to the relevant references and simulation analysis of the laser tracker measurement, it was concluded that the accuracy of three-dimensional coordinate transformation was a family of ellipsoid distribution in space. The distribution of common point has great impact on the transformation accuracy, which is reflected in two aspects, the number and intensity of common points. In the simulation analysis, the 3–7 common points were selected in order, and the transformation accuracy for the point p was calculated to acquire the impact of the number of common points on the transformation accuracy. When the number of common points was 3, the transformation error was bigger. With increasing of the number of common points, the transformation error of the measured point was lower. When the number of common points was more 6, the increasing of common points would not significantly improve the transformation accuracy, meanwhile the measurement efficiency was lower [6]. Two groups of common points in space were selected. One group was relatively sparse, and the points were evenly distributed in the whole measurement space (10 m × 10 m × 3 m). Another group was denser and the points were concentrated in a region (3 m × 3 m × 3 m). The measurement errors for two groups of common points were controlled at the same level and the transformation accuracy was analyzed. Then the impact of the common point density on the transformation accuracy could be determined. The simulation analysis showed that the more dispersed the common points, the higher the transformation accuracy. The analysis showed that when the measuring accuracy was at the same level, the accuracy of the coordinate transformation of the sparse common points was higher than that of the dense common points. When the measurement was conducted, a highly accurate measurement control field was built by the laser tracker system. The common points should be distributed around the objects to be measured and in the whole measuring space.

A technological process experiment was conducted for RVS antenna measurement of the Chang'E-3 lunar lander. In the measurement space of simulated the lunar lander, 6 common points were set up, and three-dimensional coordinates of 6 common points were acquired from two stations A and B by laser tracker and theodolite respectively. The coordinates of common points at different stations were transformed to same reference coordinate system. At first, the theodolite was set up at A station to measure the 3 pin holes on a platform and establish the coordinate system of simulated the lunar lander, and transform to the coordinate system of reference cubic prism. Then the theodolite was setup again, 4 holes A, B, C and D on simulated antenna end surface were measured directly in the coordinate system of reference cubic prism, and the normal was fitted to acquire angle between the normal and the coordinate system of simulated the lunar lander. At the same time, the position of 6 targets fixed on the test platform was measured and recorded in the coordinate system of simulated the lunar lander. The laser tracker system was set up again to measure the 6 targets and 4 holes A, B, C and D on simulated antenna end surface, as shown in Fig. 13.7. The coordinates of 6 targets in the coordinate system of simulated the lunar lander were input and matched with the coordinates of the second measurements to establish the coordinate system of simulated the lunar

Fig. 13.7 Technological process experiment

lander again. Then the coordinates of A, B, C and D were converted to the coordinate system of simulated the lunar lander, and the normal equation and vector angle were calculated. By analysis of experiment data, the maximum error was 0.017°, which satisfied the measurement requirement.

13.4.2 Laser Scanning for Interface Ring Based on General USMN Adjustment

1. Requirements

The interface ring between the lunar lander and the launch vehicle supported the total weight of the lunar lander, so its structural strength played an important role in maintaining the accuracy of the lunar lander. In order to improve reliability of the lunar lander, during the system assembly, it was necessary to measure the shape of the bottom flange of interface ring in the reference coordinate system of the lunar lander so as to verify whether the structure could meet the requirement.

2. Technical Design and Principle

A great many point cloud data is necessary for shape measurement. The point measurement methods based on theodolite intersection or laser tracking could not meet the requirement. Therefore, the laser 3-D scanning method was applied. However, the laser 3-D scanning system could not directly measure the reference coordinate system of the lunar lander. The reference prism coordinate system should be measured by theodolite at first. How to integrate measurement data of two different measurement systems was a challenge. Therefore, the general USMN adjustment method was applied.

The general USMN technology is applied to solve the problem of unified spatial metrology network adjustment for various spherical coordinate measuring systems, in order to implement the spatial positioning and orientation of various measuring instruments, and provide the unified position and orientation reference for the spatial coordinate measurement of various instruments.

Laser radar scanning technology was applied in measurement of interface ring of the lunar lander. The laser radar system is shown in Fig. 13.8. The laser radar combined the FM laser ranging technology with the positioning mechanism of the precision mirror [7], whose principle is shown in Fig. 13.9.

A laser interferometry system was applied to measure the distance R between the infrared laser spot center and the origin of laser radar measurement coordinate system. The positioning mechanism could accurately measure the angle between the laser beam and the coordinate axis of the laser radar coordinate system, including the pitching angle E_{I} and the yaw angle A_{Z}. The 3-D coordinates of the measured point were calculated by the distance and encoder angle information. The equation is as follows [8]:

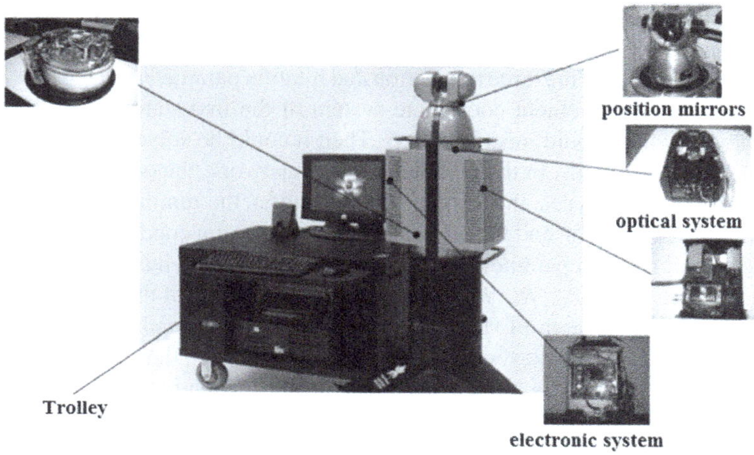

Fig. 13.8 Laser radar scanning measurement system

Fig. 13.9 Measurement principle of laser radar

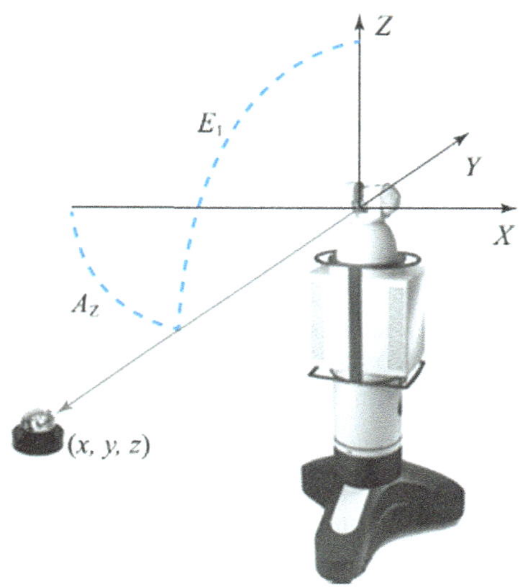

$$\begin{cases} x = -R \sin E_I \cos A_Z \\ y = -R \sin E_I \sin A_Z \\ z = -R \cos E_I \end{cases} \tag{13.2}$$

It was assumed that m (m $\geq$ 1) laser radars L_i were applied to observe k ($k \geq 3$) orientation points. At the same time, n theodolites were applied to measure p ($p \geq 1$) reference rulers by aiming at each other mutually. The reference ruler length is S_0. And q orientation points were observed in order by each theodolite. Then the number of position parameters of the whole metrology network was $t = 3q + 6m + 6n + 6p$, and the total number of observation equations was $n = 3mq + 2nq + 2C_n^2 + 4np + p$. In the adjustment calculation, the three translation and rotation parameters of the first station are all 0 in the measurement coordinate system of the first station, and the whole metrology network should meet $n \geq t-6$. Then it could be solved according to the principle of least squares. In the whole metrology network, the error equation could be classified into four types: the azimuth error equation, the zenith error equation, the slant range error equation and the reference length constraint condition equation. The translation and rotation parameters of station i relative to the measuring coordinate system were $(X_{0i}, Y_{0i}, Z_{0i}, R_{xi}, R_{yi}, R_{zi})$, and the coordinates of the orientation point k in the coordinate system of station I were (X_{ik}, Y_{ik}, Z_{ik}), and the coordinates in the measurement coordinate system were (X_k, Y_k, Z_k). When the angle, distance and mutual aiming value were applied to formulate the error equation, it was necessary to transform directional point k from the station i coordinate system to the measurement coordinate system [9].

$$\begin{bmatrix} X_{ik} \\ Y_{ik} \\ Z_{ik} \end{bmatrix} = \begin{bmatrix} a_{i1} & b_{i1} & c_{i1} \\ a_{i2} & b_{i2} & c_{i2} \\ a_{i3} & b_{i3} & c_{i3} \end{bmatrix} \cdot \begin{bmatrix} X_k - X_{0i} \\ Y_k - Y_{0i} \\ Z_k - Z_{0i} \end{bmatrix} \tag{13.3}$$

3. Analysis of Accuracy

By focusing infrared laser, the laser radar could implement fast scanning and uninterrupted focus scanning. The fastest scanning speed could be 1000 points/s with accuracy better than 0.2 mm. The measurement speed for accurate scanning and focusing on every point could be up to 20 points/s with accuracy better than 0.05 mm. When every point was scanned and focused with high accuracy and the optimal feedback signal was searched, the measurement speed could be 0.5–2 point/s with accuracy better than 0.025 mm. Setting multiple polygonal area simultaneously could help to implement automated scanning of the surface of large size component, as shown in Fig. 13.10.

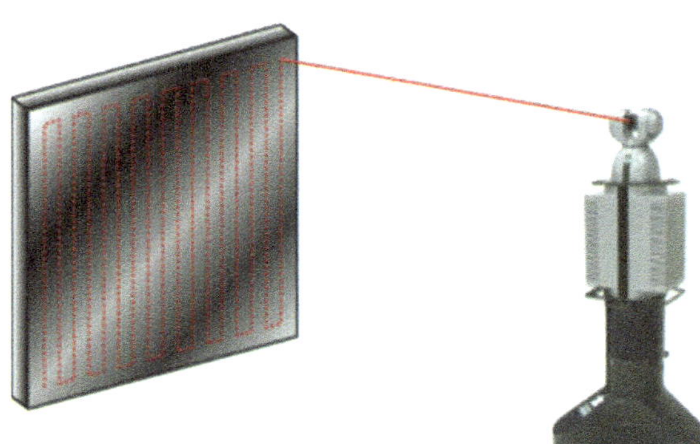

Fig. 13.10 Principle of scanning

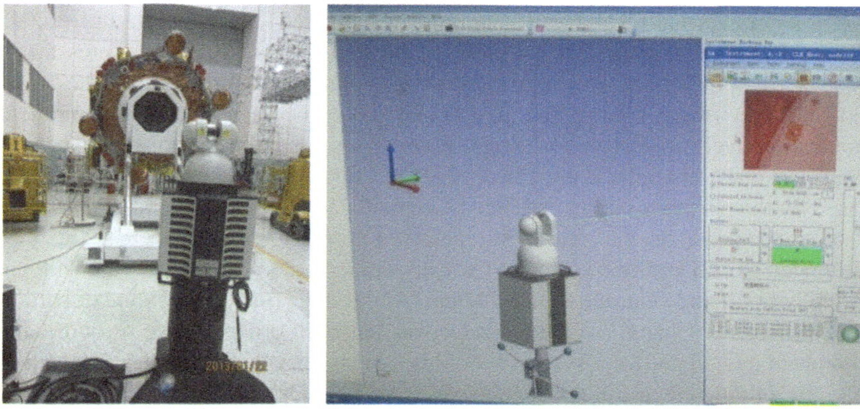

Fig. 13.11 3D scanning of the bottom flange of interface ring of the Chang'E-3 lunar lander

The bottom flange of interface ring of the Chang'E-3 lunar lander was measured by laser radar, as shown in Fig. 13.11. Based on the adjustment model, the point cloud on the end surface was measured by laser radar and theodolites, and the normal direction was fitted. Compared with result of theodolites, the maximum error of laser radar measurement was 0.0284°. Since the requirement was 0.05°, the laser radar measurement method could meet practical accuracy requirement. When the laser radar measurement method was applied, every point could be focused on by scanning path planning and accurate scanning. The measurement accuracy of single point could be improved to 0.05 mm, so the accuracy of scanning and axis measurement could be improved further [10].

13.4.3 High Accuracy and Automated Leak Test of Propulsion Subsystem

1. Requirements

For the leak test of propulsion subsystem of the lunar lander, the pressure was high, the structure was complex and requirement for gas load and vent rate was high. The safety and reliability of requirements of load and vent equipment for leak test should be very high. Because there might be much more leak in the subsystem and the leak rate requirement was high, it was necessary to develop automated leak test equipment with high accuracy to accomplish the leak test.

2. Technical Design and Principle

The high accuracy and automated leak test technology of propulsion subsystem included automated gas fill and vent technology, automated system leak rate test technology, automated comparison measurement technology with reference container introduced, and automated sampling technology.

The automated gas fill and vent technology was applied for safe, stable and reliable fill and vent of tracer gas in leak test of the propulsion subsystem. Because pressure in the leak test of the Chang'E-3 lunar lander propulsion subsystem was high, the safety and reliability of gas fill and vent equipment should be very high. The equipment should effectively control the gas fill and vent rate by automatic control of the opening of proportional valve, while the efficiency of gas fill and vent should be appropriate. Based on the principle and experience of manual gas fill and vent, and the test data of simulated products, the control parameters of proportional valve were set up. The automated gas fill and vent system effectively implemented gas fill and vent with the accuracy of fill and vent rate 0.01 MPa/min. At the same time, the volume of helium gas used for the Chang'E-3 lunar lander was large. In order to save resource, the automated gas fill and vent system should be able to recycle helium gas [11]. Up to date, the automated gas fill and vent system could achieve recovery of 80% helium.

The basic principle of non-vacuum accumulation leak test technology by helium Mass Spectrometer is as follows. At first, the propulsion subsystem of the lunar lander to be tested was filled with helium and the pressure should be same as on orbit. Through effective single point leak test technology, the leakage rate of single point should meet the designed requirement. Then the lunar lander would be put into non-vacuum container (collection room or leak test container) and sealed. With the helium Mass Spectrometer leak detector, the concentration of helium in the container would be examined. After stand still for 24 h, the change of helium concentration in the container would be examined. For the calibration of leak test system, a definite quantitative of gas should be filled into the container to examine the change of helium concentration in the container.

The leak test results should be compared with the calibration data to calculate the leak and leakage rate of the tested propulsion subsystem of the lunar lander.

The automated total leak detector was the core to complete the total leakage rate test, which was a new developed automated leakage rate detector based on sampler of laser micro hole and standard Mass Spectrometer leak detector for total leak rate test of spacecraft. The sampler based on laser micro hole was the key to automation of the leak detector, which consisted of valve, micro hole in fixed size, standard gas solenoid valve and sample gas solenoid valve. The leak test port was in the middle of the valve. The upper part of the main path was connected with the leak test port and there was a micro hole. The sampling port channels that were connected with the leak test part were respectively connected to standard gas solenoid valve and sample gas solenoid valve. On both sides of the valve, there were one input port and one output port for both of standard gas solenoid valve and sample gas solenoid valve. They were integrated into one channel and connected to the corresponding standard gas solenoid valve and sample gas solenoid valve. Compared with original technology, the leak detector could implement the switch between the standard gas and the sample gas during the total leak test by solenoid valve, and the flow limitation by micro hole in fixed size, so as to avoid the randomness caused by human operation. The air clear valve was added between the standard gas solenoid valve and the sample gas solenoid valve to solve the issue of tailing phenomenon in multi-channel gas mass

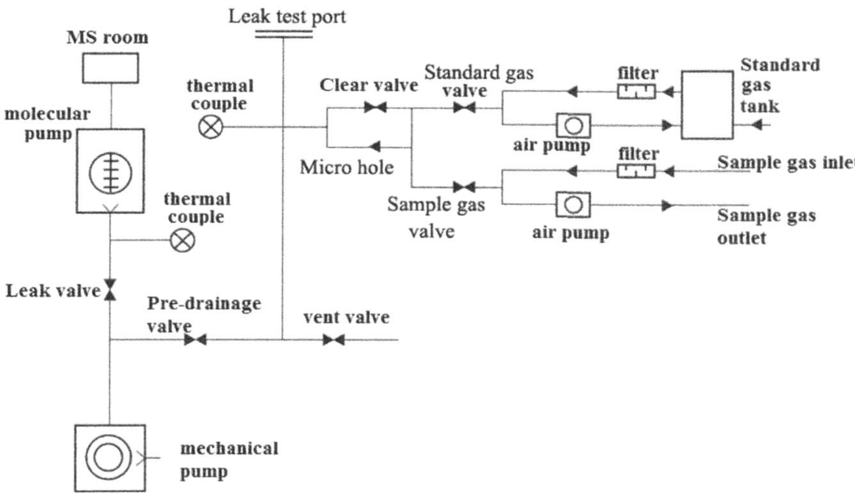

Fig. 13.12 Principle of automated total leak detector

spectrometry when standard gas and sample gas were switched, which could improve the test accuracy. The principle of automated total leak detector is shown in Fig. 13.12.

The automatic measurement technology of reference container was introduced to solve the impact of environmental background on the total leak test. The helium Mass Spectrometer leak detector was highly sensitive, but poor long-term stability. In order to avoid the instability of the leak detector, it was necessary to calibrate the leak detector with standard gas during leak test. Standard gas was man-made mixture of helium and air with same concentration of helium in the air. For calibration, before leak test, helium Mass Spectrometer leak detectors were applied to test same standard gas and the output of leak detectors was adjusted to be same. By calibrating leak detectors with standard gas, the effect of leak detector baseline drift on total leak test that helium was accumulated within 24 h in the collection chamber could be eliminated. However, the helium concentration of the standard gas should be same as the concentration of helium in the air which is generally 5×10^{-6}. It was difficult to make standard gas and costs too much to be automated. After the reference container was introduced, the natural air could be filled into the reference container and sealed for storage. The calibration and automatic comparison measurement of the leak detector could be implemented with the circulation system of the leak detector. The reference container should be designed to be consistent with performance of the collection container for the lunar lander. The reference container should be able to be sealed in order to meet the requirement that the pressure drop in 24 h was not greater than 300 Pa under 2 kPa pressure relative to external air in the container. At the same time, in order to satisfy the requirements for the atmospheric pressure change caused by external environment during stand still process, as well as the pressure difference change between internal and external of the container caused by temperature, the container should be able to adapt to the pressure difference between the internal and external of the container, which was −2–5 kPa. The inside of the reference container should be clean and smooth, and there should not be material which could adsorb helium.

The automatic sampling and calibration system was applied to implement the calibration of the leak test system. Generally, the standard leak hole is applied for sampling calibration of the leak test system. When the standard leak hole was applied, the sampling status should be exactly same as the measuring status. However, because of the long sampling period and the uncertainty of the standard leak hole itself (The uncertainty of constant pressure method was still 5%. Because of the lack of constant pressure method, the uncertainty was usually about 18% in China), the uncertainty of test result was big. In the leak test of the Chang'E-3 lunar lander, the standard gas volume was used, whose uncertainty was less than 1%. When the standard gas volume was measured, the helium was slowly filled into the collecting chamber, while the helium was instantaneously filled into the collecting container when the sample was used. Because of the difference between the measurement statuses, the adsorption of helium on the surface material of collection chamber and the lunar lander was different, which would theoretically cause error in leak test result. The relevant analysis and experimental results showed that adsorption of helium on the surface material of collecting chamber and the lunar lander would not affect the leak test result. A new automatic sampling system was developed for the Chang'E-3 lunar lander. The sampling process of conventional mode depended on experience of operator, equipment, environment and tool preparation, which could possibly lead to failure of sampling. Compared with sampling system in traditional mode, the technical process of newly developed calibration system for sampling was greatly simplified, the accuracy of sampling was raised to 0.5% FS of the regulation pressure, the sampling time was shortened to 5 min, and the efficiency of sampling was increased by more than 2 times. Meanwhile, the requirements for environment and personnel operation experience were reduced. The automatic control could be implemented by program. In the technical process of total leak test for the lunar lander container, which was based on operation mode of automatic sampling control system, there were equipment and tool preparation, logon and system parameter setup, automatic program control sampling, leakage rate test and vent.

3. Analysis of Accuracy

The application of automated leak test technology of propulsion subsystem greatly improved the automation level of leak test for the Chang'E-3 lunar lander, so as to improve the sensitivity, accuracy, safety and reliability of leak test.

In the implementation of automated leak test, the uncertainty of the leak test system was evaluated [12]. Through analysis, the factors on the uncertainty of non-vacuum accumulation leak test method by helium Mass Spectrometer are shown in Fig. 13.13.

After analysis and experiment of all factors, the uncertainty of general spacecraft leakage caused by system effect and random effect was analyzed. The uncertainty of synthetic standard of the leak test system was determined (Seen in Table 13.1).

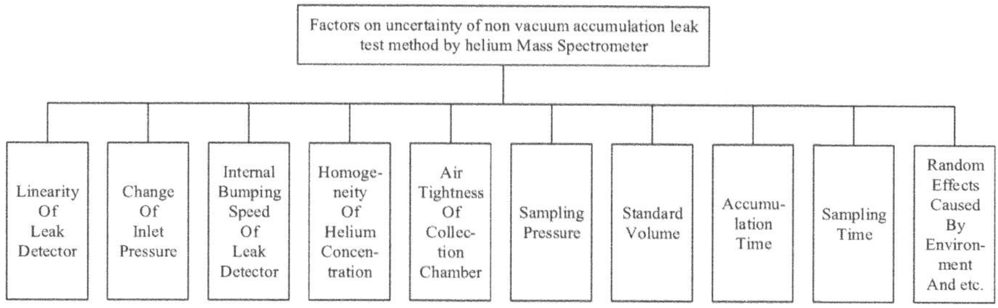

Fig. 13.13 Factors on uncertainty of non-vacuum accumulation leak test method by helium Mass Spectrometer

Table 13.1 Uncertainty components caused by different factors

Magnitude/(Pa•m3/s)	Uncertainty caused by system effect (%)	Uncertainty caused by random effect (%)	Synthetic standard uncertainty (%)
1×10^{-4}~1×10^{-5}	7.38	6.86	10.08
1×10^{-3}~1×10^{-4}	7.64	16.21	17.92
1×10^{-2}~1×10^{-3}	7.64	8.24	11.24

13.4.4 Quick Leak Positioning Technology for Propulsion Subsystem

1. Requirements

The conventional leak test in system assembly process is to conduct single point leak test, such as weld, screw connection and etc., so as to ensure that single point leak meet the requirement and then conduct total leak test. In conventional leak test, if single point leak test is qualified, the total leakage rate is generally qualified. However, if there is the leak caused by raw material and parts that is unforeseen leak in system assembly and it is missed before system assembly, there will be total leakage rate of the system over requirement level. Specially, when the leakage rate is close to the total leakage rate of system, it will be very hard to position the leak. Because propulsion subsystem of the Chang'E-3 lunar lander was complex and it was difficult to position the leak. However, there was requirement for quick positioning of the leak point during development of the lunar lander. It was necessary to develop a quick leak positioning technology. The Flexible Helium Hood method for quick leak positioning was suitable for complex system that could not be tested by bubble method.

2. Technical Design and Principle

The flexible helium hood method for quick leak positioning was a technology between the single point leak test and the total leak test of the spacecraft. It was suitable for the search and identification of the leak below the 10^{-4}Pa•M3/s.

The difference between single point leak test and system leak test was the volume and size of the covered space. When the covered space was less than 2 mL, the helium mass spectrometer sniffer method could implement quantitative test of the single point leakage rate, and the inlet pressure of the leak detector was irrelevant to volume of covered space. However, the volume of covered space and the inlet pressure of the leak detector would affect the reaction time of the leak test system. When the inlet pressure was higher and the volume of cover space was smaller, the leak test time by helium mass spectrometry sniffer method will be shortened. The leak test equipment of helium mass spectrometer sniffer consisted of helium mass spectrometer leak detector, sniffer, positive pressure standard leak and single point cover tool. Helium mass spectrometer leak detector was the core of single point leak test. In order to achieve that the leak test sensitivity of the helium mass spectrometer sniffer for single point leak test was better than 10^{-8}Pa•m3/s, the sensitivity of the leak detector should be better than 10^{-11}Pa•m3/s, and leak detector should be reliable and stable.

The sensitivity of non-vacuum accumulation leak test technology by helium mass spectrometer was affected by two factors which were the container volume and the accumulation time of hardware in stand still [13]. The smaller the volume of collection container was and the longer the accumulation time was, the higher the sensitivity of leak test system was. Whereas, the larger the volume of collection container was and the shorter the accumulation time was, the lower the sensitivity of leak test system was. At the same time, the volume of collection container was limited by spacecraft size and the accumulation time was limited by development cycle, various factors should be considered to establish the leak test system. Up to date, for 100 m^3 volume of collection container which was sealed and stood still for 24 h under atmospheric environment, its leak detection sensitivity could reach 10^{-5}Pa• m3/s.

The flexible helium hood method for quick leak positioning was applied for complex system that could not be dissembled and the leak was uncertain. When the accumulation leak test was conducted, the complex system should be partitioned and flexible coating was applied to cover and formulate a simple hood. By use of flexibility, the covered space was reduced as far as possible. The helium mass spectrometer sniffer was applied to test covered space. If there was leak, the concentration of helium in the measured space would gradually increase to indicate that there was leak point in the covered space. Otherwise, the leak point was not within the covered space. Then space with leak was continued to be partitioned until the leak point is positioned. The reverse idea was applied in such technology to solve the issue of leak positioning when the total leakage rate didn't meet the requirement. It was suitable for search and identification of unpredicted leak when the leakage rate was lower than 10^{-3}Pa•m^3/s.

3. Analysis of Accuracy

The flexible helium hood method for quick leak positioning was successfully applied to implement leak positioning in system assembly process of the Chang'E-3 lunar lander, with the leakage rage of 1×10^{-3}Pa•m^3/s. Such technology had also been successfully applied in other spacecraft integration and assembly process in China.

13.4.5 Mass Property Test of Lunar Lander with Multiple DOF Transformation

1. Requirements

The traditional mass property test of spacecraft, generally bracket in L shape was applied in the test of longitudinal mass center and inertia. During test, there were complex operation processes including lift of spacecraft and docking with L shape bracket. Compared with general spacecraft, the configuration of the lunar lander was complex and the size of envelope was large. Therefore, the traditional L shape bracket could not meet requirements of mass property test. For example, if the L shape bracket was applied in test of longitudinal mass center of the lunar lander, it was impossible to install deployable mechanisms such as landing gears and solar panels. The final mass property should be calculated further, which could seriously affect test accuracy. Meanwhile, the test of the inertia could be implemented with L shape bracket.

The linear three-coordinate converter could implement the mass property test with multiple Degree of Freedom transformation, which could meet the test requirements and avoid the complex operation processes of re-lifting and docking.

2. Technical Design and Principle

The mass property test technology of multiple DOF transformation was to establish the position relationship between the lunar lander coordinate system and the test coordinate system by the three-coordinate converter instead of L shape bracket. The longitudinal mass center and inertia of spacecraft could be tested by rotation adjustment of the three-coordinate converter.

The structure of the three-coordinate converter is shown in Fig. 13.14, which was composed of lathe bed, rotation table, inclined platform, lower bracket and ballast mechanism. The rotation table, inclined platform, lower bracket and inclined shaft were referred to as slide table. The slide table was connected with the lathe bed by guide track and can move horizontally along the guide track on the lathe bed. The lunar lander was installed on the rotation table and could be rotate around the center of inclined platform at 360°. The inclined table could be inclined at a certain angle along the inclined shaft. The lathe bed could support the weight of the lunar lander, slide table and ballast mechanism, whose bottom was docked with the mass property test bed.

The idea of mass property test technology with multiple DOF transformation was to compare the change of lateral mass center of the lunar lander in different status, so as to calculate the longitudinal mass center of the lunar lander.

The specific testing process is as follows [14].

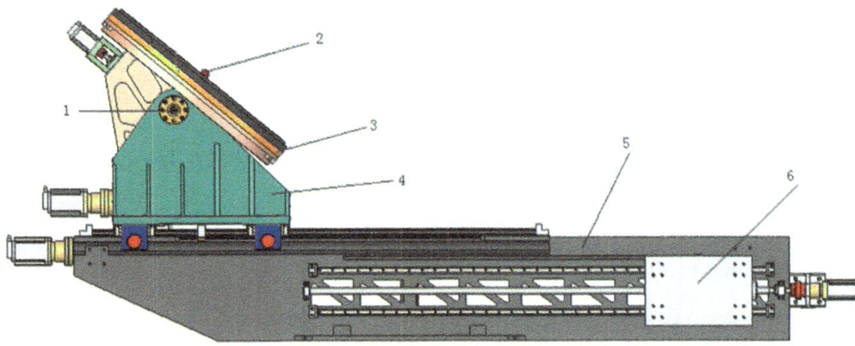

Fig. 13.14 Three coordinate converter, 1-Inclined shaft; 2-Rotation table; 3-Inclined table; 4-Lower bracket; 5-Lathe bed; 6-Ballast mechanism

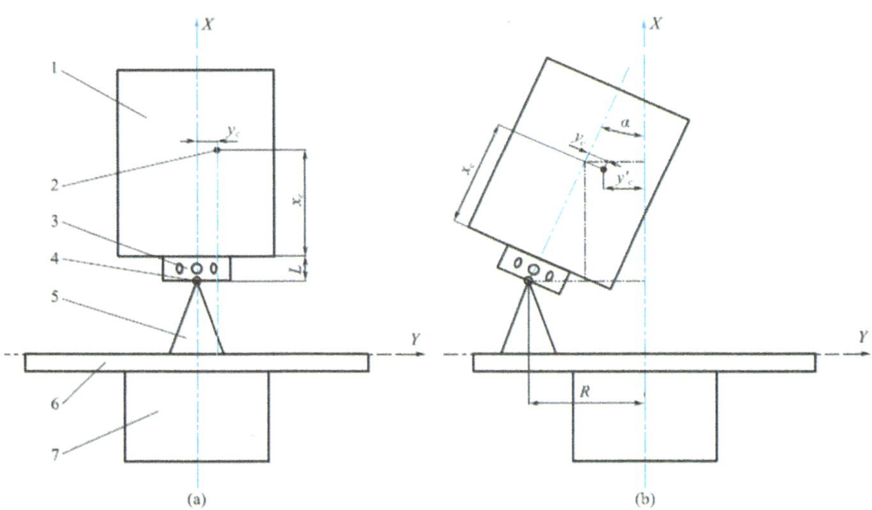

Fig. 13.15 Mass center test process 1-Lunar lander; 2-Mass center; 3-Lateral support; 4-Incline center; 5-Lower bracket; 6-Lathe bed; 7-Mass center test bed

(1) The test coordinate system was established, whose origin was the geometric center of the test bed. The X axis was upward, through the origin and vertical to the horizontal plane. The Y axis was toward the ballast mechanism, through the origin and parallel to the guide track. The Z axis formulated the right-handed system with X axis and Y axis.

(2) When the lateral mass center was tested, the inclined table was kept horizontal. The lateral mass center o Y_c was acquired directly by mass center test bed, as shown in Fig. 13.15a.

(3) When the longitudinal mass center was tested, the slide table was shifted to R and the inclined table was inclined at α around the inclined shaft. The test value Y_c' was given by mass center test bed. The longitudinal mass center X_c could be calculated, as shown in Fig. 13.15b.

The calculation process is as follows: when $\alpha = 0$, the mass center Y_c was measured; when $\alpha \neq 0$, the longitudinal mass center Xc could be calculated by Eq. (13.4).

$$X_c = \frac{R - \left(Y_c' - Y_c \cos\alpha\right)}{\sin\alpha} \tag{13.4}$$

When the inertia was calculated, the inertia of rigid body around a fixed point was calculated by Eq. (13.5).

$$\begin{aligned} I = & I_x \cos^2\alpha + I_y \cos^2\beta + I_z \cos^2\gamma - 2I_{xy} \cos\alpha \cos\beta \\ & - 2I_{yz} \cos\beta \cos\gamma - 2I_{zx} \cos\gamma \cos\alpha \end{aligned} \tag{13.5}$$

When the angles between the measurement axis and the lunar lander coordinate axis X, Y and Z were α, β and γ, and $\alpha = 0°$, $\beta = 90°$ and $\gamma = 90°$, the direction cosine was (1,0,0) and the I_x could be measured. When $\alpha \neq 0$, the 5-direction cosine was set by the three-coordinate converter as $(C_\alpha, C_\beta, 0)$, $(C_\alpha, -C_\beta, 0)$, $(C_\alpha, 0, C_\gamma)$, $(C_\alpha, 0, -C_\gamma)$, $(C_\alpha, C_\beta, C_\gamma)$. The inertia I_i of the lunar lander was measured by inertia test bed for above direction cosine respectively. Then the inertia I_x, I_z and I_{xy}, I_{xz}, I_{yz} could be calculated by Eq. (13.5). The results were inertia through the inclined center and parallel to 3 axis of the lunar lander coordinate system. Then the inertia in mass center coordinate system could be calculated with axis theorem formula.

3. Analysis of Accuracy

The accuracy of mass property test was determined by the accuracy of mass property test bed [15]. The accuracy of current mass center bed was generally 0.5 kg. The accuracy of lateral mass center test included the test accuracy of test bed and positioning accuracy of the three-coordinate converter. Through calibration of mass center bed and installation/adjustment of the three-coordinate converter, the test accuracy of lateral mass center of the lunar lander could be controlled within 0.5 mm. Longitudinal mass center was an indirect test and its accuracy should be calculated by following Equation.

$$\mathrm{d}x = \sqrt{\left(\frac{\partial X_c}{\partial L}\mathrm{d}L\right)^2 + \left(\frac{\partial X_c}{\partial Y_c}\mathrm{d}Y_c\right)^2 + \left(\frac{\partial X_c}{\partial Y_c}\mathrm{d}Y_c\right)^2 + \left(\frac{\partial X_c}{\partial \alpha}\mathrm{d}\alpha\right)^2} \tag{13.6}$$

According to error analysis of lateral mass center, assuming that the maximum value of $\mathrm{d}Y_c'$ and $\mathrm{d}Y_c$ was 0.5 mm, the value of Y_c' and Y_c was 100 mm, and reference value of $\mathrm{d}\alpha$ is 0.02°, the test errors caused by different α are listed in Table 13.2.

(1) The relative error of longitudinal axis inertia test: the test error of the test bed was about 0.1%.

Table 13.2 Analysis of mass center error

$\alpha/(^\circ)$	$\frac{\partial X_c}{\partial L}\mathrm{d}L$/mm	$\frac{\partial X_c}{\partial Y_c}\mathrm{d}Y_c$/mm	$\frac{\partial X_c}{\partial Y_c}\mathrm{d}Y_c$/mm	$\frac{\partial X_c}{\partial \alpha}\mathrm{d}\alpha$/mm	$\mathrm{d}X_c$/mm
20	0.002924	0.584761	0.5494955	1.4766	1.68
30	0.002	0.4	0.3464102	0.941305	1.08
40	0.001556	0.311145	0.2383507	0.650681	0.76

Table 13.3 Effect of k and α on inertia of lateral axis

$\alpha/(^\circ)$ k	10 (%)	20 (%)	30 (%)	40 (%)
0.8	0.58	1.57	2.61	0.57
1.0	6.53	1.61	0.70	0.38
2.0	0.67	0.43	0.31	0.24

(2) The relative error of horizontal axis inertia test:

$$\frac{\mathrm{d}I_y}{I_y} = \frac{2(I_1 + I_2 - 2I_x)\tan\beta \cdot \mathrm{d}\beta + \mathrm{d}I_1 + \mathrm{d}I_2 + 2\sin^2\beta \cdot \mathrm{d}I_x}{I_1 + I_2 - 2I_x\sin^2\beta} \tag{13.7}$$

It was assumed that the value of I_1 and I_2 was k times of I_x, the value of $\mathrm{d}I_1$ and $\mathrm{d}I_2$ was k times of $\mathrm{d}I_x$, and $\mathrm{d}\alpha = 0.02^\circ$. According to the actual test accuracy $\mathrm{d}I_x/I_x = 1/1000$ of inertia I_x, the Eq. (13.7) could be simplified as follows under above assumptions.

$$\frac{\mathrm{d}I_y}{I_y} = \frac{2(k-1)\cot\alpha \times \mathrm{d}\alpha + \left(k + \cos^2\beta\right) \times 0.001}{k - \cos^2\alpha} \tag{13.8}$$

The effect of k and α on inertia of lateral axis is listed in Table 13.3.

13.4.6 Testing Data Correction Technology for Propellant Load

1. Requirements

For the lunar lander to land on lunar surface, a great many propellant was necessary to be load. In the technical process of the lunar lander development, propellant load strategy different from other spacecraft was applied, i.e., the propellant was loaded after the lunar lander was docked with the test bed. Such design brought new issue to ensure test accuracy of propellant load in mass property test of the lunar lander.

The accuracy of mass property test consisted of two parts: one was the measurement accuracy of the test bed, which could not be improved furthermore and only a small percent of total accuracy; the other one was the positioning accuracy of the lunar lander on the test bed. Because the lunar lander was docked with the test bed at first, the lunar lander to be tested was changed to a complex of the lunar lander and the docking bracket, which was 1700 mm in height. As analysis of three-coordinate test principle, mechanical structural parameters of the docking bracket, such as flatness, parallelism and coaxiality could directly affect the mass center test accuracy.

2. Technical Design and Principle

There was no direct solution for positioning of the lunar lander on the mass center bad. It could only be achieved by correction of test data. The principle of such method was to establish the transformation matrix of the measurement coordinate system of mass center bed through reference transfer technology of accuracy test, and then test the mass center of the lunar lander in test bed coordinate system during propellant filling in real time, so as to calculate the mass center in real time by transfer matrix. The process is as follows.

(1) Before the lunar lander was docked with the mass center bed, a transitional cubic prism was fixed in the appropriate location in facility. The theodolite was applied to establish the angle and position relationship between transitional cubic prism and the coordinate system of the mass center bed. During measurement, the position of the transitional cubic prism should be stable.
(2) After the lunar lander was docked with the mass center bed, the angle and position relationship between the mechanical coordinate system of the lunar lander and the transitional cube prism was precisely measured. The relationship between the mechanical coordinate system of the lunar lander and the coordinate system of the mass center bed was calculated, including the angles between the coordinate axis of the lunar lander coordinate system and the coordinate axis, as well as the coordinates of the origin of the lunar lander coordinate system in the mass center bed coordinate system. The results of accuracy measurement are listed in Table 13.4.

If the coordinates of the lunar lander mass center in the coordinate system of mass test bed was $(X_1{}', Y_1{}', Z_1{}')$ and the coordinates in the lunar lander coordinate system is (X_1, Y_1, Z_1), then

Table 13.4 Accuracy test results

Test reference Test item and status		Coordinate system of mass center bed		
		X	Y	Z
Mechanical coordinate system of the lunar lander	X'	0.0499°	89.961°	89.9689°
	Y'	90.0388°	0.1845°	90.1804°
	Z'	90.0289°	89.8414°	0.1612°
	Coordinates of center	X_1	Y_1	Z_1

Table 13.5 Comparison of test data after correction

Mass center of the lunar lander	X_c	Y_c	Z_c
Data by test bed/mm	1767.72	−0.42	−0.82
Data after correction/mm	1768.35	−0.09	0.05
Difference/mm	0.63	0.33	0.87

$$\begin{bmatrix} X_1 & Y_1 & Z_1 \end{bmatrix} = \begin{bmatrix} X_1' - X_1 & Y_1' - Y_1 & Z_1' - Z_1 \end{bmatrix} \cdot \begin{bmatrix} \cos 0.049 & \cos 89.961 & \cos 89.9689 \\ \cos 90.0388 & \cos 0.1845 & \cos 90.1804 \\ \cos 90.0289 & \cos 89.8414 & \cos 0.1612 \end{bmatrix} \quad (13.9)$$

The test data of mass center during propellant filling process was substituted in Eq. (13.9) in turn. The test data could be corrected in real time and the mass center data of the lunar lander during propellant filling could be monitored. The core of such technology was to replace the accuracy of mechanical positioning with the accuracy in accuracy test, which could greatly improve the accuracy of mechanical positioning and provide a good solution to the issue of test accuracy for the lunar lander during propellant filling test. Such technology would continue to play a role in spacecraft test technology.

3. Practical Application

Here is an example of test data for the Chang'E-3 lunar lander during propellant filling, which can show the practice result of test data correction technology for propellant load. The parameters of mass and mass center given by the mass center bed was substituted in Eq. (13.9) for data correction in a moment during propellant filling. The test data correction results are listed in Table 13.5.

According to data comparison, it could be concluded that if the test result of mass center (traditional method) is applied, it would not be able to meet the requirement of ±0.5 mm test accuracy of Y_c and Z_c of mass center of the lunar lander. The correction effect was significant.

13.4.7 Integrated Mass Center Parameter Correction Technology by Propellant and Ballast

1. Requirements

In order to ensure the mission success, the accuracy and stability of landing for the lunar lander, there were strict requirements for mass center of the lunar lander. In order to eliminate the mass center changes in different development phases of the lunar lander, before launch, it was necessary to correct mass center of the lunar lander to achieve the ideal design status, so as to ensure the mission success. The conventional method was to adjust the mass centroid by additional ballast on outside of later plates.

For the lunar lander to be landed on lunar surface, a great many propellant was necessary to be load. When the launch mass was certain, application of integrated mass center parameter correction technology by propellant and ballast was an effective approach to increase the load of propellant as far as possible.

At the same time, the 4 parallel propellant tanks of the lunar lander were distributed horizontally, which was different from other spacecraft. Therefore, it was possible to apply the integrated mass center parameter correction technology by propellant and ballast.

2. Technical Design and Principle

The configuration of propellant tanks on the lunar lander is shown in Fig. 13.16, where $Y1$ and $Y2$ were tanks for oxidizer, $R1$ and $R2$ were tanks for methyl hydrazine. For example, in order to adjust the Y coordinate of mass center, more propellant should be loaded into $Y2$ and $R1$ other than $Y1$ and $R2$. So, the mass center would move toward $+Y$ axis and the mass center correction was achieved.

There were three methods that could correct mass center parameters of the lunar lander, including conventional ballast, propellant correction and integrated correction by ballast and propellant. The advantages of conventional ballast correction was uniform propellant load and easy to control load for each propellant tank. After propellant load, the good mass property could be achieved through small and partial ballast. The disadvantage was reduction of propellant by ballast. The advantage of propellant correction was that it could increase propellant load as much as possible. The disadvantages included: the propellant load for each tank was different and the propellant load must be calculated in advance; there would be a limit for mass center adjustment; the propellant filling process was irreversible and the accuracy of mass center adjustment was low. The integrated ballast and propellant correction method combined the advantages of two methods, which gave a solution to the issue of ballast outside the limit range of propellant correction.

The major process of integrated ballast and propellant correction was as follows: (1) According to test data before propellant filling, the risk of the limit range of propellant correction method should evaluated; (2) The mass center of the lunar lander was adjusted to the range of propellant correction method by traditional ballast correction method; (3) The propellant load for each tank should calculated and then

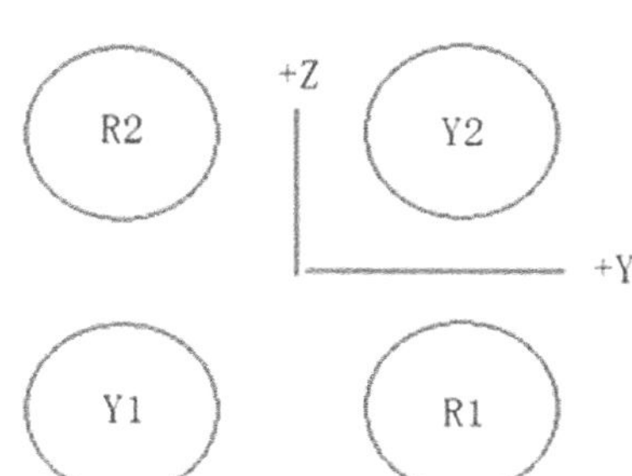

Fig. 13.16 Mass center test process

propellant is filled; and (4) According to test data after propellant filling and mission requirement, the best mass center could be achieved by fine-tuning of ballast.

3. Practical Application

Here is an example of the process and result of integrated ballast and propellant correction in the Chang'E-3 lunar lander. The dry mass of the lunar lander was adjusted by ballast. The Y_c is −8.53 mm before ballast was installed. After ballast was installed, the Y_c is −0.10 mm. By correction of the lunar lander mass center with conventional ballast, the risk beyond the mass center limit of propellant correction method was eliminated and it was prepared for propellant correction. The filling process of methyl hydrazine was as follows: (1) Each tank was filled in 425 kg. The Y_c was −0.27 mm; (2) The R1 was filled in 436.5 kg and 433.5 kg for R2. The Y_c was 1.53 mm, which meant that the mass center was changed significantly and the mass center correction by propellant instead of ballast was achieved. For the filling of oxidizer, it was filled uniformly. The test result of mass center did not change significantly. The Y_c is 0.5 mm. After propellant was loaded, the mass center of the lunar lander was corrected by traditional ballast again. The Y_c was 0.00 mm before ballast was installed, and Y_c was −0.61 mm after ballast was installed, which was optimal by design.

13.5 Summary

During development of the Chang'E-3 lunar lander, some professional metrology technologies were developed, which improved the accuracy, reliability and safety of metrology. For the alignment metrology, laser tracker, laser radar and theodolite integrated measurement technology were applied. The fusion of laser tracker and theodolite coordinate system was implemented by common point transformation, which was applied in normal direction measurement of RVS antenna end surface by laser tracker in the lunar lander coordinate system. At the same time, scanning measurement technology with laser radar and theodolite based on generalized USMN adjustment were applied in 3-D shape measurement of the interface ring in the lunar lander coordinate system to verify the stability of the interface ring under load. In the system leak test, the automated leak test technology and the flexible helium hood technology for quick leak positioning were applied. By application of highly accurate automatic leak test technology, automatic gas fill and vent, automatic leakage rate test, automatic comparison measurement with standard container and automatic sampling were implemented, which improved the efficiency and reliability of pipeline sealing performance test. At the same time, the quick leak positioning technology by flexible helium hood was applied to improve the accuracy of total leak test and the efficiency of quick leak positioning in complex conditions were realized, which gave a solution to the issue of leak positioning in complex system which could not apply bubble method. In the mass property test, technologies were applied including mass property test based on multiple DOF transformation, test data correction for propellant load, integrated ballast and propellant correction method, which gave a solution to inertia measurement of the lunar lander, and improved test accuracy and efficiency.

Professional metrology technology had played an important role in success of the Chang'E-3 lunar lander. Along with continuous development of mission requirements, the accuracy, efficiency and reliability of professional metrology technology will also be improved continuously.

References

1. Yan Z, Huang S (1998) Satellite manufacturing technology[M]. China Astronautic Publishing House, Beijing
2. Huang G, Qin G (2007) Measurement methods of the large scale 3-D coordinate and its system[J]. J Astronaut Metrol Meas 27(4):15–19
3. Wang H, Xu Z, Zhao K et al. (2011) Recent advances of mass property measuring technology for spacecraft[J]. Spacecraft Environ Eng 28(2):171–174
4. Jianxin LIU, Guowei RUAN (2012) Comparative analyses of the measurement schemes for the test of hang-staying, slow landing and obstacle-shunning of "CE" landing module[J]. Aerospace Manuf Technol 1(6):34–41
5. Yang Z (2007) An analysis of measurement technology in spacecraft assembly[J]. Spacecraft Environ Eng 24(6):390–392
6. Zhang H, Lin J, Zhu J (2012) Three-dimensional coordinate transformation accuracy and its influencing factors[J]. Opto-Electron Eng 39(10):26–31
7. Zhou Na, An Zhiyong, Li Yonghao (2011) Large-sized three-dimensional profile measurement technology based on laser radar[J]. Infrared and Laser Eng 40(12):2465–2468
8. Tao L, Yang Z, Ruan G et al. (2013) Alignment precision measurement method of satellite engine based on laser radar[J]. Spacecraft Environ Eng 30(3):328–331
9. Yang Linghui, Yang Xueyou, Lao Dabao et al (2010) Large-scale coordinate measurement method based on intersection of optical planes[J]. Infrared and Laser Eng 39(6):1105–1109
10. Chen Q, Sun G, Pang H et al. (2012). Large-size topography scanning with laser radar[J]. Spacecraft Environ Eng 29(2):215–219
11. Yan Z, Huang S (1997) A study of leakage detecting with mass spectrometry under non vacuum environment[J]. Chin Space Sci Technol 17(5):41–45
12. Yan Zhiping, Huang Shuying (2001) A study of leakage detecting with mass spectrometry under non vacuum environment[J]. Chin Space Sci Technol 21(3):42–46
13. Yan Z, Huang S (2002) Sensitivity estimation formula of leak detection method[J]. Chin Space Sci Technol 22(1):59–64
14. Du C, Chen M (2004) Research on new test method of mass properties of spacecraft[J]. Spacecraft Environ Eng 21(3):11–15
15. Chen Mian, Du Chen (2005) New technology of celerity measure satellite 3 Axis Centre of mass[J]. Spacecraft Environ Eng 22(6):358–359

Chapter 14
Electrical Testing Technology of Lunar Lander

14.1 Introduction

After system assembly and integration of the lunar lander was completed, the system electrical integration test should be conducted. The integration test was an important step in development of the lunar lander, which ran through the whole process from system integration to launch. The integrated system development was composed of system design, instrument development, assembly and integration and system electrical test.

The contents of electrical testing included testing requirement proposal, testing plan design, ground support system development, testing method design, testing implementation and testing result evaluation. In this chapter, the focus is on the contents in testing design phase, and the design methods and rationales are introduced. The contents include technology development status, requirements and technical characteristics, electrical testing plan, ground support system and electrical testing methods, and etc.

14.2 Development Status

Along with progress of space science and technology, the integrated test technology of spacecraft is developing gradually all over the world. Combined with test technology for missile, ship, airborne equipment and other equipment, an integrated test platform of cross-domain and cross-arms has been developed. Typical examples are the SCOS2000 of ESA [1], the CASS of the United States Navy, the LM STAR of the United States Air Force for the stealth fighter, as well as the next generation of automatic test (NxTest) [2] developed by the United States Department of Defense (DoD) and etc. The characteristics of integrated test technology development are listed as follows.

D.-Y. Yu et al., *Technology of Lunar Soft Lander*, Space Technology Library 38,
https://doi.org/10.1007/978-981-15-6580-9_14

(a) The extensive application of modular, synthetic and virtual instrument gradually replaces the test system composed of traditional desktop instruments [3]. In particular, with application of new data bus technologies such as LXI and PXI, flexible and reusable test equipment has been developed by combination of modular hardware platform and a great many test application, which makes the test system miniaturized, versatile, and highly automated.
(b) Standardized, portable, repeatable test languages make the development of Test Program Set (TPS) more efficient and accurate. The test language can describe the test process and express the external function of the test equipment, while it does not involve internal hard driver and does not depend on specific equipment model. Application of standardized test language in test program can be transplanted between different test systems, so as to speed up the test development and promote unify of test methods [4].
(c) The development of test standard provide foundation for modular recombination of test equipment and popularization of test language application, including the Interchangeable Virtual Instrument (IVI) series standard, the A Broad-Based Environment for Test (ABBET) standard, Automatic Test Markup Language (ATML) standard and the Automatic Test System (ATS) of the relevant international standards.

Along with development of space technology in China, especially deep space exploration technology, spacecraft electrical test technology has made great progress. The networked test system, with Overall CheckOut Equipment (OCOE) as the core structure, has been established to implement distributed collection and centralized management of data [5]. The extensive application of general and standardized instruments improves the integration and generalization level of the test system to be adapted to the test of data management protocol of CCSDS standard. The development and application of automated test software makes automatic test, automatic monitoring and automatic interpretation to be possible, and improves ability of test data analysis and mining, while improving test efficiency and saving labor cost.

14.3 Requirements and Technical Characteristics

14.3.1 Requirements

The core task of the electrical test was to fully test and verify all functions and specifications of the lunar lander including the follows.

(a) To verify the correctness and compatibility of electrical interface of all subsystems with structure, configuration and cables.
(b) To verify whether the functions and performance of the lunar lander and all subsystems meet the design requirements.
(c) To verify the correctness and rationality of flight procedures.

(d) To verify the correctness and compatibility of electrical interface design between the lunar lander and the lunar rover.
(e) To check and identify defects in the hardware and software design of the lunar lander and all subsystem.

For above tasks, the design and implementation of electrical test was conducted as the following principles and requirements.

(a) To conduct design of electrical test plan and develop ground support equipment in accordance with principles of safety, standardization, comprehension, accuracy, integrity, reliability and automation.
(b) To provide electric energy for the lunar lander with solar cells simulator.
(c) To receive telemetry data and payload data downlink, and send telecommand.
(d) To establish ground support system to unify data processing, distribution, display, storage, query and analysis, as well as control and manage ground test equipment centrally.
(e) To implement processing of data, image resolution and closed-loop test of subsystems.
(f) To meet the requirements of lunar lander test as well as test of the complex of the lunar lander and the lunar rover.
(g) To meet the requirements of local test and remote test.

14.3.2 Technical Characteristics

The challenges to electrical test of the lunar lander are listed as follows.

(a) The test process was complex. The electrical test of the lunar lander was classified into two phases: Electrical Module (EM) test and Flight Module (FM) test. For the FM test, besides electrical performance test in AIT facility, there were also tests including test in environment tests and test at launch site. At the same time, there were also joint test for complex of the lunar lander and the lunar rover, release and separation test. The test process at launch site was different from traditional spacecraft, which put forward new requirements for the design and optimization of the test process.
(b) There were lots of test modes. In order to adapt to test requirements in all phases, a variety of test modes should be designed. During environment test and test at launch site, the remote test mode was necessary, which should support the remote transmission of multiple frequency signals such as X band, S band and UHF band. In addition, there were also test modes for the lunar lander and the complex of the lunar lander with the lunar rover in the test process. All above modes and status were new in electrical test.
(c) The test methods were innovative. In order to cover the verification of all functions and performance specifications of the lunar lander, new test plan should be designed. For example, there were flight simulation test of powered descent, telecommunication function test between the lunar lander and the

lunar rover, hibernation and awakening function test and so on, which were new for electrical test and should be considered and designed.

(d) The ground support equipment was complex. There are many information sources and a great many data of the lunar lander. The data interface and information flow between ground support equipment and the lunar lander were also complex. It was necessary to develop a highly-automatic ground test system to meet the test requirements of all phases and test modes, which should be highly automated and implement test process with central control of terminal equipment.

(e) The requirement for test data interpretation was high. The test data of the Lander in all phases of development process were strictly interpreted, and the test data in different time and phases should be compared and analyzed. Based on the characteristics of the lunar lander, the test data interpretation criterion should be clearly defined, as well as the timing and interpretation rules of all data. Meanwhile, more flexible requirements were come up for automatic interpretation tools. Besides the traditional data query function, the data classification and extraction, characteristic value statistics, integrity analysis and other statistical analysis functions should also be accomplished.

14.4 Electrical Testing Plan

14.4.1 Testing Process

The electrical test of the lunar lander was divided into two phases: Electrical Module (EM) test and Flight Module (FM) test. In the EM test, there were electrical test in AIT facility and EMC test. In the FM test, there were electrical test in AIT facility, electrical test in environment tests and vacuum thermal test, interface tests with other engineering systems, electrical test at launch site. The test process of FM electrical test in AIT facility of the Chang'E-3 lunar lander test is shown in Fig. 14.1.

Unlike ordinary spacecraft, the lunar lander was usually designed to be a complex of multiple probes in order to accomplish complicated lunar missions. To ensure that the electrical test was completed within specified time in the project schedule, the maximum parallel principle was applied in the design of the test process: independent tests in parallel were conducted for different probes, including interface test, subsystem function and performance test, even parts of flight simulation test. After the test of single probe, complex test of multiple probes would be conducted.

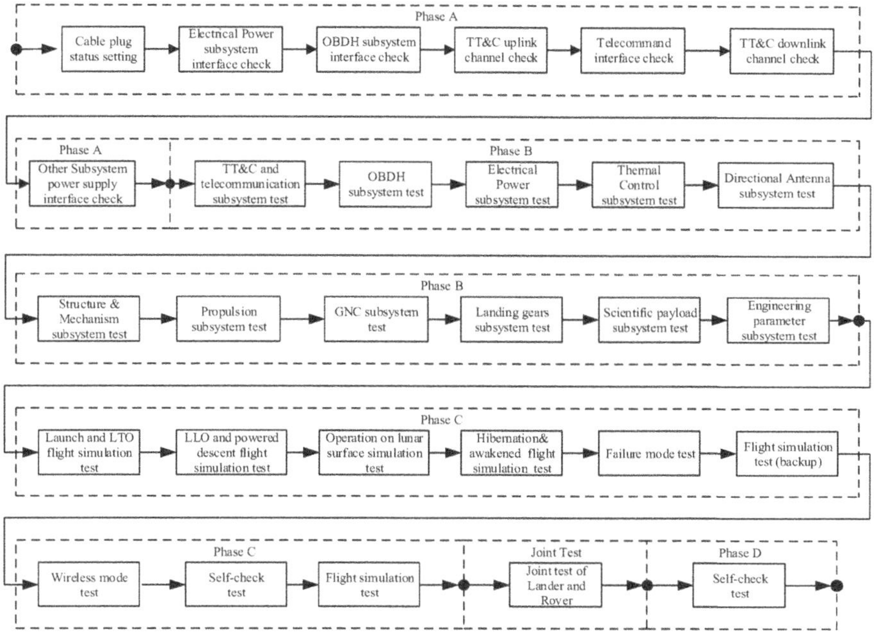

Fig. 14.1 Relationship between the orbit of the Moon and the magnetic tail of the Earth

14.4.2 Testing Mode

The testing mode for the lunar lander could be classified as single probe test or joint test, wired test or wireless test, and local test or remote test.

1. Single Probe Testing Mode

There was no electrical connection between the lunar lander and the lunar rover, which were tested separately. In the test process, single probe was tested at first, and then joint test could be conducted after all function of single probe was verified.

2. Joint Testing Mode

A complex of the lunar lander and the lunar rover was set up through cable connection. There were electrical interfaces between probes including power supply, command, signal acquisition and communication. The two probes were usually connected by separation plug. The joint test mode was applicable to the joint test in AIT facility, EMC test, mechanical environment test and thermal test of the lunar lander.

3. Wired Testing Mode

In the wired test, the transponder of the lunar Lander was connected with EGSE through high frequency ground cable instead of with the uplink/downlink antenna and the solid amplifier of the lunar lander. The mode had the following advantages.

(1) The high frequency signal could be tested to accurately determine the operation status of onboard instruments.
(2) It could avoid the interference of electromagnetic environment of the test site to the wireless signal.
(3) It could reduce the influence of electromagnetic radiation on human body.
Based on above advantages, wired test mode was applied in general test of the lunar lander, except the special conditions when wireless test mode should be applied.

4. Wireless Testing Mode

In the wireless test, the transponder and instruments of TT&C of the lunar lander were connected with the antenna and signal transmission between the lunar lander and EGSE was implemented through wireless channel. Generally, the signal quality was only monitored but not measured. Wireless test mode was applied under following situations.

(1) To verify wireless channel.
(2) To examine electromagnetic compatibility in EMC test.
(3) Unable to conduct wired test in system assembly process.
Wireless test mode was applied in wireless channel test, EMC test, mechanical environment test and test at launch site.

5. Local Testing Mode

When the lunar lander was close to ground test equipment and the signal transmission between the lunar lander and ground system was implemented by high and low frequency cables, it was the local test. It was usually applied for the test in AIT facility.

6. Remote Testing Mode

In the remote test, the lunar lander was far away from the test room and it was impossible to connect the lunar lander with ground test equipment through the test cable directly. Except that the power supply ground equipment and little of test equipment were placed close the lunar lander, OCOE equipment, TT&C equipment and display terminals were placed in the test room. Network signal and TT&C high frequency signal were transmitted by optical fiber link as shown in Fig. 14.2. Such mode could shorten test preparation time and improve test efficiency. In environment test and test at launch site, remote test mode was preferred.

The testing mode of the lunar lander in different test phases is shown in Table 14.1.

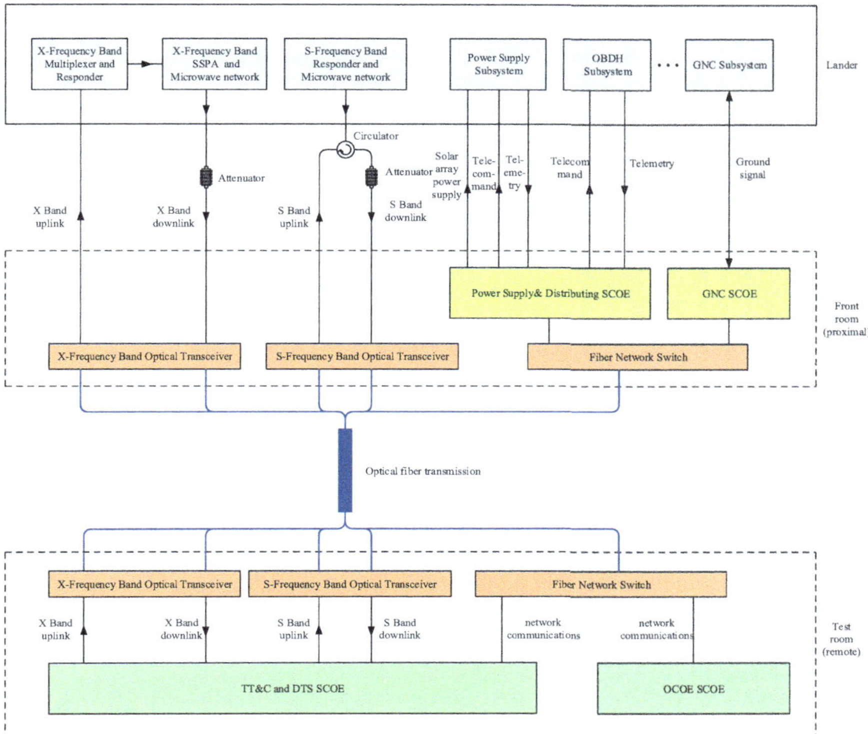

Fig. 14.2 Remote test mode

Table 14.1 Test modes in different test phases for the Chang'E-3 lunar lander

Test phase	Single probe	Complex	Wired	Wireless	Local	Remote
Single probe test	√		√	√	√	
Complex test		√	√		√	
EMC test	√	√		√		√
Mechanical Test		√		√		√
Thermal test	√	√	√			√
Electrical test in technical area at launch site	√	√	√	√	√	
Electrical test at launch pad		√		√		√

14.4.3 Design of Test Coverage

To verify the correctness of the function and performance of the lunar lander, the test coverage should be fully designed to ensure that instruments, subsystem, lunar lander system and interfaces to other engineering systems should meet the requirements.

The test coverage design of the lunar lander system is shown in Table 14.2.

During each test phase, the test coverage design of primary and backup instruments of the lunar lander should also be considered. The test modes were managed as follows.

(1) All the operation modes with primary instruments should be covered to verify normal operation modes on orbit.
(2) The backup instruments should be powered on at least once in all tests.
(3) For instruments or subsystem with hardware interface, the cross combination tests of primary and backup should be conducted.
(4) Optimization of test modes and test process should be conducted as far as possible to reduce movement of mechanisms and improve the test efficiency.

For example, in scientific payload test of the Chang'E-3 lunar lander, there were 12 test modes including OBDH computer primary and backup, 1553B bus primary and backup, data transmission channel primary and backup, primary and backup of payload electric controller, and 3 scientific instruments. The test coverage design is shown in Table 14.3.

Table 14.2 Test coverage design of lunar lander system

Test phase	Test			
	Interface check	Subsystem test	Flight simulation test	Self-check test
Electrical test in AIT facility	√	√	√	√
EMC test			√	√
Mechanical test				√
Thermal test			√	√
Interface tests with engineering systems			√	
Electrical test in technical area at launch site		√	√	√
Electrical test at launch pad				√

Table 14.3 Test coverage design for scientific payload

Test mode	OBDH computer	1553B bus	Data transmission channel	Payload controller	Payload operation mode
Mode 1	Backup	Backup	Primary	Primary	Payload 1 in operation alone, payload 2 and payload 3 in operation together
Mode 2	Backup	Backup	Primary	Primary	Payload 3 in operation alone, payload 2 and payload 1 in operation together
Mode 3	Backup	Backup	Primary	Primary	Payload 2 in operation alone, payload 3 and payload 1 in operation together
Mode 4	Primary	Primary	Primary	Primary	Payload 1, payload 2 and payload 3 in operation together (1)
Mode 5	Backup	Backup	Primary	Primary	Payload 1, payload 2 and payload 3 in operation together (1)
Mode 6	Backup	Backup	Primary	Primary	Payload 1, payload 2 and payload 3 in operation together (1)
Mode 7	Backup	Primary	Backup	Backup	Payload 1 in operation alone, payload 2 and payload 3 in operation together
Mode 8	Backup	Primary	Backup	Backup	Payload 3 in operation alone, payload 2 and payload 1 in operation together
Mode 9	Backup	Primary	Backup	Backup	Payload 1 in operation alone, payload 2 and payload 3 in operation together

(continued)

Table 14.3 (continued)

Test mode	OBDH computer	1553B bus	Data transmission channel	Payload controller	Payload operation mode
Mode 10	Primary	Backup	Backup	Backup	Payload 1, payload 2 and payload 3 in operation together (2)
Mode 11	Primary	Backup	Backup	Backup	Payload 1, payload 2 and payload 3 in operation together (2)
Mode 12	Primary	Backup	Backup	Backup	Payload 1, payload 2 and payload 3 in operation together (2)

(1) There were some differences in operation mode of payload 1, 2 and 3
(2) There were some differences in operation mode of payload 1, 2 and 3

14.4.4 Test Data Analysis Technology

During the spacecraft testing process, there will be a great many complicated test data from various channels, including telemetry data, payload data, ground test equipment data and etc. Interpretation and analysis of test data is the key basis to examine the function and performance of spacecraft.

The design for test data interpretation should be finished before implementation of test. According to the characteristics of object to be tested, a defined test data interpretation criterion should be formulated, including interpretation parameters, timing, method, normal range, data comparison requirements and interpretation report format. On the basis of interpretation criteria formulation, development of test data automatic interpretation system was also necessary to identify the abnormal data autonomously in time during tests and help the tester to discover problem. The script language for description of parameter range and corresponding data interpretation engine should be included in the test data automatic interpretation system. The script language could support sophisticated syntax structure and provided conditional judgment, loops, functions and logic control. All the interpretation scripts were stored in the background knowledge base. During the test, the data interpretation engine read the interpretation knowledge and deduced the test data. Then it distributed the interpretation results to the test monitoring terminals and stored them in the database. Automatic interpretation system could implement real-time interpretation of test data, effective data extraction of complex process, multi-parameter correlation interpretation, historical data statistical analysis and etc. It would improve the portability of interpretation knowledge, the easy use of interpretation tools and the accuracy of interpretation results, so that efficient and accurate test methods could be available for development and test of the lunar lander.

14.5 Ground Test System

The ground test system was the support system for the electric in different development phases, environment tests and tests at launch site of the lunar lander, which could not only implement the power supply and command control of the lunar lander, as well as collection, processing, display and interpretation of downlink data, but also meet the requirements for remote testing and joint testing. Reliable, universal equipment hardware configuration and test software were applied in ground test system. In the design and development of ground test system, there were the design of power supply relationship, information flow, the composition and verification of ground test equipment [6].

14.5.1 Ground Power Supply

In the test of earlier deep space probes, DC voltage stabilized power supply and solar cell simulator were applied. Up to date, only solar cell simulator could be applied for the lunar lander. The lunar lander was powered and charged by solar cell simulator through the umbilical connector, as well as wired command transmission and parameter collection function by ground equipment. The power supply relationship between the lunar lander and the ground equipment is shown in Fig. 14.3.

The following four aspects should be considered in design.

(1) Safety design. Measures such as isolation and protection to ensure safety of the lunar lander and onboard instruments should be applied. When the ground equipment was short circuit, control failure and so on, it should not cause damage to on board instruments.
(2) Electric power requirement. The ground power supply equipment should provide enough energy for the lunar lander that met the requirements of its maximum operation mode. It could simulate the performance of solar cell while the loss of long cable transmission should be considered.
(3) Automation. In order to meet the requirements of automatic test, the test equipment should have the program-controlled interface. Through the automation test software, it could implement unified distribution control of power supply, wired command transmission, wired parameter acquisition and storage function, and respond to the setting command of OCOE.
(4) Self-check. A power controller simulator of the lunar lander could provide self-check function for ground equipment, so that the operation status of ground equipment could be examined quickly and accurately after preparation of equipment and cable.

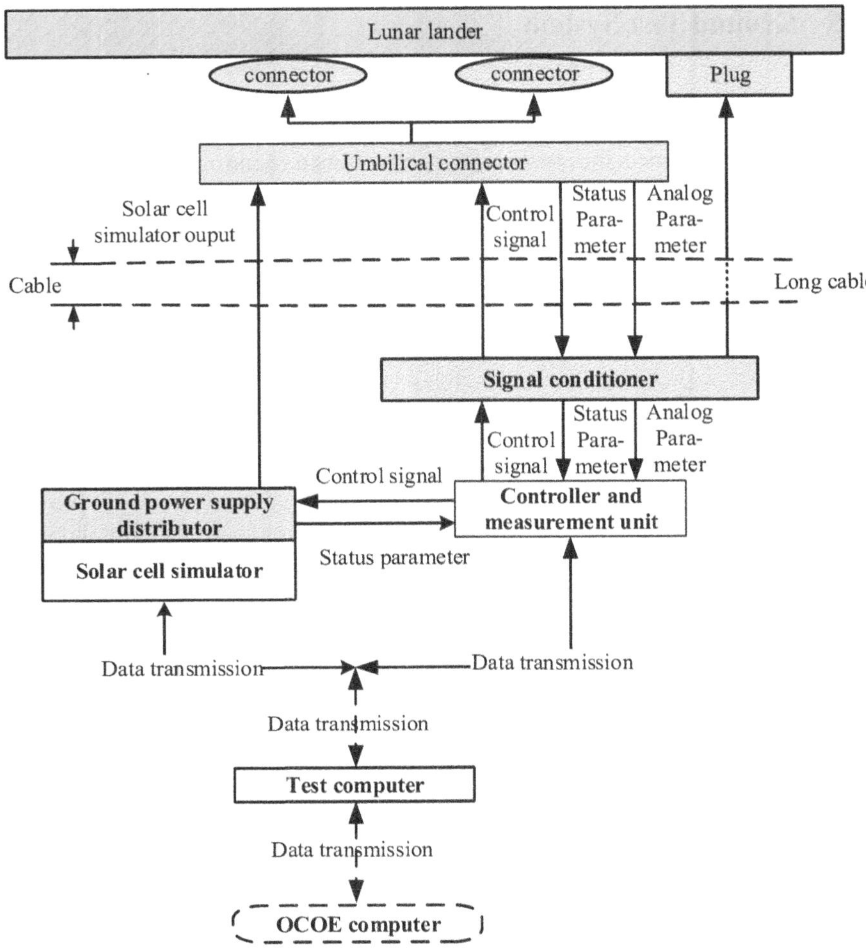

Fig. 14.3 Power supply relationship between lunar lander and ground equipment

14.5.2 *Information Flow Between Lunar Lander and Ground Equipment*

Because the joint mode of the lunar lander and the lunar rover was applied in most of flight phases, the processing of the test data both in single probe test mode and joint mode should be considered in design of integrated test. The information flow and information processing mechanism between the Chang'E-3 lunar lander and ground test system with following characteristics is shown in Fig. 14.4.

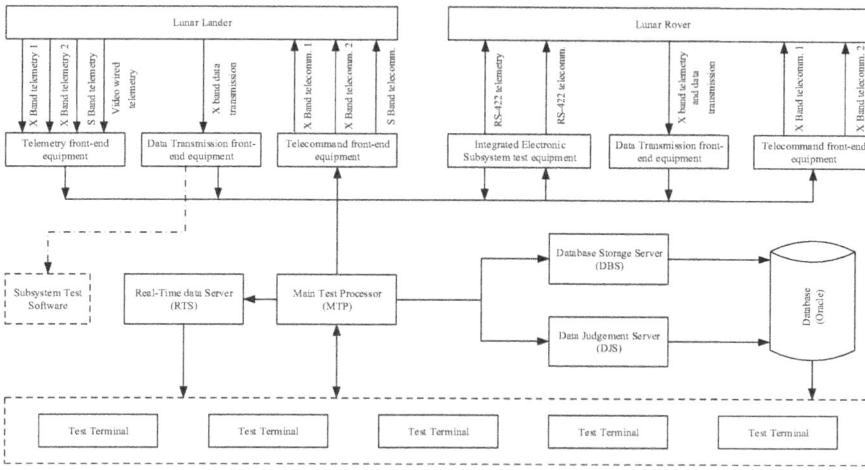

Fig. 14.4 Design of information flow between lunar lander and ground equipment

(1) The downlink test data flow of the lunar lander and the lunar rover was uniformly designed and planned. A unified test data processing management system was applied to implement archiving, processing, distribution and storage of the telemetry data of two probes. There were 4 telemetry channels of the lunar lander: two for X band and two for S band, as well as one telemetry downlink channel and one data transmission downlink channel of the lunar rover. During the single probe test, data could be monitored and queried respectively. During the joint test, processing of telemetry and data transmission data of the lunar rover forwarded to the lunar lander by RS-422 channel and UHF relay channel could be implemented, and test data of two probes could be uniformly monitored.

(2) The uplink telecommand of two probes was uniformly designed. The standard naming rules of the command were applied to distinguish command of two probes effectively, so as to ensure uniqueness of command and improve the test security. For the complexity of structure and mechanism subsystem, diversity of data injection command type and format, the configurable and flexible command generation & verification system should be developed, which could generate data injection file by command format and verify the correctness. On the one hand, it was applied to generate all data injection instructions in system test. On the other hand, it was applied to verify data injection during flight control.

(3) The unified data access interface was established. Since test data including engineering parameters measurement data and images, scientific payload exploration data and images were received centrally, the OCOE should interact with multiple Special CheckOut Equipment (SCOE). Therefore, unified data access interface for telemetry and data transmission, as well as archive format,

were established in integrated test. All kinds of test data could be accessed and acquired by subsystem software, which made it easy to analyze test data.

14.5.3 Configuration of Ground Test System

The ground test system was distributed local network system with two levels management. The OCOE was connected with special checkout equipment (SCOE) of subsystem via Ethernet. Graphic data display terminals were configured to monitor test data. Test data interpretation terminals were configured, and test database provided database storage service to all subsystems. The configuration of ground test system of the Chang'E-3 lunar lander is shown in Fig. 14.5.

The OCOE was the management center of test process for data processing, control command management, automatic test implementation, analysis of test data, command and control. It could implement the remote control of front-end equipment of power supply and distribution subsystem, TT&C and telecommunication subsystem, the reception, processing and display of telemetry and data transmission data, the generation of telecommand and data injection, and the support for storage, query and interpretation service of test data.

There were 5 types of SCOE as follows.

(1) Test equipment of power supply and distribution subsystem: to implement ground power supply, wired command sending and wired parameter acquisition for the Lander.

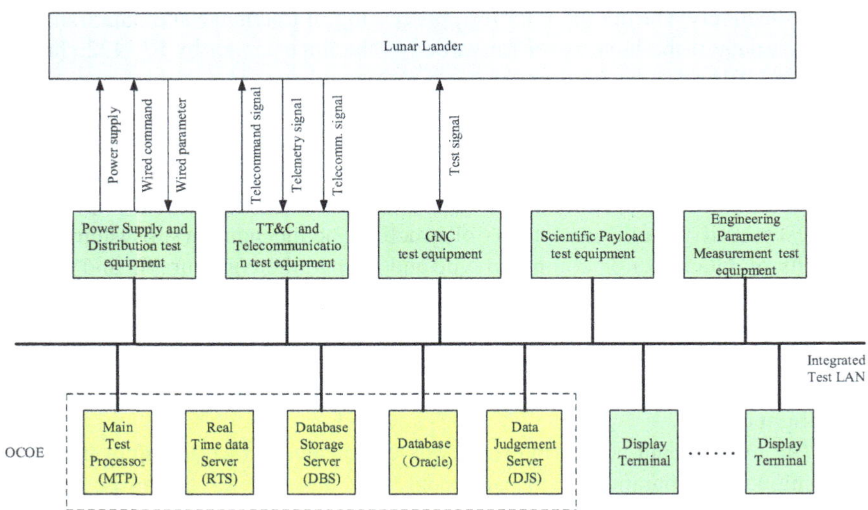

Fig. 14.5 Configuration of ground test system of the Chang'E-3 lunar lander

(2) Test equipment of TT&C and Telecommunication subsystem: to implement signal reception and demodulation of telemetry and payload transmission data, signal modulation and transmission of telecommand data, and the automatic test of the subsystem performance.
(3) Test equipment of GNC subsystem: to test the functions, performances and control accuracy in open-loop and closed-loop test of GNC subsystem.
(4) Test equipment of scientific payload subsystem: to process data and implement the test of scientific payload subsystem instruments such as cameras, lunar based observatory, Ultra Violet Camera and etc.
(5) Test equipment of engineering parameter measurement subsystem: to process data and test the monitor camera and lunar dust detector.

14.5.4 Verification of Ground Test Equipment

Before the system test of the lunar lander in all phases, the ground test equipment should be calibrated. The SCOE should be tested and debugged according to the test plan of subsystem ground test equipment. The interface of OCOE to MTP, database server, test terminals, DJS and display terminals and operation status should be examined. After internal joint test of SCOES was finished, system-level joint test between subsystem SCOEs as well as OCOE and SCOE, should be conducted according to the test plan. Generally there were joint tests including the joint test of OCOE, power supply ground equipment and the lunar lander simulator, the joint test of OCOE and telemetry front-end, telecommand front-end, data transmission front-end, the joint test of OCOE and subsystem SCOEs including GNC, propulsion, engineering data measurement and scientific payload, the joint test of data transmission front-end and subsystem SCOEs of GNC, engineering data measurement and scientific payload. For the test with test sequence, the verification of sending sequence should be conducted before the lunar lander was powered on. In the remote test mode, the optical terminals and optical links should be verified before the test began.

14.6 Electrical Testing Methodology

14.6.1 Tests

The electrical test content of the lunar lander included interface checkout, subsystem detail test of function and performance, flight simulation test and basic status test, as shown in Table 14.4. The purpose of interface checkout was to verify the correctness and compatibility of the interface between subsystems and instruments in order to ensure the test safety and establish the test state of the lunar lander. The purpose of subsystem detail test was to verify the correctness of function and performance of the subsystem and the coordination between subsystems. The purpose of flight

Table 14.4 Electrical test content of lunar lander

Classification	Test content
Interface test	Grounding network checkout
	Power supply and distribution subsystem interface checkout
	Power supply interface checkout of OBDH subsystem
	TT&C uplink channel test
	Telecommand interface checkout
	Downlink channel test
	Power supply interface checkout of other subsystems
Subsystem detail test	Detail test of power supply and distribution subsystem (including hibernation and awakening test)
	Detail test of TT&C and telecommunication subsystem (including communication test between the lunar lander and the lunar rover)
	Detail test of OBDH subsystem
	Detail test of thermal control subsystem
	Detail test of engineering parameter measurement subsystem
	Detail test of scientific payload subsystem
	Detail test of directional antenna subsystem
	Detail test of structure and mechanism subsystem
	Detail test of landing gear subsystem
	Detail test of GNC subsystem
	Detail test of propulsion subsystem
Flight simulation test	Wireless mode test
	Flight simulation test for launch and LTO phase
	Flight simulation test for circumlunar phase and powered descent phase
	Flight simulation test for operation on lunar surface
	Flight simulation test for hibernation and awakening
	Flight simulation test for all mission phases
	Failure mode test
Basic status test	Self-test of lunar lander

simulation test was to verify flight events in all flight phases according to the flight procedure, and to verify the reasonableness of flight procedure design. The purpose of basic status test was to determine whether the lunar lander was in good status by subsystem self-test.

On the basis of heritage of electrical testing methodology for previous deep space probes, in order to verify new technology and new instrument function of the lunar lander, new test methodology was designed for new tests. For example, there are examples including flight simulation test for all mission phases, flight simulation test for powered descent phase, communication function test between the lunar lander and the lunar rover, function test for hibernation and awakening as follows.

14.6.2 Flight Simulation Test for All Mission Phases

The ground flight simulation test design was as complex as mission profile of the Chang'E-3 lunar lander, including different modes such as LTO and powered descent in complex status with the lunar rover, release and separation on lunar surface, operation on lunar surface and etc. For both single probe test and all mission phases test in complex status, modularized design idea was applied. The flight procedure was classified into different units, which were designed separately with different test procedures and sequence. When the test was implemented, test units were combined according to test requirements to form test cases in order to improve the reuse level of test cases and the efficiency of test design.

For example, in the flight simulation test before release and separation process on lunar surface of Change'E-3 lunar lander, there were 16 units as listed in Table 14.5. The unit test process is shown in Fig. 14.6. After one test unit was accomplished, it could be chosen to enter the next unit or finish the test after the status recovery. The modularized design could greatly improve the flexibility of test management and the efficiency of test implementation.

Table 14.5 Electrical test content of the lunar rover

Unit no	Test content
Unit 1	Status setting before launch
Unit 2	Flight simulation test for launch phase
Unit 3	Flight simulation test of Trajectory Maneuver Correction (TCM)-1
Unit 4	Flight simulation test of TCM-3
Unit 5	Flight simulation test for Lunar Orbit Insertion
Unit 6	Flight simulation test for orbit change in 100 km × 15 km LLO
Unit 7	Flight simulation test for powered descent phase
Unit 8	Flight simulation test for initialization after landing
Unit 9	Flight simulation test for unlocking of directional antenna, transfer mechanism and the rover wheel
Unit 10	Flight simulation test for first camera imaging
Unit 11	Flight simulation test for antenna deployment and data transmission
Unit 12	Flight simulation test for propellant depletion
Unit 13	Flight simulation test for data playback
Unit 14	Flight simulation test for propulsion subsystem power-off and gas vent
Unit 15	Flight simulation test for TT&C after GNC power-off
Unit 16	Status recovery

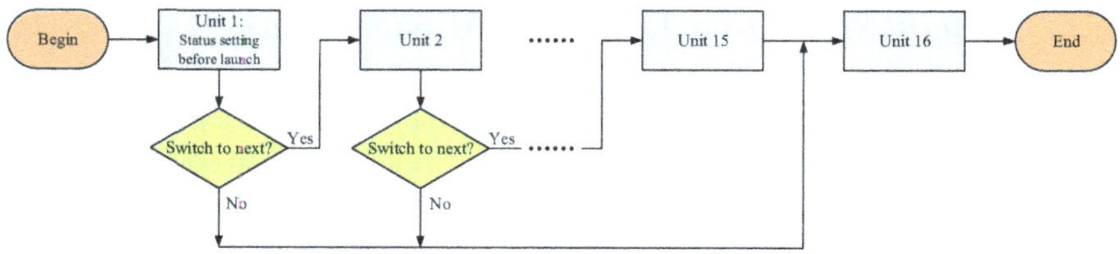

Fig. 14.6 Test process of unit test

Table 14.6 Flight simulation test for powered descent phase

System test phase	Test content
Phase B	Flight simulation test for GNC subsystem in powered descent
Phase C	Primary and back flight simulation test for all mission phases with the lunar lander
Complex test	Primary and back flight simulation test for all mission phases with the complex
EMC test	Primary and back flight simulation test for all mission phases with the complex
Thermal test	Flight simulation test for powered descent with the lunar lander
Test at launch site	Primary and back flight simulation test for all mission phases with the complex

14.6.3 Flight Simulation Test for Powered Descent Phase

One of the key processes of the Chang'E-3 lunar lander was the powered descent, which should be tested for many times at different phases of electrical test, as shown in Table 14.6.

In order to implement the flight simulation test for powered descent, several kinds of sensor simulators should be developed, including echo simulator, laser ranging simulator, optical simulator and etc. These simulators would simulate signals of microwave, distance signal and image stimulation, so that the GNC central control unite of the lunar lander could automatically complete adjusting, approaching, hovering, obstacle avoidance, slow descending and other activities according to these inputs, and finally be softly landed on lunar surface. In addition, test procedures and commands of flight simulations test were compiled strictly with the time sequence according to the flight procedures on orbit. The test equipment configuration for flight simulation test for powered descent is shown in Fig. 14.7.

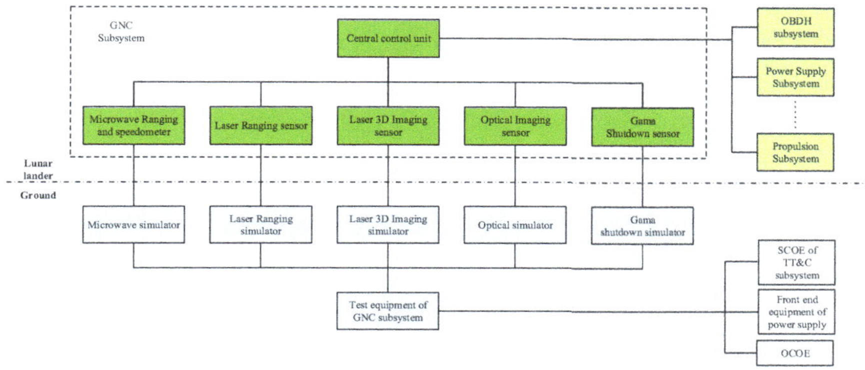

Fig. 14.7 Test equipment configuration for flight simulation test for powered descent

14.6.4 Test of Communication Function Between the Lunar Lander and the Lunar Rover

In order to implement telecommunication between the lunar lander and the lunar rover, there were UHF receiver, transmitter and antenna, the telemetry and scientific payload data of the lunar rover could be transmitted through the UHF channel between the lunar lander and the lunar rover, and then sent back to the ground by data transmission channel of the lunar lander after it was multiplexed.

For example, the communication function tests between the lunar lander and the lunar rover in integrated test phase were classified into 4 categories.

(1) Performance test of the receiver and transmitter. The test mode was wired test with the lunar lander, where the UHF receiver and transmitter of the lunar lander were connected with test equipment the TT&C and Telecommunication subsystem by ground cable. The technical specifications such as demodulation threshold of the receiver and the output power of the transmitter, should be tested.

(2) Data sending and reception function test of UHF band. The test mode was wired test with the lunar lander, where the simulated data from ground test equipment was sent to verify the function correctness of UHF receiver, OBDH data multiplexing and forwarding. At the same time, the correctness of the data format would be tested by reception data of ground test equipment from the UHF transmitter.

(3) Wired communication function test at UHF band between the lunar lander and the lunar rover. The test mode was wired joint test with the complex of the lunar rover and the lunar rover, where the UHF transmitter of one probe was connected with the UHF receiver of another probe by ground cable and attenuator. The data of the lunar rover would be forwarded to ground test equipment through data transmission channel of the lunar lander after it was sent to the lunar rover through UHF channel between probes. Meanwhile, the

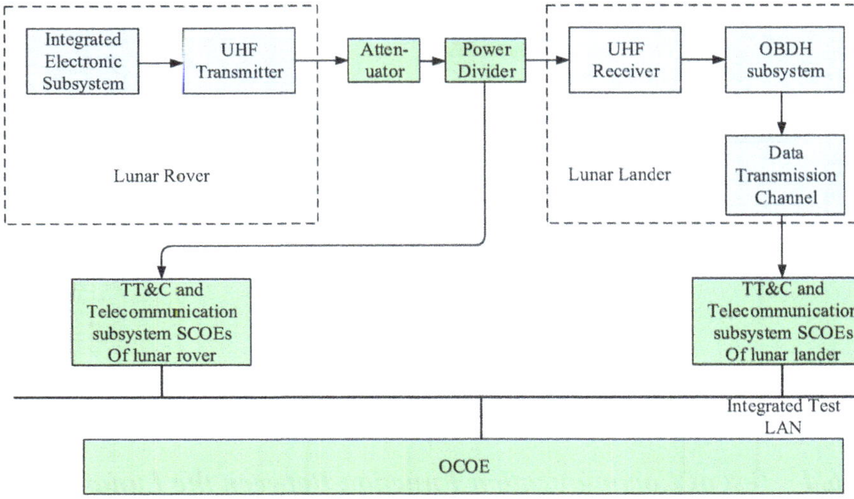

Fig. 14.8 Communication function test between the lunar rover and the lunar rover

same data would be directly sent to ground test equipment of the lunar rover, as shown in Fig. 14.8. So, the correctness of communication function between probes could be verified by comparison of data from two different channels.

(4) Wireless communication function test at UHF band between the lunar lander and the lunar rover. The test mode was wireless joint test with the complex of the lunar rover and the lunar rover, where the UHF signal was transmitted and received through wireless channel, and the UHF signal was received by ground test equipment at the same time. The wireless test process was same as that in wired mode. By the wireless test, the correctness of the UHF antennas and the interface between the antenna and other instruments could be verified.

14.6.5 Function Test of Hibernation and Awakening

During lunar day, the lunar lander was powered by the solar panel to supply electric power to the load. However, there was no sunlight during lunar night and the battery packs onboard could not maintain the minimum power requirement of the lunar lander platform, so it was necessary to enter the hibernation status by power-off of all instruments safely and reliably. After there was sunlight, the solar cell would gradually supply power to the platform load and resume to the normal operation mode in lunar day.

During the integrated test, according to real illumination condition on lunar surface, the solar cell simulator was designed to simulate the power supply curve and simulate the characteristics of solar cell power supply under different illumination conditions, as shown in Fig. 14.9. During transition process from lunar day to

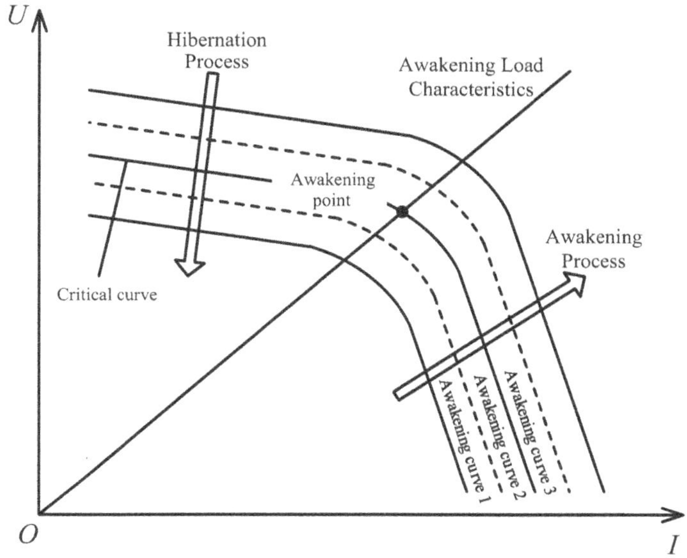

Fig. 14.9 Power Supply Curve of solar cell simulator in function test of hibernation and awakening

lunar night, the gradual reduce of ground power supply would make the lunar lander transfer to hibernation status. During transition process from lunar night to lunar day, the gradual increasing of ground power supply would awaken the lunar lander autonomously.

14.7 Summary

The development and the successful applications of electrical test technology of the lunar lander could promote the development of spacecraft electrical test technology furthermore, and accumulated helpful experience for development and test of future deep space probes. The complexity and innovation of future deep space probes will introduce new topics for electrical test technology and promote progress of test technology to great extent. Deep space telecommunication is extending to Ka and higher bands [7], which makes that the ground test system shall adapt to the multi-band, multi-system and multi-channel test requirements. The application of channel coding technologies, such as Turbo code and Low Density Parity Check (LDPC) code, requires that the corresponding encoding and decoding functions shall be developed for the test equipment. Some new requirements for ground test equipment and test methods are put forwarded due to the characteristics of long distance, low signal-to-noise ratio and low bit rate in deep space telecommunication. When the packet telemetry and telecommand system is applied to the data management of Advanced Orbiting System (AOS), the ground test software shall be able to fully and effectively verify the correctness of the virtual channel scheduling and packet scheduling strategy. The design of complicated information flow, such as multi-probe telecommunication and relay telecommunication [8], will result in higher requirements for the timeliness and integrity of test data acquisition, processing and storage. And it is a brand new subject to test the nuclear power system in future deep space probes [9], which make it necessary to develop the ground power supply equipment that can replace and simulate the characteristics of the nuclear power supply. In term of testing mode and process, multi-probe complex test is an important topic faced in of deep space exploration. In addition, analysis and verification on test coverage of complicated system shall be deeply researched, as well as the design and implementation of intelligent test and verification methodology. Along with development of test technology, the theory and methodology of ground virtual test and verification will also be introduced into the electrical test of deep space probes.

References

1. Robert M, Catherine L, Johan S et al (2002) SCOS-2000 MultiSat: the multi-satellite extension of a mission controls system: SpaceOps. SpaceOps, USA
2. Marvin R (2005) NxTest and the development of synthetic instrumentation. Def Electron 2:6–10
3. Michael JS, David D (2003) ATE open system platform. IEEE-P1552 structured architecture for test systems (SATS). IEEE, USA
4. Nestor P (2003) A vision for the future ESA Ground Operations Software: 16th Flight Dynamics Workshop. 16th Flight Dynamics Workshop, USA
5. Wang Q (2008) Electrical test technology of spacecraft. China Science and Technology Press, Beijing
6. Jong-oh P, Jong-yoen C et al (2002) Electrical Ground Support Equipment (EGSE) design for small satellite. Astron Space Sci 19(3):215–224
7. Simons RN, Wintuchy EG (2011) Demonstration of multi-Gbps data rates at Ka-Band using software-defined modem and broadband high power amplifier for space communications. Glenn Research Center, USA
8. Jiao J, Yang Z, Zhou J et al (2011) Design of relay file delivery protocol in deep space communication. J Spacecraft TT&C Technol 30(ZL):32–36
9. Cai S, He S (2004) Retrospect of development for radioisotope power systems in space and its prospect of application in new century. Chin J Nuclear Sci Eng 24(2):97–104

Chapter 15
System Testing and Validation Technology of Lunar Lander

15.1 Introduction

In term of environmental adaptability and environment test, there were so many similarities between the lunar lander and Earth orbit spacecraft. However, there were still some differences, especially in the particularity of soft landing environment and lunar surface environment, which made it necessary to solve a series of specific technical difficulties in ground environment simulation tests, build and improve the relevant environmental test facilities, and conduct adequate tests to identify the potential problems and deficiencies of the design or manufacture quality.

According to operation environment during the whole mission life cycle of the lunar lander, under the premise of synthetically consideration of benefit, schedule and feasibility, it was necessary to verify the design of the product, the correctness and reasonableness of the technological process on the ground as far as possible, which would identify the defects in manufacturing quality, material and process of the product, eliminate the early failure, and improve the operational reliability during working life.

In this chapter, the special system-level ground validation tests during development of the lunar lander are introduced, including test purpose, methodology, test system design and practical application. The conventional system-level ground validation tests (such as sound test, sinusoidal vibration test, satellite/launch vehicle dock and separation test, thermal test, integrated electrical performance test and EMC test) will not be discussed, which can be referred to the related articles of spacecraft environmental test.

D.-Y. Yu et al., *Technology of Lunar Soft Lander*, Space Technology Library 38,
https://doi.org/10.1007/978-981-15-6580-9_15

15.2 Development Status

Before implementation of the Chang'E-3 mission, only the United States and the Soviet Union had succeeded in lunar soft landing and conducted systematic test and validation on the ground. Therefore, in this section, the system-level validation test programs by the United States and the Soviet Union in lunar exploration are introduced.

15.2.1 System Test Implemented in the United States

In the 1960s and 1970s, during development of Apollo lunar exploration program, systematic and comprehensive ground test and validation for lunar landing mission were conducted, including the Lunar Landing Research Facility (LLRF), the Lunar Landing Research Vehicle (LLRV), landing stability test, landing impact test and plume impact test.

1. LLRF

The LLRF is shown in Fig. 15.1. Its function was to simulate gravity on lunar surface to acquire flight dynamic characteristics of the lunar module, and verify the control characteristics of servo control system of the lunar module in low gravity. Meanwhile, it could provide the astronaut with long term simulation training for lunar landing control.The LLRF was composed of gantry structure, low gravity simulation system, traveling bridge crane and test vehicle. Simulation of low gravity was implemented by an overhead suspension system, which consisted of rope for gimbal spreader, lift unit, crane and corresponding servo control system. One end of the rope was connected with gimbal spreader and the gimbal spreader was connected with the test

Fig. 15.1 LLRF [1]

vehicle through two connection points. The line between gimbal spreader and two connection points passed through mass center of the test vehicle. The gimbal spreader could provide pith, yaw and roll DOF. The other end of the rope was connected with the lift unit of the crane. The lift unit could control the rope to move upward or downward by servo control system. The lift force of the rope could be maintained by force sensor and server control unit, so as to ensure that the lift force was 5/6 weight of the test vehicle. The crane could move in two DOF on the horizontal plane provided by the gantry structure, which could ensure that the lift force should be always vertical and upward by tracking position change of the test vehicle with angle sensor on the rope [1].

2. LLRV

The LLRV was designed to simulate final flight trajectory of lunar soft landing process, and it could also help to train the astronaut capability to control the lunar module during soft landing and lift-off on lunar surface. The LLRV could simulate the landing process when the altitude was about 457.5 m, including hovering and horizontal motion, as well as lift-off on lunar surface, which could be applied to train the astronaut to control the lunar module to hover down and find landing site. The LLRV is shown in Fig. 15.2.

The LLRV consisted of a pyramid-shaped structural frame with four truss-type landing legs, lunar gravity simulation system and attitude control system. The landing legs could support the LLRV. The simulation of lunar gravity was implemented by two independent jet propulsion systems: A jet engine that was mounted vertically on the gimbal bracket at vehicle center was applied to balance the 5/6 weight of LLRV; Two hydrogen-peroxide rocket engines which were mounted on main frame of vehicle were applied to balance the other 1/6 weight of LLRV. Two rocket engines were symmetrical on both sides of the jet engine. The pilot could control the vehicle to maneuver horizontally by two rocket engines and inclination of vehicle. The attitude control was implemented by 16 thrusters which could provide 80 ~ 400 N force. In addition, in order to simulate lunar environment, automatic air dynamic compensation system was applied for the LLRV and the air dynamic coefficients

Fig. 15.2 LLRV

were acquired by 1:0.3 model in wind tunnel. The LLRV could simulate gravity environment on lunar surface very well. However, the system was complex, expensive and of high test cost, and there was potential hazard. During tests, there were accidents in which the vehicle went out of control and exploded [2, 3].

3. Landing Stability Test

In order to verify the landing gears performance and landing stability of Apollo lunar module, the 1/6 geometric scale model test under Earth gravity and full-scale model test under simulation lunar gravity were conducted in the United States The test results of landing gear stroke, acceleration at mass center and attitude motion acquired by two methods were in good consistency. The full-scale model could perfectly reproduce the dynamic process under symmetrical landing condition, which could be applied to study the stability during landing process.

1) 1/6 scale model test

The 1/6 scale model could satisfy test requirements for structure, size and mass. The scale model was by material with same yield strength as the full-scale model. When the 1/6 scale model was tested, it was installed on the release device at the bottom of single pendulum, as shown in Fig. 15.3. In the test, when the release device passed through the lowest point of single pendulum, the 1/6 scale test model was released by release device. Then the test model would move in a parabola at predefined horizontal velocity and contact with the landing surface to simulate soft landing process [4].

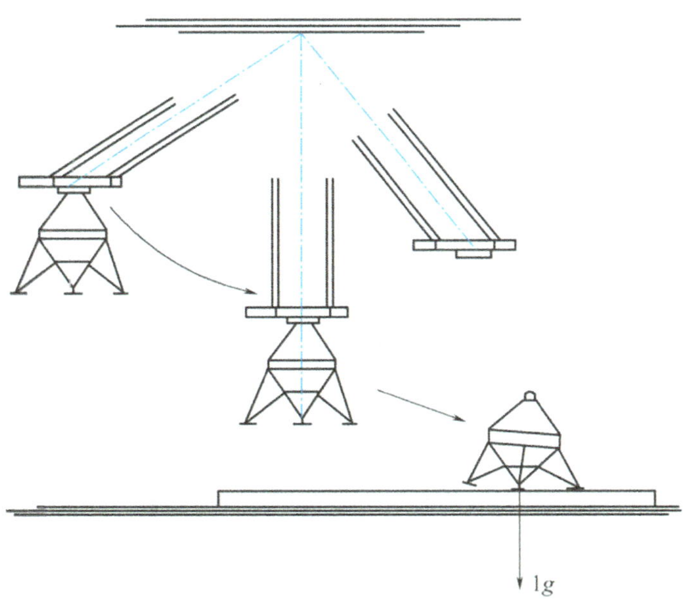

Fig. 15.3 Principle of 1/6 scale model test [4]

2) Full-scale model test with lunar gravity simulator

In order to simulate the soft landing process more practically and comprehensively, the landing stability test with full-scale model could be conducted by the lunar-gravity simulator. The lunar-gravity simulator consisted of inclined landing surface, trolley and track and a fixed-length cable which supported the test model from the trolley. The normal vector of static force on landing surface by the test model which was supported by the cable was equal to its lunar weight, as shown in Fig. 15.4 [4, 5].

4. Landing Impact Test

The landing impact test was conducted to reproduce landing impact dynamic process, so as to verify structural adequacy for lunar landing under all harsh conditions.

In the landing impact test of Apollo program, the test article was LTA-3 (Lunar Module test article 3) with flight model of landing legs (the footpad was test article). In the test, LTA-3 was released at a predefined height to drop to inclined landing platform (which could simulate landing surface environment), so as to simulate the landing impact kinematic process. The initial drop height was approximate 210.82

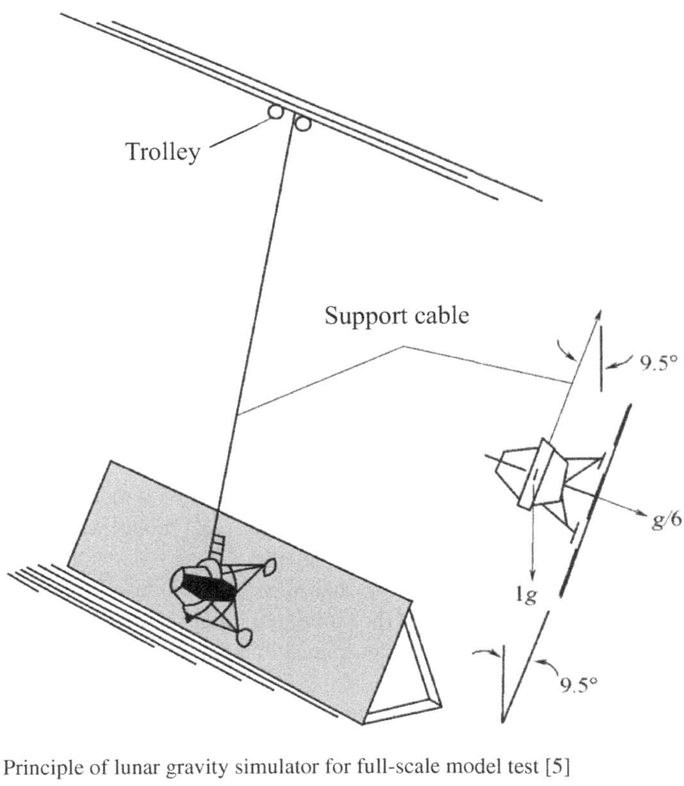

Fig. 15.4 Principle of lunar gravity simulator for full-scale model test [5]

~ 1066.8 mm, which could provide an impact velocity range of 2.04–4.57 m/s. Total 16 tests were conducted for LTA-3. Six of them were reduced-level tests. The reduced-level test was accomplished with a special test landing gear that used cartridges with crush characteristics equal to 75% of flight hardware [6–8].

5. Plume Effect Test

In the Apollo mission, the plume of the 16 attitude control engines in the service module would act on the surface of the service module and the lunar module. The plume of the land engine during landing would act on the landing gears and the lower surface of the descent module. Therefore, special plume protection was necessary for the surface of the service module and the lunar module, and the landing gears. The plume impact tests were applied to verify the effectiveness of the protection.

During development of Apollo, in the full-scale attitude control thrust system (RCS) plume impact test conducted in the Space Environment Simulation Laboratory (SESL), the full-scale lunar module system validation test was completed. Meanwhile, a full-scale descent profile was applied for the plume impact test to acquire the plume heating flux and pressure data [9].

15.2.2 System Test Implemented in the Soviet Union

In order to verify the strength and dynamics of the lunar lander, in the Soviet Union lunar exploration program, drop experiment to simulate the landing conditions on lunar surface based on the principle of energy equivalence was conducted. Under Earth gravity, it could simulate the landing process of the lunar lander with same landing energy as the lunar landing process. In the experiment, the attitude parameters and velocity parameters of the lunar lander were simulated when the lunar lander touched lunar surface.

The drop test platform for the lunar lander is shown in Fig. 15.5. It consisted of following parts.

(1) Landing platform: The angle between the landing platform and the horizontal plane could be adjusted, and the maximum slope of the landing surface could be simulated at 20°.
(2) Needle shaped steel frame structure: It consisted of winch and steel cable, which can be applied to provide the lunar lander test model with initial landing velocity parameters and attitude parameters.
(3) Simulated lunar soil: The medium hardness soil was used to simulate lunar soil, and the bearing strength of the simulated lunar soil was 34.3 ~ 49 kPa/cm.
(4) Electric lock: It was applied to release the lunar lander simulation at predefined position.
(5) Camera: It was used to record the test process.

The needle shaped steel frame structure consisted of winch and steel cable with electric lock, by which the initial drop height of the lunar lander test model could

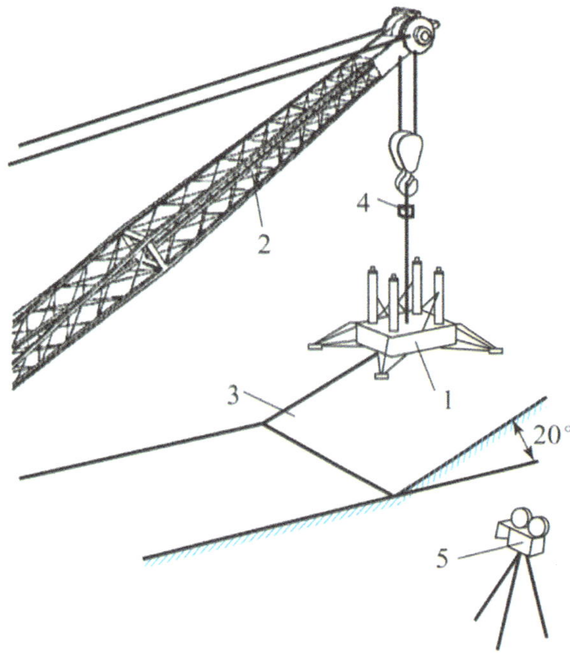

Fig. 15.5 Lunar lander drop test **1** Lunar lander test model; **2** Needle shaped steel frame structure; **3** Landing platform; **4** Electric lock; **5** Camera

be adjusted to get the required vertical landing velocity. In order to simulate the horizontal motion of the lunar lander in contact with the simulated lunar soil, the needle shaped steel frame structure could be rotated according to the required velocity relative to the vertical axis. The test platform could simulate the landing process with the vertical landing velocity of 5 m/s and the horizontal landing velocity of 3 m/s. The drop test process of Luna-21 lunar lander is shown in Fig. 15.6.

15.3 Requirement

15.3.1 Principles of Test Planning

The basic principles should be followed when the test systems for system level ground validation of the lunar rover was planned.

(1) It should be able to fully verify the test object.
(2) The experiment boundary condition should not be lower than the real environment condition of the lunar lander.
(3) On the basis of test objective satisfied, the test method should be as simple, safe and reliable as possible.

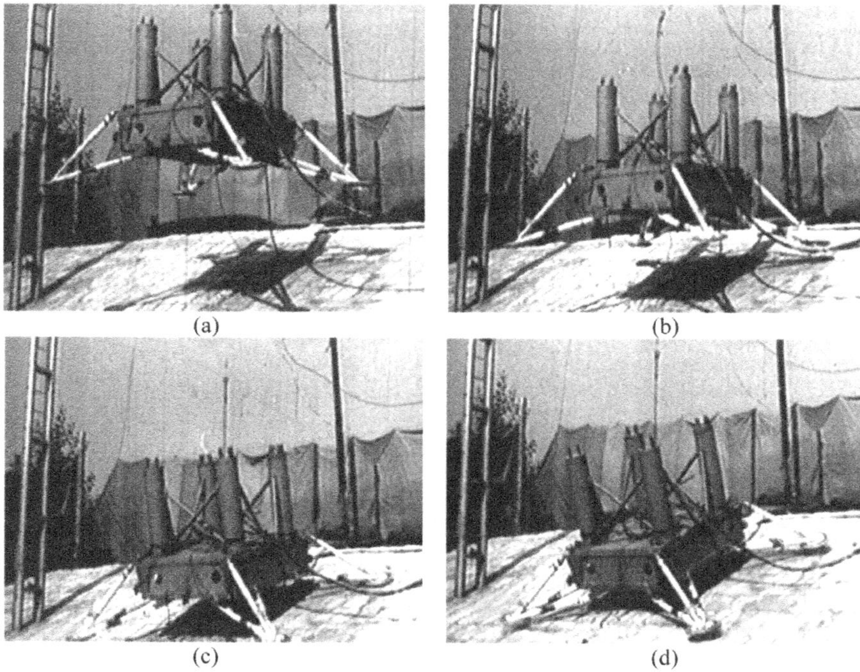

Fig. 15.6 Drop test process of Luna-21 lunar lander

(4) On the basis of test coverage of operation condition on orbit, the test conditions should be as little as possible.

15.3.2 Validation Requirements

In this section, only the requirements for the special system-level ground validation tests that should be conducted during development of the lunar lander are analyzed. For conventional tests such as mechanical and thermal tests (sinusoidal vibration test, acoustic test, spacecraft and launch vehicle separation test, vacuum thermal test) and EMC tests will not be discussed.

1. Hovering, Obstacle Avoidance and Slow Descending Test

Before the soft landing, the lunar lander would be inserted into elliptical orbit from circular lunar orbit at first. At the predefined altitude on lunar surface (15 km for the Chang'E-3 lunar lander), the powered descent was initiated. In the powered descent phase, there were main deceleration, quick attitude adjustment, approaching, hovering, hazard avoidance, and slow descending. Because of the complex processes of ranging, velocity measurement, terrain identification, autonomous navigation control of navigation sensors, coordination of control and propulsion subsystem in powered descent phase, it was necessary to test and verify the process of powered descent on the ground. However, it was almost impossible to verify the whole process of powered descent from 15 km to soft landing under current technical conditions, and it was also unnecessary. In the powered descent phase, the key activities such as deceleration, terrain identification, obstacle avoidance, hovering and shutdown should be completed by the lunar lander. With consideration of the limitation of cost, development schedule and ground simulation methodology, it was necessary to conduct the ground test for the hovering, the obstacle avoidance and the slow descending so as to verify the whole process of powered descent.

2. Landing Impact Test

After the powered descent, the lunar lander will be landed softly on lunar surface. During the final soft landing, the lunar lander would experience the landing impact load environment, as well as the instruments which would be operated on lunar surface (scientific payload, the lunar rover). During the soft landing, the lunar lander was affected by factors such as landing mass, landing velocity, landing attitude, lunar terrain topography, mechanical characteristics of landing surface and landing cushion characteristics of landing gears, and etc. And the force transfer path of the lunar lander during landing impact was that the landing impact load was transferred to the main structure connection through the landing gear, and then extended to the structural plates, and then to onboard instruments. Both the force transfer path and the load characteristics were different from the load environment during launch.

In order to examine the impact resistance of the main structure of the lunar lander during soft landing, and acquire the impact response of the key parts and the key instrument connection of the lunar lander, the landing impact process should be tested on the ground.

3. Landing Stability Test

The most important task of the lunar lander was to complete lunar soft landing. Whether the lunar lander could be landed on lunar surface with anticipated attitude and direction would determine the mission success. The landing stability of the lunar lander was affected by factors such as landing mass, landing velocity, landing attitude, lunar terrain topography, mechanical characteristics of landing surface and landing cushion characteristics of landing gears. In order to implement successful lunar soft landing, the landing stability should be verified on the ground, so as to examine the impact of the initial conditions on landing stability of the lunar lander.

4. Test of Engine Thrust Pulsation Effect on GNC Sensor

When the large liquid rocket engine is firing, there will be thrust pulsation which is determined by its inherent combustion characteristics and generally the thrust pulsation can be controlled within 3% of average thrust. The bipropellant liquid rocket engine (7500 N throttling engine for the Chang'E-3) was generally used for Lunar Orbit Insertion and deceleration during powered descent. In the powered descent, the mechanical environment caused by thrust pulsation of the engine would have a certain effect on measurement accuracy of the high-precision navigation sensors of GNC subsystem, as well as the pulsation response of structure on camera imaging. In order to determine the pulsation environment of the lunar lander and its effects on GNC sensors, it was necessary to conduct test validation on the ground.

5. Joint Firing Test of Throttling Engine and GNC CCU

According to the result of ranging and speedometer, by autonomous formulation of control strategy, the GNC CCU would control the engine in real time. Because of the high control frequency, the engine should respond quickly, which should be realized by joint work of GNC subsystem and propulsion subsystem. In order to examine joint work performance, it was necessary to conduct special test and validation on the ground.

6. Test of Plume Effect on Thermal Control

Large throttling engine (7500 N throttling engine for the Chang'E-3) was generally applied to achieve soft landing deceleration. When the lunar lander was approaching lunar surface, the plume of the engine would interact with the lunar surface and be reflected to the bottom of the lunar lander, which would affect the thermal control material at the bottom of the lunar lander, instruments on the bottom plate and the landing gears. In addition, the force of plume on the lunar lander would also affect its landing attitude. In order to determine the thermal effect of engine plume on the lunar lander, it was necessary to test the plume effect on the lunar lander.

15.4 Hovering, Obstacle Avoidance and Slow Descent Tests

15.4.1 Test Objective

The objective of ground validation test for hovering, obstacle avoidance and slow descending was to simulate the actual flight process of hovering, obstacle avoidance and slow descending, so as to fully examine the correctness of flight procedure design. At the same time, the landing kinematics of the lunar lander could be simulated to verify the autonomous navigation control algorithm based on navigation sensors, as well as to simulate the reflection characteristics of lunar surface and motion of the lunar lander so that the dynamic performance of navigation sensors could be examined.

15.4.2 Test Constraints and Methodology

In ground environment, the thrust of throttling engine could not balance weight of the lunar lander, and it was difficult to satisfy the test requirements for control of velocity and attitude of the lunar lander by GNC. The focus in the test was to establish the 1/6 g gravity environment on lunar surface. There were different solutions as follows.

1. Helicopter

The idea was to suspend the lunar lander with a cable by helicopter. During the test, the helicopter lifted the lunar lander to the test height, and then the helicopter was hovering. After it was ready, the cable would balance part of weight of the lunar lander by the force sensor and control system on the helicopter, while the engine on the lunar lander was firing to conduct test.

The disadvantage of this solution was that the hovering stability of the helicopter could not meet the requirements, and the pilot's operational response time was much longer than the GNC control period of the lunar lander. In addition, there was potential hazard.

2. Balloon

The idea was similar to the helicopter solution, with helium balloon instead of helicopter. The disadvantage of this solution was that the balloon volume would be greater than 1000 m^3 and the air drag on the lunar lander during descent would be greater than 4000 N. Therefore, it could not meet the test requirements.

3. Scale Model

The idea was to develop a dynamic scale model of the lunar lander whose mass and engines were determined by the similar theory to meet the test requirements. The disadvantages of this solution were: (1) It was difficult to develop a scaled model according by the similar theory. For example, the thrust of attitude control engine should be consistent with the scale. (2) The technical status of the scale model was different from real lunar lander, which made it difficult to verify the real lunar lander.

4. Aero engine

The idea was to use an aero jet engine to balance part of weight of the lunar lander. In the test, the jet engine was applied to lift the lunar lander up to the test height, and then the thrust was reduced. At the same time, the engine of the lunar lander was fired and the test was started. The disadvantages of this solution were: (1) because of the structure limitation of the lunar lander, it was difficult to install the aero engine and the throttling engine at the same time, which would result in change of lunar lander configuration. (2) After the aero engine was installed, it was necessary to be controlled and cooperate with throttling engine, which would increase the difficulty of control. (3) There was potential hazard in the test. Once there was accidental explosion of LLRV happened in the United States.

Fig. 15.7 Gantry test

5. Gantry

The idea was to lift the lunar lander by gantry. In the test, the part of weight of the lunar lander was balanced by the rope to meet the test requirements. Meanwhile, the lifting rope could follow the horizontal motion of the lunar lander in real time to avoid the angle deviation of the lifting rope to affect the test, as shown in Fig. 15.7.

In order to simulate the real process of hovering, obstacle avoidance and slow descending and examine performance of the lunar lander, in the ground test, the related instruments on the lunar lander should work well at design status. Therefore, it was necessary to consider the constraints in the test system design. The specific constraints are as follows.

(1) The forces upon the lunar lander during slow descending should be simulated for velocity and acceleration simulation.
(2) The gantry should provide the DOF of the lunar lander motion and enough space for movement. The attitude DOF of the lunar lander should be ensured.
(3) The flight hardware of GNC subsystem and propulsion subsystem should participate in the test. In the test, the GNC subsystem should coordinate with the propulsion engine or thrusters according to the real control law, and it should implement control of velocity and attitude of the lunar lander.
(4) In the test, the mass properties of the lunar lander such as mass center and inertia should be consistent with the real lunar lander on lunar surface.
(5) In the hovering, the GNC subsystem should accomplish the tasks of Obstacle identification and safe area determination. The lunar terrain topography on lunar surface should be simulated. In the obstacle avoidance and the slow

descending, the height and velocity of the lunar lander were respectively measured by laser ranging sensor and microwave ranging and velocity sensors. The test site should provide condition for sensors and simulate the reflection characteristics of the laser and microwave on lunar surface.

(6) The choice of test facilities and test windows should not interfere with the test system.

In order to satisfy above constraints, special attitude tracking system and translational motion tracking system should be designed to simulate free flight conditions on orbit, and simulation landing field should be designed to simulate lunar terrain topography, reflection characteristics of lunar surface, and illumination characteristics. Besides, it was also necessary to build a test gantry system to provide adequate support and test space.

15.4.3 Design of Test System

In the hovering, obstacle avoidance and slow descending tests, it was necessary to simulate the force state, motion DOF, lunar terrain, reflection characteristics of lunar surface, coordination between GNC subsystem and propulsion subsystem, and autonomous navigation control algorithms. The entire system was composed of the lunar lander test model subsystem, gantry subsystem, simulated lunar surface subsystem and test control subsystem.

1. Force State and Translational Motion Simulation

The gravity environment on the ground was different from that of lunar surface. The maximum output thrust of the engine was less than the weight of the lunar lander test model, so the lunar lander test model could not be hovered. Therefore, in the design of the test system, simulation of the force state of the lunar lander test model should be considered at first, which was also one of the cores of the test system.

The test gantry was built up. The lunar lander test model was lifted by a rope, which would balance part of weight of the lunar lander test model by control the rope to maintain a constant tension with servo control system in the test. With throttling of the engine output thrust, the resultant force on the lunar lander test model in the test would be consistent with that in lunar soft landing, so as to simulate the kinematics characteristics of the lunar lander in real landing process. The rope could also be able to follow horizontally. In the test, the rope would follow up the horizontal motion of the lunar lander test model in real time. The direction of rope tension on the lunar lander test model should always be vertical and up, so as to avoid the horizontal component force of the rope tension to interfere the test when the lunar lander test model was moving horizontally. The test gantry structure is shown in Fig. 15.8.

2. Simulation of Attitude DOF

In order to simulate the attitude DOF of the lunar lander, a gimbal spreader system should be designed. The spreader was connected to the lunar lander test model by

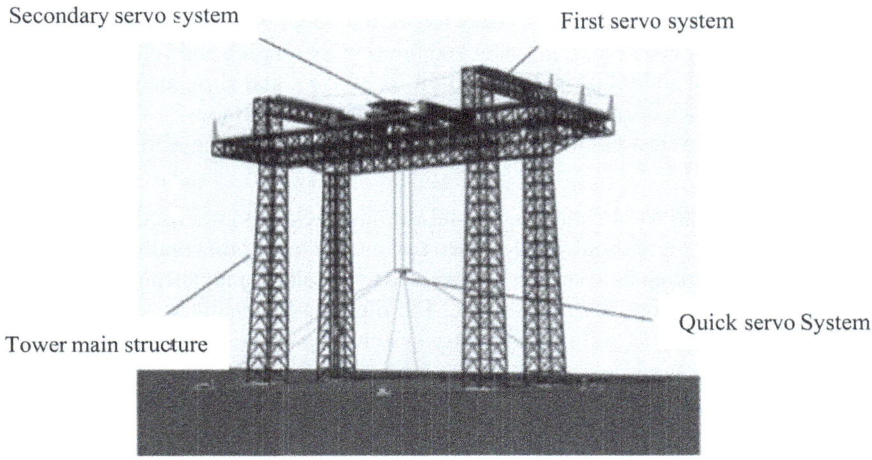

Fig. 15.8 Test gantry structure

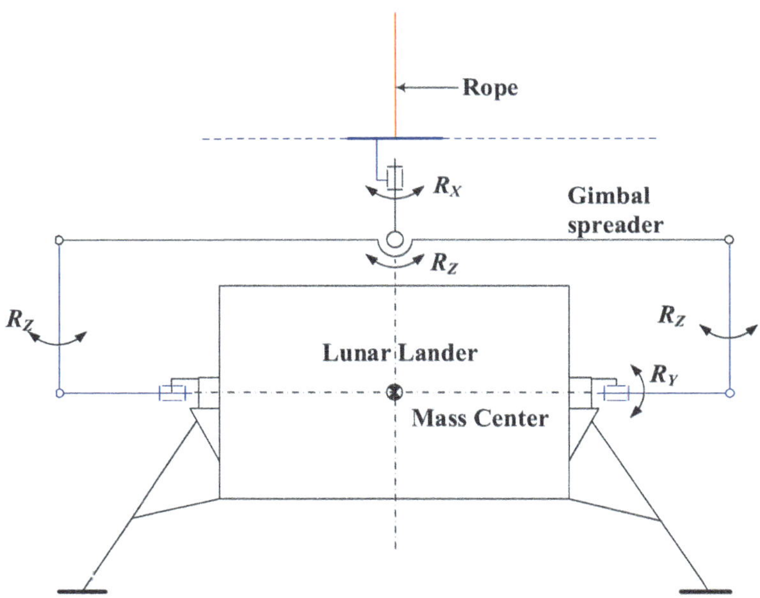

Fig. 15.9 Principle of gimbal spreader

two interfaces that the connection line between them passed the mass center of the lunar lander test model. The other end was connected to the rope of the test gantry. In the test, the gimbal spreader could provide pitch, roll and yaw DOF of the lunar lander test model. The principle is shown in Fig. 15.9.

3. Simulation of Lunar Terrain and Reflection Characteristics of Lunar Surface

In order to simulate lunar terrain and reflection characteristics of lunar surface, the simulated lunar surface system should be built. The material with similar reflectivity as lunar surface could be painted on the test site, which could simulate lunar surface reflection characteristics of laser and microwave so as to satisfy operation conditions of GNC navigation sensors and implement ranging and velocity measurement in the test.

4. Test Control

In the test, the lunar lander test model and the follow-up system of the test gantry system were all controlled by ground control equipment. The respective ground control equipment was managed and controlled by the main control center of the test system. In the test, the main control center of the test system could collect the data information generated by the lunar lander test model and the gantry system and external measurement data by ground equipment, so that the test results could be comprehensively analyzed.

15.4.4 Test Example

During development of the Chang'E-3 lunar lander, the hovering, obstacle avoidance and slow descending test and the test system were designed and developed. The hovering, obstacle avoidance and slow descending test was conducted, as shown in Fig. 15.10.

Fig. 15.10 Hovering, obstacle avoidance and slow descending test of the Chang'E-3 lunar lander

15.5 Landing Impact Tests

15.5.1 Test Objective

The objective of landing impact test was to examine the impact resistance of main structure of the lunar lander under the typical landing conditions during the soft landing. At the same time, the impact response of key parts of the lunar lander and the connection of key instrument could be determined, which could provide references for formulation of instrument environment conditions.

15.5.2 Test Constraints and Methodology

In order to simulate real soft landing impact of the lunar lander and acquire the response of the lunar lander, there were constraints to be considered when the test system was designed as follows.

(1) The mass property, load transfer characteristics and force transfer path of the lunar lander should be same as actual landing process.
(2) The cushion characteristics of the landing gear and the impact load environment of the lunar lander during soft landing should be the same as actual.
(3) The landing surface should be able to simulate the terrain features of lunar surface and mechanical property of lunar soil.
(4) The velocity and attitude when touched landing surface should be same as actual landing on lunar surface.
(5) The lunar lander should be free without any additional constraints during landing.

When the test case was designed, simulation analysis of all conditions should be conducted. Simulation conditions should cover all landing modes, initial parameters and landing surface characteristics. The initial parameters should cover the mass property of the lunar lander, velocity, and attitude. The landing surface characteristics should cover the friction characteristics, load bearing characteristics, and slope of the landing surface. The most severe typical landing conditions should be chosen to be tested.

15.5.3 Design of Test System

In the landing impact test, it should be simulated including the lunar lander energy, motion DOF, lunar terrain and mechanical property of lunar soil during soft landing. The test system was composed of the lunar rover test model subsystem, lift and release device, simulated landing surface and ground measurement subsystem, as shown in Fig. 15.11.

1. Test Model Subsystem of the Lunar Lander

In order to acquire real landing impact response of the lunar lander, it was necessary to simulate the force transfer path and the force state of the lunar lander during soft landing. Therefore, the main structure and landing gears of the lunar lander test model was real, while other instrument could be structural analog model.

2. Lift, Release and Separation Device

For the lunar lander to get the energy during soft landing, the lunar lander test model should be lifted to predefined height by lift, release and separation device. After the initial parameters were setup, the release and separation device was started. The lunar

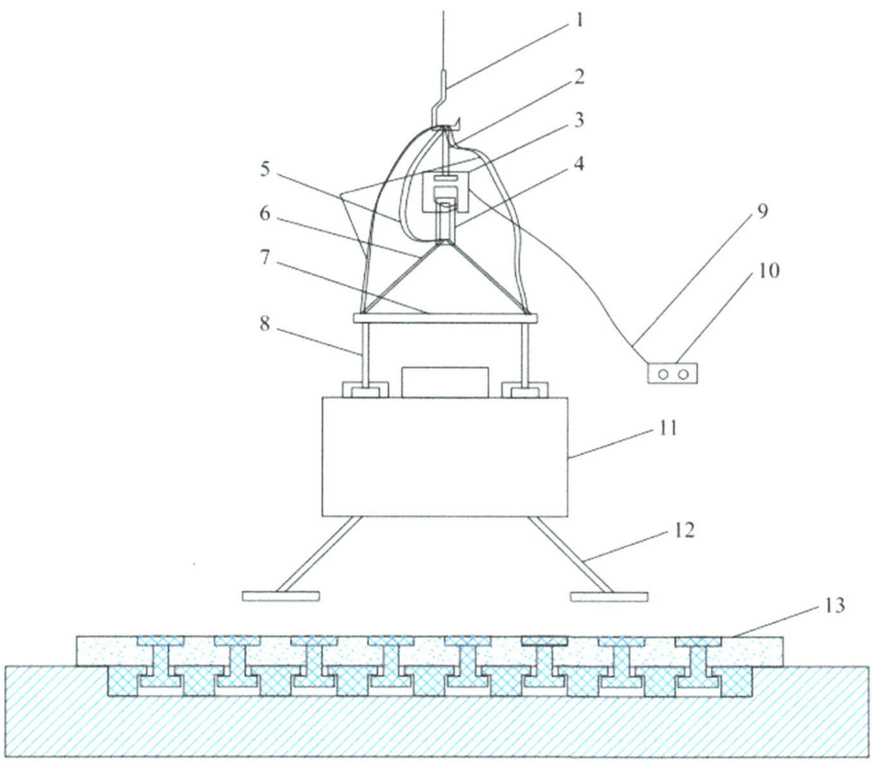

Fig. 15.11 Configuration of test system for landing impact test **1** Crane hook; **2** Release and separation device rope; **3** Release and separation device; **4** Ring; **5** Limit protection device; **6** Hanging beam rope; **7** Hanger; **8** lunar Lander rope; **9** Control cable of Release and separation device; **10** Control switch of Release and separation device; **11** Lunar lander test model; **12** Landing gear; **13** Simulated landing surface

lander test model would be separated from lift, release and separation device. Then the lunar lander test model would free fall and touch the landing surface.

3. Simulated Landing Surface

In order to simulate the contact characteristics between the landing gear foot pad of the lunar lander and the landing surface during soft landing, simulated landing surface should be designed. Simulation of landing surface could simulate typical lunar terrain and mechanical property of lunar surface. The parameters should be simulated were including friction factor, load bearing strength, slope, and etc.

4. Ground Measurement Subsystem

The ground measurement subsystem consisted of a high-speed camera system and an acceleration/strain measurement system. During the test, the high-speed camera system would measure the parameters of interested points with time, such as displacement, cushion stroke, landing velocity, attitude angle, angular rate, and angular acceleration, and the acceleration/strain measurement system would measure the time history of acceleration and strain response.

15.5.4 Test Example

During development of the Chang'E-3 lunar lander, the landing impact test and the test system were designed and developed. The landing impact test was conducted, as shown in Fig. 15.12.

15.6 Landing Stability Tests

15.6.1 Test Objective

The objective of the landing stability test was to examine influence on landing stability of the lunar lander caused by landing attitude, landing velocity, landing surface terrain and mechanical property of lunar soil, to verify landing stability under extreme conditions and correct landing stability simulation model.

15.6.2 Test Constraints and Methodology

When the test system for landing stability test was designed, the following constraints should be considered:

Fig. 15.12 Landing impact test of the Chang'E-3 lunar lander

(1) It could simulate a 1/6 g gravity environment during soft landing.
(2) The tilt DOF of the lunar lander should be free without any additional constraints during landing.
(3) The energy of the lunar lander during soft landing should be same as landing on lunar surface.
(4) The kinematic parameters such as velocity and attitude of the lunar lander during soft landing should be same as landing on lunar surface.
(5) It should be able to simulate the landing cushion characteristics during soft landing, as well as the interface load between the landing gear subsystem and the main structure.
(6) The shape and mass property of the lunar lander should be same as the flight model, as well as the landing gears.
(7) The landing surface should simulate lunar surface terrain and mechanical property of lunar soil.

In order to satisfy above constraints, the simulated main structure and real landing gears were applied in the lunar lander test model. The low-gravity simulation system was designed to simulate 1/6 g gravity environment on lunar surface. The follow-up system was designed to simulate the free motion of the lunar lander. The gantry was designed to support the lunar lander test model and provide enough test space. The lift and release subsystem was designed to set up the initial landing parameters. The simulated landing surface was designed to simulate lunar terrain and mechanical properties of lunar soil.

15.6.3 Design of Test System

The landing stability test system was composed of the lunar rover test model subsystem, low-gravity simulation subsystem, gantry, one-dimensional follow-up subsystem, lift and release subsystem, simulated landing surface and ground measurement subsystem, as shown in Fig. 15.13.

1. Test Model Subsystem of the Lunar Lander

The lunar lander test model subsystem consisted of a simulated main structure of the lunar lander and landing gears, which could simulate the mass property (mass center and inertia) of the lunar lander during soft landing. The main structure of the test model was in same physical dimensions as the real structure, and it could simulate the connection interface with landing gears. The landing gears were flight model.

2. Low-Gravity Simulation Subsystem

Inclined landing surface was applied to simulate the lunar gravity. When the angle between inclined landing surface and the horizontal plane was 80.5°, the gravity component of the lunar lander in the normal direction of inclined landing surface was the gravity upon the test model when landing on lunar surface. Therefore, the low-gravity lunar gravity could be simulated in the normal direction of inclined landing surface. In the test, the gravity component of the test model along inclined landing surface was balanced by the rope of lift and release device.

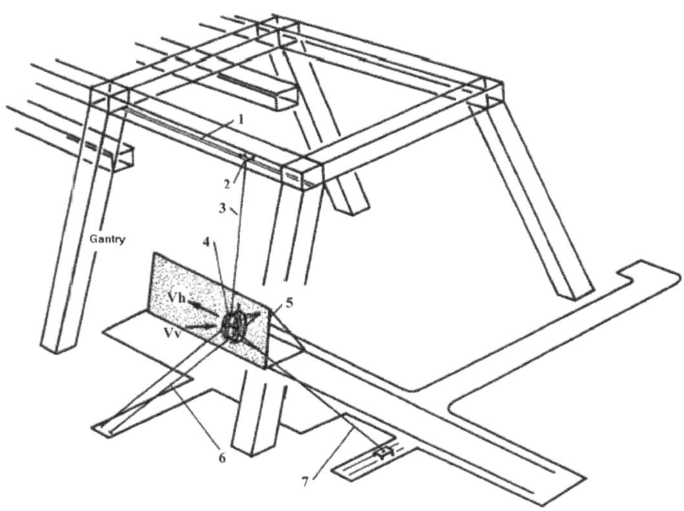

Fig. 15.13 Configuration of landing stability test system **1** One-dimensional follow-up track; **2** One-dimensional follow-up subsystem; **3** Rope; **4** Lunar lander test model; **5** Simulated landing surface; **6** Sling rope; **7** Horizontal rope

3. Test Gantry and One-Dimensional Follow-Up Subsystem

Test gantry could support the one-dimensional follow-up subsystem and provide sufficient test space. The one-dimensional follow-up system was generally designed in passive mode. In order to minimize the interference force caused by follow-up lag to the test model, the mass and follow-up friction factor of the follow-up subsystem should be as small as possible. The follow-up capability of the one-dimensional follow-up subsystem should cover the maximum horizontal landing velocity of the lunar lander with certain margin.

4. Lift and Release Device

The lift and release device was composed of lift device, rope, sling rope, release device of sling rope, length adjusting device of sling rope, horizontal rope, release device of horizontal rope and length adjusting device of horizontal rope. The specific functions of each part are as follows.

(1) Lift device: to connect the rope and implement the one-dimensional follow-up of the rope to the test model by the follow-up device.
(2) Rope: to connect the test model with lift device. During the test, the gravitational component of the lunar lander could be balanced. In order to minimize the rope angle change caused by displacement changes of test model mass center caused by attitude change and cushion stroke, the length of the rope should be designed specially.
(3) Sling rope: to achieve the initial attitude of the lunar lander and the vertical landing velocity.
(4) Release device of sling rope: to release sling rope at the set position by triggering of pyrotechnic.
(5) Length adjusting device of sling rope: to adjust the length of sling rope. The initial attitude angle of the test model could be achieved by adjusting the relative length of sling rope.
(6) Horizontal rope: to achieve the horizontal landing velocity of the test model parallel to the guide track.
(7) Release device of horizontal rope: to release horizontal rope at the set position by triggering of pyrotechnic.
(8) Length adjustment device of horizontal rope: to adjust the length of the horizontal rope to achieve different horizontal landing velocity.

5. Simulated Landing Surface

The simulated landing surface could simulate typical lunar terrain and mechanical property of lunar soil, including friction factor, load bearing strength and slope of landing surface.

When the size of simulated landing surface was designed, it was necessary to consider the space motion of the test model and the slip during contact with landing surface. Generally, it was determined by theoretical calculation and simulation analysis. The stiffness and strength constraints should be comprehensively considered when simulated landing surface was designed. The fundamental frequency should be kept away from the primary modal frequency of test model in order to avoid interference caused by resonance.

6. Ground Measurement Subsystem

The ground measurement subsystem consisted of high-speed camera system and load measurement equipment. In the test, the time history of the displacement, cushion stroke, velocity, attitude angle, angular rate and angular acceleration at interested points was measured by high-speed camera, and the time history of acceleration and load was measured by load measurement equipment.

15.6.4 Test Example

During development of the Chang'E-3 lunar lander, the landing stability test and the test system were designed and developed. The landing stability test was conducted, as shown in Fig. 15.14.

Fig. 15.14 Landing stability test of the Chang'E-3 lunar lander

15.7 Test of Engine Thrust Pulsation Effect on GNC Sensors

15.7.1 Test Objective

The test objective of engine thrust pulsation effect on GNC sensors was to determine the influence of engine thrust pulsation on GNC landing navigation sensors, so as to verify adaptability of GNC sensors such as accelerometer to the thrust pulsation environment.

15.7.2 Test Constraints and Methodology

When the test system for engine thrust pulsation effect on GNC sensors was designed, the following constraints should be considered.

(1) It could simulate the free boundary of the lander when the engine was firing on orbit.
(2) It could simulate the pulsation thrust condition when the engine was firing on orbit.
(3) It could simulate the interface and mode of pulsation thrust when the engine was firing on orbit.
(4) It could simulate the transfer path and transfer characteristics of pulsation thrust on the lunar lander.
(5) It could simulate the mass property of the lunar lander.
(6) The real main structure and real landing gears should be applied in the test.
(7) The flight hardware of GNC landing navigation sensors should be applied and powered on in the test.

In order to satisfy the above constraints and simulate the mechanical impact of the engine thrust pulsation on orbit environment on the lunar lander as realistically as possible, the real main structure and the real landing gear should be applied to simulate the force transfer path and load transfer characteristic. A gravity compensation system of near zero stiffness spring should be designed to simulate the free boundary of the lunar lander. A pulsation thrust simulation method should be developed to simulate the pulsation thrust. The pulsation thrust input was simulated through an exciter.

When the test plan was designed, the time domain data of the pulsation loads under all condition in the ground vacuum firing test of the engine should be statistically analyzed at first. The time slice with the largest time domain load in nominal thrust and throttling thrust should be selected as the input excitation of the test, and the closed-loop control method should be applied in the test.

15.7.3 Design of Test System

The test system for engine thrust pulsation effect on GNC sensors was composed of the lunar rover test model, gravity compensation subsystem of near zero stiffness spring, pulsation thrust exciter and input control system, and measurement subsystem, as shown in Fig. 15.15.

1. Test Model of the Lunar Lander

The real main structure and real landing gears were applied and the mass property of the lunar lander during landing on lunar surface should be simulated. The flight models of concerned GNC navigation sensors should be on the test model.

2. Gravity Compensation Subsystem of Near Zero Stiffness Spring

The metal spring is general elastic element and can be used as a gravity compensation unit to constitute gravity compensation subsystem. The basic principle of gravity compensation with springs was: With the spring's stiffness characteristics, under the load gravity, the elastic recovery force by elastic deformation of the spring would balance the load gravity; at the same time, the load mass and the spring constitute

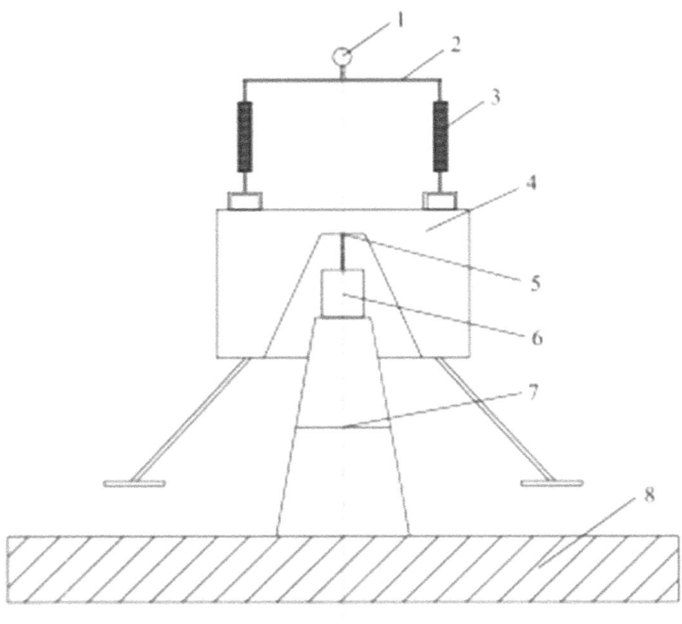

Fig. 15.15 Configuration of test system for engine thrust pulsation effect on GNC sensors **1** Ring; **2** Beam; **3** Gravity compensation subsystem of near zero stiffness spring; **4** the lunar rover test model; **5**-Pulsation thrust interaction interfaces; **6** Pulsation thrust exciter and input control system; **7** Exciter support: **8** Ground.

a spring-mass oscillator model with a certain natural frequency. The key of gravity compensation by spring in pulsation tests was that the spring should meet the load capacity and have a low enough natural frequency.

The gravity compensation subsystem of near zero stiffness spring was applied to simulate the free boundary conditions of the lunar lander when it was excited by pulsation thrust on orbit. As long as the support stiffness was low enough, the natural oscillation caused by elastic recovery in the test had negligible influence on the test system.

3. Pulsation Thrust Exciter and Input Control System

In the test of dynamic response of pulsation excitation to the lunar lander and the test of transfer path, it was necessary to use the pulsation source of the lunar lander as the excitation source. However, in order to determine the response and transfer characteristics of the lunar lander structure, it was inconvenient to install the pulsation excitation source on the structure and simulate by firing of engine. It was necessary to use external excitation to simulate the pulsation excitation. In the ground test, the exciter was a good choice. The amplitude and frequency range of the pulsation load should be considered when the exciter was selected.

4. Measurement Subsystem

The measurement subsystem was composed of acceleration measurement system, angle measurement system and GNC sensors ground test system. The acceleration measurement system and the angle measurement system were applied to measure the acceleration response and angle response at the installation interface of the sensor and critical position of the lunar lander. Because the pulsation load was in small amplitude and wide frequency range, generally sensors with low-range, high-sensitivity and wide frequency bandwidth were preferred. In the test, the interested GNC landing navigation sensors should be powered on and its performance should be monitored, so as to analyze the influence of pulsation load.

15.7.4 Test Example

During development of the Chang'E-3 lunar lander, the test system for 7500 N engine thrust pulsation effect on GNC sensors were designed and developed. The pulsation thrust test was conducted, as shown in Fig. 15.16.

Fig. 15.16 Test for 7500 N engine thrust pulsation effect on GNC sensors of the Chang'E-3 lunar lander

15.8 Joint Firing Test of Throttling Engine and GNC CCU

15.8.1 Test Objective

The objective of joint firing test of throttling engine and GNC CCU was to verify the control coordination and compatibility of CCU with throttling engine, and measure the average specific impulse during frequent thrust throttling in the approaching and hovering.

15.8.2 Test Constraints and Methodology

When the test system for joint firing test of throttling engine and CCU was designed, the following constraints should be considered.

(1) It could simulate the vacuum environment while the engine was firing on orbit.
(2) It could measure the actual thrust of the throttling engine.
(3) It could simulate the throttling control system.
(4) It could cover the firing condition of the throttling engine on orbit.

In order to satisfy above constraints, vacuum test chamber and engine testbed was necessary to simulate the vacuum environment. Real engine should be applied to simulate the thrust characteristics in the test. Real CCU and throttling control system should also be applied in the test.

When the test conditions were determined, all conditions on orbit should be comprehensively regarded. In all flight phases of the lunar lander, the throttling engine should be fired for many times. The first one was for TCM in LTO, then LOI, orbit change on circumlunar orbit, and powered descent at last. In all of this, the throttling condition during powered descent was the most complicated process, with high-frequency throttling, large throttling range and high-demand for thrust throttling response. At the same time, the response of propulsion subsystem to throttling command of GNC subsystem should be quickly and performed correctly. Therefore, the test for powered descent was generally selected for examination.

15.8.3 Design of Test System

The test system for joint firing test of throttling engine and CCU was composed of throttling engine, throttling control subsystem, vacuum simulation subsystem, parameter measurement subsystem, test control subsystem and engine testbed pipeline, as shown in Fig. 15.17.

1. 7500 N Throttling Engine

As controlled target of the test system, in order to simulate output load characteristics of the throttling engine, real engine should be applied in the test.

2. Throttling Control Subsystem

The throttling control subsystem was applied to control the thrust of throttling engine in the test. The throttling control subsystem was composed of real controller, ground test equipment and test cables.

3. Vacuum Simulation Subsystem

The vacuum simulation subsystem was applied to simulate vacuum environment in the test, which consisted of vacuum tank, pre-evacuation system, ejection pipeline and ejection system.

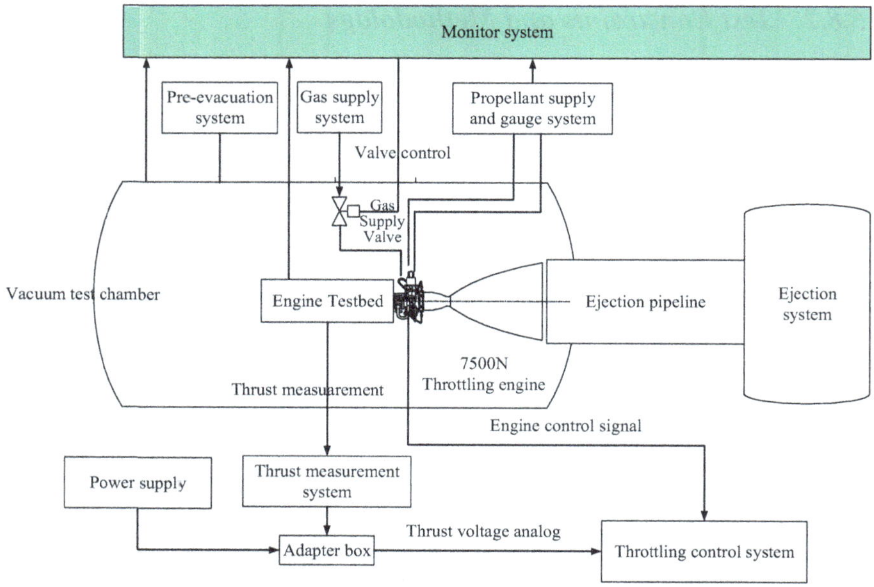

Fig. 15.17 Configuration of test system for joint firing test of throttling engine and CCU

4. Parameter Measuring Subsystem

The parameter measuring subsystem was applied to measure and record the parameters of the test system in the test, including vacuum chamber pressure, vacuum chamber temperature, propellant pressure, propellant flow rate, propellant temperature, propellant density, heat flux of the combustion chamber body of the engine, engine thrust, and etc.

5. Test Control Subsystem

The test control system consisted of test monitoring module, control and schedule module, network transmission and time system module, and test data processing module. The test monitoring module was applied to monitor and record the image information of the operation and test process of the test system. The control and schedule module was applied to schedule the operation of the test system and send remote operation instructions to the test system. The network transmission and time system module was responsible for data transmission and time uniformity for the test system. The test data processing module was applied to classify and store various test data and information generated in the test, and comprehensively analyze the test data generated at the same time.

6. Engine Testbed Pipeline

The engine testbed pipeline was used to supply propellant and the gas for valve control. There should be filters on the pipeline to avoid foreign object.

15.9 Test of Plume Effect on Thermal Control

15.9.1 Test Objective

The test objective of plume effect on thermal control was to acquire the resistant ability to ablation and erosion of the engine plume for the thermal control material on landing gears, bottom plate and protrusions on bottom plate during soft landing by simulation of plume condition, and to provide important reference for thermal design of landing gears and bottom plate of the lunar lander.

15.9.2 Test Constraints and Methodology

When the test system for plume effect on thermal control was designed, the following constraints should be considered.

(1) It could simulate the vacuum environment while the engine was firing on orbit.
(2) It could simulate heat flow while the engine was firing on orbit.
(3) It could simulate the real thermal state
(4) It could simulate the effect of engine plume on thermal control.

Due to the limitation of vacuum test conditions, the plume of large engine was difficult to simulate on the ground. Therefore, small engine could be applied for the equivalent heat flow simulation. The small engine was firing in high vacuum simulation system, while test articles covered with thermal control material was placed at appropriate position in the plume of small engine in order to simulate the plume effect of large engine on thermal control material.

15.9.3 Design of Test System

The test system of plume effect on thermal control was composed of vacuum simulation subsystem, three-dimensional moving subsystem, engine, test article with thermal control material and parameter measurement subsystem, as shown in Fig. 15.18.

1. Vacuum Simulation Subsystem

The vacuum simulation subsystem was applied to simulate the vacuum environment when the engine was firing on orbit.

2. 3-D Moving Subsystem

The 3-D moving subsystem was applied to fix the test article and adjust the position of test article so that the heat flux in the plume field could satisfy the test requirement.

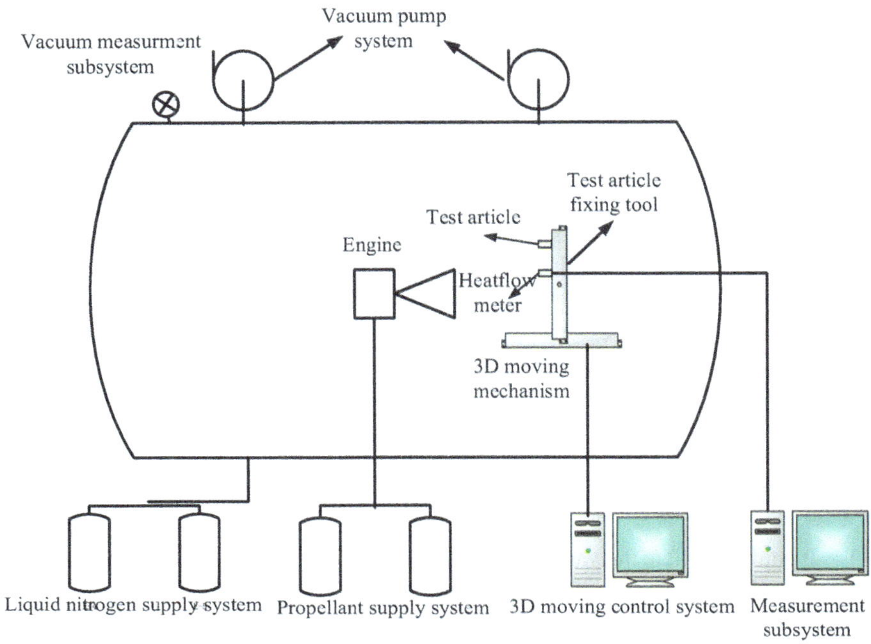

Fig. 15.18 Configuration of test system for plume effect on thermal control

3. Engine

The engine should be flight model. When the engine was installed, the engine vector line should be perpendicular to the tested surface of test article (installation surface). The axis of the engine nozzle was coaxial with the axis of the heat flow meter installation tool.

4. Test Article with Thermal Control Material

The test article was the thermal control coating material used in the landing gears, bottom plate and protrusion instrument on bottom plate.

5. Parameter Measurement Subsystem

The test parameter measurement subsystem was applied to measure and acquire data of heat flux and temperature in the test, which was composed of heat flow sensor, pressure sensor, thermocouple, vacuum gauge and resistance gauge.

Fig. 15.19 Test of plume effect on thermal control for the Chang'E-3 lunar lander

15.9.4 Test Example

During development of the Chang'E-3 lunar lander, the test of plume effect on thermal control was conducted, as shown in Fig. 15.19.

15.10 Summary

In this chapter, the test methodology and test process for typical ground validation conducted by the United States and the Soviet Union in the lunar exploration mission are systematically introduced. In term of mission characteristics for lunar soft landing exploration, the environment and its impact on the lunar lander is analyzed. The corresponding ground test requirements are analyzed.

For the analysis of requirements, special system-level validation tests of the lunar lander that were different from conventional spacecraft are introduced, including hovering, obstacle avoidance and slow descending test, landing impact test, landing stability test, test of engine thrust pulsation effect on GNC sensors, joint firing test of throttling engine and GNC CCU and test of plume effect on thermal control. The comprehensive and systematic introduction are given on the necessity of test, test objective, test constraints and methodology, and the composition of test system. Test examples during development of the Chang'E-3 lunar lander are also given.

The adequacy and effectiveness of ground validation test are the key to mission success. How to create simulated space mission environment on earth as realistic a as possible is an important task for system design of the lunar lander. In the future, along with development of deep space exploration method, the requirements for test methodology and test system are also more and more complicated and continuous optimization in terms of accuracy, convenience and cost-effective of simulation is also necessary.

References

1. Thomas CO, Donald EH (1967) Operational features of the Langley Lunar landing research facility[R]. National Aeronautics and Space Administration, Washington DC
2. Donald RB, Gene JM (1965) Design and operational characteristics of a Lunar-landing research vehicle[R]. National Aeronautics and Space Administration, Washington DC
3. Caluin RJ (1966) Operational experience with the electronic flight control systems of a Lunar-landing research vehicle[R]. National Aeronautics and Space Administration, Washington DC
4. Ulysse JB (1968) Evaluation of a full-scale Lunar-gravity simulator by comparison of landing-impact tests of a full-scale and a 1/6-scale Model[R]. National Aeronautics and Space Administration, Washington DC
5. Ulysse JB (1965) Model investigation of technique for conducting full-scale landing-impact tests at simulated lunar gravity[R]. National Aeronautics and Space Administration, Washington DC
6. William FR (1972) Apollo experience report Lunar module landing gear subsystem[R]. National Aeronautics and Space Administration, Washington DC
7. Ulysse JB (1969) Full-scale Dynamic landing-impact investigation of a prototype Lunar module landing gear [R]. National Aeronautics and Space Administration, Washington DC
8. Stanley PW (1973) Apollo experience report Lunar module structural subsystem [R]. National Aeronautics and Space Administration, Washington DC
9. Thomas JT (1972) Apollo experience report thermal protection from engine-plume environments[R]. National Aeronautics and Space Administration, Washington DC

Chapter 16
Scientific Instrument Technology of Lunar Lander

16.1 Introduction

The scientific instrument of the lunar lander was the instrument for scientific or engineering exploration [1, 2].

In term of different objectives, the scientific instruments can be classified as scientific payload or engineering payload. The scientific payload can acquire scientific data and its objective is to acquire the characteristics of objects interested by scientists. The engineering payload can acquire interested engineering data including environment and status, and its objective is to acquire interested engineering data to improve engineering design or support for successful engineering implementation.

In term of exploration method, scientific instrument can be classified into direct sensing instrument and remote sensing instrument. The direct sensing instrument can touch the target by sensors directly to acquire status property of the target to be studied. The remote sensing instrument can study the target status such as light and electromagnetic wave by sensors, which is an indirect way.

In deep space exploration, there are direct sensing instruments such as high-energy particle detector, low-energy charged particles detector, ion detector, dust detect and magnetometer, remote sensing instruments such as radio astronomy instrument, optical imager, polariscope, photometer, infrared radiometer, microwave radar and laser radar. In the scientific instruments of the Chang'E-3 lunar lander, there were remote sensing instruments including descent camera, lunar terrain camera, lunar-based ultraviolet telescope, extreme ultraviolet camera and monitoring camera, as well as direct sensing instrument such as lunar dust detector.

In this chapter, the typical instruments and their achievements in lunar landers of the United States and the Soviet Union are introduced. The requirements for scientific instruments of the Chang'E-3 lunar lander are analyzed. The design of scientific instruments on the Chang'E-3 lunar lander is introduced, as well as special tests and test methodology for the instruments such as calibration.

D.-Y. Yu et al., *Technology of Lunar Soft Lander*, Space Technology Library 38,
https://doi.org/10.1007/978-981-15-6580-9_16

16.2 Development Status

Up to date, only the United States, the Soviet Union and China had implemented successful robotic lunar soft landing [3–8].

From May 1966 to Jan. 1968, 7 robotic Surveyor lunar landers were launched by the United States There were 5 lunar landers achieved successful soft landing, including Surveyor 1, Surveyor 3, Surveyor5, Surveyor 6 and Surveyor 7. Scientific instruments such as television cameras, surface samplers, α scattering instrument and the magnets onboard the lunar lander achieved designed exploration tasks, including acquiring high definition images, mechanical property of lunar soil, soil compositions and chemical compound of lunar soil [9–15].

From Jan. 1963 to Dec. 1966, 13 robotic Luna landers were launched by the Soviet Union. There were 2 lunar landers achieved successful soft landing, including Luna 9 and Luna 13. In Nov. 1970 and Jan. 1973, 2 lunar landers with lunar rovers (Luna 17 and Luna 21) were launched by the Soviet Union and achieved successful lunar landing. In Sept. 1970, Feb. 1972 and Aug. 1976, 3 robotic lunar landers for lunar sampling and return (Luna 16, Luna 20 and Luna 24) were launched by the Soviet Union and achieved successful sampling and return.

In the early lunar soft landing missions of the United States and the Soviet Union, especially for the first mission, the main objective was to demonstrate engineering technology and acquire environment parameters for subsequent missions. Due to technology limitation at that time, there were few types and quantities of instruments on the lunar lander. However, they had achieved designed exploration and acquired valuable information for engineers and scientists, which provided important reference for humankind acknowledgement of lunar surface characteristics.

16.2.1 Instruments on Surveyor Lunar Landers and Major Achievements

From 1966 to 1968, there were 7 robotic lunar landers (Surveyor 1 to Surveyor7) launched by the United States [9–15].

Surveyor 1 and Surveyor 2 were launched in May and September of 1966 respectively, and the primary mission objectives were to demonstrate technology for landing softly on the Moon and acquire information of lunar surface. There were TV camera and Strain gauges on the lunar landers. The TV camera was applied to take image of lunar surface. There were zoom lenses operated at either end of its range resulting in 25 mm and 100 mm focal-lengths, resulting in 1 mm highest resolution. The TV camera was mounted under a mirror that could be moved in azimuth (360°) and elevation (from +40° above the plane normal to the camera's axis to –65° below this plane). Surveyor 1 transmitted 11,350 still photos of the lunar surface to the Earth. Strain gauges were mounted on each leg shock absorber to record the peak axial forces at landing impact of the lunar lander. They were designed to accept a force of approximately 800 kg. Because of engine failure, Surveyor 2 failed.

Surveyor 3 was launched in April 1967. The specific objectives for this mission were to select potential landing site for Apollo program and obtain information on lunar surface radar reflectivity, thermal properties and bearing strength. Besides TV camera, there was lunar surface sampler on the lunar lander. The specification of TV camera was the same as that of Surveyor 1. The camera returned 6326 pictures to the Earth. The sampler was mounted below the television camera, which could be extended about 1.5 m and dig to 0.5 m depth. The scoop sampler was mounted on a pantograph arm. The arm could also be moved from an azimuth of –72° to +40°. The scoop was capable of holding a maximum of 100 cm^3 of granular material. The surface sampler performed seven bearing tests, four trench tests, and 13 impact tests. A 175 mm depth was reached during trenching operations. According to above exploration results, scientists implied that the density of lunar soil was identical to sand on Earth and the bearing strength is 0.7 kg/cm^2, which was enough to support Apollo lunar module. Because the landing radar was powered off, the radar reflectivity of lunar soil was not measured. In Nov. 1969, two astronauts of Apollo 12 visited the spacecraft and removed about 10 kg of parts from the spacecraft, including the TV camera and the sampler, for later examination back on Earth to study lunar surface environment effects in two and half years.

Surveyor 4 was launched in July 1967. There were instruments same as that of Surveyor 3, with additional lunar soil magnetic sensor inside the sampler for iron abundance measurement in lunar surface soil. Approximately 2.5 min before touchdown, the radio signals from the probe ceased during the terminal-descent phase and the mission failed.

Surveyor 5 was launched in Sept. 1967. The specific objectives for this mission were to perform a soft landing on the Moon in Mare Tranquillitatis and obtain post landing television pictures of the lunar surface, determine the relative abundances of the chemical elements in the lunar soil he mechanical properties of lunar soil, conduct a vernier engine erosion experiment, obtain touchdown dynamics data, and obtain thermal and radar reflectivity data. The instrumentation for Surveyor 5 was similar to that of the previous Surveyors and included the television camera and lunar surface sampler. A α scattering instrument and a small bar magnet attached to one footpad were included. The spacecraft returned 20018 pictures. On 13 Sept., the main engine was fired for 0.55 s to study interference to lunar surface. The image showed that there was no pit and obvious dust cloud. The α scattering instrument was mounted inside the sampler with six α sources (^{242}Cm) to radiate area in 100 mm diameter. There were two detection systems in the α scattering instrument. One detection system with two sensors was applied to investigate α particle energy spectrum scattered by lunar surface. Another detection system with four sensors was applied to investigate proton energy spectrum generated by lunar surface reaction. The α scattering instrument worked continuously for 83 h in first lunar day and high quality data was acquired. It worked continuously for 22 h in second lunar day. The α scattering results indicated that the oxygen, silicon and aluminum in soil composition was over 50%.

Surveyor 6 was launched in Nov. 1967. The mission objectives and instruments onboard were identical to Surveyor 5. With another mirror, the TV camera could observe lunar surface under the lunar lander. The original mirror was modified to help the TV camera to take pictures in 360° azimuth and –60°~90° pitch. The camera returned 29952 images. The powered lift-off experiment was conducted by Surveyor 6 on lunar surface. Three vernier engines were fired for 2.5 s, causing Surveyor 7 to lift off the lunar surface about 4 meters and land about 2.5 meters west of its original position. Viewing of the original landing site by camera provided further acknowledgement for mechanical properties of lunar soil. The results of the experiments showed that the surface had a basaltic composition, similar to that found at the Surveyor 5 landing site.

Surveyor 7 was launched in Jan. 1968. The mission objectives were to acquire images of different terrain and conduct measurement of lunar soil. The instruments onboard were identical to Surveyor 6. The probe returned 21,274 images. The lunar surface sampler conducted multiple digging experiments, over 16 measurement of bearing strength, 7 digging trenches and 2 impacts. The α scattering instrument investigated 3 lunar samples, undisturbed lunar surface, one rock and one new trench. The TV camera recorded laser beam emitted from Earth observatory for two times. As the target, the Surveyor 7 lander also successfully detected laser beams transmitted from Earth to conduct precise measurement of Earth-Moon distance for the first time.

16.2.2 Instruments on Luna Probes and Major Achievements

From 1966 to 1976, there were 6 successful robotic lunar landers launched by the Soviet Union, including luna 9, luna 16, luna 17, luna 20, luna 21 and luna 24.

Luna 9 was launched in Jan. 1966, and was the first spacecraft to achieve a lunar soft landing. The mission objectives were to demonstrate lunar landing technology and explore lunar surface environment. The instruments onboard included TV camera and radiometer. The TV camera acquired panorama images of landing site by rotation in azimuth and pitch. Luna 9 transmitted three series of TV pictures under different illumination conditions back to Earth. The radiation dose by the radiometer was 0.3 mGy (1 Gy = 1 J/kg). The Luna 9 proved that the lunar surface could support the weight of a lunar lander and an object would not sink on lunar surface.

The Luna 13 was launched in Dec. 1966, and its shape was similar to Luna 9. The mission objectives were to demonstrate lunar landing technology and explore lunar surface environment furthermore. The instruments onboard included TV camera, soil penetrometer, radiation densitometer, backscatter densitometer, three-axis accelerator and infrared radiometer. The TV camera took five panorama images. The soil penetrometer was applied to measure mechanical properties of lunar soil. The radiation densitometer was applied to measure cosmic radiation reflected by lunar surface. The three-axis accelerator was applied to measure impact and stroke when the probe was landing on lunar surface and the measurement results indicated that the Luna 13 sunk down to a depth of 20–30 cm. The backscatter densitometer was applied to measure density of lunar regolith. The infrared radiometer was applied to measure lunar surface temperature. The measurement results indicated that the temperature at noon was 117 ± 3 °C. The measurement of radiation densitometer indicated that radiation levels on lunar surface would be less than hazardous for humans.

Luna 17 and Luna 21 were launched in November 1970 and January 1973. The mission objectives were to be landed on lunar surface and implement the lunar rover deployment on lunar surface. There was no instrument on the lunar lander. Both missions were successful.

Luna 16, Luna 20 and Luna 24 were launched in September 1970, February 1972, and August 1973 separately. The mission objectives of the lunar lander were to be landed on lunar surface and take sample of lunar soil and support lift-off on lunar surface. There were drill samplers on Luna 16, Luna 20 and Luna 24. About 101, 55 and 170 g lunar soil samples were returned to Earth in the three missions. Luna 16 was the first robotic probe to land on the Moon and return sample back to Earth. the Soviet Union scientists studied returned lunar sample with infrared spectrometer and found that there were 0.1% water in lunar soil, which was published in 1978.

16.3 Requirements

16.3.1 Requirements for Scientific Exploration

The four scientific objectives are designed in the second phase of CLEP as follows.

(1) To survey on the morphology and geological structure of the landing site and roving area.
(2) To analyze mineral composition and chemical composition in the landing site and roving area.
(3) To explore internal structure of the Moon.
(4) To explore the environment of Earth-Moon space and lunar surface, and perform optical astronomy observation on lunar surface.

After optimization of the scientific payloads, the Chang'E-3 lander and the rover would jointly implement above four scientific objectives. The four scientific exploration tasks should be implemented by the Chang'E-3 lunar lander as follows.

(1) To in situ take image of lunar surface.
(2) To investigate lunar crater in high resolution.
(3) To take extreme ultraviolet image of Earth plasmasphere.
(4) To perform the optical astronomy observation on lunar surface.

Where, task (1) and (2) were designed for the investigation of lunar terrain and geological structure of the landing site; task (3) and (4) were designed for exploration of Earth-Moon space and optical astronomy observation based on the Moon.

Table 16.1 Scientific exploration tasks

Scientific objective	Task	Measurement parameter or data
Survey on the morphology and geological structure of the landing area	To in situ take image of lunar surface To investigate lunar crater in high resolution	Lunar surface image of landing site
Environment exploration of Earth- Moon space and lunar-based optical astronomy observation	To take extreme ultraviolet image of Earth plasmasphere To perform optical astronomy observation on lunar surface	Extreme ultraviolet image of Earth plasmasphere Optical image of interested celestial area

The scientific exploration tasks of the Chang'E-3 lunar lander are listed in Table 16.1.

After analysis, proposed instruments on the Chang'E-3 lunar were as follows: one descent camera for taking image of landing site during powered descent, one morphology camera for taking high definition image of lunar surface after landing, one extreme ultraviolet (EUV) camera for taking image of Earth plasmasphere, and one Lunar-based Ultraviolet Telescope (LUT) for observing interested celestial area.

16.3.2 Requirements for Engineering Measurement

For the first soft landing mission in China, there were both the scientific objectives and engineering objectives. The engineering objectives were to acquire first-hand surface engineering environment parameters, which would greatly benefit to design of subsequent lunar probes. In the Chang'E-3 mission, engineering environments and parameters were focus as follows.

(1) To take images of operation status of the lunar lander and the lunar rover. The images would provide the visual information to ground operators for making decisions and evaluations.
(2) To take images during powered descent. The images would provide the information for path planning of the lunar rover and help to monitor the attitude of the lunar lander for analysis of subsystems status including GNC and propulsion.
(3) To conduct measurement including landing impact, lunar surface thermal environment, mechanical properties of lunar soil, lunar dust, meteoroid environment, radiation environment, and etc. It would acquire first-hand engineering environment data and help to improve design of subsequent lunar missions.
(4) To determine effects of lunar natural environment on performance of specific material. It would provide effect parameter of lunar surface environment on specific material, which would provide fundamental data for design of future lunar base.

The measurement requirements for engineering environment parameter are listed in Table 16.2.

Table 16.2 Measurement requirements for engineering environment parameter

Engineering environment	Requirements	Measurement parameter or data
Status of the lunar rover	To monitor the lock/unlock of the lunar rover on the lunar lander To monitor the movement and position of the lunar rover on the lunar lander To monitor the release and separation of the lunar rover on the lunar lander	Images of monitoring camera
Descending process and terrain of landing site	To survey rock distribution and terrain of landing site, provide data for path planning of the lunar rover To observe attitude of the lunar lander during descending for analysis of propulsion and GNC	Video taken during descending by descent camera
Landing impact feature	To study landing impact, optimize structural and mechanism design of future lander with mechanical properties of lunar soil	3D acceleration of landing impact
Lunar dust properties on lunar surface	To study lunar dust accumulation with time To study lunar dust accumulation with movement of the lunar rover To study effect of lunar dust on output power of solar cell To study effect of lunar dust on optical instrument for optimization of future instrument design	Accumulation of lunar dust
Thermal environment on lunar surface	To study lunar surface temperature of landing site, radiative heat flux on all spectrum band on lunar surface, heat conductivity of lunar soil To analyze thermal environment on lunar surface, its variation in all life cycle of the lunar lander, so as to optimize thermal design of future lander	Temperature, radiative heat flux on lunar surface and conductivity of lunar soil

(continued)

Table 16.2 (continued)

Engineering environment	Requirements	Measurement parameter or data
Mechanical properties of luna soil.	To study bearing capacity and shearing strength of lunar soil To study stress status at different depth To study hardness, strength, friction of lunar soil To optimize locomotion design of future lunar rover To optimize structural and mechanism design of future lander with landing impact feature	bearing capacity, shearing strength, stress at different depth
Electrical property of shallow lunar soil	To study permittivity of shallow lunar soil To study dielectric loss tangent of shallow lunar soil To optimize telecommunication system design between probes for future missions	Complex permittivity
Lunar meteoroid environment	To study the flux of meteoroid with time To study velocity, mass of meteoroid on lunar surface To analyze and evaluate potential effects of meteoroid on the lunar lander.	Relationship of flux, velocity and mass of meteoroid
Radiation environment on lunar surface	To study optimized radiation protection design for robotic and manned lunar lander	Equivalent radiation dose with protection
Long term environment effect	To study performance change of specific material in long term space weathering	Specific test material

Above engineering and environment parameters on lunar surface, would be important basis for design of future robotic and manned lunar exploration missions in China, which played an important role in sustainable development of China lunar exploration.

The first priority of Chagn'E-3 lunar lander was successful landing. Then priority should be given to the very important and lightweight engineering instruments. After proposal and deliberately analyzing, it was determined that there would be three monitoring cameras to monitor the status of the lunar rover, and one lunar dust detector to investigate lunar dust accumulation on lunar surface.

16.3.3 Analysis of Target Feature

The descent camera was designed to take images for acquisition of lunar surface information during powered descent, which could show terrain and geological feature of lunar surface at landing site. The major target features included lunar surface albedo and contrast, solar elevation and azimuth angle. The lunar morphology camera was mounted on a mast with two motion dimensions, which was able to take images to survey lunar terrain. The major target features were the same as the descent camera. The extreme ultraviolet (EUV) camera was on the top plate of the lunar lander, which would conduct system and long term observation of radiation generated by Earth plasmasphere in 30.4 nm wavelength. The major target features included Earth plasma irradiation and background. The LUT would continuously observe brightness change of celestial bodies in interested area. The major target features included stellar spectrum, irradiation intensity, background noise and stray lights. Three monitoring cameras would take images when the lunar rover was deployed. The major target features included albedo of observed area, solar elevation and azimuth, and shadow. The lunar dust detector would investigate lunar dust accumulation on lunar surface. The major target feature was the accumulation of lunar dust with time.

According to the task characteristics, analysis of target features should cover following aspects.

1. Albedo of Lunar Surface

The predefined landing site of the Chang'E-3 mission was Sinus Iridum, a lunar mare area on near side of the Moon. The albedo of lunar mare is relatively low. The albedo in different band is listed in Table 16.3.

For optical instruments, the lunar surface target is relatively simple. The major material on lunar surface is lunar soil in lunar mare and lunar highland, with relatively low albedo. The reflection of rock is relatively high (0.4~0.6). The vast majority of rock is dispersed on the lunar soil. Compared with Earth, the reflectivity on lunar surface is lower. Optical instrument should adapt to changes in solar elevation on lunar surface. The low albedo and low solar elevation would bring some difficulties to take good images in visible light.

Table 16.3 Albedo of lunar mare

Band/nm	Albedo		
	Range	Concentration range	Average
500	0.055~0.135	0.070~0.110	0.083
600	0.060~0.167	0.080~0.130	0.091
700	0.060~0.198	0.080~0.160	0.100
750	0.060~0.200	0.080~0.165	0.110
800	0.083~0.225	0.085~0.180	0.110

Compared with lunar highland, the contrast of lunar mare area is relatively low. Therefore, the sensitivity of camera should be high enough, or the image quality would be poor.

2. Solar Incident Angle

The light cycle is about 28 days on most area of lunar surface, i.e., there will be continuous 14 days in lunar day, and the rest of the time in lunar night. The maximum of solar incident angle (angle between sunlight and the normal of lunar surface) is about 90°, and the minimum of solar incident angle is up to the local latitude. The smaller the solar incident angle is, the more sufficient sunlight there is. The minimum solar incident angles in different landing site are listed in Table 16.4.

The albedo of lunar surface is only 1/6 of Earth. With solar incident angle listed in above table, the aperture of lens or detector size should be as large as possible so that the image quality will be good.

In a lunar day, the solar incident angles change greatly. In addition, the installation position of instrument might cause direct sunlight to the optical lens. Therefore, the exposure time of camera should be able to be automatically adjusted, so as to adapt to the target brightness under different solar elevation. At the same time the precision and sensitivity of camera should be higher and it should be able to eliminate stray light, to ensure anticipated image quality.

3. Brightness and Optical Properties of Stars

In the astronomy, the brightness of celestial body is generally measured by apparent magnitude. The change of apparent magnitude means that the celestial body may be born, disappear or change by something. The brighter an object appears, the lower its magnitude value (i.e. inverse relation). The apparent magnitude is up to distance to Earth, natural brightness of celestial body and occultation of interstellar dust. The apparent magnitude of the Sun is –26.74.

In the astronomy, three wavelengths are generally applied for observation. (1) About 360 nm. It is applied to measure near violet composition of celestial body in violet magnitude. (2) About 440 nm. It is applied to measure blue composition of

Table 16.4 Minimum solar incident angles in different landing site

Landing site	Time	Minimum solar incident angle/(°)
Sinus Iridum	Jan. 2012 to Dec. 2012	44
Mare Nectaris	Jan. 2012 to Dec. 2012	14.5
Mare Humorum	Jan. 2012 to Dec. 2012	24
Kepler Crater	Jan. 2012 to Dec. 2012	8
Aristillus Crater	Jan. 2012 to Dec. 2012	34

celestial body in blue magnitude. (3) About 550 nm. It is applied to measure yellow and green composition of celestial body in apparent magnitude.

The main interference to observation is the stray light from the Sun. Therefore, the control of stray light is a very important factor in lunar-based optical observation.

4. Shapes and Radiation of Earth Plasmasphere

The plasmasphere, or inner magnetosphere, is a region of the Earth's magnetosphere. It is located above the ionosphere. It consists of low energy plasma. The major composition includes H, He and O. About 90% of its composition is H^+ and He^+ is secondary. When the distance to Earth is further, the plasma density is decreasing. When the plasma density is lower than one decimal, the boundary is called outer boundary of plasmasphere, which is also known as the plasmapause. The biggest radius of the plasmasphere is about 5 times of Earth radius.

The shape of Earth plasmasphere is affected by change of solar wind. When there is solar flare, the shape of Earth plasmasphere will change greatly. The H^+ in Earth plasmasphere doesn't generate radiation. The main characteristic wavelength of He^+ is 30.4 nm. The radiation intensity of top plasmasphere is about 0.1 RayL (1RayL = 1 Pa•s/m). The maximum radiation intensity of central region is about 10 RayL. Generally, the change of Earth plasmasphere is observed by observing change of radiation characteristic at 30.4 nm.

There are two kinds of background interference in adjacent He^+ characteristic radiation wavelength. The one is 58.5 nm radiation generated by inner He atom at the edge of Earth plasmasphere, which is below 2000 km altitude; the maximum radiation intensity is about 700 RayL; the average radiation intensity is about 60 RayL. The other is 121.6 nm radiation generated between Earth plasmasphere and interplanetary space, with radiation intensity at about 35000 RayL.

5. Lunar Dust

Lunar dust is the finest lunar soil grain with diameter less than 1 mm. Its electrical conductivity is very low. So, it is easy to be charged in sunlight and maintain for a long time. After charged by solar wind radiation, the lunar dust would float and move. When the lunar day and lunar night is alternated, the lunar dust will be periodically lifted because of charged effect by solar ultraviolet radiation. With landing impact and engine firing of the lunar lander, movement of the lunar rover and impact of small bodies on lunar surface, lunar dust will be activated. The charged lunar dust may stick to any surface it comes in contact with.

16.4 Design Principles

16.4.1 Scientific Instrument Configuration

According to the requirements for scientific objectives, current instrument technology, system resource constraints and target features, after proposal and deliberately analyzing, it was determined that there would be scientific instruments on the lunar lander as follows. One descent camera was installed inside a module of the lunar lander, which was applied for imaging of landing site during powered descent. One morphology camera was installed on a mast for imaging surrounding lunar surface in high resolution after landing. One lunar-based ultraviolet telescope (LUT) was installed in a module of the lunar lander for optical observation of interested area after landing. One extreme ultraviolet (EUV) camera was installed on the top plate of the lunar lander for imaging of Earth plasmasphere. The scientific instrument configuration on the Chang'E-3 lunar lander is listed in Table 16.5.

1. Function of Scientific Instruments

The function of descent camera was to continuously obtain images for the landing site during powered descent, adjust exposure, compress image data and eliminate stray light.

The function of morphology camera was to obtain continuous and still images of the landing site, monitor the moving the lunar rover on lunar surface and deployment of the lunar rover, acquire colorful image, adjust exposure, compress image data, eliminate stray light, and be capable of anti-lunar-dust.

The function of LUT was to observe the celestial bodies in near-ultraviolet band, point to specific celestial area, adjust exposure, compress image data and eliminate stray light.

The function of EUV camera was to observe Earth plasmasphere, point to Earth, adjust exposure and be capable of anti-lunar-dust.

2. Specification of Scientific Instruments

Table 16.5 Scientific instruments configuration

Scientific objective	Scientific instrument configuration	
	Scientific instrument	Quantity
To take image of landing site for study of lunar terrain	Descent camera	1
To take image of landing site during descending, for analysis of landing site terrain and lunar crater	Morphology camera	1
To conduct lunar-based optical observation	LUT	1
To conduct EUV Imaging of Earth plasmasphere	EUV camera	1

The specifications of descent camera included: band range, normal imaging distance, effective pixels number, Field of View (FoV), frame frequency, quantitative value, MTF (Modulation Transfer Function), signal to noise ratio, maximum data rate, data compression ratio, power consumption and mass.

The specifications of morphology camera included: band range, color, imaging modes, normal imaging distance, effective pixels number, FoV, frame frequency, quantitative value, MTF, signal to noise ratio, maximum data rate, data compression ratio, power consumption and mass.

The specifications of LUT included: band range, FoV, observation limitation, observation uncertainty, effective pixels number, image spot energy concentration, quantitative value, maximum data rate, data compression ratio, power consumption and mass.

The specifications of extreme ultraviolet cameras were central frequency, bandwidth, FoV, angular resolution, time for single measurement, observed intensity range, sensitivity, maximum data rate, power consumption and mass.

16.4.2 Engineering Instrument Configuration

According to the requirements for engineering exploration, current instrument technology, system resource constraints and target features, after proposal and deliberately analyzing, it was determined that there would be engineering instruments on the lunar lander as follows. Three monitoring cameras were installed on the top plate of the lunar lander for imaging the lunar rover status of locking/unlocking and deployment. One lunar dust detector was installed for measurement of lunar dust accumulation, which had two heads, one was solar cell detector on the top plate of the lunar lander and another was the stick quartz crystal microbalance (SQCM) in a module of the lunar lander which had a cover to be opened and closed. The engineering instruments on the Chang'E-3 lunar lander are listed in the Table 16.6.

1. Functions of Engineering Instruments

The function of monitoring camera A was to monitor the lunar rover locking/unlocking status (the front wheel of the lunar rover) and movement to the transfer mechanism. The function of monitoring camera B was to monitor the lunar rover locking/unlocking status (the rear wheel of the lunar rover), release status of

Table 16.6 Engineering instruments configuration

Engineering objective	Engineering instrument configuration	
	Engineering instrument	Quantity
To monitor the deployment of the lunar rover on the lunar lander	Monitoring camera A, B, C	3
To measure lunar dust accumulation	Lunar dust detector	1

the separation plug states, and movement to the transfer mechanism. The function of monitoring camera C was to monitor transfer and deployment to lunar surface of the lunar rover, as well as movement on lunar surface. All the monitoring cameras should be able to acquire colorful continuous and still images, adjust exposure, compress image data and be capable of anti-lunar-dust.

The function of lunar dust detector was to measure lunar dust accumulation by landing impact and natural floating on lunar surface, and compress data.

2. Specification of Engineering Instruments

The specifications of monitoring cameras included: color, band range, imaging modes, normal imaging distance, effective pixels number, FoV, frame frequency, quantitative value, signal to noise ratio, MTF, maximum data rate, data compression ratio, power consumption and mass.

The specifications of lunar dust detector included: resolution, measurement range, effective detection area, maximum data rate, power consumption and mass.

16.4.3 Functions, Performances and Implementation

In term of design and development, the function and performance requirements for all scientific and engineering instruments were analyzed. The detail implementation methods are listed in Table 16.7.

16.5 Calibration and Special Testing

Besides conventional tests including cyclic aging test, running-in test, mechanical environment and thermal environment, the calibration tests and relevant special tests should be conduct for the scientific and engineering instruments.

The purpose of calibration test is to test the accuracy of the payload instruments by standard measurement instrument, and determine whether it meets the requirements and corrects to the expected state. The purpose of the payload instrument calibration was to confirm that the performances should meet the requirements of technical specifications or design requirements. If the requirements were not met, modifications shall be made until the requirements were met.

Table 16.7 Implementation methods for instrument function and performance

Instrument	Function requirement	Performance requirement	Implementation methods in design and development
Descent camera	To continuously obtain lunar surface image of landing site during powered descent	Band range	It was achieved by spectral of lens coating and spectral of detector
		Normal imaging distance	When the optical lens was designed, the best imaging distance was determined with the larger F number and other means to make the camera depth meet the requirement
		Effective pixels number	It was achieved by selection of detector that met the requirement
		FoV	It was achieved by optical design. In the optical lens calibration test, the focal length should be controlled in certain accuracy range
		Frame frequency	It was achieved by selection of detector which met the frame frequency requirement. The exposure time and image compression time should be designed reasonably
		Quantitative value	It was achieved by selection of detector that met the requirement
		MTF	In optical design, the optical system should achieve the diffraction limit imaging quality. The lens shape accuracy and central thickness error should be controlled. In optical lens assembly, the lens interval should be adjusted to ensure the optical imaging quality

(continued)

Table 16.7 (continued)

Instrument	Function requirement	Performance requirement	Implementation methods in design and development
		Signal to noise ratio	The photonic flow density under different conditions should be analyzed. According to the detector features, the optical system should be analyzed to achieve anticipated signal to noise ratio
	To adjust exposure	Solar incident angle range	The automatic exposure adjustment function should be designed to adapt different solar incident angle
	To compress image data	Image data compression ratio	Wavelet algorithm was applied to achieve better image compression
	To eliminate stray light		In design of optical system, the optical field should be set, which could effectively suppress the non-imaging beam entering the system. The outer cover was applied to suppress the stray light greatly, and the inner wall was equipped with multiple baffle rings. The clutter light bar should be set. All mechanical parts should be able to eliminate stray light. The inner wall of the optical system was black and had a dull thread
Morphology camera	To acquire colorful image	band range	It was achieved by spectral of lens coating and spectral of detector
		Color	It was achieved by selection of detector which met the color requirement
	To acquire continuous and still image	Imaging modes: continuous and still	The imaging mode could be changed by telecommand to select still imaging or the continuous imaging, or switch to each other

(continued)

Table 16.7 (continued)

Instrument	Function requirement	Performance requirement	Implementation methods in design and development
	To acquire lunar surface image of landing site, as well as transfer and movement of the lunar rover	Normal imaging distance	When the optical lens was designed, the best imaging distance was determined with the larger F number and other means to make the camera de
		Effective pixels number	It was achieved by selection of detector that met the requirement
		FoV	It was achieved by optical design. In the optical lens calibration test, the focal length should be controlled in certain accuracy range
		Frame frequency	It was achieved by selection of detector which met the frame frequency requirement. The exposure time and image compression time should be designed reasonably
		Quantitative value	It was achieved by selection of detector that met the requirement
		Signal to noise ratio	The photonic flow density under different conditions should be analyzed. According to the detector features, the optical system should be analyzed to achieve anticipated signal to noise ratio
		MTF	In optical design, the optical system should achieve the diffraction limit imaging quality. The lens shape accuracy and central thickness error should be controlled. In optical lens assembly, the lens interval should be adjusted to ensure the optical imaging quality

(continued)

Table 16.7 (continued)

Instrument	Function requirement	Performance requirement	Implementation methods in design and development
	To compress image data	Maximum data rate	The maximum data rate requirements should be satisfied by reasonable selection of pixel number, image compression ratio and frame frequency
		Image data compression ratio	Wavelet algorithm was applied to achieve better image compression
EUV camera	To acquire EUV image	Band range	It was achieved by EUV band multilayer coating reflector, UV band filter, UV band photon counting imaging detector together
		FoV Resolution	It was achieved by design of camera aperture, FoV and equivalent pixel number together
		Sensitivity	It was achieved by determination of parameters such as into pupil area, resolution, reflectivity of the reflector, transmittance of the filter and quantum efficiency of the detector
	To point to Earth	Adjustment of azimuth and pitch	After the landing site position, landing attitude and Earth-Moon relation were analyzed, the adjustment range of azimuth and pitch of two-dimensional mechanism was determined
LUT	To observe celestial body of brightness higher than specific value in near ultraviolet wavelength	Band range	It was achieved by near ultraviolet film and near ultraviolet CCD
		Observation limitation, Observation uncertainty, FoV	It was achieved by design of telescope aperture, focal length, transmittance of the films, elimination of stray light, high-efficiency detector, low-noise and low-dark-current electronics together

(continued)

Table 16.7 (continued)

Instrument	Function requirement	Performance requirement	Implementation methods in design and development
		Energy concentration	The transfer function of optical system should be as high as possible and defocusing should be conducted according to requirement
	To point to interested area	–	The range of azimuth and pitch of the turntable should cover observation area
	To adjust exposure	–	Multiple exposure time could be set
	To compress image data	–	The effective information should be acquired and lower data rate as far as possible
Monitoring camera A/B/C	To monitor the status of front wheel of the lunar rover by A To monitor the status of rear wheel of the lunar rover by B To monitor transfer process on the lunar lander and movement on lunar surface of the lunar rover	Normal imaging distance, Effective pixels number FoV Frame frequency Quantitative value Signal to noise ratio MTF	In optical design, according to target position and features as well as sunlight, reasonable imaging distance and aperture should be determined. With larger F number, the camera should have good depth of field. The effective pixel number, quantization value and frame frequency of the detector should meet the requirements. Based on analysis of exposure time and image compression time, the camera frame frequency was determined. The anticipated signal-to-noise ratio should be achieved. The surface and central thickness error of the lens should be controlled in manufacture. In optical lens assembly, the lens interval was adjusted to ensure the optical imaging quality
	To acquire colorful image	Color, Band range	It was achieved by spectral of lens coating and spectral of detector

(continued)

Table 16.7 (continued)

Instrument	Function requirement	Performance requirement	Implementation methods in design and development
	To acquire continuous and still image	Frame frequency	The imaging mode could be changed by telecommand to select still imaging or the continuous imaging, or switch to each other
	To adjust exposure	–	The automatic exposure adjustment function should be designed
	To compress image data	Image data compression ratio, Maximum data rate	Wavelet algorithm was applied to achieve better image compression
Lunar dust detector	To measure lunar accumulation by landing impact and natural floating respectively	–	The lunar dust detector consisted of having two heads. One was solar cells for measurement of lunar dust accumulation by landing impact, and another one was the SQCM. The lunar dust accumulation of natural floating could be measured by two heads
	To monitor lunar dust with time	Resolution, Measurement range, Effective detection area	According to the requirements such as resolution, measurement range and effective detection area of probes, the crystal and solar cell should be selected to be probes. According to the probe features, the electric circuit was designed
	To compress data	Maximum data rate	The data sampling period should be controlled reasonably, and the data should be smoothed (mean) to control maximum data rate

16.5.1 Calibration and Special Testing of Descent Camera

1. Tests

The calibration and special testing of the descent camera are listed in Table 16.8.

2. Test Principle and Method
1) Principle and method of flat field calibration

Table 16.8 Calibration and special tests of descent camera

Tests	Purpose
Flat fielding calibration	Camera dark current detection and relative radiation calibration
Absolute radiation calibration	To establish the quantitative relationship between the camera output signal and the radiance of the target
Interior orientation elements calibration	To measure the focal length and main point location
Video test	To evaluate the image quality of dynamic camera at different lighting and imaging distance
Automatic exposure test	To evaluate the function of automatic exposure imaging under different light intensity
Image compression test	To evaluate image quality difference and data rate before and after compression

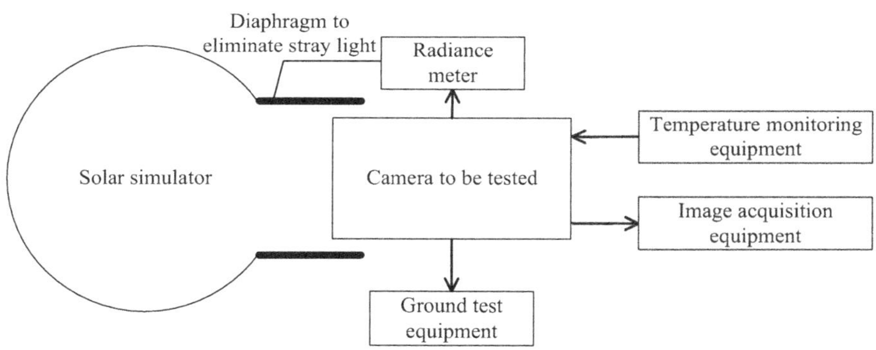

Fig. 16.1 Principle diagram of flat field calibration

There were two tests in flat field calibration, camera dark current detection and relative radiation calibration.

(1) The dark current was the output of the camera in the dark condition. In the dark current detection test, the camera was placed in a dark room environment, and the output image data was recorded at different exposure times, which was the dark current of the camera.

(2) Relatively radiation calibration was to test the camera outputs of each pixel under different uniform radial brightness of light, and through certain mathematical processing method, to obtain the relative calibration coefficient for changing un-uniform output to uniform output. In the relative radiation calibration test, the camera was placed in the uniform brightness environment, and the output image data was recorded in the fixed exposure time, and the calibration coefficient was calculated.

The testing steps for calibration of the flat field are as follows:

(1) In the darkroom, the test instrument was connected as shown in Fig. 16.1. The position of the tested camera was adjusted, so that the entrance pupil center was consistent with the radiation opening center of the solar simulator; the optical bar and the camera temperature monitoring equipment were set up, and the cable between the camera and the computer and the image acquisition equipment was connected.
(2) Set the camera exposure time, collect the dark current image data of the camera, and the average of the multi-group dark current images was the camera dark current.
(3) Turn on the power of the solar simulator, and the power current should not be less than 70% of the rated current value of the light source, and then measured the spectral radiance of the radiation opening of the solar simulator after stabilization.
(4) Set the camera exposure time and collected camera image data.
(5) Check the temperature monitoring value of the camera and determine whether to collect the next set of calibration data.
(6) Change the radiance level of the solar simulator, repeated steps (3)~(5) until all the calibration data of the radiance level are collected.

The image data was obtained by subtracting the dark current from the image under the different radiance illumination conditions, and the relative calibration coefficients of each pixel were obtained by using the least square method.

2) Principle and method of absolute radiation calibration

The absolute radiation calibration is the quantitative relationship between the output of the camera and the input radiation brightness under uniform irradiation.

In absolute calibration, the test method was the same as shown in Fig. 16.1. Under the conditions of different exposure time and different radiance, the image data were collected, and the absolute calibration coefficient was calculated after the calibration of the flat field.

3) Principle and method for interior orientation elements calibration

The element of interior orientation calibration was to measure the camera's focal length and the main point internal parameters.

As shown in Fig. 16.2, the test method of element of interior orientation calibration was shown by taking the space feature point (standard grid board) and using the direct

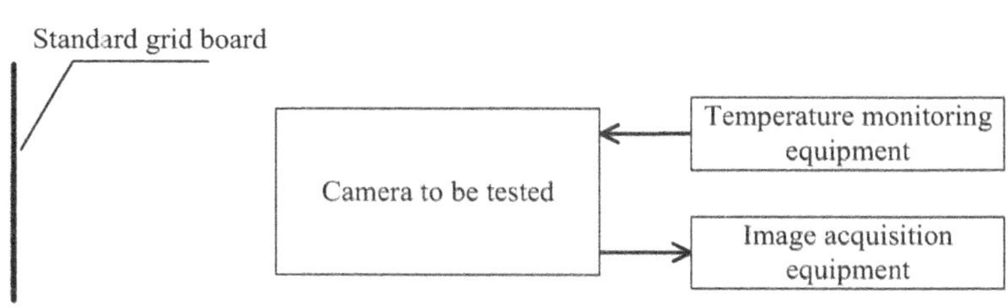

Fig. 16.2 Principle diagram of interior orientation elements calibration

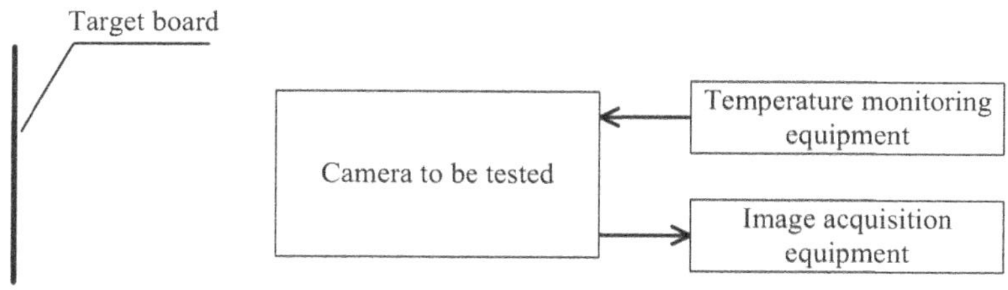

Fig. 16.3 Principle diagram of video test

linear transformation algorithm (DLT) to calculate the element of interior orientation of the camera.

4) Principle and method of video test

Video test method is shown in Fig. 16.3. Under atmosphere and light environment, the camera could image the target board at different distances, and the image sharpness and image compression quality were evaluated manually.

5) Principle and method of automatic exposure test

The automatic exposure test was conducted under the different radiance light. The camera could automatically expose the imaging target board, obtain the images under different radiance conditions, and judge the image sharpness manually.

The test method of the automatic exposure was similar as shown in Fig. 16.3 and different brightness conditions were realized by using the light source.

6) Principle and method of image compression test

Image compression test was to analyze and evaluate the image before and after compression, and verify the performance of image compression algorithm. The analysis and evaluation were combined with subjective evaluation (artificial evaluation) and objective evaluation.

Image compression test method was the same as shown in Fig. 16.1, to evaluate the video images before and after the compression. The subjective evaluation content was the image definition, and the objective evaluation could use the peak signal-to-noise ratio.

16.5.2 Calibration and Special Testing of Morphology Camera

1. Tests

The calibration and special testing of the morphology camera are listed in Table 16.9.

2. Test Principle and Method

Table 16.9 Calibration and special tests of morphology camera

Tests	Purpose
Flat fielding calibration	Camera dark current detection and relative radiation calibration
Absolute radiation calibration	To establish the quantitative relationship between the camera output signal and the radiance of the target
Interior orientation elements calibration	To measuring the focal length and main point location
Image test	Still imaging test: to evaluate still image quality at different distance Sampling video test: to evaluate the sample image quality of video at different distance Open window video test: to evaluate the open window image quality of video at different distance
Automatic exposure test	To evaluate the function of automatic exposure image under different light intensity
Image compression test	To evaluate image quality difference and data rate before and after compression
Lunar dust occlusion test	To evaluate the imaging quality under different lunar dust condition
Color calibration	To obtain white balance correction matrix and color correction matrix

1) Principle and method of flat field calibration

The flat field calibration method of morphology camera was the same of the descent camera. When the dark current was subtracted from the images under different radial brightness, R (red), G (green) and B (blue) channel image data would be obtained respectively, and then the calibration coefficients of each channel would be obtained by the least squares fitting method.

2) Principle and method of absolute radiation calibration

The absolute radiation calibration method of morphology camera was the same of descent camera. But the light should be monochromatic light source, R, G, B channel image data should be adopted in different exposure time and under different radial brightness respectively, after the flat field calibration, the R, G, B channel absolute calibration coefficient would be obtained.

3) Principle and method of interior orientation elements calibration

The calibration method of the interior orientation elements of morphology camera was the same of descent camera.

4) Principle and method of video test

The video tests of morphology camera included: still imaging, sampling video and open window video.

The video test method of morphology camera was the same of descent camera, as shown in Fig. 16.3. The video test could be conducted under the natural atmosphere.

The images were obtained at different distance to target board, artificial judging image color, image resolution, and image compression quality.

5) Principle and method of automatic exposure test

The automatic exposure test of morphology camera was the same of descent camera.

The imaging target board was placed at the best distance, and different brightness was realized by using the light source.

6) Principle and method of image compression test

The image compression test of morphology camera was the same of descent camera. Target board placed was the best distance. The images were obtained under the same light brightness to evaluate the images before and after the compression. The subjective evaluation content was the color authenticity and image sharpness; the objective evaluation could use the peak signal-to-noise ratio.

7) Principle and method of lunar dust occlusion test

The dust occlusion test was to analyze the camera image quality in different occlusion of lunar dust.

Method of lunar dust occlusion test is shown in Fig. 16.4. The plain glass which was covered with lunar dust was close to the camera lens, with different amount of covered dust for different test. The target board was placed in the best distance, under the same light brightness, respectively to obtain the separately imaging and continuously imaging, analyze and evaluate the images by experts.

8) Principle and method of color calibration

The purpose of color calibration was to make the color output of the camera close to the color perception of the human eye, and the color calibration generally included two processes: color correction and white balance.

The method of color calibration test is shown in Fig. 16.5.

In color calibration, 24 color standard board and D65 standard light source were used generally.

In the white balance calibration, the neutral color (black, white, grey) diffuse standard plate and D65 standard light source were used generally.

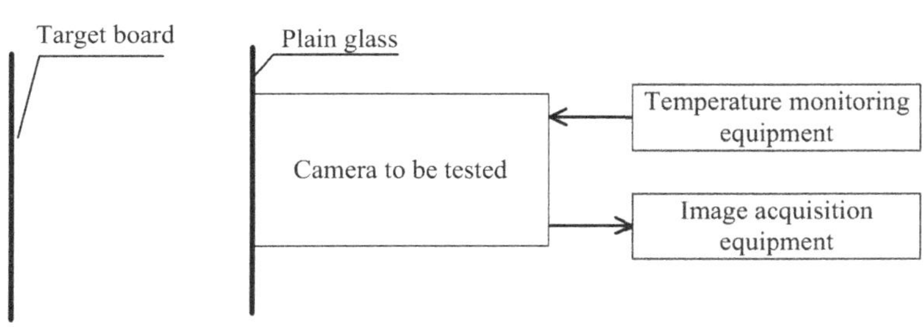

Fig. 16.4 Principle diagram of lunar dust occlusion test

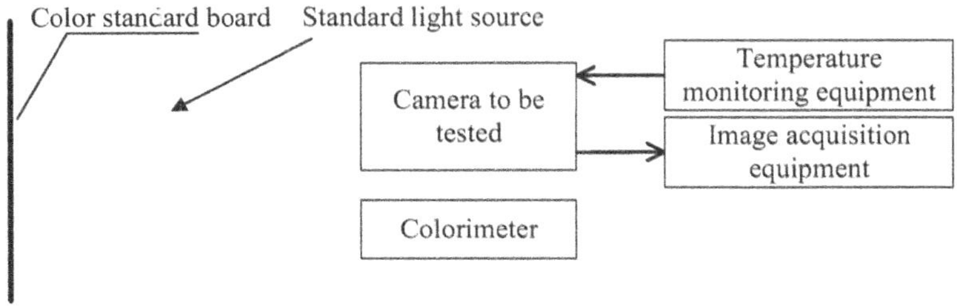

Fig. 16.5 Principle diagram of color calibration

Table 16.10 Calibration and special tests of LUT

Tests	Purpose
Flat fielding calibration	Camera dark current detection and relative radiation calibration
Spectral response test	To measure the spectral response of the camera system
Focal spot test	To measure the size of the focal spot
Full field imaging test	To test the full field imaging effect
Ultimate detection capability test	To detect the ultimate detection capability

16.5.3 Calibration and Special Testing of LUT

1. Tests

The calibration and special testing of LUT are listed in Table 16.10.

2. Test Principle and Method
1) Principle and method of flat field calibration

The flat field calibration method of LUT was the same of descent camera. The CCD should be cooled to the predetermined temperature before the dark current detection and relative radiation calibration. Ultraviolet light source should be used for relative radiation calibration.

2) Principle and method of spectral response test

The spectral response test was to test the response characteristic of the telescope in the working band.

The method of spectral response was shown. With different wavelengths of monochromatic light source and under the condition of same radial brightness, the CCD was refrigerated to predetermined temperature and images were obtained

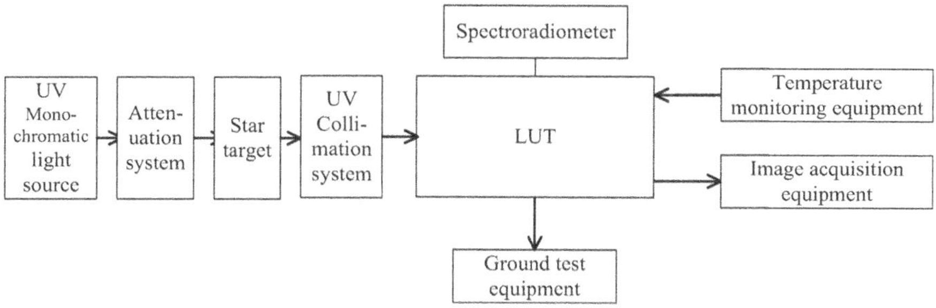

Fig. 16.6 Principle diagram of color calibration

respectively. After the flat field calibration, the response characteristics of LUT in different wavelength light source should be calculated.

3) Principle and method of focal spot test

The method of the focal spot test is shown in Fig. 16.6. The star target was placed in the focal plane of the collimation system (parallel optical tube).

The CCD was cooled to the predetermined temperature, and the image was collected to measure the shape and size of the spot.

4) Principle and method of full-field imaging test

The method of the full-field imaging test is shown in Fig. 16.6, and the star-target plate was replaced by the stars-target matrix board.

The CCD as cooled to predetermined temperature, the image was collected and the imaging effect of CCD array was analyzed.

5) Principle and method of ultimate detection capability test

The method of the ultimate detection capability test is shown in Fig. 16.6. UV monochromatic light source was changed to UV light source, and the light source brightness adjustment system was added at the focal plane of the collimating system (parallel optical tube).

The CCD was refrigerated to predetermined temperature and the light source brightness was adjusted to test light radiation brightness and obtain images. After light radiation brightness being converted to the corresponding magnitude, the limits of detection ability of the telescope were analyzed.

16.5.4 Calibration and Special Testing of EUV Camera

1. Tests

The calibration and special testing of EUV camera are listed in Table 16.11 [16–18].

2. Test Principle and Method
1) Principle and method of geometry scaling

The detector (sensor) used in EUV camera was a non-plane detector, and the geometry scaling of the detector should be the first test. The geometry scaling of EUV camera was to complete the geometry scaling of the detector (sensor) and then complete the camera system geometry scaling. In the test, UV light source was used, and there was no filter coating in the optical system.

The method of geometry scaling of the detector is shown in Fig. 16.7. Equipment including UV source, UV collimation system, pinhole array metal plate and detector were placed in the vacuum chamber. The pinhole array metal plate was attached to the detector. The vacuum in the vacuum chamber was better than 6×10^{-4}Pa. The detector obtained the pinhole array image and the calibration matrix of the detector according to the information such as pinhole size and pinhole spacing.

The method of geometry scaling of camera system is shown in Fig. 16.8. Equipment including UV light source, UV collimation system, single pinhole metal plate and EUV camera were placed in the vacuum chamber. The single pinhole metal plate was placed in the focal plane of the UV collimation system (parallel optical tube).The vacuum in the vacuum chamber was better than 6×10^{-4} Pa, and EUV camera captured the pinhole image. In practice, images should be obtained in the camera view of symmetrical directions. The first step was to complete the geometry calibration of sensor, then to establish geometry calibration matrix of camera system. The geometry calibration matrix was used to realize point by point calibration by interpolation method.

2) Principle and method of spatial resolution test

The spatial resolution test as to test the spatial resolution of EUV camera. Since it was difficult to get a EUV plane/point light source, in the test, the EUV camera was installed in a two-dimensional rotating platform, using an UV light source to

Table 16.11 Calibration and special tests of EUV camera

Tests	Purpose
Geometry scaling	Geometry scaling of sensor and camera system
Spatial resolution test	To measure spatial resolution of the camera system
Relative radiation calibration	To correct response uniformity of camera system
Absolute radiation calibration	To establish the quantitative relationship between the camera output signal and the radiance of the target
Suppression test at 58.4 nm	To test the suppression capability of camera at 58.4 nm

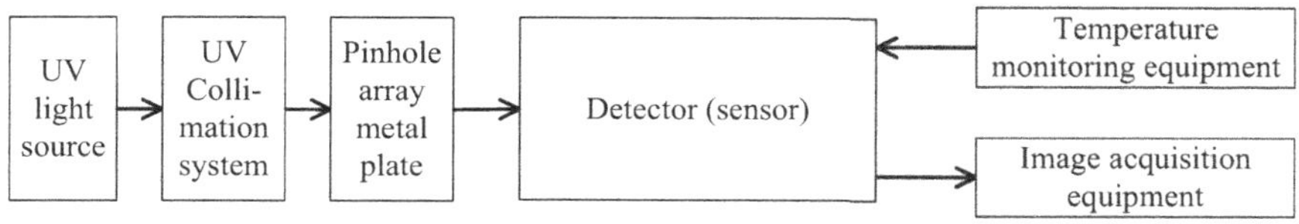

Fig. 16.7 Principle diagram of geometry scaling test for EUV camera detector

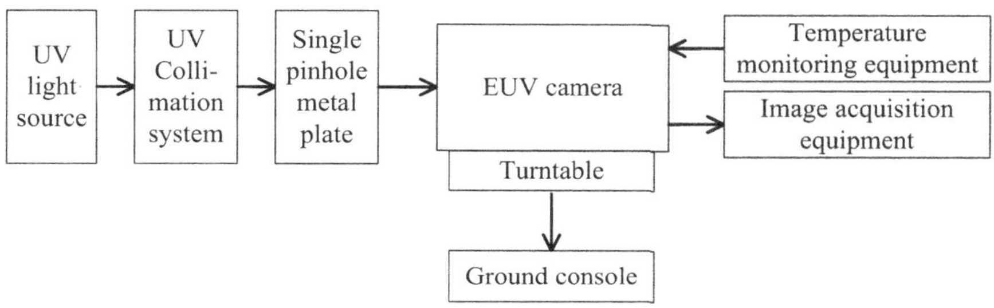

Fig. 16.8 Principle diagram of geometry scaling test for EUV camera

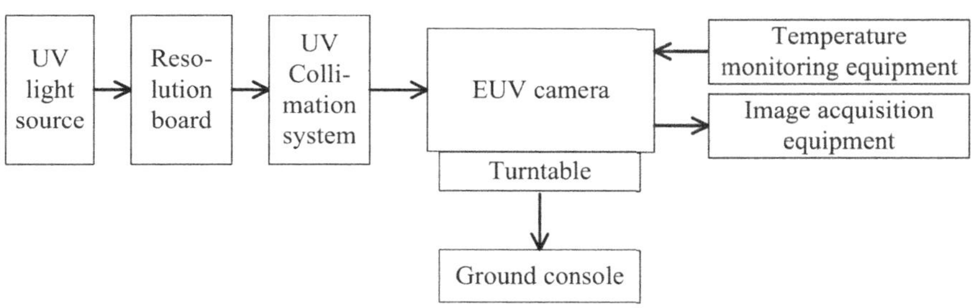

Fig. 16.9 Principle diagram of spatial resolution test for EUV camera

accomplish the measurement of the spatial resolution, and there was no extreme UV film in optical system.

Spatial resolution test method is shown in Fig. 16.9, The resolution board was at UV collimation system (parallel light pipe) focal plane. Equipment including UV light source, resolution board, UV collimation system, EUV camera and turntable were placed in a vacuum chamber that could adjust temperature.

The vacuum in vacuum chamber should be better than 6×10^{-4} Pa. The temperature of EUV should be kept at predetermined temperature and camera obtained images. By distinguishing line of resolution board in images, combining with the information such as the focal length of the camera system, the spatial resolution of EUV camera under different temperature conditions could be evaluated.

3) Principle and method of relative radiation calibration

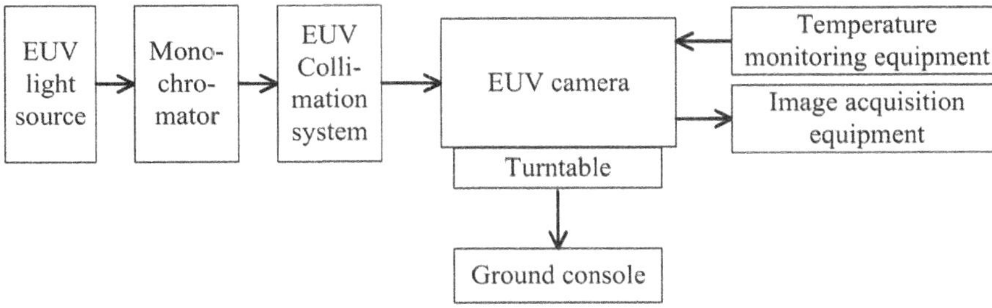

Fig. 16.10 Principle diagram of relative radiation calibration for EUV camera

The relative radiation calibration was to test and calibrate the response uniformity of the extreme ultraviolet camera system.

The relative radiation calibration method of EUV camera is shown in Fig. 16.10. Equipment including EUV light source, monochromator (which can let out 30.4 nm $\pm$ 5 nm EUV light), EUV collimating system, EUV camera and turntable was placed in vacuum environment. The relative radiation calibration matrix could be obtained by EUV camera to image the light source at the same radiance. Because the working spectrum of EUV camera was 30.4 nm $\pm$ 5 nm, it was difficult to generate a plane light source; the point light source was used. A point-by-point imaging of the sensor of EUV camera would be very expensive in cost and time. In practice, the images would be obtained in several symmetrical directions in the camera field, and then establish relative radiation calibration matrix of point by point by interpolation method.

4) Principle and method of absolute radiation calibration

The absolute radiation calibration method was the same as shown in Fig. 16.10.

A point-by-point imaging of the sensor of the extreme ultraviolet camera would be very expensive in cost and time. In practice, the images would be obtained in several symmetrical directions in the camera field, under different temperature, with different exposure time, and under the different radial brightness. After the relative radiation calibration, relative radiation calibration matrix of point by point by interpolation method was established.

5) Principle and method of suppression test at 58.4 nm

The test method was the same as shown in Fig. 16.10.

With monochromator respectively through 30.4 nm and 58.4 nm EUV light, the images were obtained at two kinds of light by EUV camera and standard instrument respectively. The responding image values of EUV camera were obtained at two kinds of light in the same irradiation intensity.

Table 16.12 Calibration and special tests of monitoring camera

Tests	Purpose
Interior orientation elements calibration	To measure the optical distortion coefficient, pixel position, focal length and optical axis offset
Flat fielding calibration	To measure inconsistent response of pixel and channel
Absolute radiation calibration	To establish the quantitative relationship between the camera output signal and the radiance of the target
Sill imaging	To evaluate the still image quality at different lighting and imaging distance
Video test	To evaluate the image quality of dynamic camera at different lighting and imaging distance
Automatic exposure test	To evaluate the function of automatic exposure image under different light intensity
Image compression test	To evaluate image quality difference and data rate before and after compression
Lunar dust occlusion test	To evaluate the imaging quality under different dust cover
Color calibration	To obtain white balance correction matrix and color correction matrix

16.5.5 Calibration and Special Testing of Monitoring Camera

1. Tests

The calibration and special testing of monitoring camera are listed in Table 16.12.

2. Test Principle and Method

The principle and method of monitoring camera calibration and special tests cold refer to similar tests of descent camera and morphology camera.

16.5.6 Calibration and Special Testing of Lunar Dust Detector

1. Tests

The calibration and special testing of lunar dust detector are listed in Table 16.13.

2. Test Principle and Method
1) Principle and method of SQCM detector calibration

The SQCM detector calibration was to measure the frequency and drift range of SQCM detector under different lunar dust accumulation mass, and to establish the relationship between the crystal oscillator frequency and the lunar dust accumulation mass.

Table 16.13 Calibration and special tests of lunar dust detector

Tests	Purpose
SQCM detector calibration	To measure the frequency and drift range of SQCM detector under different lunar dust accumulation mass
Solar cell detector calibration	To measure the output current and drift range of solar cell detector under different lunar dust accumulation mass
SQCM mucosa stability test	To measure the stability of the SQCM mucosa

Lunar dust particles were spread evenly on SQCM detector surface, under the condition of vacuum (better than that of 6×10 Pa), several tests should be conducted from no dust to full dust. The curve of crystal frequency and dust accumulation mass could be determined.

2) Principle and method of Solar cell calibration

Solar cell detector calibration was to measure the output current of solar cell detector and its drift under different dust accumulation mass, and determine the relationship between output current and dust accumulation mass.

Lunar dust particles were spread evenly in the surface of solar cell detector. Under the condition of vacuum (better than that of 6×10 Pa), the surface of solar cell detector was vertically illuminated by fixed light intensity, to respectively test no load to full load between multiple test points and measure the output current probe of solar cell detector and its drift. Then the curve of output current and dust accumulation mass under certain light intensity could be determined.

3) Principle and method of SQCM mucosa stability test

Under vacuum condition, the SQCM detector was vertically illuminated by solar simulator. The SQCM detector would continue to run longer (generally more than 100 h). If the frequency change of SQCM detector was far less than detection range resolution, it implied that SQCM mucosa was stable and without apparent volatile, or vice versa.

16.6 Summary

In lunar soft landing missions and other deep space exploration activities, mission objectives are fundamental requirements for instruments. In order to achieve different mission objectives, the different instruments would be configured. In addition, technical conditions and resource constraints would affect the final configuration of instruments.

In the early stage of the United States and the Soviet Union lunar landing exploration, especially in the first mission, the major objectives were to demonstrate engineering technologies or investigate environment parameter that would affect engineering implementation. Then scientific exploration was gradually conducted.

Limited by technology development status at that time, there were only few instruments. But the successful exploration of those instruments provided valuable information for engineers and scientists, which was also important for humankind to know more about the Moon.

The Chang'E-3 lunar lander was the first spacecraft that was softly landed on extraterrestrial body in China. There were four scientific instruments on the Chang'E-3 lunar lander, including descent camera, morphology camera, LUT and EUV camera, which could perform in situ investigation of Lunar surface, imaging of Earth plasmasphere and lunar-based astronomy observation. There were also 3 monitoring cameras and one lunar dust detector, which could monitor deployment of the lunar rover and measure the floating lunar dust and its accumulation. All the instruments should adapt to flight environments, such as mechanics, heat, vacuum and radiation, as well as the target features. When the instruments were designed, above factors should be comprehensively considered. In order to confirm the correctness of instrument design, conventional environment tests such as mechanics, thermal and vacuum should be conducted, as well as calibration and special tests to confirm that the instruments could obtain anticipated exploration data under specific environment, and implement designed function and performance.

In subsequent missions to the Moon or Mars, more attention will be paid to\lightweight, miniaturization, low power consumption and high performance of the instruments to implement more exploration. The type of instruments will be diversified. For example, small camera can investigate the terrain. Descent camera can help to monitor descent process. Ray spectrometer can investigate chemical element, optical spectrometer for mineral composition investigation, mass spectrometer for gas composition investigation, radar for soil structure, force gauge for mechanical properties, and so on.

References

1. Chu G, Zhang H (2007) Technology of lunar probes. China Science and Technology Press, Beijing
2. Ouyang Z (2005) Introduction to Lunar Science. China Astronautic Publishing House, Beijing
3. Chu G, Shijun MA (2002) Introduction to astronautic technology. China Astronautic Publishing House, Beijing
4. Tan W, Hu J (2009) Spacecraft system engineering. China Science and Technology Press, Beijing
5. Peng C (2011) Spacecraft system design. China Science and Technology Press, Beijing
6. Ouyang Z (2004) Scientific objectives of Chinese Lunar Exploration Project and development strategy. Adv Earth Sci 19(3):351–355
7. Ouyang, Z (2005) Review of lunar exploration and introduction of Chinese Lunar Exploration Project. Chinese J Nat 27(4):187–190
8. Ye P, Sun Z, Rao W (2007) Research and development of Chang'E-1. Spacecraft Eng 16(6):9–15
9. Jet Propulsion Laboratory (1966) Surveyor 1 Mission Report:Part I. Mission description and performance. California Institute of Technology
10. Jet Propulsion Laboratory (1967) Surveyor 2 Mission Report. California Institute of Technology

11. Jet Propulsion Laboratory (1967) Surveyor 3 Mission Report: Part I. Mission description and performance. California Institute of Technology
12. Jet Propulsion Laboratory (1968) Surveyor 4 Mission Report. California Institute of Technology
13. Jet Propulsion Laboratory (1968) Surveyor 5 Mission Report: Part I. Mission description and performance. California Institute of Technology
14. Jet Propulsion Laboratory (1968) Surveyor 6 Mission Report: Part I. Mission description and performance. California Institute of Technology
15. Jet Propulsion Laboratory (1969) Surveyor 7 Mission Report: Part I. Mission description and performance. California Institute of Technology
16. Chen B, He F (2011) Optical design of moon-based Earth's plasmaspheric extreme ultraviolet imager. Opt Precis Eng 19(9):2057–2062
17. Chen B, Gao L, Kan S et al (2007) An evaluation method of angular resolution for extreme ultraviolet telescope. Opt Precis Eng 15(11):1644–1648
18. He F, Chen B, Zhang X (2010) Moon-based imaging of earth plasmaspheric extreme ultraviolet radiation. Opt Precis Eng 18(12):2564–2573

Chapter 17
Future Prospects of Lunar Lander Technology

17.1 Introduction

Although there were over 100 lunar probes that had visited the Moon and even 12 astronauts who had set their feet on lunar surface [1–3], human knowledge about the moon is still limited. There are lots of unsolved mysteries such as origin and evolution of the Moon, whether there is large amount of water or ice on the Moon, lunar internal structure and etc. At the same time, utilization of lunar resource became the focus of all space countries because the contradiction between the increasing demand for resources and the depletion of the Earth's resources is becoming increasingly prominent.

The prospect to requirements of future development of lunar exploration and trend of lunar lander technology is given in this chapter.

17.2 Trends in Future Development of Lunar Exploration

The basic technology for lunar exploration has been mastered by human being after 60 years practice. China will become one of the advanced countries in lunar exploration technology after implementation of CLEP orbiting, landing and sample return. The future development of lunar exploration can be summarized on three trends.

(1) Lunar exploration is still focus of deep space exploration for all countries. Lunar missions were planned in the future 30 years strategic plan of aerospace industry of the United States, Russia, Japan and ESA. Although the hottest spot of deep space exploration is now Mars exploration for space countries such as the United States, the importance of lunar exploration cannot be ignored yet because that the Moon is a convenient test bed for technology demonstration and the potential of lunar resources utilization is enormous, which makes lunar exploration be another focus of deep space exploration. After 1990s, 13 lunar

D.-Y. Yu et al., *Technology of Lunar Soft Lander*, Space Technology Library 38,
https://doi.org/10.1007/978-981-15-6580-9_17

missions were launched by the United States Japan, China, India and ESA in second climax of lunar exploration [5, 9]. The lunar exploration plans with lunar base construction as ultimate goal for future strategic plan were come up [3].

(2) The important trend is fine survey of resources and technology demonstration of resources utilization. The lunar resources can be summarized as position resources, environment resources and material resources. The position resources of the Moon mean that the Moon can serve as a 'high ground' because the near side is toward the Earth forever, long term and large scale observation of the Earth and its proximity can be conducted based on the Moon. The slow rotation rate of the Moon makes it possible to observe specific space region continuously. The environment resources including natural lunar environment such as high vacuum, low gravity and weak magnetic field make the Moon a good place for scientific research, manufacture and refinement of special material. The material resources are water ice, Helium-3 [4, 5] and other rare elements on the Moon which may be raw material for future solution of energy crisis of human being.
The Moon is the nearest celestial body of the Earth which makes resources utilization feasible. Major space countries are all conducting fine survey of lunar resources urgently in order to gain advantage. The Moon might be a contest target for all countries in the future.
The Moon may be a favorable place for the key technology demonstration of future deep space exploration because of its position and environment. For example, technologies such as long term electric propulsion, autonomous navigation and control, thermal control under extreme environment and long range telecommunication were validated in ESA SMART-1 lunar mission. Most of deep space exploration technologies can be applied in future deep space exploration after demonstration in lunar missions.

(3) The long term goal is lunar base construction for lunar exploration [10]. In the middle and of later twentieth centuries, proposals to construct long term lunar facilities and implement efficient lunar resources utilization were come up by the United States and Russia. Some concepts of lunar base are shown in Fig. 17.1. The proposals of lunar base are developed and improved by different countries recently. China also initiated primitive research on robotic lunar scientific research station and manned lunar missions.

Construction of lunar base [7] can help to implement long term and continuous research and manufacture, Earth observation and resources mining. Key technologies including space power, thermal control, telecommunication, navigation, structure and configuration, life support, space environment protection and maintenance can be tested and validated by construction of lunar base which has function of power generation, telecommunication, transportation, experiment and etc., which may prepare for subsequent manned lunar missions and Mars missions as well as improve technology and capability of deep space exploration [6, 7]. The research on life science, Earth science, astronomy, material science and space science can be

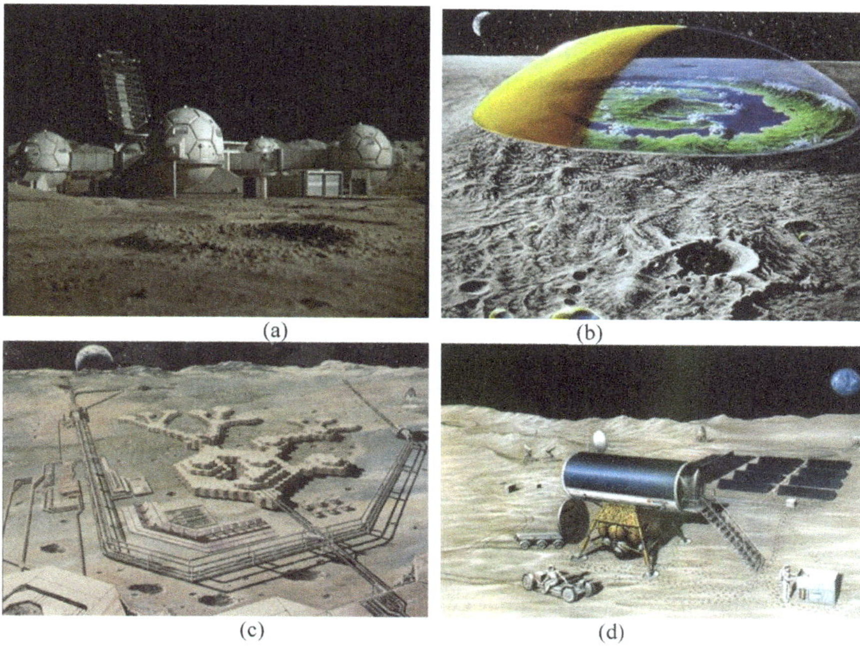

Fig. 17.1 Proposed lunar base concept

conducted on lunar base. The observatory center for astronomy and Earth observation can conduct high accuracy monitoring of changing of Earth climate and space environment, especially all-around and continuous astronomical observation which may monitor Potential Hazardous deep space small bodies for Earth.

The construction of the lunar base which is a facility of long term robotic operation and short term operated by human for scientific research and manufacture is an inevitable trend along with further survey of lunar resources and readiness of technology development.

17.3 New Challenges for Lunar Lander Technology

In a foreseeable future, lunar landing is absolutely necessary for lunar sample return, lunar scientific research station construction and manned lunar missions. The new challenges of such exploration for lunar lander technology can be summarized as following six aspects [8, 9].

(1) Pin-point landing technology. The major water ice resources are in lunar polar region. The lunar terrain, light condition and thermal environment are complex in lunar polar region. The accuracy of lunar landing shall be improved to ensure

the landing at concerned region. For lunar base construction, landing sites shall be close to each other for different lunar landers in order to cooperate. The landing accuracy can be improved to higher than 100 m by terrain matching and beacon guiding, which makes landing at requirement feasible.

(2) Technology adapting to complex lunar terrain. The stable attitude after landing is required for subsequent tasks. The capability of lunar lander adapting to different lunar terrain shall be improved for steeper slope, more rocks and pits in polar region. Technology such as lunar terrain identification and obstacles avoiding, landing cushion are involved.

(3) Power supply technology in shadow region. The major water ice resources are in Permanent Shadow Region (PSR) in the craters of lunar polar region. The lunar lander will be commanded to operate in PSR to mining water ice resources. The power supply is crucial for long term operation. Technologies such as wireless power transmission and RTG may be the solution.

(4) Resources utilization technology. The important aim of lunar exploration is to utilize lunar resources for human being. Technology shall be studied continuously such as how to implement resources utilization by deep drill, heating and melting, how to construct basic components with lunar regolith, how to construct high accuracy and large telescope on lunar surface, and etc.

(5) Environment protection technology. There are severe effects on structure, material and etc. of lunar lander by radiation, gravity and lunar dust when the lunar lander is long term operated or astronauts work for a long time on lunar surface. Therefore, environment protection technology shall be studied.

(6) Reusable and mobile lunar lander technology. Now the exploration range is limited for the lunar lander because it can be landed once and cannot move. The reusable and mobile lunar lander shall be developed to implement extensive exploration in different location and acquire more benefits in the future.

More accurate, intelligent and adaptable lunar lander technology will be developed along with increasing requirement. At the same time, the improvement of lunar lander technology may be directly applied in landing missions on Mars, asteroids and other celestial bodies, which will prompt further deep space exploration of human being.

References

1. Ouyang Z et al (2002) Moon: the first stop of the humanity towards space exploration[M]. Tsinghua University Press, Beijing
2. Chu G, Zhang H (2007) Technology of lunar probes[M]. China Science and Technology Press, Beijing
3. Zhang H (2007) Soaring the empire: from artificial satellite to lunar probe[M]. Shanghai Scientific and Technological Education Publishing House, Shanghai
4. Wang J (2007) Beyond the moon: fascinating prospect of lunar exploration[M]. Shanghai Scientific and Technological Education Publishing House, Shanghai
5. Ouyang Z (2005) Introduction to lunar science[M]. China Astronautic Publishing House, Beijing

6. Wu W, Yu D (2014) Development of deep space exploration and its future key technologies[J]. J Deep Space Explor 1(1):5–17
7. Wu W (2007) To the moon[M]. China Astronautic Publishing House, Beijing
8. Xi X, Zeng G, Ren X et al (2001) Orbit design of lunar probe[M]. National Defense Industry Press, Beijing
9. Wu W, Liu W, Tang Y et al (2013) Several key technologies and development trends for deep space exploration[J]. Space Int 420:45–51
10. Yu D, Ge Z, Wang N et al (2012) Supposal for structure form of lunar base[J]. J Astronaut 33(12):1840–1844

Appendix

See Figs. A.1, A.2, A.3, A.4, A.5, A.6, A.7, A.8, A.9, A.10, A.11, A.12, A.13, A.14, A.15 and A.16.

Fig. A.1 First image taken by Chang'E-3 lunar lander after landing

D.-Y. Yu et al., *Technology of Lunar Soft Lander*, Space Technology Library 38,
https://doi.org/10.1007/978-981-15-6580-9

Fig. A.2 The Yutu rover after deployment, ready for exploration

Fig. A.3 Image of Chang'E-3 lunar lander taken by panorama camera of Yutu rover

Fig. A.4 Image of Chang'E-3 lunar lander taken by panorama camera of Yutu lunar rover when it moved to east of the lander

Fig. A.5 Image of Chang'E-3 lunar lander taken by panorama camera of Yutu lunar rover when it moved to south-east of the lander

Fig. A.6 Image of Chang'E-3 lunar lander, taken by panorama camera of Yutu rover when it moved to south of the lander

Fig. A.7 Image of Yutu lunar rover, taken by panorama camera of Chang'E-3 lunar lander

Fig. A.8 Image of Chinese national flag on the Chang'E-3 lunar lander, taken by panorama camera of Yutu lunar rover

Fig. A.9 Image of landing leg footpad of Chang'E-3 lunar lander, taken by panorama camera of Yutu lunar rover

Fig. A.10 Image of Chang'E-3 lunar lander taken by panorama camera of Yutu lunar rover

Fig. A.11 Image of Yutu lunar rover when it was roving, taken by panorama camera of Chang'E-3 lunar lander

Fig. A.12 Image of roving Yutu lunar rover when it was roving, taken by panorama camera of Chang'E-3 lunar lander

Fig. A.13 Chang'E-3 lunar lander docked with the launch vehicle at launch site

Fig. A.14 Landing impact test of Chang'E-3 lunar lander

Fig. A.15 Mechanical test of Chang'E-3 lunar lander

Fig. A.16 Hovering, obstacle avoidance and slow descending test of Chang'E-3 lunar lander

The manufacturer's authorised representative in the EU is Springer Nature Customer Service Centre GmbH, Europaplatz 3, 69115 Heidelberg, Germany. If you have any concerns regarding our products, please contact ProductSafety@springernature.com

Printed and bound by CPI Group (UK) Ltd, Croydon, CR0 4YY
30/09/2026
02225320-0018